W9-CUK-171

D/MF

CONCEPTS AND DESIGN OF
CHEMICAL REACTORS

Chemical Engineering: Concepts and Reviews

A series edited by

Jaromir J. Ulbrecht,
Center for Chemical Engineering
National Bureau of Standards
Washington, D.C.

Volume 1   **MIXING OF LIQUIDS BY MECHANICAL AGIT-
ATION**
Edited by Jaromir J. Ulbrecht and Gary K. Patterson

Volume 2   **DYNAMICS OF NONLINEAR SYSTEMS**
Edited by Vladimir Hlavacek

Volume 3   **CONCEPTS AND DESIGN OF CHEMICAL RE-
ACTORS**
Edited by Stephen Whitaker and Alberto E. Cassano

Other volumes in preparation

ISSN 0734-1644

CONCEPTS AND DESIGN OF CHEMICAL REACTORS

Edited by

STEPHEN WHITAKER
University of California at Davis

and

ALBERTO E. CASSANO
INTEC, Argentina

GORDON AND BREACH SCIENCE PUBLISHERS
New York · London · Paris · Montreux · Tokyo

© 1986 by Gordon and Breach Science Publishers S.A., P.O. Box 161, 1820 Montreux 2, Switzerland. All rights reserved.

Gordon and Breach Science Publishers

P.O. Box 786
Cooper Station
New York, NY 10276
United States of America

5379 8843

P.O. Box 197
London WC2E 9PX
England

58, rue Lhomond
75005 Paris
France

14-9 Okubo 3-chome,
Shinjuku-ku,
Tokyo 160
Japan

Library of Congress Cataloging in Publication Data

Concepts and design of chemical reactors.

(Chemical engineering concepts and reviews,
ISSN 0734-1644 ; v. 3)
  1. Chemical reactors--Design and construction.
I. Cassano, Alberto E., 1935-   . II. Whitaker,
Steven, 1932-   . III. Series.
TP157.C63   1986   660.2'81   86-7614
ISBN 2-88124-118-2

CONTENTS

SERIES EDITOR'S INTRODUCTION

There has long been a need for monographs that discuss, in detail, specific subjects in chemical engineering. The aim of this series is to publish such monographs, each of which comprises articles developed from critical reviews of the relevant literature. The articles are written by experts whose contributions to the field under discussion are well known and significant. Articles written by more than one author still aim to present a single concept, although the exposition will certainly reflect the authors' somewhat differing opinions.

It is my belief that any review article worth reading must be not only informative, but also stimulating to the point of provocation. Thus, the reader will never see an "objective unbiased" review in this series (or, indeed, anywhere else—the mythical beast never has existed).

The authors are encouraged to structure their articles around their own and other closely related work. This approach, of course, involves the risk that the products will be noncritical self-serving summaries. It is to the credit of both the authors and the volume editors that no such thing has happened.

This present volume, *Concepts and Design of Chemical Reactors*, was preceded by *Mixing of Liquids by Mechanical Agitation* and *Dynamics of Nonlinear Systems*. Further volumes on topics ranging from *Chemical Process Diagnostics* to *Biochemical Reactors* are under preparation. Suggestions for future volumes are welcome.

J. J. Ulbrecht

PREFACE

This book contains ten chapters covering a broad spectrum of topics associated with the design of chemical reactors. These chapters were prepared for presentation at the "25th Conicet Anniversary Reactor Design Conference" at Santa Fe, Argentina. The topics range from fundamental treatments of the basic transport equations for reacting systems to practical applications in the design of multi-phase reactors. While every aspect of reactor design is not covered in this monograph, it does contain reviews of the current state-of-the-art of the most important areas of reactor design, along with the most recent developments and applications of experts in the field.

Stephen Whitaker
Alberto E. Cassano

Chapter 1

Transport Processes with Heterogeneous Reaction

STEPHEN WHITAKER

Department of Chemical Engineering, University of California, Davis, California 95616, U.S.A.

Abstract

In this paper we consider one of the threads of analysis that leads from axioms to applications in the design of chemical reactors. The development is restricted to isothermal systems and begins with the continuum axioms for the mass and momentum of a species body. From the axioms we obtain a set of transport equations that can be used, in conjunction with the appropriate constitutive equations and jump conditions, to determine the concentration fields of reacting species. From these fields overall rates of reaction can then be determined.

Since most reactor design problems involve multiphase systems with interfacial regions that can only be determined in an average or statistical sense, a detailed knowledge of the concentration fields is not possible. This situation naturally leads to a search for average concentrations and the governing differential equations for averaged quantities. Use of the method of volume averaging provides a precise route from the well-known point equations to the design equations for chemical reactors. In this paper the method is first explored in a detailed treatment of bulk diffusion and reaction in porous media. The analysis is then extended to diffusion and reaction in a micropore–macropore system, and finally the general transport problem involving bulk diffusion, Knudsen diffusion, and Darcy flow is considered. As a final illustration of the method of volume averaging, the design equations for a two-phase reactor are derived and the general closure problem is considered.

1. INTRODUCTION

The design of chemical reactors represents the focal point of the chemical engineering educational process since it draws upon all aspects of that process. It is a process that begins with fluid

1

mechanics, thermodynamics, quantum mechanics, and the usual measure of applied mathematics. Upon these subjects one fortifies the structure with heat transfer, mass transfer, and chemical kinetics, and with these tools in hand one is prepared to enter the arena of reactor design. As the focal point of the educational process, it is not a special discipline (sometimes referred to as "chemical reaction engineering") but an integral part of the whole of chemical engineering education.

A pictorial representation of this point of view is given in Fig. 1-1. The axioms of continuum mechanics refer to the axioms associated with the classical field theories as described by Truesdell and Toupin [45]. The term *transport phenomena* refers to fluid mechanics, heat transfer, and mass transfer while *thermodynamics* refers to what some identify as "rational thermodynamics" [4, 46]. The latter includes thermostatics as a special and very important case.

The complete structure of a mathematical theory of physics could follow the outline given by Truesdell [46] which suggests that any theory should be based on:

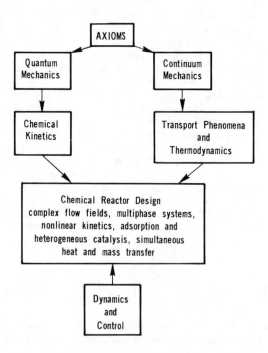

Figure 1-1   Foundations of chemical reactor design.

1. *Primitive quantities*, defined only in terms of their mathematical properties.
2. *Definitions* of other quantities in terms of the primitives.
3. *Axioms* stated as mathematical relations between the primitives and the defined quantities.
4. Proved *theorems*.
5. *Constitutive equations*.

Most often we think of the primitives and the defined quantities as being self-evident; however, this is not at all true and in this paper we find it worthwhile to devote a small amount of space to the primitive concept of a *species body*. We identify a species body in the following manner:

SPECIES BODY

A species body (specifically species A) occupies a region in space designated by $\mathcal{V}_A(t)$. Associated with every species body is a non-negative scalar m_A called the mass of the species body and a scalar $\mathcal{R}_A$ called the mass rate of production of species A. The region occupied by a species body is not exclusive.

The concept of a species body occupying a region in space that is simultaneously occupied by other species bodies is a central idea in the continuum approach to multicomponent transport processes. This idea is illustrated in Fig. 1-2 for a two-component system.

The axioms for the mass of multicomponent systems can be stated as

$$\left\{ \begin{array}{l} \text{time rate of change of the mass} \\ \text{of a species body} \end{array} \right\} = \left\{ \begin{array}{l} \text{mass rate of production} \\ \text{of the species} \end{array} \right\}$$

(1-1)

$$\left\{ \begin{array}{l} \text{total mass rate of} \\ \text{production of all species} \end{array} \right\} = 0$$

(1-2)

In mathematical terms we can express the first of these axioms as

$$\frac{dm_A}{dt} = \mathcal{R}_A$$

(1-3)

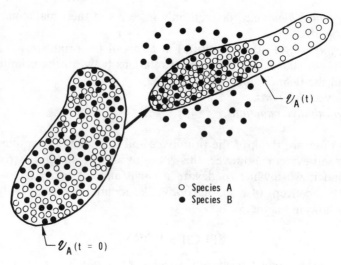

Figure 1-2   Motion of a species body.

however, the second axiom is not an unambiguous statement. To remove this ambiguity we extend the mathematical characteristics of our species body to require that the mass, m_A, and the mass rate of production, $\mathscr{R}_A$, are absolutely continuous functions of the volume. This means that density functions exist and Eq. (1-3) can be expressed as

$$\frac{d}{dt} \int_{\mathscr{V}_{A(t)}} \rho_A \, dV = \int_{\mathscr{V}_{A(t)}} r_A \, dV \qquad (1\text{-}4)$$

Here ρ_A is the mass density of species A and r_A is the mass rate of production of species A per unit volume. An unambiguous alternative to Eq. (1-2) is given by

$$\sum_{A=1}^{A=N} r_A = 0 \qquad (1\text{-}5)$$

Even though neither of these axiomatic statements has been verified experimentally, the *proved* theorems resulting from Eqs. (1-4) and (1-5) have been [30, page 1] and we tend to accept them as a reasonable starting point for the analysis of mass transport and chemical reaction.

One is inclined to think of Eq. (1-5) as representing the concept that "mass is neither created nor destroyed by chemical reaction"

and there is reason to think of Eq. (1-5) as a "weak form" of the axiom associated with mass and chemical reaction. To develop an alternate form of this axiom we define the molar rate of production of species A by

$$R_A = r_A / M_A \qquad (1\text{-}6)$$

in which M_A is the molecular weight of species A. If we define $N_{\alpha A}$ as

$$N_{\alpha A} = \left\{ \begin{array}{l} \text{number of } \alpha\text{-type atoms} \\ \text{associated with species } A \end{array} \right\} \qquad (1\text{-}7)$$

the molecular weight of species A is given by

$$M_A = \sum_{\alpha=1}^{\alpha=Q} N_{\alpha A} M_\alpha \qquad (1\text{-}8)$$

Here M_α is the atomic weight of α-type atoms and we think of Q as representing the total quantity of atomic types. An alternate axiomatic statement that replaces Eq. (1-5) is given by

$$\sum_{A=1}^{A=N} R_A N_{\alpha A} = 0, \quad \alpha = 1, 2, \ldots, Q \qquad (1\text{-}9)$$

and we tend to think of this result as representing the concept that

"Atomic species are neither created nor destroyed by chemical reaction".

It is appropriate to think of Eq. (1-9) as the "strong form" of the axiom associated with mass and chemical reaction since Eq. (1-5) can be derived directly from Eq. (1-9). The converse is not true however, since one must assume that the reaction rates are independent of atomic weight in order to derive Eq. (1-9) from Eq. (1-5). The result given by Eq. (1-9) has considerable appeal since it can be used to immediately deduce the stoichiometric coefficients for simple reactions; however, we will base our study of mass transport in multicomponent systems on Eqs. (1-4) and (1-5) since the latter can be utilized directly in the analysis of momentum transport.

One can use the general transport theorem [51, Sec. 3.4] to express Eq. (1-4) as

$$\int_{\mathscr{V}_A(t)} \left[\frac{\partial \rho_A}{\partial t} + \nabla \cdot (\rho_A \mathbf{v}_A) - r_A \right] dV = 0 \qquad (1\text{-}10)$$

Here $\mathbf{v}_A$ represents the continuum velocity of species A, thus the surface of $\mathscr{V}_A(t)$ moves at a velocity equal to $\mathbf{v}_A$. From Eq. (1-10) we can extract the *species continuity equation* and we list the point equations that describe the mass of multicomponent systems as

$$\frac{\partial \rho_A}{\partial t} + \nabla \cdot (\rho_A \mathbf{v}_A) = r_A \qquad (1\text{-}11)$$

$$\sum_{A=1}^{A=N} r_A = 0 \qquad (1\text{-}12)$$

One must think of Eq. (1-11) as the governing differential equation for ρ_A while Eq. (1-12) represents an axiomatic constraint on the chemical reaction rates that will normally be imposed in terms of stoichiometric coefficients.

In order to solve Eq. (1-11) to produce the ρ_A-field we need:

1. a quantum mechanical constitutive equation for r_A in terms of $\rho_A, \rho_B, \ldots, \rho_N$
2. the species velocity field, $\mathbf{v}_A$

In this development we assume that the *chemical kinetic constitutive equation* will be discovered experimentally. The velocity field is another matter, for velocity fields are determined by the laws of mechanics and it should be clear that additional axioms are required if we are to determine ρ_A and thus r_A.

We present two of the *laws of mechanics* for multicomponent systems as

I. $\quad \dfrac{d}{dt} \displaystyle\int_{\mathscr{V}_A(t)} \rho_A \mathbf{v}_A \, dV = \int_{\mathscr{V}_A(t)} \rho_A \mathbf{b}_A \, dV + \int_{\mathscr{A}_A(t)} \mathbf{t}_A \, dA$

$$+ \int_{\mathscr{V}_A(t)} \sum_{B=1}^{B=N} \mathbf{P}_{AB} \, dV + \int_{\mathscr{V}_A(t)} r_A \bar{\mathbf{v}}_A \, dV$$

$$(1\text{-}13)$$

II. $$\frac{d}{dt}\int_{\mathcal{V}_{A(t)}}\mathbf{r}\times\rho_A\mathbf{v}_A\,dV=\int_{\mathcal{V}_{A(t)}}\mathbf{r}\times\rho_A\mathbf{b}_A\,dV+\int_{\mathcal{A}_{A(t)}}\mathbf{r}\times\mathbf{t}_A\,dA$$

$$+\int_{\mathcal{V}_{A(t)}}\sum_{B=1}^{B=N}\mathbf{r}\times\mathbf{P}_{AB}\,dV$$

$$+\int_{\mathcal{V}_{A(t)}}\mathbf{r}\times r_A\bar{\mathbf{v}}_A\,dV \qquad (1\text{-}14)$$

The first of these represents the linear momentum principle for a species body and it indicates that the time rate of change of momentum is balanced by forces acting on the body and momentum sources caused by the presence of other molecular species. In Eq. (1-13) the term $\mathbf{b}_A$ represents the *body force* vector that plays a crucial role in the motion of ionic species in the presence of an electric field. The species stress vector is identified by $\mathbf{t}_A$ and it accounts for the contact force that acts on the surface of the body which is represented by $\mathcal{A}_A(t)$. One can follow the usual development [51, Chap. 4] to represent $\mathbf{t}_A$ in terms of the species stress tensor according to

$$\mathbf{t}_A=\mathbf{n}_A\cdot\mathbf{T}_A \qquad (1\text{-}15)$$

There are two momentum source terms that appear in Eq. (1-13) that do not appear in Euler's laws of mechanics [47, Chap. II] for single component systems. The term $\mathbf{P}_{AB}$ represents the force per unit volume that species B exerts on species A and we naturally require that $\mathbf{P}_{CC}$ be zero. Since the process of diffusion represents a situation in which the individual species move at different velocities, we expect the momentum source $\mathbf{P}_{AB}$ to be important when A and B are interdiffusing. The last term in Eq. (1-13) represents a momentum source owing to the production of species A by chemical reaction. The velocity of species A produced by chemical reaction has been designated by $\bar{\mathbf{v}}_A$ with the thought that it need not be equal to the continuum velocity $\mathbf{v}_A$. In writing the angular momentum principle given by Eq. (1-14), we have assumed that all torques are the moments of forces so that the effect of intrinsic angular momentum has been neglected [2, Sec. 5.13].

There are two other axioms associated with the laws of mechanics for multicomponent systems and we state the first of these as

IIIa. $$\sum_{A=1}^{A=N}\sum_{B=1}^{B=N}\mathbf{P}_{AB}=0, \quad \text{weak form} \qquad (1\text{-}16)$$

IIIb. $\mathbf{P}_{AB} = -\mathbf{P}_{BA}$,   strong form   (1-17)

The second of these appears to be Newton's law of action and reaction, but in terms of an axiomatic formulation of the laws of continuum mechanics for multicomponent systems either Eq. (1-16) or (1-17) must stand as a separate axiom. The fourth axiom is given by

IV. $$\sum_{A=1}^{A=N} r_A \bar{\mathbf{v}}_A = 0$$ (1-18)

and this requires that there be no net momentum source owing to homogeneous reaction.

The derivation of the species stress equations of motion parallels that for single component systems [41, Chap. 2] and leads to

$$\rho_A\left(\frac{\partial \mathbf{v}_A}{\partial t} + \mathbf{v}_A \cdot \nabla \mathbf{v}_A\right) = \rho_A \mathbf{b}_A + \nabla \cdot \mathbf{T}_A + \sum_{B=1}^{B=N} \mathbf{P}_{AB} + r_A(\bar{\mathbf{v}}_A - \mathbf{v}_A)$$

(1-19)

$$\mathbf{T}_A = \mathbf{T}_A^T$$ (1-20)

If we define the following *mass average* quantities

$$\rho = \sum_{A=1}^{A=N} \rho_A$$ (1-21)

$$\mathbf{v} = \sum_{A=1}^{A=N} \left(\frac{\rho_A}{\rho}\right)\mathbf{v}_A$$ (1-22)

$$\mathbf{T} = \sum_{A=1}^{A=N} \mathbf{T}_A - \rho_A \mathbf{u}_A \mathbf{u}_A$$ (1-23)

$$\mathbf{b} = \sum_{A=1}^{A=N} \left(\frac{\rho_A}{\rho}\right)\mathbf{b}_A$$ (1-24)

in which $\mathbf{u}_A$ is the mass diffusion velocity given by

$$\mathbf{u}_A = \mathbf{v}_A - \mathbf{v}$$ (1-25)

we can sum Eqs. (1-11) and (1-19) to obtain

$$\frac{\partial \rho}{\partial t} + \nabla \cdot (\rho \mathbf{v}) = 0$$ (1-26)

$$\rho\left(\frac{\partial \mathbf{v}}{\partial t} + \mathbf{v} \cdot \nabla \mathbf{v}\right) = \rho \mathbf{b} + \nabla \cdot \mathbf{T} \tag{1-27}$$

$$\mathbf{T} = \mathbf{T}^{T} \tag{1-28}$$

Here we have used the axioms given by Eqs. (1-16) and (1-18) to eliminate any possible *diffusion momentum source* or *chemical reaction momentum source* in the mass average stress equations of motion.

The Stefan–Maxwell Equations

In traditional studies of fluid mechanics one can find processes in which each term in Eq. (1-27) is significant. If one considers hydrostatics, viscometric flows, boundary layer flows, and accelerated flows, one finds examples of processes in which all four terms in Eq. (1-27) must be taken into account. With the species momentum equation given by Eq. (1-19), the situation is quite different and in the analysis that follows we show how Eq. (1-19) simplifies to the Stefan–Maxwell equations. We begin with a decomposition that is appropriate for fluids and express the species stress tensor as

$$\mathbf{T}_{A} = -p_{A}\mathbf{I} + \boldsymbol{\tau}_{A} \tag{1-29}$$

Here p_A represents the partial pressure of species A and use of Eq. (1-29) in Eq. (1-19) yields

$$\rho_{A}\left(\frac{\partial \mathbf{v}_{A}}{\partial t} + \mathbf{v}_{A} \cdot \nabla \mathbf{v}_{A}\right) = -\nabla p_{A} + \rho_{A}\mathbf{g} + \nabla \cdot \boldsymbol{\tau}_{A}$$

$$+ \sum_{B=1}^{B=N} \mathbf{P}_{AB} + r_{A}(\bar{\mathbf{v}}_{A} - \mathbf{v}_{A}) \tag{1-30}$$

Here we have replaced $\mathbf{b}_A$ with the gravitational body force $\mathbf{g}$ so that we are now restricted from considering the diffusion process of charged species in the presence of an electrostatic or electromagnetic field. Our next step is also restrictive since we draw upon dilute gas kinetic theory [12, page 109] to represent the diffusion momentum source as

$$\mathbf{P}_{AB} = px_{A}x_{B}(\mathbf{v}_{B} - \mathbf{v}_{A})/\mathscr{D}_{AB} \tag{1-31}$$

Here x_A and x_B are the mole fractions of species A and B respectively and $\mathscr{D}_{AB}$ represents the binary diffusion coefficient [7, Chap. 16]. Although this result comes from dilute gas kinetic theory, we expect it to be a suitable form for the diffusion momentum source in liquid mixtures in which the binary diffusivity is an experimentally determined parameter.

We now substitute Eq. (1-31) into Eq. (1-30) and summarize our problem of diffusion and reaction as

$$\frac{\partial \rho_A}{\partial t} + \nabla \cdot (\rho_A \mathbf{v}_A) = r_A \tag{1-32}$$

$$\sum_{A=1}^{A=N} r_A = 0 \tag{1-33}$$

$$\rho_A \left(\frac{\partial \mathbf{v}_A}{\partial t} + \mathbf{v}_A \cdot \nabla \mathbf{v}_A \right) = -\nabla p_A + \rho_A \mathbf{g} + \nabla \cdot \boldsymbol{\tau}_A$$

$$+ p \sum_{B=1}^{B=N} \frac{x_A x_B (\mathbf{v}_B - \mathbf{v}_A)}{\mathscr{D}_{AB}} + r_A (\bar{\mathbf{v}}_A - \mathbf{v}_A)$$

$$\tag{1-34}$$

$$\boldsymbol{\tau}_A = \boldsymbol{\tau}_A^T \tag{1-35}$$

One should think of Eq. (1-32) as the governing differential equation for ρ_A with a constitutive equation for r_A determined by experiment or quantum mechanics. Although Eq. (1-33) has already been used in our theoretical analysis, its main contribution (in the form given by Eq. (1-9)) resides with the constraint that it imposes on stoichiometric coefficients. Equation (1-34) represents the governing differential equation for the species velocity, $\mathbf{v}_A$, that is required in order to solve Eq. (1-32), and Eq. (1-35) provides only a constraint on the form of the viscous stress tensor for species A.

While the determination of the species velocity by means of Eq. (1-34) would appear to be overwhelmingly complex, most of the terms in the species momentum equation are negligibly small and Eq. (1-34) generally reduces to the Stefan–Maxwell equations. The process of simplifying Eq. (1-34) begins with an order of magnitude estimate [52, Sec. 2.9] of the diffusion momentum source term. Use of Eq. (1-25) leads to

$$p \sum_{B=1}^{B=N} \frac{x_A x_B (\mathbf{v}_B - \mathbf{v}_A)}{\mathscr{D}_{AB}} = px_A \sum_{B=1}^{B=N} \frac{x_B (\mathbf{u}_B - \mathbf{u}_A)}{\mathscr{D}_{AB}} \tag{1-36}$$

and order-of-magnitude analysis yields the estimate

$$p \sum_{B=1}^{B=N} \frac{x_A x_B (\mathbf{v}_B - \mathbf{v}_A)}{\mathscr{D}_{AB}} = \mathbf{O}\left(\frac{p x_A u}{\mathscr{D}}\right) \tag{1-37}$$

Here u is an appropriate characteristic diffusion velocity and $\mathscr{D}$ is an appropriate measure of the diffusivity. It is convenient to restrict this analysis to gases and express the total pressure in terms of the *speed of sound* [51, Sec. 10.3] and write

$$p = \mathbf{O}(\rho C^2) \tag{1-38}$$

in which C represents the speed of sound for the N-component mixture. Use of this result in Eq. (1-37) leads to

$$p \sum_{B=1}^{B=N} \frac{x_A x_B (\mathbf{v}_B - \mathbf{v}_A)}{\mathscr{D}_{AB}} = \mathbf{O}\left(\frac{x_A \rho C^2 u}{\mathscr{D}}\right) \tag{1-39}$$

and in order to obtain an estimate based on quantities that are known *a priori* we need a useful representation for the diffusion velocity. We obtain this estimate from a result to be proved as part of this analysis, i.e. Fick's law for binary diffusion. This can be written as [7, Sec. 16.2]

$$\rho_A \mathbf{u}_A = -\rho \mathscr{D}_{AB} \nabla(\rho_A/\rho) \tag{1-40}$$

and leads to an estimate of the diffusion velocity given by

$$u = \mathbf{O}(\mathscr{D}/L_{\mathscr{D}}) \tag{1-41}$$

Here $L_{\mathscr{D}}$ is the diffusion length and it represents the distance over which significant variations in ρ_A occur. For the evaporation process taking place in the Stefan diffusion tube illustrated in Fig. 1-3, the diffusion length is equivalent to l while for the boundary layer mass transfer process we have $L_{\mathscr{D}} = \delta$. Use of the estimate given by Eq. (1-41) in Eq. (1-39) leads to

$$p \sum_{B=1}^{B=N} \frac{x_A x_B (\mathbf{v}_B - \mathbf{v}_A)}{\mathscr{D}_{AB}} = \mathbf{O}\left(\frac{\rho_A C^2}{L_{\mathscr{D}}}\right) \tag{1-42}$$

in which we have replaced $x_A \rho$ by the species density ρ_A. This approximation is in keeping with the accuracy of the estimate given by Eq. (1-42).

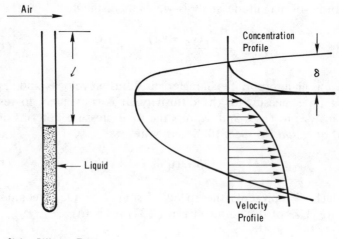

Figure 1-3   Diffusion length.

All of the terms on the right hand side of Eq. (1-42) can be estimated *a priori*, and if we can do this with the other terms in Eq. (1-34) some rational simplification can be effected. We begin with the first term in Eq. (1-34) and estimate its magnitude as

$$\rho_A \frac{\partial \mathbf{v}_A}{\partial t} = \mathbf{O}\left(\frac{\rho_A \Delta v_A}{t^*}\right) \tag{1-43}$$

in which Δv_A represents the change in $\mathbf{v}_A$ that takes place during the characteristic time t^*. We can *overestimate* the local acceleration of species A according to

$$\rho_A \frac{\partial \mathbf{v}_A}{\partial t} = \mathbf{O}\left(\frac{\rho_A v}{t^*}\right) \tag{1-44}$$

where v represents an estimate of the mass average velocity. In order to neglect local acceleration in Eq. (1-34) we require that

$$\rho_A \frac{\partial \mathbf{v}_A}{\partial t} \ll p \sum_{B=1}^{B=N} \frac{x_A x_B (\mathbf{v}_B - \mathbf{v}_A)}{\mathscr{D}_{AB}} \tag{1-45}$$

where it is understood that the inequality refers to the magnitude of the vectors involved. In terms of the estimates given by Eqs. (1-42) and (1-44) we have

$$\frac{\rho_A v}{t^*} \ll \frac{\rho_A C^2}{L_{\mathscr{D}}} \tag{1-46}$$

and we can express this result as

$$\frac{t^* C^2}{v} \gg \mathrm{Re}\left(\frac{L_{\mathscr{D}}}{L_\mu}\right), \quad \begin{array}{l} \text{local acceleration} \\ \text{is negligible} \end{array} \tag{1-47}$$

Here L_μ represents the *viscous length* discussed in detail elsewhere [50, Sec. 1.1.2.5] and the Reynolds number is defined by

$$\mathrm{Re} = v L_\mu / v \tag{1-48}$$

For a typical gas phase process we might choose

$$v \sim 10^{-5}\,\mathrm{m}^2/\mathrm{s}, \quad t^* \sim 1\,\mathrm{s}, \quad C \sim 300\,\mathrm{m/s}$$

so that $t^* C^2 / v$ is given by

$$\frac{t^* C^2}{v^2} \sim 10^{10} \tag{1-49}$$

and it is clear from this result that the local acceleration in Eq. (1-34) can be neglected under all but the most unusual circumstances. It is important to remember that this is not at all true for the mass average momentum equation given by Eq. (1-27), and it should be clear from this analysis that the species momentum equation has rather different characteristics than the mass average momentum equation.

In order to neglect convective acceleration we require

$$\rho_A \mathbf{v}_A \cdot \nabla \mathbf{v}_A \ll p \sum_{B=1}^{B=N} \frac{x_A x_B (\mathbf{v}_B - \mathbf{v}_A)}{\mathscr{D}_{AB}} \tag{1-50}$$

Our estimate of the convective acceleration follows a development given elsewhere [50, Sec. 1.1.25], and the inequality given by Eq. (1-50) is equivalent to the restriction

$$M^2 (L_{\mathscr{D}} / L_\rho) \ll 1, \quad \begin{array}{l} \text{convective acceleration} \\ \text{is negligible} \end{array} \tag{1-51}$$

Here L_ρ is the *inertial length* and M is the Mach number. For most processes $L_{\mathscr{D}} \le L_\rho$, thus the convective acceleration term in Eq.

(1-35) can be neglected when the Mach number is small compared
to one.

At this point we need to spend some time thinking about the fact
that Eq. (1-34) is a *vector equation* and the results given by Eqs.
(1-47) and (1-51) are *scalar constraints*. For the boundary layer
diffusion process illustrated in Fig. 1-3 we are normally interested in
using Eq. (1-34) to determine the component of the diffusion
velocity, $\mathbf{u}_A$, in a direction *orthogonal* to the *mean flow*. However,
the constraint given by Eq. (1-51) represents a comparison of the
convective acceleration *parallel* to the mean flow with the diffusion
momentum source. A little thought will indicate that the component
of the convective acceleration in a direction orthogonal to the mean
flow is generally much smaller than the component parallel to the
mean flow, thus Eq. (1-51) is an overly severe constraint in terms of
the practical application of Eq. (1-34) to determine the diffusion
velocity.

We now move from the left hand side of Eq. (1-34) to the right
hand side and note that the body force term can be neglected when

$$\rho_A \mathbf{g} \ll p \sum_{B=1}^{B=N} \frac{x_A x_B (\mathbf{v}_B - \mathbf{v}_A)}{\mathscr{D}_{AB}} \tag{1-52}$$

and this leads to the constraint

$$\frac{g L_{\mathscr{D}}}{C^2} \ll 1 \quad \begin{array}{l}\text{negligible gravitational}\\ \text{effects}\end{array} \tag{1-53}$$

For all realistic values of $L_{\mathscr{D}}$ this constraint indicates that gravitational effects are *always* negligible in the species momentum
equation, a conclusion that is in stark contrast to the role of gravity
in the mass average momentum equation. The last term in Eq.
(1-34) represents a momentum source owing to the possibility that
molecules of species A created by chemical reaction may have an
average velocity different from $\mathbf{v}_A$. This effect may be neglected
when

$$r_A (\bar{\mathbf{v}}_A - \mathbf{v}_A) \ll p \sum_{B=1}^{B=N} \frac{x_A x_B (\mathbf{v}_B - \mathbf{v}_A)}{\mathscr{D}_{AB}} \tag{1-54}$$

If we accept the plausible hypothesis that

$$\bar{\mathbf{v}}_A - \mathbf{v}_A = \mathbf{O}(u) \tag{1-55}$$

we can use Eq. (1-39) and the approximation $x_A \rho \sim \rho_A$ to obtain the constraint

$$r_A \ll \rho_A C^2/\mathscr{D} , \quad \begin{array}{l}\text{negligible reaction}\\\text{momentum source}\end{array} \qquad (1\text{-}56)$$

To be definitive we could consider the ubiquitous first-order irreversible reaction and write

$$r_A = k\rho_A \qquad (1\text{-}57)$$

so that Eq. (1-56) leads to a constraint on the reaction rate coefficients given by

$$k \ll C^2/\mathscr{D} , \quad \begin{array}{l}\text{negligible reaction}\\\text{momentum source}\end{array} \qquad (1\text{-}58)$$

For typical gas phase conditions

$$C \sim 300 \text{ m/s} , \quad \mathscr{D} \sim 10^{-5} \text{ m}^2/\text{s}$$

we find that Eq. (1-59) yields

$$k \ll 10^{10} \text{ s}^{-1} \qquad (1\text{-}59)$$

For liquids these constraints cannot be expressed in the convenient form given by Eqs. (1-47), (1-51), (1-53), and (1-58); however, an examination of the original inequalities given by Eqs. (1-45), (1-50), (1-52), and (1-54) indicates that all of these constraints are satisfied under normal circumstances (this excludes non-linear and viscoelastic fluids) and Eq. (1-35) can be simplified to

$$0 = -\nabla p_A + \nabla \cdot \boldsymbol{\tau}_A + p \sum_{B=1}^{B=N} \frac{x_A x_B (\mathbf{v}_B - \mathbf{v}_A)}{\mathscr{D}_{AB}} \qquad (1\text{-}60)$$

In the next section we will consider the problem of diffusion and reaction in the porous system illustrated in Fig. 1-4. The κ-phase represents a rigid solid and the γ-phase represents a gas or liquid in which the diffusion of reactants and products takes place. If we designate the mean free path of species A by l_A, then we know that bulk diffusion occurs when $l_A \ll l_\gamma$. Under these circumstances we can estimate the term $\nabla \cdot \boldsymbol{\tau}_A$ by analogy with the comparable term in the mass averaged stress equations of motion. For a Newtonian fluid this leads to [50, Sec. 1.1.2.5]

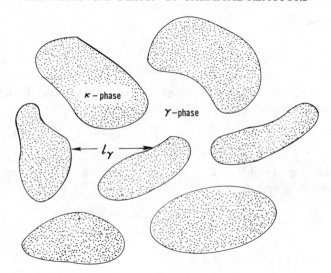

Figure 1-4   Diffusion and reaction in porous media.

$$\boldsymbol{\nabla} \cdot \boldsymbol{\tau}_A = \mathbf{O}\left(\frac{\mu_A v_A}{L_\mu^2}\right) \tag{1-61}$$

in which μ_A represents the "viscosity" of species A. Given that the viscosity of a gas mixture is approximated by [21, page 160]

$$\mu = \frac{1}{3} \sum_{A=1}^{A=N} \rho_A C_A l_A \tag{1-62}$$

where C_A is the mean speed, we represent μ_A as

$$\mu_A \sim \rho_A C_A l_A \sim \rho_A C l_A \tag{1-63}$$

We can use this result in Eq. (1-61) and identify the viscous length as l_γ to obtain

$$\boldsymbol{\nabla} \cdot \boldsymbol{\tau}_A = \mathbf{O}\left(\frac{\rho_A C l_A v_A}{l_\gamma^2}\right) \tag{1-64}$$

A comparison of this estimate with that given by Eq. (1-42) indicates that we can neglect $\boldsymbol{\nabla} \cdot \boldsymbol{\tau}_A$ in Eq. (1-60) when

$$\left(\frac{l_A}{l_\gamma}\right)\left(\frac{L_{\mathscr{D}}}{l_\gamma}\right)\left(\frac{v_A}{C}\right) \ll 1, \quad \begin{array}{l} \text{negligible viscous} \\ \text{effects} \end{array} \tag{1-65}$$

Table 1-1   Restrictions Leading to the Stefan–Maxwell Equations

---

(1) $\dfrac{t^* C^2}{\nu} \gg \mathrm{Re}\left(\dfrac{L_{\mathscr{D}}}{L_{\mu}}\right)$,   negligible local acceleration

(2) $M^2\left(\dfrac{L_{\mathscr{D}}}{L_{\rho}}\right) \ll 1$,   negligible convective acceleration

(3) $\dfrac{g L_{\mathscr{D}}}{C^2} \ll 1$,   negligible gravitational effects

(4) $r_A \ll \rho_A C^2 / \mathscr{D}$,   negligible reaction momentum source

(5) $\left(\dfrac{l_A}{L_{\mu}}\right)\left(\dfrac{L_{\mathscr{D}}}{L_{\mu}}\right)\left(\dfrac{v_A}{C}\right) \ll 1$,   negligible viscous effects

---

Under these circumstances Eq. (1-60) simplifies to

$$0 = -\nabla p_A + p \sum_{B=1}^{B=N} \frac{x_A x_B (\mathbf{v}_B - \mathbf{v}_A)}{\mathscr{D}_{AB}}, \quad A = 1, 2, \ldots, N$$

$$(1\text{-}66)$$

and this result can be simplified to the Stefan–Maxwell equations after we consider the role of the total momentum equation in the solution of diffusion problems.

A summary of the restrictions that led from Eq. (1-34) to the above result is given in Table 1-1, and, in general, these restrictions are easily satisfied for a wide variety of diffusion and reaction processes. However, there is one familiar case for which the inequality given by Eq. (1-65) is reversed. This occurs when $l_A \gg l_{\gamma}$ and the diffusion momentum source becomes negligible relative to the rate of momentum exchange between species A and the κ-phase illustrated in Fig. 1-4. The analysis for liquids indicates that the viscous term in Eq. (1-60) is always negligible relative to the diffusion momentum source and Knudsen diffusion is strictly a gas-phase phenomenon. In Sec. 4 we will consider Eq. (1-60) in detail and develop a theory for bulk diffusion, Knudsen diffusion, and Darcy flow in porous catalysts.

Role of the Total Momentum Equation

Before going on to the application of Eq. (1-66) we note that because of Axiom III we can sum Eq. (1-66) over all species to obtain

$$0 = -\nabla p \qquad (1\text{-}67)$$

in which the total pressure, p, is the sum of all the partial pressures. This appears to be an overly severe restriction on the total pressure field that is not at all consistent with the mass average momentum equation which can be written as

$$\rho \frac{\partial \mathbf{v}}{\partial t} + \mathbf{v} \cdot \nabla \mathbf{v} = -\nabla p + \rho \mathbf{g} + \nabla \cdot \boldsymbol{\tau} \qquad (1\text{-}68)$$

A comparison of Eq. (1-67) and (1-68) suggests that some of the restrictions listed in Table 1-1 cannot be valid and perhaps Eq. (1-66) is not to be trusted as the basis for predicting either the species velocity or the species diffusion velocity. This paradox can be explained by returning to Eq. (1-34) and imposing the restrictions given in Table 1-1. Under these circumstances we can define a "small" vector $\boldsymbol{\varepsilon}_A$ as

$$\boldsymbol{\varepsilon}_A = \rho_A \left(\frac{\partial \mathbf{v}_A}{\partial t} + \mathbf{v}_A \cdot \nabla \mathbf{v}_A \right) - \rho_A \mathbf{g} - \nabla \cdot \boldsymbol{\tau}_A - r_A (\bar{\mathbf{v}}_A - \mathbf{v}_A)$$

$$(1\text{-}69)$$

and express Eq. (1-34) in the form

$$\boldsymbol{\varepsilon}_A = -\nabla p_A + \sum_{B=1}^{B=N} \mathbf{P}_{AB} \qquad (1\text{-}70)$$

Here it should be clear that by "small" we mean that $\boldsymbol{\varepsilon}_A$ is small relative to the terms on the right hand side of Eq. (1-70). In any given species momentum equation, $\boldsymbol{\varepsilon}_A$ can be neglected and Eq. (1-70) reduces to Eq. (1-66). However, while there are N independent equations of the form given by Eq. (1-70) there are *only* $N-1$ independent equations of the form given by Eq. (1-66). This means that the N-component diffusion problem *cannot be solved* solely in terms of the Stefan–Maxwell equations and the species continuity equation given by Eq. (1-11). The closure is obtained either by the employment of the total or mass average momentum equation given by Eq. (1-68) or by the imposition of some plausible constraint. The most common of these are:

1. The velocity of one species is zero (the "stagnant film theory" applied to mass transfer processes)

2. The fluxes are constrained by energy considerations ("equi-molar counter diffusion" applied to distillation processes)
3. The fluxes are constrained by stoichiometric considerations (commonly applied in reactor design problems)
4. The dilute solution approximation

In each of these cases the mass average momentum equation is circumvented (and therefore violated) by means of a plausible intuitive hypothesis [8, page 4]. Given that Eq. (1-68) is a weak member of the set of N independent momentum equations, the use of a reasonable constraint, such as one of those listed above, in order to circumvent the total momentum equation represents a judicious course of action. The use of the dilute solution approximation is a particularly intriguing example of the manner in which we commonly violate the total momentum equation.

The total momentum equation can be used to accomplish the final simplification of Eq. (1-66) by using Eq. (1-70) and Axiom III to obtain

$$\boldsymbol{\varepsilon} = \sum_{A=1}^{A=N} \boldsymbol{\varepsilon}_A = -\nabla p \tag{1-71}$$

This result can be used, along with the following expression for the partial pressure

$$p_A = x_A p \tag{1-72}$$

to express Eq. (1-70) as

$$\boldsymbol{\varepsilon}_A = -p\nabla x_A - x_A \boldsymbol{\varepsilon} + p \sum_{B=1}^{B=N} \frac{x_A x_B (\mathbf{v}_B - \mathbf{v}_A)}{\mathscr{D}_{AB}} \tag{1-73}$$

It is consistent with the simplifications that were used in the development of Eq. (1-66) to neglect $x_A \boldsymbol{\varepsilon}$ and $\boldsymbol{\varepsilon}_A$ in Eq. (1-73) in order to arrive at the traditional form of the Stefan–Maxwell equations

$$0 = -\nabla x_A + \sum_{B=1}^{B=N} \frac{x_A x_B (\mathbf{v}_B - \mathbf{v}_A)}{\mathscr{D}_{AB}}, \quad A = 1, 2, \ldots, N \tag{1-74}$$

Molar Forms

It is well established that the molar form of the species continuity equation is preferred for practical applications, thus we divide Eq.

(1-12) by the molecular weight of species A to obtain

$$\frac{\partial c_A}{\partial t} + \nabla \cdot \mathbf{N}_A = R_A \qquad (1\text{-}75)$$

The molar rates of reaction, R_A, are constrained by

$$\sum_{A=1}^{A=N} R_A N_{\alpha A} = 0, \quad \alpha = 1, 2, \dots, Q \qquad (1\text{-}76)$$

while the Stefan–Maxwell equations can be expressed in terms of the molar fluxes to give

$$0 = -\nabla x_A + \sum_{B=1}^{B=N} \frac{x_A \mathbf{N}_B - x_B \mathbf{N}_A}{c\mathscr{D}_{AB}} \qquad (1\text{-}77)$$

The dilute solution approximation mentioned above results when†

$$x_A \ll 1, \quad \mathbf{N}_B = \mathbf{O}(\mathbf{N}_A), \quad B = 1, 2, \dots, N \qquad (1\text{-}78)$$

and it allows us to write Eq. (1-73) as

$$0 = -\nabla x_A - \left\{ \frac{1}{c} \sum_{\substack{B=1 \\ B \neq A}}^{B=N} \frac{x_B}{\mathscr{D}_{AB}} \right\} \mathbf{N}_A \qquad (1\text{-}79)$$

If the temperature is taken to be constant and the mixture diffusivity is defined as

$$\frac{1}{\mathscr{D}_{AM}} = \sum_{\substack{B=1 \\ B \neq A}}^{B=N} \frac{x_B}{\mathscr{D}_{AB}} \qquad (1\text{-}80)$$

one can use Eq. (1-79) to represent the flux by

$$\mathbf{N}_A = -\mathscr{D}_{Am} \nabla c_A \qquad (1\text{-}81)$$

Substitution of this result into Eq. (1-75) leads to

$$\frac{\partial c_A}{\partial t} = \nabla \cdot (\mathscr{D}_{Am} \nabla c_A) + R_A \qquad (1\text{-}82)$$

---

†This approximation is most often applied to particular components of the molar flux vectors.

and one can proceed to solve for the concentration of species A without recourse to the total momentum equation.

In this section we have seen the role of the species momentum equation in formulating diffusion and reaction problems. It should be clear that most problems of this type are solved by the imposition of some judicious constraint on the species fluxes rather than by satisfaction of the entire set of the laws of mechanics for multicomponent systems. Having acquired a degree of understanding of the general problem of diffusive transport, we are ready to move on to the influence of boundary conditions and the complexities of multiphase transport processes.

2. BULK DIFFUSION AND REACTION IN POROUS MEDIA

In this section we consider the simple process of bulk diffusion in a catalyst pellet with a heterogeneous, first-order, irreversible reaction. This analysis will serve to illustrate the principles of the method of volume averaging [1, 29, 40, 53] and provide some insight into the closure problem that accompanies every application of the method of volume averaging.

The system under consideration is illustrated in Fig. 1-4 and the governing equations are given by

$$\frac{\partial c_A}{\partial t} + \mathbf{\nabla} \cdot \mathbf{N}_A = 0 \qquad (2\text{-}1)$$

$$0 = -\mathbf{\nabla} c_A + \sum_{B=1}^{B=N} \frac{x_A \mathbf{N}_B - x_B \mathbf{N}_A}{\mathscr{D}_{AB}} \qquad (2\text{-}2)$$

and we plan to avoid the use of the total momentum equation by restricting the analysis to either of the two cases:

1. $x_A \ll 1$, $\mathbf{N}_B = \mathbf{O}(\mathbf{N}_A)$, $B = 1, 2, \ldots, N$
2. equimolar counterdiffusion

These restrictions are not at all consistent with reacting systems; however, they lead to a tractable problem from which we can learn a great deal. These restrictions lead to

$$\mathbf{N}_A = -\mathscr{D} \mathbf{\nabla} c_A \qquad (2\text{-}3)$$

in which $\mathscr{D}$ is $\mathscr{D}_{AB}$ for the binary system or is given by

$$\frac{1}{\mathscr{D}} = \sum_{\substack{B=1 \\ B \neq A}}^{B=N} \frac{x_B}{\mathscr{D}_{AB}} \tag{2-4}$$

for the dilute solution approximation. Use of Eq. (2-3) in Eq. (2-1) yields the governing differential equation for c_A

$$\frac{\partial c_A}{\partial t} = \mathbf{\nabla} \cdot (\mathscr{D} \mathbf{\nabla} c_A) \tag{2-5}$$

and we express the boundary condition at the γ–κ interface as

$$\mathbf{n}_{\gamma\kappa} \cdot \mathbf{N}_A = \frac{\partial \hat{c}_A}{\partial t} + \hat{k} \hat{c}_A , \qquad \text{at } A_{\gamma\kappa} \tag{2-6}$$

Here $\mathbf{n}_{\gamma\kappa}$ represents the unit normal vector pointing from the γ-phase into the κ-phase, and $\hat{c}_A$ represents the surface concentration of species A. In writing Eq. (2-6) we have neglected surface diffusion and assumed that the reaction is first order and irreversible.

While we are definitely interested in transient diffusion and reaction processes, it is quite plausible that the surface is in a quasi-steady condition, i.e.

$$\frac{\partial \hat{c}_A}{\partial t} \ll \hat{k} \hat{c}_A \tag{2-7}$$

and Eq. (2-6) simplifies to

$$\mathbf{n}_{\gamma} \cdot \mathbf{N}_A = \hat{k} \hat{c}_A , \qquad \text{at } A_{\gamma\kappa} \tag{2-8}$$

The inequality given by Eq. (2-7) will be satisfied when

$$t^* \hat{k} \gg 1 \tag{2-9}$$

in which t^* represents a characteristic time during which significant changes in $\hat{c}_A$ take place.

In order to complete our statement of the boundary condition, we require a constitutive equation for the flux and it is in keeping with the simplistic nature of this problem to choose a linear relation of the form

$$\mathbf{n}_{\gamma\kappa} \cdot \mathbf{N}_A = k_1 c_A - k_2 \hat{c}_A , \qquad \text{at } A_{\gamma\kappa} \tag{2-10}$$

Here k_1 is the adsorption rate constant and k_2 is the desorption rate constant. Use of Eq. (2-10) in conjunction with Eq. (2-8) leads to

$$\hat{k}\hat{c}_A = \left(\frac{k_1 \hat{k}}{\hat{k} + k_2} \right) c_A , \qquad \text{at } A_{\gamma\kappa} \qquad (2\text{-}11)$$

We can use this result along with Eq. (2-3) in the boundary condition given by Eq. (2-8) to obtain

$$-\mathbf{n}_{\gamma\kappa} \cdot \mathscr{D}\nabla c_A = k c_A , \qquad \text{at } A_{\gamma\kappa} \qquad (2\text{-}12)$$

Here k is a pseudo rate constant defined in the obvious manner. In terms of k, the quasi-steady constraint given by Eq. (2-9) takes the form

$$t^* \left(\frac{k_2 k}{k_1 - k} \right) \gg 1 \qquad (2\text{-}13)$$

and from this we see that k, k_1, and k_2 must be known before one can determine if the surface is in a quasi-steady state. A more conservative form of Eq. (2-13) is given by

$$t^* k / K \gg 1 \qquad (2\text{-}14)$$

where K is the equilibrium coefficient, k_1/k_2. In this form one needs to know only the pseudo rate constant and the equilibrium constant in order to determine the state of the surface.

Volume Averaging

The method of volume averaging begins with the boundary value problem

$$\frac{\partial c_A}{\partial t} = \nabla \cdot (\mathscr{D}\nabla c_A) , \qquad \text{in the } \gamma\text{-phase} \qquad (2\text{-}15)$$

B.C.1 $\qquad\qquad -\mathbf{n}_{\gamma\kappa} \cdot \mathscr{D}\nabla c_A = k c_A , \qquad \text{at } A_{\gamma\kappa} \qquad (2\text{-}16)$

B.C.2 $\qquad\qquad c_A = f(\mathbf{r}, t) , \qquad \text{at } \mathscr{A}_{\gamma e} \qquad (2\text{-}17)$

in which $\mathscr{A}_{\gamma e}$ represents the bounding surface of the γ-phase as illustrated in Fig. 2-1. Since the location of the γ–κ interface is unknown, Eqs. (2-15) through (2-17) cannot be solved to determine

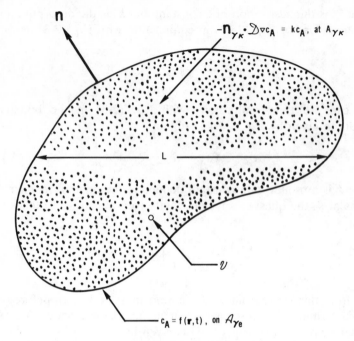

Figure 2-1   Boundary conditions for diffusion and reaction in a porous medium.

c_A and the overall rate of reaction. As an alternative, we seek to determine an *average concentration* with the hope that this will lead us to a reasonably accurate prediction of the overall rate of reaction. To accomplish this we need to develop a volume-averaged transport equation and a method of predicting the coefficients that appear in that equation.

With every point in space we associate an averaging volume $\mathcal{V}$ that can be expressed as

$$\mathcal{V} = V_\gamma + V_\kappa \qquad (2\text{-}18)$$

Here V_γ represents the volume of the γ-phase contained within the averaging volume and the volume fraction of the γ-phase (often identified as the *void fraction*) is given by

$$\varepsilon_\gamma = V_\gamma / \mathcal{V} \qquad (2\text{-}19)$$

The *phase average* concentration is defined by

$$\langle c_A \rangle = \frac{1}{\mathscr{V}} \int_{V_\gamma} c_A \, dV \qquad (2\text{-}20)$$

and this concentration must be continuous and differentiable if the method of volume averaging is to be successful. If the radius of the averaging volume is r_0 we normally think of the pertinent length scales as being constrained by [54]

$$l_\gamma \ll r_0 \ll L \qquad (2\text{-}21)$$

however, a more recent study [11, Sec. 2] suggests that even more severe constraints on the length scales may be required, and these are given by

$$\frac{r_0}{L} \ll \frac{l_\gamma}{r_0} \ll 1 \qquad (2\text{-}22)$$

These constraints result from order-of-magnitude estimates and a recent computation study by Tsing [48] suggests that they are overly severe. Further detailed computations are required before we can be certain about the nature of the length scale constraints.

A crucial tool in the method of volume averaging is the spatial averaging theorem [1, 29, 40, 53] which can be stated as

$$\langle \nabla \psi \rangle = \nabla \langle \psi \rangle + \frac{1}{\mathscr{V}} \int_{A_{\gamma\kappa}} \mathbf{n}_{\gamma\kappa} \psi \, dV \qquad (2\text{-}23)$$

in which ψ is any quantity associated with the γ-phase. We can form the average of Eq. (2-15) to obtain

$$\frac{1}{\mathscr{V}} \int_{V_\gamma} \left(\frac{\partial c_A}{\partial t} \right) dV = \frac{1}{\mathscr{V}} \int_{V_\gamma} \nabla \cdot (\mathscr{D} \nabla c_A) \, dV \qquad (2\text{-}24)$$

and since V_γ is not a function of time, the left hand side of this result becomes

$$\frac{1}{\mathscr{V}} \int_{V_\gamma} \left(\frac{\partial c_A}{\partial t} \right) dV = \frac{\partial}{\partial t} \left\{ \frac{1}{\mathscr{V}} \int_{V_\gamma} c_A \, dV \right\} = \frac{\partial \langle c_A \rangle}{\partial t} \qquad (2\text{-}25)$$

We now write Eq. (2-24) as

$$\frac{\partial \langle c_A \rangle}{\partial t} = \langle \nabla \cdot (\mathscr{D} \nabla c_A) \rangle \qquad (2\text{-}26)$$

and use the averaging theorem to obtain

$$\frac{\partial \langle c_A \rangle}{\partial t} = \nabla \cdot \langle \mathscr{D} \nabla c_A \rangle + \frac{1}{\mathscr{V}} \int_{A_{\gamma\kappa}} \mathbf{n}_{\gamma\kappa} \cdot \mathscr{D} \nabla c_A \, dA \qquad (2\text{-}27)$$

If one is willing to neglect variations of the diffusivity within the averaging volume, the averaging theorem can be used again along with the boundary condition given by Eq. (2-16) to write Eq. (2-27) in the form

$$\frac{\partial \langle c_A \rangle}{\partial t} = \nabla \cdot \left\{ \mathscr{D} \left[\nabla \langle c_A \rangle + \frac{1}{\mathscr{V}} \int_{A_{\gamma\kappa}} \mathbf{n}_{\gamma\kappa} c_A \, dA \right] \right\} - \frac{1}{\mathscr{V}} \int_{A_{\gamma\kappa}} k c_A \, dA$$

$$(2\text{-}28)$$

Here we have neglected variations in $\mathscr{D}$ over distances on the order of r_0; however, we retain the capability of allowing $\mathscr{D}$ to change over distances on the order of L as illustrated in Fig. 2-1. In the study of multiphase transport phenomena, the average of choice is usually the *intrinsic phase average* which is defined as

$$\langle c_A \rangle^{\gamma} = \frac{1}{V_{\gamma}} \int_{V_{\gamma}} c_A \, dV \qquad (2\text{-}29)$$

The reason for this choice is that when c_A is a constant, the *intrinsic phase average* is exactly equal to that constant whereas the *phase average* is not. These two averages are related by

$$\langle c_A \rangle = \varepsilon_{\gamma} \langle c_A \rangle^{\gamma} \qquad (2\text{-}30)$$

and use of this result in Eq. (2-28) yields

$$\varepsilon_{\gamma} \frac{\partial \langle c_A \rangle^{\gamma}}{\partial t} = \nabla \cdot \left\{ \mathscr{D} \left[\varepsilon_{\gamma} \nabla \langle c_A \rangle^{\gamma} + \langle c_A \rangle^{\gamma} \nabla \varepsilon_{\gamma} + \frac{1}{\mathscr{V}} \int_{A_{\gamma\kappa}} \mathbf{n}_{\gamma\kappa} c_A \, dA \right] \right\}$$

$$- \frac{1}{\mathscr{V}} \int_{A_{\gamma\kappa}} k c_A \, dA \qquad (2\text{-}31)$$

If we think of this result as the governing differential equation for $\langle c_A \rangle^{\gamma}$ then it becomes clear that we must eliminate the point concentration c_A from Eq. (2-31).

It is important to note that in going from Eqs. (2-15) and (2-16) to the volume-averaged transport equation given by Eq. (2-31), we have accepted the *loss of information* associated with the use of the average concentration, $\langle c_A \rangle^\gamma$. Nevertheless, Eq. (2-31) can be thought of as an *exact result* in terms of the initial problem statement which was, of course, an approximate description of diffusion and reaction. We can eliminate the point concentration by use of the decomposition [16]

$$c_A = \langle c_A \rangle^\gamma + \tilde{c}_A \tag{2-32}$$

in which $\tilde{c}_A$ is referred to as a spatial deviation. Use of this relation in Eq. (2-31) yields

$$\varepsilon_\gamma \frac{\partial \langle c_A \rangle^\gamma}{\partial t} = \nabla \cdot \left\{ \mathscr{D} \left[\varepsilon_\gamma \nabla \langle c_A \rangle^\gamma + \langle c_A \rangle^\gamma \nabla \varepsilon_\gamma + \frac{1}{\mathscr{V}} \int_{A_{\gamma\kappa}} \mathbf{n}_{\gamma\kappa} \langle c_A \rangle^\gamma \, dA \right. \right.$$

$$\left. \left. + \frac{1}{\mathscr{V}} \int_{A_{\gamma\kappa}} \mathbf{n}_{\gamma\kappa} \tilde{c}_A \, dA \right] \right\} - \frac{1}{\mathscr{V}} \int_{A_{\gamma\kappa}} k \langle c_A \rangle^\gamma \, dA$$

$$- \frac{1}{\mathscr{V}} \int_{A_{\gamma\kappa}} k \tilde{c}_A \, dA \tag{2-33}$$

It is shown elsewhere [11, Sec. 2] that the simplification

$$\frac{1}{\mathscr{V}} \int_{A_{\gamma\kappa}} \mathbf{n}_{\gamma\kappa} \langle c_A \rangle^\gamma \, dA = \left\{ \frac{1}{\mathscr{V}} \int_{A_{\gamma\kappa}} \mathbf{n}_{\gamma\kappa} \, dA \right\} \langle c_A \rangle^\gamma \tag{2-34}$$

is acceptable when $(r_0/L)^2 \ll 1$ and the averaging theorem can be used to show that

$$\frac{1}{\mathscr{V}} \int_{A_{\gamma\kappa}} \mathbf{n}_{\gamma\kappa} \, dA = -\nabla \varepsilon_\gamma \tag{2-35}$$

Use of these two results in Eq. (2-32) provides some simplification leading to

$$\varepsilon_\gamma \frac{\partial \langle c_A \rangle^\gamma}{\partial t} = \nabla \cdot \left\{ \mathscr{D} \left[\varepsilon_\gamma \nabla \langle c_A \rangle^\gamma + \frac{1}{\mathscr{V}} \int_{A_{\gamma\kappa}} \mathbf{n}_{\gamma\kappa} \tilde{c}_A \, dA \right] \right\}$$

$$- \frac{k}{\mathscr{V}} \int_{A_{\gamma\kappa}} \langle c_A \rangle^\gamma \, dA - \frac{k}{\mathscr{V}} \int_{A_{\gamma\kappa}} \tilde{c}_A \, dA \tag{2-36}$$

Here we have treated the pseudo rate constant in the same manner as the diffusivity, i.e. variations within the averaging volume illustrated in Fig. 2-1 are neglected but we retain the possibility that k

undergoes significant changes over the distance L. At this time, there is no *analysis* of the type associated with Eq. (2-34) that supports the simplification

$$\frac{1}{\mathcal{V}} \int_{A_{\gamma\kappa}} \langle c_A \rangle^{\gamma} \, dA = \left\{ \frac{1}{\mathcal{V}} \int_{A_{\gamma\kappa}} dA \right\} \langle c_A \rangle^{\gamma} \qquad (2\text{-}37)$$

however, it seems that this is reasonable when $(r_0/L)^2 \ll 1$ and we use this result in Eq. (2-36) to obtain

$$\varepsilon_{\gamma} \frac{\partial \langle c_A \rangle^{\gamma}}{\partial t} = \nabla \cdot \left\{ \mathcal{D} \left[\varepsilon_{\gamma} \nabla \langle c_A \rangle^{\gamma} + \frac{1}{\mathcal{V}} \int_{A_{\gamma\kappa}} \mathbf{n}_{\gamma\kappa} \tilde{c}_A \, dA \right] \right\}$$

$$- a_v k \langle c_A \rangle^{\gamma} - a_v k \langle \tilde{c}_A \rangle_{\gamma\kappa} \qquad (2\text{-}38)$$

The term a_v represents the interfacial area per unit volume of the porous medium and is given explicitly by

$$a_v = A_{\gamma\kappa}/\mathcal{V} \qquad (2\text{-}39)$$

Care needs to be taken when identifying an "area per unit volume" for in some cases this refers to an "interfacial area per unit volume of fluid" while in other cases it refers to the quantity defined by Eq. (2-39).

When a population or a field is averaged, one is usually committed to a loss of information. In the method of volume averaging this occurs in two steps. The first step is the averaging process in which one loses information in going from the c_A-field to the $\langle c_A \rangle^{\gamma}$-field, and the second step is in the development of a closure scheme for $\tilde{c}_A$. Information loss is greatest if one simply postulates a constitutive equation for $\tilde{c}_A$ and this information can be retrieved only through experimental studies. Information loss is minimized if the closure is obtained in terms of a governing differential equation and boundary conditions for $\tilde{c}_A$. This is the route to closure that we will follow in this section and the subsequent one; however, the route using a plausible constitutive equation will be illustrated in Sec. 4.

The boundary condition for $\tilde{c}_A$ at the $\gamma-\kappa$ interface can be obtained by use of the decomposition given by Eq. (2-32) in the boundary condition given by Eq. (2-16). This leads to

B.C.1 $-\mathbf{n}_{\gamma\kappa} \cdot \mathcal{D} \nabla \tilde{c}_A - k \tilde{c}_A = \mathbf{n}_{\gamma\kappa} \cdot \mathcal{D} \nabla \langle c_A \rangle^{\gamma} + k \langle c_A \rangle^{\gamma}$, at $A_{\gamma\kappa}$

$$(2\text{-}40)$$

Not only does this result provide a boundary condition for the $\tilde{c}_A$-field, but Ryan et al. [37] have shown that it can be used to produce the estimate

$$\tilde{c}_A = \mathbf{O}\left\{ \frac{\phi^2 + \mathbf{O}(L/l_\gamma)}{\phi^2 + \mathbf{O}(L/l_\gamma)^2} \right\} \langle c_A \rangle^\gamma \qquad (2\text{-}41)$$

Here ϕ represents the Thiele modulus

$$\phi = L\sqrt{a_v k/\mathcal{D}} \qquad (2\text{-}42)$$

and the length l_γ illustrated in Fig. 1-4 has been defined explicitly as

$$l_\gamma = a_v^{-1} \qquad (2\text{-}43)$$

For cases of practical interest, one finds that $\phi^2 \leqslant L/l_\gamma$ and Eq. (2-41) reduces to

$$\tilde{c}_A = \mathbf{O}\left\{ \left(\frac{l_\gamma}{L}\right) \langle c_A \rangle^\gamma \right\} \qquad (2\text{-}44)$$

and we see that $\tilde{c}_A$ is small compared to $\langle c_A \rangle^\gamma$. Although it does not occur in practice, it is interesting to note that the condition $\phi^2 \geqslant (L/l_\gamma)^2$ leads to

$$\tilde{c}_A = \mathbf{O}(\langle c_A \rangle^\gamma), \qquad \text{for } \phi^2 \geqslant (L/l_\gamma)^2 \qquad (2\text{-}45)$$

For very rapid reactions the concentration at the $\gamma-\kappa$ interface tends to zero and the decomposition given by (2-32) indicates that $\tilde{c}_A$ and $\langle c_A \rangle^\gamma$ are the same order of magnitude. Under these circumstances the reaction is diffusion controlled and the use of a porous catalyst is inappropriate. Thus, the practical considerations of reactor design dictate that

$$\tilde{c}_A \ll \langle c_A \rangle^\gamma \qquad (2\text{-}46)$$

and both Eqs. (2-38) and (2-40) can be simplified to

$$\varepsilon_\gamma \frac{\partial \langle c_A \rangle^\gamma}{\partial t} = \mathbf{\nabla} \cdot \left\{ \mathcal{D}\left[\varepsilon_\gamma \mathbf{\nabla} \langle c_A \rangle^\gamma + \frac{1}{\mathcal{V}} \int_{A_{\gamma\kappa}} \mathbf{n}_{\gamma\kappa} \tilde{c}_A \, dA \right] \right\} - a_v k \langle c_A \rangle^\gamma$$

$$(2\text{-}47)$$

B.C.1 $-\mathbf{n}_{\gamma\kappa}\cdot\mathcal{D}\nabla\tilde{c}_A = \mathbf{n}_{\gamma\kappa}\cdot\mathcal{D}\nabla\langle c_A\rangle^\gamma + k\langle c_A\rangle^\gamma$,   at $A_{\gamma\kappa}$

$$(2\text{-}48)$$

One must remember that the volume-averaged reaction rate is given *exactly* by the last two terms in Eq. (2-38) provided the pseudo rate constant undergoes negligible variation within the averaging volume. Thus, when Eq. (2-46) is applicable we can accurately describe the overall rate of reaction in terms of the average concentration $\langle c_A\rangle^\gamma$, and what is required at this point is a reliable transport equation for $\langle c_A\rangle^\gamma$.

The Closure Problem

In order to obtain a governing differential equation for $\tilde{c}_A$ we return to Eq. (2-15) and use the decomposition given by Eq. (2-32) to obtain†

$$\frac{\partial\langle c_A\rangle^\gamma}{\partial t} - \nabla\cdot(\mathcal{D}\nabla\langle c_A\rangle^\gamma) = -\left[\frac{\partial\tilde{c}_A}{\partial t} - \nabla\cdot(\mathcal{D}\nabla\tilde{c}_A)\right] \quad (2\text{-}49)$$

The intrinsic phase average form of this result is given by

$$\left\langle\frac{\partial\langle c_A\rangle^\gamma}{\partial t} - \nabla\cdot(\mathcal{D}\nabla\langle c_A\rangle^\gamma)\right\rangle^\gamma = -\left\langle\frac{\partial\tilde{c}_A}{\partial t} - \nabla\cdot(\mathcal{D}\nabla\tilde{c}_A)\right\rangle^\gamma$$

$$(2\text{-}50)$$

and it is shown elsewhere [11, Sec. 2] that Eq. (2-50) can be expressed as

$$\frac{\partial\langle c_A\rangle^\gamma}{\partial t} - \nabla\cdot(\mathcal{D}\nabla\langle c_A\rangle^\gamma) = -\left\langle\frac{\partial\tilde{c}_A}{\partial t} - \nabla\cdot(\mathcal{D}\nabla\tilde{c}_A)\right\rangle^\gamma \quad (2\text{-}51)$$

when $(r_0/L)^2 \ll 1$. Since we are already committed to this condition by means of Eq. (2-21) or (2-22) we can always treat intrinsic phase average quantities as constants with respect to the averaging process. Use of Eq. (2-51) in Eq. (2-49) yields a governing equation for $\tilde{c}_A$.

$$\frac{\partial\tilde{c}_A}{\partial t} - \nabla\cdot(\mathcal{D}\nabla\tilde{c}_A) = \left\langle\frac{\partial\tilde{c}_A}{\partial t} - \nabla\cdot(\mathcal{D}\nabla\tilde{c}_A)\right\rangle^\gamma \quad (2\text{-}52)$$

It should be obvious that there are many transient diffusion and

---

†This approach was suggested by Guillermo H. Crapiste.

reaction processes of interest [22] and for those processes one cannot neglect the first term in Eq. (2-47). Transient effects are important when the characteristic time t^* is comparable to or smaller than $L^2/\mathscr{D}$, i.e.,

$$\mathscr{D}t^*/L^2 \leqslant 1, \qquad \text{for a transient process} \qquad (2\text{-}53)$$

Although we must retain the transient term in the volume-averaged transport equation, this is *not* true of Eq. (2-52) since the $\tilde{c}_A$-field will be quasi-steady whenever

$$\mathscr{D}t^*/l_\gamma^2 \gg 1, \qquad \text{for a quasi-steady process} \qquad (2\text{-}54)$$

Because l_γ is so small relative to L this constraint is easily satisfied for all practical cases and Eq. (2-52) reduces to

$$\boldsymbol{\nabla} \cdot (\mathscr{D} \boldsymbol{\nabla} \tilde{c}_A) = \frac{1}{V_\gamma} \int_{V_\gamma} \boldsymbol{\nabla} \cdot (\mathscr{D} \boldsymbol{\nabla} \tilde{c}_A)\, dV \qquad (2\text{-}55)$$

While we wish to retain the possibility that the diffusivity $\mathscr{D}$ varies over the domain of interest shown in Fig. 2-1, this does not prevent us from using the obvious inequality

$$\mathscr{D} \nabla^2 \tilde{c}_A \gg \boldsymbol{\nabla} \mathscr{D} \cdot \boldsymbol{\nabla} \tilde{c}_A \qquad (2\text{-}56)$$

to write Eq. (2-55) in the form

$$\nabla^2 \tilde{c}_A = \frac{1}{V_\gamma} \int_{V_\gamma} \nabla^2 \tilde{c}_A\, dV \qquad (2\text{-}57)$$

Use of the divergence theorem along with Eq. (2-48) leads to

$$\nabla^2 \tilde{c}_A = -\frac{1}{V_\gamma} \int_{A_{\gamma\kappa}} \mathbf{n}_{\gamma\kappa} \cdot \boldsymbol{\nabla} \langle c_A \rangle^\gamma\, dA - \frac{1}{V_\gamma} \int_{A_{\gamma\kappa}} \frac{k \langle c_A \rangle^\gamma}{\mathscr{D}}\, dA$$

$$+ \frac{1}{V_\gamma} \int_{A_{\gamma e}} \mathbf{n}_\gamma \cdot \boldsymbol{\nabla} \tilde{c}_A\, dA \qquad (2\text{-}58)$$

Here $A_{\gamma e}$ represents the area of entrances and exits for the volume V_γ and $\mathbf{n}_\gamma$ represents the outwardly directed unit normal vector at the entrances and exits. We can follow the development given by Eqs. (2-34) and (2-35) to express Eq. (2-58) in the form

$$\nabla^2 \tilde{c}_A = \varepsilon_\gamma^{-1} \nabla \varepsilon_\gamma \cdot \nabla \langle c_A \rangle^\gamma - \left(\frac{a_v k}{\varepsilon_\gamma \mathscr{D}} \right) \langle c_A \rangle^\gamma + \frac{1}{V_\gamma} \int_{A_{\gamma e}} \mathbf{n}_\gamma \cdot \nabla \tilde{c}_A \, dA$$

(2-59)

At this point we return to the averaging theorem as given by Eq. (2-23) and make use of the divergence theorem to obtain

$$\nabla \langle \psi \rangle = \frac{1}{\mathscr{V}} \int_{A_{\gamma e}} \mathbf{n}_\gamma \psi \, dA$$

(2-60)

This is the form of the averaging theorem originally presented by Slattery [40] and it can be used with the last term in Eq. (2-59) to obtain the final form of the governing differential equation for $\tilde{c}_A$.

$$\nabla^2 \tilde{c}_A = \varepsilon_\gamma^{-1} \nabla \varepsilon_\gamma \cdot \nabla \langle c_A \rangle^\gamma - \left(\frac{a_v k}{\varepsilon_\gamma \mathscr{D}} \right) \langle c_A \rangle^\gamma + \varepsilon_\gamma^{-1} \nabla \cdot \langle \nabla \tilde{c}_A \rangle$$

(2-61)

The boundary conditions for $\tilde{c}_A$ are given by

B.C.1 $-\mathbf{n}_{\gamma\kappa} \cdot \mathscr{D} \nabla \tilde{c}_A = \mathbf{n}_{\gamma\kappa} \cdot \mathscr{D} \nabla \langle c_A \rangle^\gamma + k \langle c_A \rangle^\gamma,$ at $A_{\gamma\kappa}$

(2-62)

B.C.2 $\tilde{c}_A = g(\mathbf{r}, t),$ at $\mathscr{A}_{\gamma e}$ (2-63)

Aside from the fact that this boundary value problem is quasi-steady, it has little to recommend it over the original problem given by Eqs. (2-15) through (2-17). However, it should be intuitively appealing that the $\tilde{c}_A$-field is not influenced by the boundary condition imposed at $\mathscr{A}_{\gamma e}$ and this suggests the existence of a *local problem*.

The presence of the nonhomogeneous terms in Eqs. (2-61) and (2-62) involving $\nabla \langle c_A \rangle^\gamma$ and $\langle c_A \rangle^\gamma$ suggests a solution of the form

$$\tilde{c}_A = \mathbf{f} \cdot \nabla \langle c_A \rangle^\gamma + s \langle c_A \rangle^\gamma + \psi$$

(2-64)

in which ψ might be zero or negligible. If we consider ψ to be completely arbitrary, we are free to specify $\mathbf{f}$ and s in any manner we choose and we do so in terms of the following boundary value problems

Problem I

$$\nabla^2 \mathbf{f} = \varepsilon_\gamma^{-1} \nabla \varepsilon_\gamma \tag{2-65}$$

B.C.1 $\qquad -n_{\gamma\kappa} \cdot \nabla \mathbf{f} = \mathbf{n}_{\gamma\kappa}, \qquad \text{at } A_{\gamma\kappa} \tag{2-66}$

B.C.2 $\qquad \mathbf{f} = \mathscr{G}(\mathbf{r}, t), \qquad \text{at } \mathscr{A}_{\gamma e} \tag{2-67}$

Problem II

$$\nabla^2 s = -\frac{a_v k}{\varepsilon_\gamma \mathscr{D}} \tag{2-68}$$

B.C.1 $\qquad -\mathbf{n}_{\gamma\kappa} \cdot \nabla s = k/\mathscr{D}, \qquad \text{at } A_{\gamma\kappa} \tag{2-69}$

B.C.2 $\qquad s = \mathscr{H}(\mathbf{r}, t), \qquad \text{at } \mathscr{A}_{\gamma e} \tag{2-70}$

In addition we require that the average values of $\mathbf{f}$ and s be zero

$$\frac{1}{V_\gamma} \int_{V_\gamma} \mathbf{f}\, dV = 0, \qquad \frac{1}{V_\gamma} \int_{V_\gamma} s\, dV = 0 \tag{2-71}$$

About the functions $\mathscr{G}(\mathbf{r}, t)$ and $\mathscr{H}(\mathbf{r}, t)$ we say only that they must satisfy the condition

$$g(\mathbf{r}, t) = \mathscr{G}(\mathbf{r}, t) \cdot \nabla \langle c_A \rangle^\gamma + \mathscr{H}(\mathbf{r}, t) \langle c_A \rangle^\gamma \tag{2-72}$$

When Eq. (2-64) is used in Eqs. (2-61) through (2-63) we obtain (after some algebra) the following boundary value problem for ψ.

$$\nabla^2 \psi = -2(\nabla \mathbf{f} : \nabla \nabla \langle c_A \rangle^\gamma + \nabla s \cdot \nabla \langle c_A \rangle^\gamma) - \mathbf{f} \cdot \nabla \nabla^2 \langle c_A \rangle^\gamma$$

$$- s\nabla^2 \langle c_A \rangle^\gamma + \varepsilon_\gamma^{-1} \nabla \cdot (\langle \nabla \mathbf{f} \rangle \cdot \nabla \langle c_A \rangle^\gamma + \langle \nabla s \rangle \langle c_A \rangle^\gamma + \langle \nabla \psi \rangle)$$

$$\tag{2-73}$$

B.C.1 $\qquad -\mathbf{n}_{\gamma\kappa} \cdot \nabla \psi = \mathbf{n}_{\gamma\kappa} \mathbf{f} : \nabla \nabla \langle c_A \rangle^\gamma + s\mathbf{n}_{\gamma\kappa} \cdot \nabla \langle c_A \rangle^\gamma, \qquad \text{at } A_{\gamma\kappa}$

$$\tag{2-74}$$

B.C.2 $\qquad\qquad\qquad \psi = 0, \qquad \text{at } \mathscr{A}_{\gamma e} \tag{2-75}$

Using order of magnitude estimates [52, Sec. 2.9] of the form

$$\nabla \mathbf{f} = \mathbf{O}\left(\frac{f}{l_\gamma}\right), \qquad \nabla s = \mathbf{O}\left(\frac{s}{l_\gamma}\right), \qquad \nabla \langle c_A \rangle^\gamma = \mathbf{O}\left(\frac{\nabla \langle c_A \rangle^\gamma}{L}\right)$$

$$(2\text{-}76)$$

allows us to express Eqs. (2-73) and (2-74) as

$$\nabla^2 \psi = \{\mathbf{O}(\mathbf{f} \cdot \nabla \langle c_A \rangle^\gamma / l_\gamma L) + \mathbf{O}(s \langle c_A \rangle^\gamma / l_\gamma L)\} \qquad (2\text{-}77a)$$

B.C.1 $\qquad -\mathbf{n}_{\gamma\kappa} \cdot \nabla \psi = \{\mathbf{O}(\mathbf{f} \cdot \langle c_A \rangle^\gamma / L) + \mathbf{O}(s \langle c_A \rangle^\gamma / L)\} \qquad (2\text{-}77b)$

From both the governing differential equation *and* the boundary condition we obtain the same estimate for ψ which is given by

$$\psi = \left\{\mathbf{O}\left[\mathbf{f} \cdot \nabla \langle c_A \rangle^\gamma \left(\frac{l_\gamma}{L}\right)\right] + \mathbf{O}\left[s \langle c_A \rangle^\gamma \left(\frac{l_\gamma}{L}\right)\right]\right\} \qquad (2\text{-}78)$$

From this result it is clear that ψ makes a negligible contribution to $\tilde{c}_A$ and Eq. (2-64) can be expressed as

$$\tilde{c}_A = \mathbf{f} \cdot \nabla \langle c_A \rangle^\gamma + s \langle c_A \rangle^\gamma \qquad (2\text{-}79)$$

Substitution of this result into Eq. (2-47) provides the *form*, but not the details, of the volume-averaged transport equation which is given by

$$\varepsilon_\gamma \frac{\partial \langle c_A \rangle^\gamma}{\partial t} = \nabla \cdot \left\{\varepsilon_\gamma \mathscr{D}\left[\mathbf{I} + \frac{1}{V_\gamma} \int_{A_{\gamma\kappa}} \mathbf{n}_{\gamma\kappa} \mathbf{f} \, dA\right] \cdot \nabla \langle c_A \rangle^\gamma\right\}$$

$$+ \nabla \cdot \left\{\varepsilon_\gamma \mathscr{D}\left[\frac{1}{V_\gamma} \int_{A_{\gamma\kappa}} \mathbf{n}_{\gamma\kappa} s \, dA\right] \langle c_A \rangle^\gamma\right\} - a_v k \langle c_A \rangle^\gamma$$

The effective diffusivity tensor can be defined as

$$\mathbf{D}_{\text{eff}} = \mathscr{D}\left[\mathbf{I} + \frac{1}{V_\gamma} \int_{A_{\gamma\kappa}} \mathbf{n}_{\gamma\kappa} \mathbf{f} \, dA\right] \qquad (2\text{-}80)$$

along with the vector

$$\boldsymbol{\zeta} = \frac{1}{V_\gamma} \int_{A_{\gamma\kappa}} \mathbf{n}_{\gamma\kappa} (\mathscr{D} s / k) \, dA \qquad (2\text{-}81)$$

so that Eq. (2-80) takes the form

$$\varepsilon_\gamma \frac{\partial \langle c_A \rangle^\gamma}{\partial t} = \nabla \cdot (\varepsilon_\gamma \mathbf{D}_{\text{eff}} \cdot \nabla \langle c_A \rangle^\gamma) + \nabla \cdot (\varepsilon_\gamma \zeta k \langle c_A \rangle^\gamma) - a_v k \langle c_A \rangle^\gamma$$

$$(2\text{-}82)$$

Here we see that the interaction between diffusion and reaction gives rise to a convective-like term in the volume-averaged transport equation; however, it is not difficult to show that this term is unimportant. From both the governing differential equation for s given by Eq. (2-68) and the boundary condition given by Eq. (2-69) we obtain the same estimate

$$\mathscr{D}s/k = \mathbf{O}(l_\gamma) \qquad (2\text{-}83)$$

and when used with Eq. (2-81) we find

$$\zeta = \mathbf{O}(1) \qquad (2\text{-}84)$$

From this result and Eq. (2-43) we see that

$$a_v k \langle c_A \rangle^\gamma \gg \nabla \cdot (\varepsilon_\gamma \zeta k \langle c_A \rangle^\gamma) \qquad (2\text{-}85)$$

and the practical form of Eq. (2-82) is†

$$\varepsilon_\gamma \frac{\partial \langle c_A \rangle^\gamma}{\partial t} = \nabla \cdot (\varepsilon_\gamma \mathbf{D}_{\text{eff}} \cdot \langle c_A \rangle^\gamma) - a_v k \langle c_A \rangle^\gamma \qquad (2\text{-}86)$$

From the definition of $\mathbf{D}_{\text{eff}}$ and the boundary value problem for $\mathbf{f}$ we see that the effective diffusivity is independent of the rate of chemical reaction. This tends to confirm the experiments of Balder and Peterson [5] and to reject the speculation of Wakao and Smith [49] concerning the experimental results of Otani and Smith [33]. The suggestion that the effective diffusivity might depend on the reaction rate was based on the micropore–macropore model of Wakao and Smith [49] and that model will be examined in the next section.

The definition of the effective diffusivity given by Eq. (2-80) is not universal and in the reactor design literature one most often finds the void fraction ε_γ incorporated into the effective diffusivity

---

†When the kinetics are nonlinear, numerical methods must be used to solve Eq. (2-86), and this subject is treated in Chapter 2 by Gottifredi, Gonzo and Quiroga.

according to

$$\mathbf{D}'_{eff} = \varepsilon_\gamma \mathbf{D}_{eff} \qquad (2\text{-}87)$$

In order to calculate the flux at the boundary of the porous medium illustrated in Fig. 2-1 one uses

$$\left\{ \begin{array}{c} \text{diffusive flux} \\ \text{of species } A \end{array} \right\} = -\mathbf{n} \cdot \varepsilon_\gamma \mathbf{D}_{eff} \cdot \mathbf{\nabla} \langle c_A \rangle^\gamma, \qquad \text{at } \mathscr{A}_{\gamma e}$$

$$(2\text{-}88)$$

The choice of the definition of the effective diffusivity seems to vary with the problem under consideration. In reactor design problems, one is more interested in the flux at the interface between a catalyst pellet and the surrounding fluid and $\mathbf{D}'_{eff}$ is preferred in Eq. (2-86). For processes in which the prime concern is a knowledge of $\langle c_A \rangle^\gamma$ as a function of time and space the preference is to use $\mathbf{D}_{eff}$ in Eq. (2-86).

In order to produce theoretical values of $\mathbf{D}_{eff}$ we must solve for the **f**-field, thus we must finally come to grips with the boundary value problem given by Eqs. (2-65) through (2-67). In that problem there is a *volume source* for the **f**-field given by $\varepsilon_\gamma^{-1} \mathbf{\nabla} \varepsilon_\gamma$ in Eq. (2-65) and a *surface source* given by $\mathbf{n}_{\gamma\kappa}$ in Eq. (2-66). The magnitude of **f** owing to the volume source can be estimated as

$$\mathbf{f} = \mathbf{O} \left\{ \left[\left(\frac{\Delta \varepsilon_\gamma}{\varepsilon_\gamma} \right) \frac{l_\gamma}{L} \right] l_\gamma \right\}, \qquad \text{volume source} \qquad (2\text{-}89)$$

while the surface source gives rise to the estimate

$$\mathbf{f} = \mathbf{O}(l_\gamma), \qquad \text{surface source} \qquad (2\text{-}90)$$

Clearly the volume source has a negligible effect on the **f**-field and the boundary value problem for **f** can be reduced to

$$\nabla^2 \mathbf{f} = 0 \qquad (2\text{-}91)$$

B.C.1 $\qquad\qquad -\mathbf{n}_{\gamma\kappa} \cdot \mathbf{\nabla f} = \mathbf{n}_{\gamma\kappa}, \qquad \text{at } A_{\gamma\kappa} \qquad (2\text{-}92)$

B.C.2 $\qquad\qquad \mathbf{f} = \mathscr{G}(\mathbf{r}), \qquad \text{at } A_{\gamma e} \qquad (2\text{-}93)$

It should be clear that we have no intention of solving for **f** over the entire region illustrated in Fig. 2-1. Instead we want to solve for **f** in some representative region and use the computed results in Eq. (2-80) to determine the effective diffusivity tensor. Such a region is illustrated in Fig. 2-2, and if we want to solve for **f** in that region we must be willing to abandon the boundary condition given by Eq. (2-93). That leaves us without any *long range variables* in the boundary value problem for **f** and the region illustrated in Fig. 2-2 automatically becomes a unit cell in a *spatially periodic porous medium*. Under these circumstances the boundary value problem for **f** takes the form

$$\nabla^2 \mathbf{f} = 0 \tag{2-94}$$

B.C.1 $$\qquad -\mathbf{n}_{\gamma\kappa} \cdot \nabla \mathbf{f} = \mathbf{n}_{\gamma\kappa} \,, \qquad \text{at } A_{\gamma\kappa} \tag{2-95}$$

B.C.2 $$\qquad \mathbf{f}(\mathbf{r} + \mathbf{l}_i) = \mathbf{f}(\mathbf{r}) \,, \qquad i = 1, 2, 3 \tag{2-96}$$

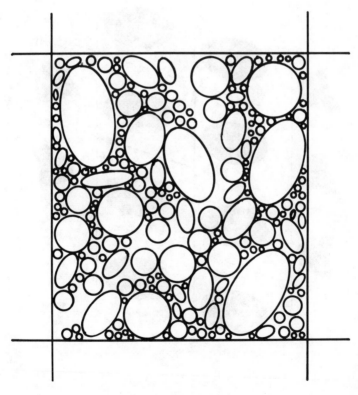

Figure 2-2   Complex unit cell.

Here $\mathbf{l}_i$ represents the three non-unique lattice vectors needed to construct a spatially periodic porous medium and the proof that $\mathbf{f}$ is spatially periodic is given by Ryan et al. [37]. A qualitative explanation is simply that Eq. (2-91) is invariant to a transformation of the type $\bar{\mathbf{r}} = \mathbf{r} + \mathbf{l}_i$ and for a spatially periodic porous medium the boundary condition given by Eq. (2-92) is also invariant to such a transformation. Couple this with the assumption that the boundary condition given by Eq. (2-93) can be ignored and one concludes that $\mathbf{f}$ is spatially periodic.

It is important to note that the *form* of the volume-averaged transport equation given by Eq. (2-86) and the *form* of the boundary value problem for $\mathbf{f}$ given by Eqs. (2-91) through (2-93) were obtained *without* imposing the restriction of either a homogeneous porous medium or a spatially periodic porous medium. The use of a unit cell, such as the one shown in Fig. 2-2, to determine $\mathbf{D}_{eff}$ does

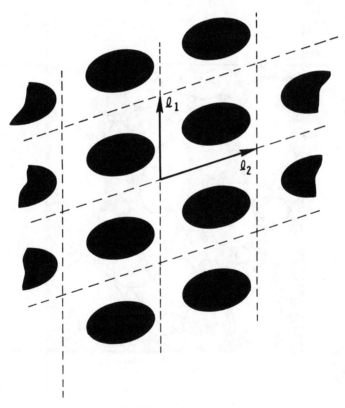

Figure 2-3   Spatially periodic porous medium.

not require that the effective diffusivity tensor be treated as a constant. If conditions vary over the region illustrated in Fig. 2-1, one need only construct unit cells that accurately model that variation in order to determine the spatial dependence of $\mathbf{D}_{\text{eff}}$. The absence of any significant long-range effects in the boundary value problem for $\mathbf{f}$ given by Eqs. (2-91) through (2-93) assures us of this fact. While the unit cell illustrated in Fig. 2-2 indicates the possibility of unlimited geometrical complexity, we are currently limited by the cost of numerical solutions to simpler models such as the one shown in Fig. 2-3. For diffusive processes, simple geometrical models appear to be satisfactory [31]; however, studies of dispersion phenomena [14] indicate that a deeper understanding of the influence of the interfacial geometry is obviously required for processes involving convective transport.

Comparison with Experiment

Our objective in the analysis of multiphase chemical reactors is to derive the correct form of the volume-averaged transport equations and to predict the coefficients that appear in those equations. This prediction excludes chemical reaction rate coefficients, thus we assume that the coefficient k in Eq. (2-86) will be determined experimentally and our task centers on a comparison between the theoretical and experimental values of $\mathbf{D}_{\text{eff}}$. To accomplish this comparison we have used the simple two-dimensional models illustrated in Fig. 2-4. The comparison between theory and experiment will be restricted to systems for which ε_γ and $\mathbf{D}_{\text{eff}}$ are constant and for which there is no reaction. Under these circumstances Eq. (2-86) takes the form

$$\frac{\partial \langle c_A \rangle^\gamma}{\partial t} = \mathbf{D}_{\text{eff}} : \nabla\nabla \langle c_A \rangle^\gamma \tag{2-97}$$

and since $\nabla\nabla \langle c_A \rangle^\gamma$ is a symmetric tensor, the appropriate defintion of $\mathbf{D}_{\text{eff}}$ is

$$\mathbf{D}_{\text{eff}} = \mathscr{D}\left[\mathbf{I} + \frac{1}{V_\gamma} \int_{A_{\gamma\kappa}} \tfrac{1}{2}(\mathbf{n}_{\gamma\kappa}\mathbf{f} + \mathbf{f}\mathbf{n}_{\gamma\kappa})\, dA \right] \tag{2-98}$$

The details of the numerical solution of Eqs. (2-94) through (2-96) are given by Ryan [37] and here we will say only that care was taken to insure that the results were independent of the parameters associated with the numerical method.

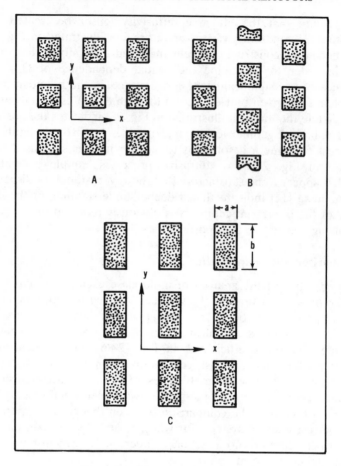

Figure 2-4   Two-dimensional models.

For the system shown in Fig. 2-4A it is easy to prove that $D_{xx} = D_{yy}$ where D_{xx} and D_{yy} represent the components of $\mathbf{D}_{\text{eff}}$. A detailed examination of the boundary value problem for $\mathbf{f}$ indicates that $D_{xy} = D_{yx} = 0$ for the system shown in Fig. 2-4A and under these circumstances this geometrical model is isotropic with respect to volume-averaged diffusion. It is interesting to note that for this particular spatially periodic model, the medium is *not isotropic* but the diffusion process *is isotropic*. A more general version of the model presented in Fig. 2-4A is illustrated by Fig. 2-4C and it should be clear that this model is capable of producing anisotropic effects.

An attractive array of experimental values of the effective diffusivity has been obtained by Currie [13] under equimolar counterdiffusion conditions. This means that the arguments leading from Eqs. (2-1) and (2-2) to Eq. (2-5) are valid and a profitable comparison between theory and experiment can be made. Currie obtained data for a wide range of void fractions and the comparison between theory and experiment is shown in Fig. 2-5. For the isotropic diffusion process, i.e. $a = b$ we see excellent agreement for nearly a dozen different systems with void fractions ranging from about 0.2 to nearly 1.0. In general the theory provides an upper bound for the experimental data for void fractions greater than one half; however, for $\varepsilon_\gamma < 0.5$ the agreement between theory and experiment is exceptionally good. In addition, the theory for $a = b$ is in good agreement with the theoretical work of Strieder and Aris [44, Eq. (2.6.15)] for a random bed of spheres.

The system having the largest deviation from the theoretical values for $a = b$ is mica, and that system is obviously anisotropic. In Currie's experiments hydrogen and air underwent an equimolar counterdiffusion process parallel to the direction of the gravity

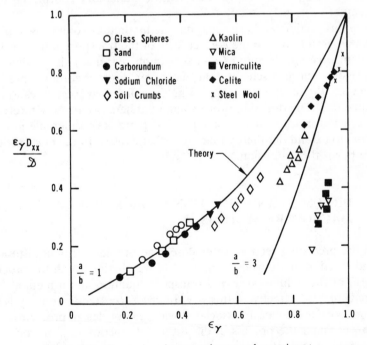

Figure 2-5 Comparison between theory and experiment.

vector. It is reasonable to assume that the mica particles tended to be orthogonal to the gravity vector, a situation that can be simulated by requiring that $a > b$ for the particles in Fig. 2-4C if we take $g_x = g$. For the case where $a/b = 3$ we see better agreement between the theory and the experimental data for mica, and the more detailed calculations of Ryan [37] suggest that a perfect fit between theory and experiment for mica is obtained for $a/b \sim 8$. Those calculations would be informative if the geometry of the mica particles were known; however, this information was not reported by Currie. Under these circumstances we can only say that the theory is capable of predicting observed effects of anisotropy using reasonable particle geometries. An interesting alternative to the cubic array shown in Fig. 2-4A is the body-centered array shown in Fig. 2-4B. The diffusion process in the x-direction for the body-centered array appears to be more "tortuous" than that encountered in the cubic array and one might guess that D_{xx} would be smaller for the body-centered array. In fact, it was found that D_{xx} and D_{yy} for the body-centered array were nearly equal and both values were within a percent or two of the value for the cubic array. Ryan et al. [37] have provided some commentary concerning this result.

In this section we have seen that the method of volume averaging can be used to derive the correct form of the transport equations for volume-average quantities and for simple systems it is possible to develop a closure scheme that allows for direct theoretical prediction of the coefficients appearing in volume-averaged transport equations. The diffusion process analyzed here occurs only rarely in practical reactor design problems, thus there is considerable motivation to extend the theory to more realistic cases. In the next section we begin this extension.

3. DIFFUSION AND REACTION IN MICROPORE–MACROPORE SYSTEMS

In the preceding section we explored the problem of bulk diffusion and reaction in a porous medium, and we obtained both the proper form of the volume-averaged transport equation and a method for predicting the effective diffusivity. In order to extend our analysis to conditions more closely associated with reactor design problems, we now consider the process of diffusion and reaction in a micropore–macropore system.

In most catalyst pellets there exists a wide range of pore diameters and the mechanism of diffusive transport ranges from bulk to Knudsen diffusion. For dilute solutions or equimolar counterdiffusion an appealing approach is to model the process of diffusion and reaction by

$$\varepsilon \frac{\partial C_A}{\partial t} = \mathscr{D}_{\text{eff}} \nabla^2 C_A - a_v k C_A \tag{3-1}$$

in which the concentration C_A is usually undefined and the effective diffusivity takes into account both bulk and Knudsen diffusion. Often catalyst pellets have a bidispersed pore size distribution so that bulk diffusion takes place in one region and Knudsen diffusion takes place in another. Under these circumstances the intuitive appeal of Eq. (3-1) is lessened and this has motivated various authors [9, 17, 30, 36] to model bidispersed pore size distributions in terms of two phases. We follow that line of attack in this section making use of the method of volume averaging and *local mass equilibrium* to develop a one-equation model of diffusion and reaction in a micropore–macropore system. Following the analysis in the preceding section, a method of closure is developed that allows for the direct theoretical determination of the effective diffusivity in the absence of adjustable parameters.

In the micropore–macropore system illustrated in Fig. 3-1 Knudsen diffusion and heterogeneous reaction take place in σ-phase, which is the micropore phase, and bulk diffusion occurs in the

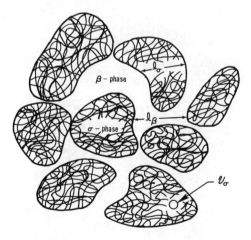

Figure 3-1   Micropore–macropore system.

β-phase. Since many porous catalysts are made by compacting small particles containing fine pores into a larger catalyst pellet, the model illustrated in Fig. 3-1 is not unreasonable. The σ-phase is actually a two-phase system and a portion of this system, identified as $\mathcal{V}_\sigma$ in Fig. 3-1, is illustrated in Fig. 3-2.

The averaging process for Knudsen diffusion in the micropore phase is described in detail in Sec. 4, and at this point we simply note that the transport process in the γ-phase is described by

$$\frac{\partial c_A}{\partial t} + \mathbf{\nabla} \cdot \mathbf{N}_A = 0 , \qquad \text{in } V_\gamma \qquad (3\text{-}2)$$

B.C. $$\mathbf{n}_{\gamma\kappa} \cdot \mathbf{N}_A = \frac{\partial \hat{c}_A}{\partial t} + \hat{k}\hat{c}_A , \qquad \text{at } A_{\gamma\kappa} \qquad (3\text{-}3)$$

This problem statement is identical to that given by Eqs. (2-1) and (2-6) of the preceding section, and we also make use of the linear flux relation given by Eq. (2-10) to write

$$\mathbf{n}_{\gamma\kappa} \cdot \mathbf{N}_A = k_1 c_A - k_2 \hat{c}_A , \qquad \text{at } A_{\gamma\kappa} \qquad (3\text{-}4)$$

When the $\gamma{-}\kappa$ interface can be treated as quasi-steady, the volume-averaged transport equation takes the form

$$\varepsilon_\gamma \frac{\partial \langle c_A \rangle^\gamma}{\partial t} = \mathbf{\nabla} \cdot (\varepsilon_\gamma \mathbf{D}^K_{\text{eff}} \cdot \mathbf{\nabla} \langle c_A \rangle^\gamma) - a_{\gamma\kappa} \left(\frac{k_1 k}{\hat{k} + k_2} \right) \langle c_A \rangle^\gamma \qquad (3\text{-}5)$$

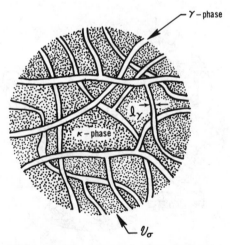

Figure 3-2 Averaging volume for the micropore phase.

Here the interfacial area per unit volume is given explicitly by

$$a_{\gamma\kappa} = A_{\gamma\kappa}/\mathcal{V}_\sigma \qquad (3\text{-}6)$$

and the Knudsen effective diffusivity tensor is given by

$$\mathbf{D}_{\text{eff}}^K = \frac{\mathbf{A}}{a_{\gamma\kappa}}\sqrt{\frac{\pi\mathcal{R}T}{8M_A}} \qquad (3\text{-}7)$$

At this time there is no explicit representation for the tensor $\mathbf{A}$ and we know only that it is of order one. It is important to note that Eq. (3-5) is not restricted to dilute solutions of species A or equimolar counter-diffusion. Arguments are given elsewhere [55] to support the idea that the boundary conditions at the σ–β interface can be expressed as

B.C.1 $\qquad\qquad \langle c_A \rangle^\gamma = (c_A)_\beta , \qquad$ at $A_{\sigma\beta}$ $\qquad$ (3-8)

B.C.2 $\quad \mathbf{n}_{\sigma\beta} \cdot \varepsilon_\gamma \mathbf{D}_{\text{eff}}^K \cdot \nabla \langle c_A \rangle^\gamma = \mathbf{n}_{\sigma\beta} \cdot \mathcal{D}_\beta \nabla (c_A)_\beta , \qquad$ at $A_{\sigma\beta}$

$$(3\text{-}9)$$

The crucial idea here is that the length scales, l_γ and l_κ, are so much smaller than the length scale for the β-phase that the *average* concentration and flux in the σ-phase can be viewed as *point* quantities relative to the β-phase.

In order to avoid confusion in the subsequent analysis, the nomenclature must be simplified and to do so we define the following quantities

$$c_\sigma = \langle c_A \rangle^\gamma \qquad (3\text{-}10)$$

$$c_\beta = (c_A)_\beta \qquad (3\text{-}11)$$

$$k = \frac{a_{\gamma\kappa}}{\varepsilon_\gamma}\left(\frac{k_1\hat{k}}{\hat{k} + k_2}\right) \qquad (3\text{-}12)$$

$$\mathbf{D}_{\text{eff}}^K = \mathbf{I}\left(\frac{A}{a_{\gamma\kappa}}\sqrt{\frac{\pi\mathcal{R}T}{8M_A}}\right) = \mathbf{I}\mathcal{D}_\sigma \qquad (3\text{-}13)$$

Here we have required that the micropore phase be isotropic with respect to diffusion, and in addition we require that it be homogeneous so that ε_γ is a constant. The mathematical statement of diffusion and reaction in a micropore–macropore system is now represented by

$$\frac{\partial c_\sigma}{\partial t} = \nabla \cdot (\mathscr{D}_\sigma \nabla c_\sigma) - kc_\sigma , \qquad \text{at } V_\sigma \qquad (3\text{-}14)$$

B.C.1 $\qquad\qquad c_\sigma = c_\beta , \qquad \text{at } A_{\sigma\beta} \qquad\qquad (3\text{-}15)$

B.C.2 $\qquad \mathbf{n}_{\sigma\beta} \cdot \varepsilon_\gamma \mathscr{D}_\sigma \nabla c_\sigma = \mathbf{n}_{\sigma\beta} \cdot \mathscr{D}_\beta \nabla c_\beta , \qquad \text{at } A_{\sigma\beta} \qquad (3\text{-}16)$

$$\frac{\partial c_\beta}{\partial t} = \nabla \cdot (\mathscr{D}_\beta \nabla c_\beta) , \qquad \text{in } V_\beta \qquad (3\text{-}17)$$

From the form of Eq. (3-17) it should be clear that we have restricted the transport process in the β-phase to dilute solutions of species A or equimolar counterdiffusion. Although reaction does occur at the σ–β interface, the contribution to the overall rate of reaction should be negligible since most of the surface area is contained within the micropore phase.

The averaging procedure for Eqs. (3-14) and (3-17) follows that given in the previous section and is described in detail elsewhere [55]. The results are given by

$$\varepsilon_\sigma \frac{\partial \langle c_\sigma \rangle^\sigma}{\partial t} = \nabla \cdot \left\{ \varepsilon_\sigma \mathscr{D}_\sigma \left[\nabla \langle c_\sigma \rangle^\sigma + \frac{1}{V_\sigma} \int_{A_{\sigma\beta}} \mathbf{n}_{\sigma\beta} \tilde{c}_\sigma \, dA \right] \right\}$$
$$+ \frac{1}{\mathscr{V}} \int_{A_{\sigma\beta}} \mathbf{n}_{\sigma\beta} \cdot \mathscr{D}_\sigma \nabla c_\sigma \, dA - \varepsilon_\sigma k \langle c_\sigma \rangle^\sigma \qquad (3\text{-}18)$$

$$\varepsilon_\beta \frac{\partial \langle c_\beta \rangle^\beta}{\partial t} = \nabla \cdot \left\{ \varepsilon_\beta \mathscr{D}_\beta \left[\nabla \langle c_\beta \rangle^\beta + \frac{1}{V_\beta} \int_{A_{\beta\sigma}} \mathbf{n}_{\beta\sigma} \tilde{c}_\beta \, dA \right] \right\}$$
$$+ \frac{1}{\mathscr{V}} \int_{A_{\beta\sigma}} \mathbf{n}_{\beta\sigma} \cdot \mathscr{D}_\beta \nabla c_\beta \, dA \qquad (3\text{-}19)$$

Here $\mathscr{V}$ represents the averaging volume illustrated in Fig. 3-3 and $A_{\sigma\beta} = A_{\beta\sigma}$ represents the interfacial area contained within the averaging volume. At this point one can develop closure schemes for Eqs. (3-18) and (3-19) separately and this has been done elsewhere [55]. However, the possibility exists that the process of diffusion and reaction can be described in terms of a *single equation* and this is the path followed here.

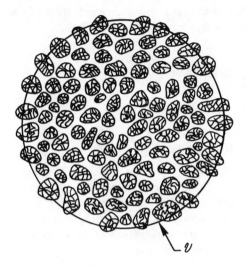

Figure 3-3   Averaging volume for two-phase transport.

One-Equation Model

In order to develop a one-equation model for diffusion and reaction in a micropore–macropore system, we follow previous analyses [31, 55] and combine Eqs. (3-18) and (3-19) in a way that eliminates the interfacial flux term. To accomplish this we multiply Eq. (3-18) by ε_γ and add the result to Eq. (3-19) to obtain

$$\frac{\partial}{\partial t}(\varepsilon_\beta \langle c_\beta \rangle^\beta + \varepsilon_\gamma \varepsilon_\sigma \langle c_\sigma \rangle^\sigma)$$

$$= \nabla \cdot \left\{ \varepsilon_\beta \mathscr{D}_\beta \left[\nabla \langle c_\beta \rangle^\beta + \frac{1}{V_\beta} \int_{A_{\beta\sigma}} \mathbf{n}_{\beta\sigma} \tilde{c}_\beta \, dA \right] \right\}$$

$$+ \nabla \cdot \left\{ \varepsilon_\gamma \varepsilon_\sigma \mathscr{D}_\sigma \left[\nabla \langle c_\sigma \rangle^\sigma + \frac{1}{V_\sigma} \int_{A_{\sigma\beta}} \mathbf{n}_{\sigma\beta} \tilde{c}_\sigma \, dA \right] \right\} - \varepsilon_\gamma \varepsilon_\sigma k \langle c_\sigma \rangle^\sigma$$

$$(3\text{-}20)$$

It is shown elsewhere [55, Sec. 4] that

$$\varepsilon_\beta \langle c_\beta \rangle^\beta + \varepsilon_\gamma \varepsilon_\sigma \langle c_\sigma \rangle^\sigma = \langle c_A \rangle \qquad (3\text{-}21)$$

where $\langle c_A \rangle$ is the spatial average concentration of species A. Use of Eq. (3-21) allows us to write Eq. (3-20) as

$$\frac{\partial \langle c_A \rangle}{\partial t} = \nabla \cdot \left\{ \varepsilon_\beta \mathscr{D}_\beta \left[\nabla \langle c_\beta \rangle^\beta + \frac{1}{V_\beta} \int_{A_{\beta\sigma}} \mathbf{n}_{\beta\sigma} \tilde{c}_\beta \, dA \right] \right\}$$

$$+ \nabla \cdot \left\{ \varepsilon_\gamma \varepsilon_\sigma \mathscr{D}_\sigma \left[\nabla \langle c_\sigma \rangle^\sigma + \frac{1}{V_\sigma} \int_{A_{\sigma\beta}} \mathbf{n}_{\sigma\beta} \tilde{c}_\sigma \, dA \right] \right\}$$

$$- \varepsilon_\gamma \varepsilon_\sigma k \langle c_\sigma \rangle^\sigma \qquad (3\text{-}22)$$

In the case of multiphase heat conduction, the one-equation model uses the spatial average temperature; however, in this case $\langle c_A \rangle$ is not the preferred average. This becomes clear if we think about the special case where the concentration of species A is a constant designated by c_A^0. For this case

$$\langle c_A \rangle = (\varepsilon_\beta + \varepsilon_\gamma \varepsilon_\sigma) c_A^0 , \qquad \langle c_\sigma \rangle^\sigma = c_A^0 , \qquad \langle c_\beta \rangle^\beta = c_A^0$$
$$(3\text{-}23)$$

and it becomes apparent that we cannot replace $\langle c_\beta \rangle^\beta$ and $\langle c_\sigma \rangle^\sigma$ by the spatial average concentration, $\langle c_A \rangle$. This difficulty is easily overcome by use of what is called an *overall intrinsic phase average* concentration. This concentration is defined by

$$\{c_A\} = \frac{\varepsilon_\beta \langle c_\beta \rangle^\beta + \varepsilon_\gamma \varepsilon_\sigma \langle c_\sigma \rangle^\sigma}{\varepsilon_\beta + \varepsilon_\gamma \varepsilon_\sigma} \qquad (3\text{-}24)$$

and it has the characteristic that $\{c_A\} = \langle c_\beta \rangle^\beta = \langle c_\sigma \rangle^\sigma$ when the point concentration of species A is a constant. Clearly the one-equation model must be expressed in terms of $\{c_A\}$, and at this point we follow the heat transfer analysis [56, Sec. 3.3] and represent the individual intrinsic phase average concentrations as

$$\langle c_\beta \rangle^\beta = \{c_A\} + \hat{c}_\beta \qquad (3\text{-}25)$$

$$\langle c_\sigma \rangle^\sigma = \{c_A\} + \hat{c}_\sigma \qquad (3\text{-}26)$$

Use of these results along with Eqs. (3-21) and (3-24) in Eq. (3-22) leads to

$$(\varepsilon_\beta + \varepsilon_\gamma \varepsilon_\sigma) \frac{\partial \{c_A\}}{\partial t} = \mathbf{\nabla} \cdot \left\{ (\varepsilon_\beta \mathcal{D}_\beta + \varepsilon_\gamma \varepsilon_\sigma \mathcal{D}_\sigma) \mathbf{\nabla} \{c_A\} \right.$$

$$+ \frac{\varepsilon_\beta \mathcal{D}_\beta}{V_\beta} \int_{A_{\beta\sigma}} \mathbf{n}_{\beta\sigma} \tilde{c}_\beta \, dA$$

$$\left. + \frac{\varepsilon_\gamma \varepsilon_\sigma \mathcal{D}_\sigma}{V_\sigma} \int_{A_{\sigma\beta}} \mathbf{n}_{\sigma\beta} \tilde{c}_\sigma \, dA \right\}$$

$$- \varepsilon_\gamma \varepsilon_\sigma k \{c_A\}$$

$$+ \mathbf{\nabla} \cdot \{ \varepsilon_\beta \mathcal{D}_\beta \mathbf{\nabla} \hat{c}_\beta + \varepsilon_\gamma \varepsilon_\sigma \mathcal{D}_\sigma \mathbf{\nabla} \hat{c}_\sigma \}$$

$$- \varepsilon_\gamma \varepsilon_\sigma k \hat{c}_\sigma \qquad (3\text{-}27)$$

The principle of local mass equilibrium requires that

$$\hat{c}_\sigma \ll \{c_A\} \qquad (3\text{-}28\text{a})$$

$$\mathbf{\nabla} \hat{c}_\beta \ll \mathbf{\nabla} \{c_A\} \qquad (3\text{-}28\text{b})$$

$$\mathbf{\nabla} \hat{c}_\sigma \ll \{c_A\} \qquad (3\text{-}28\text{c})$$

and when these restrictions are satisfied Eq. (3-27) simplifies to

$$(\varepsilon_\beta + \varepsilon_\gamma \varepsilon_\sigma) \frac{\partial \{c_A\}}{\partial t} = \mathbf{\nabla} \cdot \left\{ (\varepsilon_\beta \mathcal{D}_\beta + \varepsilon_\gamma \varepsilon_\sigma \mathcal{D}_\sigma) \mathbf{\nabla} \{c_A\} \right.$$

$$+ \frac{\varepsilon_\beta \mathcal{D}_\beta}{V_\beta} \int_{A_{\beta\sigma}} \mathbf{n}_{\beta\sigma} \tilde{c}_\beta \, dA$$

$$\left. + \frac{\varepsilon_\gamma \varepsilon_\sigma \mathcal{D}_\sigma}{V_\sigma} \int_{A_{\sigma\beta}} \mathbf{n}_{\sigma\beta} \tilde{c}_\sigma \, dA \right\}$$

$$- \varepsilon_\gamma \varepsilon_\sigma k \langle c_a \rangle \qquad (3\text{-}29)$$

It is shown elsewhere [55, Sec. 4] that the principle of local mass equilibrium is valid when the following constraints are satisfied

$$\left(\frac{l_\sigma}{d_p} \right)^2 \frac{\mathcal{D}_\beta}{\mathcal{D}_\sigma} \ll 1 \qquad (3\text{-}30)$$

$$\phi_\sigma^2 \ll 1 \qquad (3\text{-}31)$$

$$\frac{\mathscr{D}_\sigma t^*}{l_\sigma^2} \gg 1 \qquad (3\text{-}32)$$

Here d_p represents the catalyst pellet "particle diameter" and ϕ_σ is the micropore Thiele modulus defined by

$$\phi_\sigma = l_\sigma \sqrt{k/\mathscr{D}_\sigma} \qquad (3\text{-}33)$$

The first of these constraints is generally quite easily satisfied since d_p is much, much larger than l_σ, and the micropore Thiele modulus is usually small compared to one so that the inequality given by Eq. (3-31) is satisfied. Under certain circumstances the constraint given by Eq. (3-32) might be difficult to satisfy. For example, if the pores in the micropore phase are very small, the Knudsen diffusivity $\mathscr{D}_\sigma$ will also be very small and Eq. (3-32) may not be satisfied at short times.

Closure

In order to complete our analysis of diffusion and reaction in a micropore–macropore system we need to develop a closure scheme so that $\tilde{c}_\beta$ and $\tilde{c}_\sigma$ can be expressed in terms of $\{c_A\}$. The procedure follows that described in Sec. 2 and we begin by using the decomposition

$$c_\sigma = \langle c_\sigma \rangle^\sigma + \tilde{c}_\sigma \qquad (3\text{-}34)$$

in Eq. (3-14) and following the analysis from Eq. (2-49) to Eq. (2-52) in order to obtain

$$\frac{\partial \tilde{c}_\sigma}{\partial t} - \nabla \cdot (\mathscr{D}_\sigma \nabla \tilde{c}_\sigma) + k\tilde{c}_\sigma = \left\langle \frac{\partial \tilde{c}_\sigma}{\partial t} - \nabla \cdot (\mathscr{D}_\sigma \nabla \tilde{c}_\sigma) + k\tilde{c}_\sigma \right\rangle^\sigma \qquad (3\text{-}35)$$

The constraint on the micropore Thiele modulus given by Eq. (3-31) naturally leads to

$$k\tilde{c}_\sigma \ll \nabla \cdot (\mathscr{D}_\sigma \nabla \tilde{c}_\sigma) \qquad (3\text{-}36)$$

and use of the inequality

$$\mathscr{D}_\sigma \nabla^2 \tilde{c}_\sigma \gg \nabla \mathscr{D}_\sigma \cdot \nabla \tilde{c}_\sigma \qquad (3\text{-}37)$$

allows us to simplify Eq. (3-36) to

$$\nabla^2 \tilde{c}_\sigma = \frac{1}{V_\sigma} \int_{V_\sigma} \nabla^2 \tilde{c}_\sigma \, dV \qquad (3\text{-}38)$$

The same type of analysis applies to the macropore phase transport equation given by Eq. (3-17) and we express the result as

$$\nabla^2 \tilde{c}_\beta = \frac{1}{V_\beta} \int_{V_\beta} \nabla^2 \tilde{c}_\beta \, dV \qquad (3\text{-}39)$$

To develop the boundary conditions we make use of decompositions of the form given by Eq. (3-34) along with Eqs. (3-25) and (3-26) so that Eq. (3-15) takes the form

B.C.1 $\qquad \{c_A\} + \hat{c}_\sigma + \tilde{c}_\sigma = \{c_A\} + \hat{c}_\beta + \tilde{c}_\beta \,, \qquad$ at $A_{\sigma\beta}$ $\qquad (3\text{-}40)$

Obviously this reduces to

B.C.1 $\qquad \tilde{c}_\sigma + \hat{c}_\sigma = \tilde{c}_\beta + \hat{c}_\beta \,, \qquad$ at $A_{\sigma\beta}$ $\qquad (3\text{-}41)$

and we are faced with the question about the magnitude of $\hat{c}_\sigma$ and $\hat{c}_\beta$ relative to $\tilde{c}_\sigma$ and $\tilde{c}_\beta$. With a bit of algebraic effort one can use Eqs. (3-25) and (3-26), along with Eq. (3-24) to obtain

$$\hat{c}_\sigma = \frac{\varepsilon_\beta}{\varepsilon_\beta + \varepsilon_\gamma \varepsilon_\sigma} (\langle c_\sigma \rangle^\sigma - \langle c_\beta \rangle^\beta) \qquad (3\text{-}42)$$

$$\hat{c}_\beta = \frac{\varepsilon_\gamma \varepsilon_\sigma}{\varepsilon_\beta + \varepsilon_\gamma \varepsilon_\sigma} (\langle c_\beta \rangle^\beta - \langle c_\sigma \rangle^\sigma) \qquad (3\text{-}43)$$

These results can be used with previous estimates [55, Eqs. (A-14) and (A-18)] to obtain the estimates for $\hat{c}_\sigma$ and $\hat{c}_\beta$ given by

$$\hat{c}_\sigma = \mathbf{O}\left(\frac{\varepsilon_\beta}{\varepsilon_\beta + \varepsilon_\gamma \varepsilon_\sigma} \right) \left\{ \begin{array}{l} \left(\dfrac{\varepsilon_\sigma l_\sigma}{a_{\beta\sigma} \mathcal{D}_\sigma t^*} \right)\left[1 + \left(\dfrac{l_\beta}{l_\sigma} \right)\kappa \right] \\[3ex] \left(\dfrac{\varepsilon_\sigma l_\sigma}{a_{\beta\sigma} d_p^2} \right)\left[1 + \left(\dfrac{l_\beta}{l_\sigma} \right)\kappa \right] \\[3ex] \left(\dfrac{\varepsilon_\sigma l_\sigma k}{a_{\beta\sigma} \mathcal{D}_\sigma} \right)\left[1 + \left(\dfrac{l_\beta}{l_\sigma} \right)\kappa \right] \end{array} \right\} \langle c_\sigma \rangle^\sigma \qquad (3\text{-}44)$$

$$
\hat{c}_\beta = \mathbf{O}\!\left(\frac{\varepsilon_\gamma \varepsilon_\sigma}{\varepsilon_\beta + \varepsilon_\gamma \varepsilon_\sigma}\right)\!\left\{ \begin{array}{c} \dfrac{\varepsilon_\beta l_\sigma}{a_{\beta\sigma} t^* \varepsilon_\gamma \mathscr{D}_\sigma}\left[1 + \left(\dfrac{l_\beta}{l_\sigma}\right)\kappa\right] \\[3ex] \dfrac{\varepsilon_\beta l_\sigma \mathscr{D}_\beta}{a_{\beta\sigma} d_p^2 \varepsilon_\gamma \mathscr{D}_\sigma}\left[1 + \left(\dfrac{l_\beta}{l_\sigma}\right)\kappa\right] \end{array} \right\} \langle c_\beta \rangle^\beta \qquad (3\text{-}45)
$$

Here we have used κ to represent the ratio of diffusivities according to

$$
\kappa = \varepsilon_\gamma \mathscr{D}_\sigma / \mathscr{D}_\beta \qquad (3\text{-}46)
$$

For a porous medium consisting of compacted microporous particles, one can use the ideas associated with granular systems [52, Sec. 7.8] to conclude that

$$
l_\beta \leqslant l_\sigma , \qquad l_\sigma \sim a_{\beta\sigma}^{-1} \qquad (3\text{-}47)
$$

These results, along with

$$
\kappa \leqslant 1 , \qquad \varepsilon_\beta + \varepsilon_\gamma \varepsilon_\sigma \sim 1 \qquad (3\text{-}48)
$$

$$
\langle c_\sigma \rangle^\sigma \sim \langle c_\beta \rangle^\beta \sim \{c_A\} \qquad (3\text{-}49)
$$

can be used to express Eqs. (3-44) and (3-45) as

$$
\hat{c}_\sigma = \mathbf{O}\!\left\{ \begin{array}{c} \varepsilon_\beta \varepsilon_\sigma \left(\dfrac{l_\sigma^2}{\mathscr{D}_\sigma t^*}\right) \\[3ex] \varepsilon_\beta \varepsilon_\sigma \left(\dfrac{l_\sigma}{d_p}\right)^2 \\[3ex] \varepsilon_\beta \varepsilon_\sigma \phi_\sigma^2 \end{array} \right\} \{c_a\} \qquad (3\text{-}50)
$$

$$
\hat{c}_\beta = \mathbf{O}\!\left\{ \begin{array}{c} \varepsilon_\beta \varepsilon_\sigma \left(\dfrac{l_\sigma^2}{\mathscr{D}_\sigma t^*}\right) \\[3ex] \varepsilon_\beta \varepsilon_\sigma \left(\dfrac{l_\sigma}{d_p}\right)^2 \left(\dfrac{\mathscr{D}_\beta}{\mathscr{D}_\sigma}\right) \end{array} \right\} \{c_A\} \qquad (3\text{-}51)
$$

The type of analysis given by Ryan et al. [37] and referred to by Eqs. (2-40) through (2-44) has been extended to two-phase systems [31, Appendix], and it allows us to underestimate $\hat{c}_\sigma$ and $\hat{c}_\beta$ as

$$\hat{c}_\sigma = \mathbf{O}(\hat{c}_\beta) = \mathbf{O}\left[\left(\frac{l_\beta}{d_p}\right)\{c_A\}\right] \tag{3-52}$$

In order that $\hat{c}_\sigma$ and $\hat{c}_\beta$ be small compared to $\tilde{c}_\sigma$ and $\tilde{c}_\beta$ we see that the following inequalities must be satisfied

$$\varepsilon_\beta \varepsilon_\sigma \left(\frac{l_\sigma^2}{\mathscr{D}_\sigma t^*}\right) \ll \left(\frac{l_\beta}{d_p}\right) \tag{3-53a}$$

$$\varepsilon_\beta \varepsilon_\sigma \left(\frac{l_\sigma}{d_p}\right)^2 \frac{\mathscr{D}_\beta}{\mathscr{D}_\sigma} \ll \left(\frac{l_\beta}{d_p}\right) \tag{3-53b}$$

$$\varepsilon_\beta \varepsilon_\sigma \phi_\sigma^2 \ll \left(\frac{l_\beta}{d_p}\right) \tag{3-53c}$$

In general, these constraints will be more difficult to satisfy than those required for local mass equilibrium given by Eqs. (3-30) through (3-32). When the above constraints are satisfied, the boundary condition given by Eq. (3-41) simplifies to

B.C.1 $\qquad\qquad \tilde{c}_\sigma = \tilde{c}_\beta , \qquad$ at $A_{\sigma\beta}$ $\qquad\qquad$ (3-54)

Use of the representations given in Eq. (3-40) in the flux boundary condition indicated by Eq. (3-16) leads to

B.C.2

$$\mathbf{n}_{\sigma\beta} \cdot \varepsilon_\gamma \mathscr{D}_\sigma [\boldsymbol{\nabla}\{c_A\} + \boldsymbol{\nabla}\tilde{c}_\sigma + \boldsymbol{\nabla}\hat{c}_\sigma]$$

$$= \mathbf{n}_{\sigma\beta} \cdot \mathscr{D}_\beta [\boldsymbol{\nabla}\{c_A\} + \boldsymbol{\nabla}\hat{c}_\beta + \boldsymbol{\nabla}\tilde{c}_\beta] , \qquad \text{at } A_{\sigma\beta} \qquad (3-55)$$

On the basis of Eqs. (3-28), this result undergoes the obvious simplification when Eqs. (3-30) through (3-31) are in effect.

At this point we summarize the boundary value problem for $\tilde{c}_\beta$ and $\tilde{c}_\sigma$ as

$$\nabla^2 \tilde{c}_\beta = \frac{1}{V_\beta} \int_{V_{\beta\sigma}} \nabla^2 \tilde{c}_\beta \, dV \,, \quad \text{in } V_\beta \qquad (3\text{-}56)$$

B.C.1 $$\tilde{c}_\beta = \tilde{c}_\sigma \,, \quad \text{at } A_{\sigma\beta} \qquad (3\text{-}57)$$

B.C.2 $$\mathbf{n}_{\sigma\beta} \cdot \nabla \tilde{c}_\beta = \kappa \mathbf{n}_{\sigma\beta} \cdot \nabla \tilde{c}_\sigma + (\kappa - 1)\mathbf{n}_{\sigma\beta} \cdot \nabla \{c_A\} \,, \quad \text{at } A_{\sigma\beta}$$
$$(3\text{-}58)$$

$$\nabla^2 \tilde{c}_\sigma = \frac{1}{V_\sigma} \int_{V_\sigma} \nabla^2 \tilde{c}_\sigma \, dV \,, \quad \text{in } V_\sigma \qquad (3\text{-}59)$$

This problem is somewhat more complex than the problem encountered in the previous section; however, we proceed in the same manner and use the divergence theorem and Eq. (3-58) to express Eqs. (3-56) and (3-59) as

$$\nabla^2 \tilde{c}_\beta = \varepsilon_\beta^{-1} \kappa \left\{ \frac{1}{\mathscr{V}} \int_{A_{\beta\sigma}} \mathbf{n}_{\beta\sigma} \cdot \nabla \tilde{c}_\sigma \, dA \right\} - \varepsilon_\beta^{-1}(\kappa - 1)\nabla \varepsilon_\beta \cdot \nabla \{c_A\}$$
$$+ \varepsilon_\beta^{-1} \nabla \langle \nabla \tilde{c}_\beta \rangle \qquad (3\text{-}60)$$

$$\nabla^2 \tilde{c}_\sigma = \varepsilon_\sigma^{-1} \kappa^{-1} \left\{ \frac{1}{\mathscr{V}} \int_{A_{\sigma\beta}} \mathbf{n}_{\sigma\beta} \cdot \nabla \tilde{c}_\beta \, dA \right\} + \varepsilon_\sigma^{-1}\left(\frac{\kappa - 1}{\kappa}\right)\nabla \varepsilon_\sigma \cdot \nabla \{c_A\}$$
$$+ \varepsilon_\sigma^{-1} \nabla \langle \nabla \tilde{c}_\sigma \rangle \qquad (3\text{-}61)$$

The presence of the nonhomogeneous source term $\nabla\{c_A\}$ in both the governing equations and the boundary condition given by Eq. (3-58) suggests a solution of the form

$$\tilde{c}_\beta = \mathbf{f} \cdot \nabla \{c_A\} + \psi \,, \qquad \tilde{c}_\sigma = \mathbf{g} \cdot \nabla \{c_A\} + \xi \qquad (3\text{-}62)$$

Here we let ψ and ξ be completely arbitrary so that we are free to specify $\mathbf{f}$ and $\mathbf{g}$ according to the following boundary value problem

$$\nabla^2 \mathbf{f} = -\varepsilon_\beta^{-1}(\kappa - 1)\nabla \varepsilon_\beta \,, \quad \text{in } V_\beta \qquad (3\text{-}63)$$

B.C.1 $$\mathbf{f} = \mathbf{g} \,, \quad \text{at } A_{\sigma\beta} \qquad (3\text{-}64)$$

B.C.2 $$\mathbf{n}_{\sigma\beta} \cdot \nabla \mathbf{f} = \kappa \mathbf{n}_{\sigma\beta} \cdot \nabla \mathbf{g} + (\kappa - 1)\mathbf{n}_{\sigma\beta} \,, \quad \text{at } A_{\sigma\beta} \qquad (3\text{-}65)$$

$$\nabla^2 \mathbf{g} = \varepsilon_\sigma^{-1}\left(\frac{\kappa - 1}{\kappa}\right)\nabla \varepsilon_\sigma \,, \quad \text{in } V_\sigma \qquad (3\text{-}66)$$

In addition we require that the average of $\mathbf{f}$ and $\mathbf{g}$ be zero

$$\frac{1}{V_\beta} \int_{V_\beta} \mathbf{f} \, dV = 0 \,, \qquad \frac{1}{V_\sigma} \int_{V_\sigma} \mathbf{g} \, dV = 0 \qquad (3\text{-}67)$$

At this point we need estimates of ψ and ξ and these can be obtained by the use of the representations given by Eqs. (3-62) in the governing differential equations given by Eqs. (3-60) and (3-61). We make use of *complete* representations of the form

$$\nabla \tilde{c}_\beta = \nabla \mathbf{f} \cdot \nabla \{c_A\} + \mathbf{f} \cdot \nabla \nabla \{c_A\} + \nabla \psi \qquad (3\text{-}68a)$$

$$\nabla^2 \tilde{c}_\beta = \nabla^2 \mathbf{f} \cdot \nabla \{c_A\} + 2 \nabla \mathbf{f} : \nabla \nabla \{c_A\} + \mathbf{f} \cdot \nabla \nabla^2 \{c_A\} + \nabla^2 \psi \qquad (3\text{-}68b)$$

along with Eqs. (3-63) through (3-67), and from Eq. (3-60) we obtain

$$\nabla^2 \psi + \varepsilon_\beta^{-1} \varepsilon_\sigma \kappa \langle \nabla^2 \xi \rangle^\sigma - \varepsilon_\beta^{-1} \nabla \cdot (\varepsilon_\beta \langle \nabla \psi \rangle^\beta)$$

$$= -\varepsilon_\beta^{-1} \varepsilon_\sigma \kappa \langle \nabla \mathbf{g} \rangle^\sigma : \nabla \nabla \{c_A\}$$

$$\quad - 2 \nabla \mathbf{f} : \nabla \nabla \{c_A\} - \varepsilon_\beta^{-1} (\kappa - 1) \nabla \varepsilon_\sigma \cdot \nabla \{c_A\}$$

$$\quad + \varepsilon_\beta^{-1} \nabla \cdot (\varepsilon_\beta \langle \nabla \mathbf{f} \rangle^\beta) \cdot \nabla \{c_A\}$$

$$\quad + \varepsilon_\beta^{-1} \kappa \nabla \cdot (\varepsilon_\sigma \langle \nabla \mathbf{g} \rangle^\sigma \cdot \nabla \{c_A\} + \varepsilon_\sigma \langle \nabla \xi \rangle^\sigma) - \mathbf{f} \cdot \nabla \nabla^2 \{c_A\} \qquad (3\text{-}69)$$

In arriving at this result we have made use of the averaging theorem in the form given by Eq. (2-23) and in the form given by Eq. (2-60), but we have made no simplifications or approximations. Since averaged quantities undergo significant variation over the large length scale, d_p, while point quantities vary over the small length scales, l_β and l_σ, it should be clear that we can make use of inequalities of the type

$$\varepsilon_\beta^{-1} \nabla \cdot (\varepsilon_\beta \langle \nabla \psi \rangle^\beta) \ll \nabla^2 \psi \,, \qquad \mathbf{f} \cdot \nabla \nabla^2 \{c_A\} \ll \nabla \mathbf{f} : \nabla \nabla \{c_A\} \qquad (3\text{-}70)$$

A little thought will indicate that for the purpose of estimating the magnitude of ψ we can express Eqs. (3-69) as

$$\nabla^2 \psi + \varepsilon_\beta^{-1} \varepsilon_\sigma \kappa \langle \nabla^2 \xi \rangle^\sigma = \varepsilon_\beta^{-1} [\kappa \nabla \cdot (\varepsilon_\sigma \langle \nabla \mathbf{g} \rangle^\sigma) + \nabla \cdot (\varepsilon_\beta \langle \nabla \mathbf{f} \rangle^\beta)] \cdot \nabla \{c_A\}$$

$$\quad - \varepsilon_\beta^{-1} (\kappa - 1) \nabla \varepsilon_\sigma \cdot \nabla \{c_A\} \qquad (3\text{-}71)$$

From the boundary conditions given by Eqs. (3-57) and (3-58) we obtain

B.C.1 $$\psi = \xi \,, \qquad \text{at } A_{\sigma\beta} \tag{3-72}$$

B.C.2 $$\mathbf{n}_{\sigma\beta} \cdot \nabla\psi = \kappa \mathbf{n}_{\sigma\beta} \cdot \nabla\xi + (\kappa - 1)\mathbf{n}_{\sigma\beta}\mathbf{g} : \nabla\nabla\{c_A\} \tag{3-73}$$

and the analogous form of Eq. (3-71) for ξ is given by

$$\kappa\nabla^2\xi + \varepsilon_\sigma^{-1}\langle\nabla^2\psi\rangle^\beta = \varepsilon_\sigma^{-1}[\kappa\nabla\cdot(\varepsilon_\sigma\langle\nabla\mathbf{g}\rangle^\sigma) + \nabla\cdot(\varepsilon_\beta\langle\nabla\mathbf{f}\rangle^\beta)]\cdot\nabla\{c_A\}$$
$$+ \varepsilon_\sigma^{-1}(\kappa - 1)\nabla\varepsilon_\beta\cdot\nabla\{c_A\} \tag{3-74}$$

From Eq. (3-64) we know that **f** and **g** are the same magnitude, and from Eq. (3-72) we know that ψ and ξ are also the same magnitude. We can use this information to deduce from both Eqs. (3-71) and (3-74) and from Eq. (3-73) that the order of magnitude estimates of ψ and ξ are given by

$$\psi = \left\{ \mathbf{O}\left[\mathbf{f}\cdot\nabla\{c_A\}\left(\frac{l_\beta}{d_p}\right) \right] + \mathbf{O}\left[(\kappa - 1)\left(\frac{l_\beta}{d_p}\right)^2 \nabla\{c_A\} \right] \right\} \tag{3-75}$$

$$\xi = \left\{ \mathbf{O}\left[\mathbf{g}\cdot\nabla\{c_A\}\left(\frac{l_\sigma}{d_p}\right) \right] + \mathbf{O}\left[\left(\frac{\kappa - 1}{\kappa}\right)\left(\frac{l_\sigma}{d_p}\right)^2 \nabla\{c_A\} \right] \right\} \tag{3-76}$$

On the basis of Eq. (3-65) we know that **f** and **g** are on the order of the small length scales, thus ψ and ξ can be neglected in the representations given by Eqs. (3-62) and the concentration deviations can be expressed as

$$\tilde{c}_\beta = \mathbf{f}\cdot\nabla\{c_A\} \,, \qquad \tilde{c}_\sigma = \mathbf{g}\cdot\nabla\{c_A\} \tag{3-77}$$

By following the development presented in the previous section one can deduce that the volume source terms in Eqs. (3-63) and (3-66) have a negligible effect on the **f**- and **g**-fields relative to the *surface source* term in Eq. (3-65) and the boundary value problem for **f** and **g** can be expressed as

$$\nabla^2\mathbf{f} = 0 \,, \qquad \text{in } V_\beta \tag{3-78}$$

B.C.1 $$\mathbf{f} = \mathbf{g} \,, \qquad \text{at } A_{\sigma\beta} \tag{3-79}$$

B.C.2 $\qquad$ $\mathbf{n}_{\sigma\beta}\cdot\boldsymbol{\nabla}\mathbf{f}=\kappa\mathbf{n}_{\sigma\beta}\cdot\boldsymbol{\nabla}\mathbf{g}+(\kappa-1)\mathbf{n}_{\sigma\beta}$, $\qquad$ at $A_{\sigma\beta}$ $\qquad$ (3-80)

$$\nabla^2\mathbf{g}=0 , \qquad \text{in } V_\sigma \qquad (3\text{-}81)$$

The arguments presented in the previous section can also be used to construct the spatially periodic conditions given by

$$\mathbf{f}(\mathbf{r}+\mathbf{l}_i)=\mathbf{f}(\mathbf{r}) , \qquad \mathbf{g}(\mathbf{r}+\mathbf{l}_i)=\mathbf{g}(\mathbf{r}) , \qquad i=1,2,3 \qquad (3\text{-}82)$$

We are now in a position to return to the one-equation model given by Eq. (3-29) and make use of Eqs. (3-77) and (3-79) to obtain

$$(\varepsilon_\beta+\varepsilon_\gamma\varepsilon_\sigma)\frac{\partial\{c_A\}}{\partial t}=\boldsymbol{\nabla}\cdot[(\varepsilon_\beta+\varepsilon_\gamma\varepsilon_\sigma)\mathbf{D}_{\text{eff}}\cdot\boldsymbol{\nabla}\{c_A\}]-\varepsilon_\gamma\varepsilon_\sigma k\{c_A\}$$

$$(3\text{-}83)$$

Here the diffusivity is defined as

$$\mathbf{D}_{\text{eff}}=\left(\frac{\varepsilon_\beta\mathscr{D}_\beta+\varepsilon_\gamma\varepsilon_\sigma\mathscr{D}_\sigma}{\varepsilon_\beta+\varepsilon_\gamma\varepsilon_\sigma}\right)\mathbf{I}+\frac{\mathscr{D}_\beta-\varepsilon_\gamma\mathscr{D}_\sigma}{(\mathscr{V}\varepsilon_\beta\varepsilon_\gamma\varepsilon_\sigma)}\int_{A_{\beta\sigma}}\mathbf{n}_{\beta\sigma}\mathbf{f}\,dA \qquad (3\text{-}84)$$

Nozad [32] has determined the $\mathbf{f}$- and $\mathbf{g}$-fields for the system illustrated in Fig. 3-4 and although the system itself is not isotropic, the diffusion process is isotropic. If we restrict our discussion to the case for which $\mathbf{D}_{\text{eff}}$ is constant, the system is homogeneous, and the diffusion process is isotropic, Eq. (3-83) reduces to

$$(\varepsilon_\beta+\varepsilon_\gamma\varepsilon_\sigma)\frac{\partial\{c_A\}}{\partial t}=(\varepsilon_\beta+\varepsilon_\gamma\varepsilon_\sigma)D_{\text{eff}}\nabla^2\{c_A\}-\varepsilon_\gamma\varepsilon_\sigma k\{c_A\} \qquad (3\text{-}85)$$

Here the effective diffusivity is given by

$$D_{\text{eff}}=\frac{\varepsilon_\beta\mathscr{D}_\beta+\varepsilon_\gamma\varepsilon_\sigma\mathscr{D}_\sigma}{\varepsilon_\beta+\varepsilon_\gamma\varepsilon_\sigma}+\mathbf{I}:\left\{\frac{\mathscr{D}_\beta-\varepsilon_\gamma\mathscr{D}_\sigma}{\mathscr{V}(\varepsilon_\beta+\varepsilon_\gamma\varepsilon_\sigma)}\int_{A_{\beta\sigma}}\tfrac{1}{2}(\mathbf{n}_{\beta\sigma}\mathbf{f}+\mathbf{f}\mathbf{n}_{\beta\sigma})\,dA\right\}$$

$$(3\text{-}86)$$

and the theoretical values of D_{eff} are illustrated in Fig. 3-5 in terms of $(\varepsilon_\beta+\varepsilon_\gamma\varepsilon_\sigma)D_{\text{eff}}/\mathscr{D}_\beta$ as a function of κ. Other geometrical arrangements have been explored by Nozad et al. [31] and by Ryan [37] and both find that the *isotropic* effective diffusivity is dominated by the void fraction and *not* the particle geometry. It is important to note

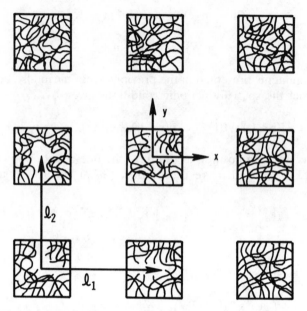

Figure 3-4　Two-dimensional spatially periodic porous medium.

that the quantity $(\varepsilon_\beta + \varepsilon_\gamma \varepsilon_\sigma) D_{\mathrm{eff}}/\mathcal{D}_\beta$ depends only on ε_β and κ and is, for all practical purposes, a linear function of κ. Any theory of diffusion in micropore–macropore systems should satisfy the two limiting conditions

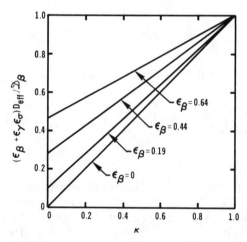

Figure 3-5　Theoretical values of the effective diffusivity for the one-equation model.

$$\frac{(\varepsilon_\beta + \varepsilon_\gamma \varepsilon_\sigma) D_{\text{eff}}}{\mathcal{D}_\beta} = \frac{\varepsilon_\gamma \mathcal{D}_\sigma}{\mathcal{D}_\beta} = \kappa, \qquad \varepsilon_\beta = 0 \qquad (3\text{-}87)$$

$$\frac{(\varepsilon_\beta + \varepsilon_\gamma \varepsilon_\sigma) D_{\text{eff}}}{\mathcal{D}_\beta} = 1, \qquad \kappa = 1 \qquad (3\text{-}88)$$

and it is clear from the results shown in Fig. 3-5 that the current theory satisfies these two constraints.

In order to use the effective diffusivity to determine the molar flux at the boundary between a catalyst pellet and the surrounding fluid phase, one makes use of the type of reasoning that led to Eq. (3-9) and which is described elsewhere [55, Sec. 2] in order to express the flux as

$$\mathbf{n} \cdot \mathbf{N}_A = -(\varepsilon_\beta + \varepsilon_\gamma \varepsilon_\sigma) D_{\text{eff}} \mathbf{n} \cdot \nabla \{c_A\} \qquad (3\text{-}89)$$

Here one should note that $\varepsilon_\beta + \varepsilon_\gamma \varepsilon_\sigma$ represents the fraction of the surface of the catalyst pellet that is fluid and a more traditional nomenclature would identify ε_β as the macropore void fraction ε_M and $\varepsilon_\gamma \varepsilon_\sigma$ as the micropore void fraction ε_μ.

In order to compare the results shown in Fig. 3-5 with the well-known model of Wakao and Smith [49], one expresses their result as [42, Eq. (11-27)]

$$(\varepsilon_\beta + \varepsilon_\gamma \varepsilon_\sigma) D_{\text{eff}} = \varepsilon_M^2 \bar{D}_M + \frac{\varepsilon_\mu^2 (1 + 3\varepsilon_M)}{(1 - \varepsilon_M)} \bar{D}_\mu \qquad (3\text{-}90)$$

The volume fractions are related by

$$\varepsilon_M = \varepsilon_\beta, \qquad \varepsilon_\mu = \varepsilon_\gamma \varepsilon_\sigma \qquad (3\text{-}91)$$

and if there is pure bulk diffusion in the macropore phase and pure Knudsen diffusion in the micropore phase we can write [42, Eqs. (11-28) and (11-29)]

$$\bar{D}_M = \mathcal{D}_{AB} = \mathcal{D}_\beta, \qquad \bar{D}_\mu = (\bar{\mathcal{D}}_K)_\mu \qquad (3\text{-}92)$$

so that Eq. (3-90) takes the form

$$(\varepsilon_\beta + \varepsilon_\gamma \varepsilon_\sigma) D_{\text{eff}} = \varepsilon_\beta^2 \mathcal{D}_\beta + \frac{(\varepsilon_\gamma \varepsilon_\sigma)^2 (1 + 3\varepsilon_\beta)}{(1 - \varepsilon_\beta)} (\bar{\mathcal{D}}_K)_\mu \qquad (3\text{-}93)$$

In order that the limiting condition given by Eq. (3-87) be satisfied we are forced to accept the interpretation

$$\varepsilon_\gamma \mathcal{D}_\sigma = (\varepsilon_\gamma \varepsilon_\sigma)^2 (\bar{\mathcal{D}}_K)_\mu \qquad (3\text{-}94)$$

so that Eq. (3-93) takes the form

$$\frac{(\varepsilon_\beta + \varepsilon_\gamma \varepsilon_\sigma) D_{\text{eff}}}{\mathcal{D}_\beta} = \varepsilon_\beta^2 + \frac{(1 + 3\varepsilon_\beta)}{(1 - \varepsilon_\beta)} \kappa \qquad (3\text{-}95)$$

Since $(\bar{\mathcal{D}}_K)_\mu$ is to be determined by a capillary tube model, it is reasonable that it should be modified by a void fraction and a "tortuosity" in order to extract an effective diffusivity. The representation given by Eq. (3-94) seems to suggest that the tortuosity is inversely proportional to $\varepsilon_\gamma \varepsilon_\sigma$. While the Wakao and Smith model can be made to satisfy Eq. (3-87) by an appropriate interpretation of the micropore diffusivity, there exist no contortions that allow Eq. (3-88) to be satisfied. This means that the model cannot be satisfactory at values of κ approaching one. At the other end of the spectrum ($\kappa = 0$), the model of Wakao and Smith can be compared with the results shown in Fig. 3-5 (or Fig. 2-5) along with the theory of Strieder and Aris [44, Eq. (2.6.15)] for a random bed of spheres. The comparison is shown in Fig. 3-6 and there we see good agreement with the theory of Ryan et al. [37] and Strieder and Aris [44] for low void fractions and good agreement with the model of Wakao and Smith [49] for high void fractions. Normally the range of interest is for $\varepsilon_\beta \leq 0.4$, and in view of the comparison between theory and experiment shown in Fig. 2-5 one must consider the work of Ryan et al. and Strieder and Aris to be reliable while the model of Wakao and Smith seriously underestimates the effective diffusivity. The exception to this generalization is the comparison between theory and experiment presented by Wakao and Smith in which they found excellent agreement between theory and experiment for values of ε_β as low as 0.09 and values of κ that were small compared to one [49, Eq. (15)].

In order to compare the theory presented in this section with experiment one requires a system in which equimolar counterdiffusion occurs and a system for which the effective diffusivity in the microporous phase is known. Experimental results of this type are currently unavailable, thus there is considerable motivation to extend the theoretical framework to more realistic cases. In the next section, a step is made in this direction.

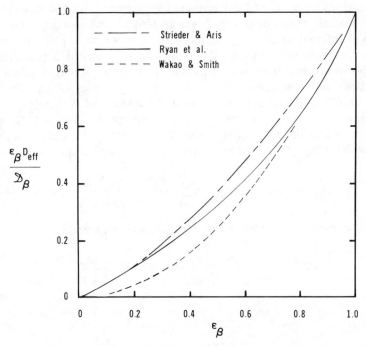

Figure 3-6   Theories for diffusion in granular porous media.

4. MASS TRANSPORT AND REACTION IN CATALYST PELLETS†

The subject of mass transport and reaction in porous catalysts is a central issue in the chemical process industries where catalytic reactors abound. The subject is discussed extensively in every text on reactor design and is the main topic of several monographs [3, 18, 38]. Since the late 1960's the interpretation of mass transport in porous catalysts has been based on the "dusty gas" model popularized by Mason et al. [23–25]. In this model one begins with the Stefan–Maxwell equations and assumes that one of the species, i.e. the "dust" species, is fixed in space. The dust species is taken to represent the rigid solid matrix of the porous medium, and the molecular diffusion coefficients are replaced with effective dif-

---

†This section is based largely on conversations with Roy Jackson along with selected reading from his monograph entitled *Transport in Porous Catalysts*.

fusivities which must be determined experimentally. The model has its origins in an attempt by Maxwell [26] to explain Graham's [15] experimental studies of gaseous diffusion in a porous plug. Maxwell adopted the point of view that "we may suppose the action of the porous material to be similar to that of a number of particles fixed in space, and obstructing the motion of the particles of the moving system". The configuration suggested by Maxwell (which is consistent with dilute gas kinetic theory) is illustrated in Fig. 4-1. Such a system can be analyzed by use of the Stefan–Maxwell equations given by Eq. (1-74) and listed here for an $N + 1$ component system.

$$0 = -\nabla x_A + \sum_{B=1}^{B=N+1} \frac{x_A \mathbf{N}_B - x_B \mathbf{N}_A}{c \mathcal{D}_{AB}} \tag{4-1}$$

For constant temperature this result can be expressed as

$$0 = -\nabla c_A + \sum_{B=1}^{B=N} \frac{x_A \mathbf{N}_B - x_B \mathbf{N}_A}{\mathcal{D}_{AB}} + \frac{x_A \mathbf{N}_{N+1} - x_{N+1} \mathbf{N}_A}{\mathcal{D}_{A,N+1}} \tag{4-2}$$

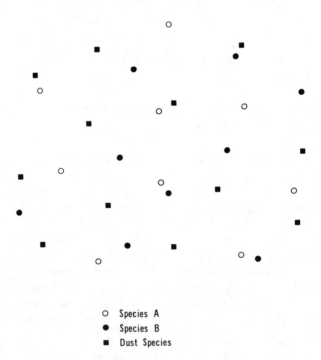

○ Species A
● Species B
■ Dust Species

Figure 4-1   Dusty gas.

We now require that the $N + 1$ species be the dust species which is fixed in space and forms the rigid solid matrix. Under these circumstances, Eq. (4-2) is expressed as

$$0 = -\nabla\langle c_A \rangle^\gamma + \sum_{B=1}^{B=N} \frac{\langle x_A \rangle^\gamma \langle \mathbf{N}_B \rangle^\gamma - \langle x_B \rangle^\gamma \langle \mathbf{N}_A \rangle^\gamma}{\mathscr{D}_{AB,\text{eff}}} - \frac{\langle \mathbf{N}_A \rangle^\gamma}{D_{A,\text{eff}}^K}$$

$$(4\text{-}3)$$

Here we have identified $\mathscr{D}_{A,N+1}$ as $\dot{D}_{A,\text{eff}}^K$ in Eq. (4-3) with the idea that this term represents the Knudsen diffusion coefficient. Since the last term in Eq. (4-2) and in Eq. (4-3) represents the momentum exchange between species A and the "solid matrix", it should not be surprising that the coefficient accounting for this exchange would be a Knudsen diffusion coefficient.

If one restricts Eq. (4-3) to a binary system and demands equimolar counterdiffusion, one obtains

$$\langle \mathbf{N}_A \rangle^\gamma = -\mathbb{D}_{\text{eff}} \nabla \langle c_A \rangle^\gamma \qquad (4\text{-}4)$$

yielding the well-known mixing rule for bulk and Knudsen diffusion

$$\frac{1}{\mathbb{D}_{\text{eff}}} = \frac{1}{\mathscr{D}_{AB,\text{eff}}} + \frac{1}{D_{A,\text{eff}}^K} \qquad (4\text{-}5)$$

That this result ignores the axioms of Sec. 1 is of little consequence, provided there is some valid motivation for

$$\langle \mathbf{N}_A \rangle^\gamma = -\langle \mathbf{N}_B \rangle^\gamma \qquad (4\text{-}6)$$

If we avoid this constraint and instead impose the condition of constant temperature and pressure we have $\nabla\langle c \rangle = 0$ and for a *binary system* Eq. (4-3) yields

$$\frac{\langle \mathbf{N}_A \rangle^\gamma}{D_{A,\text{eff}}^K} + \frac{\langle \mathbf{N}_B \rangle^\gamma}{D_{B,\text{eff}}^K} = 0 \qquad (4\text{-}7)$$

From kinetic theory Maxwell was able to argue that

$$D_{A,\text{eff}}^K \sim \frac{1}{\sqrt{M_A}}, \qquad D_{B,\text{eff}}^K \sim \frac{1}{\sqrt{M_B}} \qquad (4\text{-}8)$$

and Eq. (4-7) leads to

$$\langle \mathbf{N}_A \rangle^\gamma = -\sqrt{\frac{M_B}{M_A}} \langle \mathbf{N}_B \rangle^\gamma , \qquad \text{Graham} \qquad (4\text{-}9)$$

This was the result found experimentally by Graham and is usually referred to as the *Graham relation*. It represents a constraint on binary fluxes, under constant temperature and pressure conditions, that violates Darcy's law which would require

$$\langle \mathbf{N}_A \rangle^\gamma = -\left(\frac{M_B}{M_A}\right) \langle \mathbf{N}_B \rangle^\gamma , \qquad \text{Darcy} \qquad (4\text{-}10)$$

The fact that Graham's experiments violated Darcy's law must certainly have been a motivating factor for Maxwell's "dusty gas" model, and subsequent workers have been encouraged to add a *Darcy flux* to the flux predicted by Eq. (4-3). This "patching together" of bulk diffusion, Knudsen diffusion, and Darcy flow can be avoided by direct application of the method of volume averaging to the appropriate form of the species momentum equation.

Volume Averaging

This system under consideration is illustrated in Fig. 1-4 and the governing differential equations are

$$\frac{\partial c_A}{\partial t} + \nabla \cdot \mathbf{N}_A = 0 , \qquad A = 1, 2, \ldots , N \qquad (4\text{-}11)$$

$$0 = -\nabla p_A + \nabla \cdot \boldsymbol{\tau}_A + p \sum_{B=1}^{B=N} \frac{x_A \mathbf{N}_B - x_B \mathbf{N}_A}{c \mathscr{D}_{AB}} , \qquad A = 1, 2, \ldots , N \qquad (4\text{-}12)$$

Here the "dust species" has not been included in the species momentum equation since the momentum exchange between the diffusing molecular species and the rigid solid matrix, illustrated as the κ-phase in Fig. 1-4, will be accounted for in the traditional manner of continuum mechanics.

The form of Eq. (4-11) indicates that homogeneous reactions are ignored in this treatment, and the quasi-steady form of the flux condition at the $\gamma-\kappa$ interface is expressed as

$$\mathbf{n}_{\gamma\kappa} \cdot \mathbf{N}_A = -\hat{R}_A , \qquad \text{at } A_{\gamma\kappa} \qquad (4\text{-}13)$$

Here $\hat{R}_A$ represents the molar rate of *production* of species A owing to heterogeneous reaction. One can follow the analysis in Sec. 2 to conclude that the volume-averaged form of Eq. (4-11) is

$$\frac{\partial \langle c_A \rangle}{\partial t} + \nabla \cdot \langle \mathbf{N}_A \rangle = a_{\gamma\kappa} \langle \hat{R}_A \rangle_{\gamma\kappa} \qquad (4\text{-}14)$$

in which $a_{\gamma\kappa}$ is used in place of a_v to represent the interfacial area per unit volume. In terms of intrinsic phase averages we have

$$\varepsilon_\gamma \frac{\partial \langle c_A \rangle^\gamma}{\partial t} + \nabla \cdot (\varepsilon_\gamma \langle \mathbf{N}_A \rangle^\gamma) = a_{\gamma\kappa} \langle \hat{R}_A \rangle_{\gamma\kappa} \qquad (4\text{-}15)$$

and throughout this study we will deal only with homogeneous systems so that Eq. (4-15) can be expressed as

$$\frac{\partial \langle c_A \rangle^\gamma}{\partial t} + \nabla \cdot \langle \mathbf{N}_A \rangle^\gamma = \left(\frac{a_{\gamma\kappa}}{\varepsilon_\gamma} \right) \langle \hat{R}_A \rangle_{\gamma\kappa} \qquad (4\text{-}16)$$

The treatment of the reaction rate term follows the discussion presented in Sec. 2, and in this section our sole objective is the development of a suitable expression for the flux, $\langle \mathbf{N}_A \rangle^\gamma$. In this effort we will be guided by two experimental results. The first of these is the Graham relation for binary systems which we list as

$$\langle \mathbf{N}_A \rangle^\gamma = -\sqrt{\frac{M_B}{M_A}} \langle \mathbf{N}_B \rangle^\gamma, \qquad \nabla \langle p \rangle^\gamma = 0 \qquad (4\text{-}17)$$

and the second is Darcy's law which can be expressed as

$$\langle \mathbf{v} \rangle^\gamma = -\frac{\mathbf{K}_\gamma}{\mu} \cdot \nabla \langle p \rangle^\gamma \qquad (4\text{-}18)$$

Since Darcy's law is normally expressed in terms of the phase average velocity [54], the permeability tensor $\mathbf{K}_\gamma$ is larger by a factor of ε_γ^{-1} than the traditional Darcy's law permeability. The gravitational term in Darcy's law has been omitted since hydrostatic effects are unimportant in the catalyst pellet diffusion process. An important part of the analysis to be presented in this section is the removal of the conflict between Eqs. (4-17) and (4-18); however, at this point we ignore that conflict and proceed with the averaging of Eq. (4-12).

The phase average form of Eq. (4-12) is given by

$$0 = -\langle \boldsymbol{\nabla} p_A \rangle + \langle \boldsymbol{\nabla} \cdot \boldsymbol{\tau}_A \rangle + \left\langle p \sum_{B=1}^{B=N} \frac{x_A \mathbf{N}_B - x_B \mathbf{N}_A}{c \mathscr{D}_{AB}} \right\rangle \qquad (4\text{-}19)$$

Application of the type of development given by Eqs. (2-27) through (2-36) allows us to express the first term in Eq. (4-19) as

$$\langle \boldsymbol{\nabla} p_A \rangle = \varepsilon_\gamma \boldsymbol{\nabla} \langle p_A \rangle^\gamma + \frac{1}{\mathscr{V}} \int_{A_{\gamma\kappa}} \mathbf{n}_{\gamma\kappa} \tilde{p}_a \, dA \qquad (4\text{-}20)$$

At this time the route to the governing differential equation for $\tilde{p}_A$ has not been identified, and closure must be obtained on the basis of a plausible constitutive equation. On the basis of the analysis presented in Sec. 2, the following constitutive equation is suggested

$$\tilde{p}_A = \mathbf{f} \cdot \boldsymbol{\nabla} \langle p_A \rangle^\gamma \qquad (4\text{-}21)$$

For the special case of equimolar counterdiffusion and negligible viscous effects, this result is essentially identical to that derived in Sec. 2 and given by Eq. (2-79). In this case we have ignored any contribution to the $\tilde{p}_A$-field owing to heterogeneous reaction, and our studies in Sec. 2 support this point of view. Use of Eq. (4-21) in Eq. (4-20) leads to

$$\langle \boldsymbol{\nabla} p_A \rangle = \varepsilon_\gamma \boldsymbol{\nabla} \langle p_A \rangle^\gamma + \left\{ \frac{1}{\mathscr{V}} \int_{A_{\gamma\kappa}} \mathbf{n}_{\gamma\kappa} \mathbf{f} \, dA \right\} \cdot \boldsymbol{\nabla} \langle p_A \rangle^\gamma \qquad (4\text{-}22)$$

and we can express this result as

$$\langle \boldsymbol{\nabla} p_A \rangle = \varepsilon_\gamma \mathbf{B} \cdot \boldsymbol{\nabla} \langle p_A \rangle^\gamma \qquad (4\text{-}23)$$

where the second order tensor $\mathbf{B}$ is given by

$$\mathbf{B} = \mathbf{I} + \frac{1}{V_\gamma} \int_{A_{\gamma\kappa}} \mathbf{n}_{\gamma\kappa} \mathbf{f} \, dA \qquad (4\text{-}24)$$

From our studies in Sec. 2 we expect that $\mathbf{B}$ depends only on the geometry and is order one.

Substitution of Eq. (4-23) into (4-19) yields

$$0 = -\varepsilon_\gamma \mathbf{B} \cdot \boldsymbol{\nabla} \langle p_A \rangle^\gamma + \langle \boldsymbol{\nabla} \cdot \boldsymbol{\tau}_A \rangle + \left\langle p \sum_{B=1}^{B=N} \frac{x_A \mathbf{N}_B - x_B \mathbf{N}_A}{c \mathscr{D}_{AB}} \right\rangle \qquad (4\text{-}25)$$

Our treatment of the viscous stress term differs from that for the pressure gradient, and we simply use the divergence theorem to obtain

$$\langle \nabla \cdot \tau_A \rangle = \frac{1}{\mathcal{V}} \int_{A_{\gamma e}} \mathbf{n}_\gamma \cdot \tau_A \, dA + \frac{1}{\mathcal{V}} \int_{A_{\gamma \kappa}} \mathbf{n}_{\gamma \kappa} \cdot \tau_A \, dA \qquad (4\text{-}26)$$

Here $A_{\gamma e}$ represents the area of entrances and exists for the volume of the γ-phase contained within the averaging volume, and $\mathbf{n}_\gamma$ is the outwardly directed unit normal over these entrances and exits. It is well established from fluid mechanical studies that the viscous stresses at entrances and exists are small compared to those at solid surfaces, i.e.

$$\mathbf{n}_\gamma \cdot \tau_A \ll \mathbf{n}_{\gamma \kappa} \cdot \tau_A \qquad (4\text{-}27)$$

In addition, the nature of a porous medium requires that

$$\frac{A_{\gamma e}}{A_{\gamma \kappa}} = \mathbf{O}\left(\frac{l_\gamma}{r_0}\right) \qquad (4\text{-}28)$$

and the following inequality is assumed

$$\frac{1}{\mathcal{V}} \int_{A_{\gamma e}} \mathbf{n}_\gamma \cdot \tau_A \, dA \ll \frac{1}{\mathcal{V}} \int_{A_{\gamma \kappa}} \mathbf{n}_{\gamma \kappa} \cdot \tau_A \, dA \qquad (4\text{-}29)$$

This, along with Eq. (4-26) allows us to write Eq. (4-25) in the form

$$0 = -\varepsilon_\gamma \mathbf{B} \cdot \nabla \langle p_A \rangle^\gamma + \frac{1}{\mathcal{V}} \int_{A_{\gamma \kappa}} \mathbf{n}_{\gamma \kappa} \cdot \tau_A \, dA + \left\langle p \sum_{B=1}^{B=N} \frac{x_A \mathbf{N}_B - x_B \mathbf{N}_A}{c \mathcal{D}_{AB}} \right\rangle \qquad (4\text{-}30)$$

The central issue in the proper description of mass transport in catalyst pellets is the development of a reliable expression for $\mathbf{n}_{\gamma \kappa} \cdot \tau_A$; however, we bypass this problem for the present and go on to the last term in Eq. (4-30).

With some loss in generality, we restrict the analysis to isothermal systems and express the last term in Eq. (4-30) as

$$\left\langle p \sum_{B=1}^{B=N} \frac{x_A \mathbf{N}_B - x_B \mathbf{N}_A}{c \mathcal{D}_{AB}} \right\rangle = \frac{p}{c} \sum_{B=1}^{B=N} \frac{\langle x_A \mathbf{N}_B - x_B \mathbf{N}_A \rangle}{\mathcal{D}_{AB}} \qquad (4\text{-}31)$$

Use of decompositions of the form

$$x_A = \langle x_A \rangle^\gamma + \tilde{x}_A \qquad (4\text{-}32)$$

$$\mathbf{N}_B = \langle \mathbf{N}_B \rangle^\gamma + \tilde{\mathbf{N}}_B \qquad (4\text{-}33)$$

leads to the result

$$\langle x_A \mathbf{N}_B \rangle = \varepsilon_\gamma \langle x_A \rangle^\gamma \langle \mathbf{N}_B \rangle^\gamma + \langle \tilde{x}_A \tilde{\mathbf{N}}_B \rangle \qquad (4\text{-}34)$$

The second term on the right hand side of Eq. (4-34) represents a dispersive-like transport mechanism [53]. While this mechanism is not to be neglected when convective transport dominates, it would appear to be unimportant for the process of mass transport in catalyst pellets and we write

$$\left\langle p \sum_{B=1}^{B=N} \frac{x_A \mathbf{N}_B - x_B \mathbf{N}_A}{c \mathscr{D}_{AB}} \right\rangle = \varepsilon_\gamma \frac{p}{c} \sum_{B=1}^{B=N} \frac{\langle x_A \rangle^\gamma \langle \mathbf{N}_B \rangle^\gamma - \langle x_B \rangle^\gamma \langle \mathbf{N}_A \rangle^\gamma}{\mathscr{D}_{AB}}$$

$$(4\text{-}35)$$

With this result we can express the volume-averaged species momentum equation given by Eq. (4-30) as

$$0 = -\varepsilon_\gamma \mathbf{B} \cdot \nabla \langle p_A \rangle^\gamma + \frac{1}{\mathscr{V}} \int_{A_{\gamma\kappa}} \mathbf{n}_{\gamma\kappa} \cdot \boldsymbol{\tau}_A \, dA$$

$$+ \frac{\varepsilon_\gamma p}{c} \sum_{B=1}^{B=N} \frac{\langle x_A \rangle^\gamma \langle \mathbf{N}_B \rangle^\gamma - \langle x_B \rangle^\gamma \langle \mathbf{N}_A \rangle^\gamma}{\mathscr{D}_{AB}} \qquad (4\text{-}36)$$

For isothermal conditions p/c is constant and we can express this result as

$$0 = -\varepsilon_\gamma \mathbf{B} \cdot \nabla \langle c_A \rangle^\gamma + \frac{c/p}{\mathscr{V}} \int_{A_{\gamma\kappa}} \mathbf{n}_{\gamma\kappa} \cdot \boldsymbol{\tau}_A \, dA$$

$$+ \varepsilon_\gamma \sum_{B=1}^{B=N} \frac{\langle x_A \rangle^\gamma \langle \mathbf{N}_B \rangle^\gamma - \langle x_B \rangle^\gamma \langle \mathbf{N}_A \rangle^\gamma}{\mathscr{D}_{AB}} \qquad (4\text{-}37)$$

For equimolar counterdiffusion or the dilute solution of species A and *negligible momentum exchange* with the solid phase, Eq. (4-37) reduces to

$$\langle \mathbf{N}_A \rangle^\gamma = -\mathscr{D} \mathbf{B} \cdot \nabla \langle c_A \rangle^\gamma \qquad (4\text{-}38)$$

Here $\mathscr{D}$ is $\mathscr{D}_{AB}$ for the binary system and is given by

$$\frac{1}{\mathscr{D}} = \sum_{\substack{B=1 \\ B \neq A}}^{B=N} \frac{\langle x_B \rangle^\gamma}{\mathscr{D}_{AB}} \tag{4-39}$$

for the dilute solution case. From our studies on Sec. 2 we know that

$$\mathscr{D}\mathbf{B} = \mathbf{D}_{\text{eff}} \tag{4-40}$$

where $\mathbf{D}_{\text{eff}}$ is given explicitly by Eq. (2-80) and can be predicted by the theory described in Sec. 2. At this point our problem has been reduced to the determination of the second term in Eq. (4-37), and here we must rely on the solutions to special limiting cases and experimental results to produce a reasonable approximation for the momentum exchange at the $\gamma-\kappa$ interface.

Momentum Exchange at the Gas–Solid Interface—An Extension of Darcy's Law

We begin our discussion of momentum exchange in terms of systems for which $l_A \ll l_\gamma$, thus our initial discussion is restricted to bulk diffusion and viscous flow. In subsequent paragraphs we consider the other extreme, $l_A \gg l_\gamma$, and finally we patch the two results together to obtain a general theory of mass transport in porous catalysts.

A special case that can be extracted from Eq. (4-36) is obtained by summing over all species to provide

$$0 = -\varepsilon_\gamma \mathbf{B} \cdot \mathbf{\nabla} \langle p \rangle^\gamma + \frac{1}{\mathscr{V}} \int_{A_{\gamma\kappa}} \sum_{A=1}^{A=N} \mathbf{n}_{\gamma\kappa} \cdot \boldsymbol{\tau}_A \, dA \tag{4-41}$$

The last term in Eq. (4-36) is removed by Axiom III and the total pressure is defined by

$$p = \sum_{A=1}^{A=N} p_A \tag{4-42}$$

A comparison of Eq. (4-41) with Darcy's law in the form given by Eq. (4-18) might yield some information about the species stresses; however, that result is in conflict with Eq. (4-17) and we must remove the conflict between Darcy's law and the Graham relation before we can proceed with Eq. (4-41). It is crucial to this development to recognize that the velocity in Eq. (4-18) represents the intrinsic phase average velocity *relative to the velocity of the rigid solid matrix*. This is required in order that Eq. (4-18) be invariant to Galilean transformations. In general, when using Eq. (4-18) one assumes that the velocity of the rigid solid is zero. However, if the velocity of the rigid solid, measured relative to some inertial frame, is $\mathbf{u}_0$, then the intrinsic phase average velocity (relative to that same inertial frame) is given by

$$\langle \mathbf{v} \rangle^\gamma = -\frac{\mathbf{K}_\gamma}{\mu} \cdot \mathbf{\nabla} \langle p \rangle^\gamma + \mathbf{u}_0 \qquad (4\text{-}43)$$

Both the theoretical [54] and experimental studies leading to Eq. (4-18) are based on systems for which

$$\text{B.C.} \qquad \mathbf{v} = 0\,, \qquad \text{at the } \gamma\text{-}\kappa \text{ interface} \qquad (4\text{-}44)$$

i.e., the "no slip" condition is in effect. In 1879 Maxwell [27] presented *mean free path* arguments to indicate that the actual velocity at a gas–solid interface should be expressed as

$$\text{B.C.} \qquad \mathbf{v} = 2l\mathbf{n}_{\kappa\gamma} \cdot \mathbf{\nabla}\mathbf{v}\,, \qquad \text{at } A_{\gamma\kappa} \qquad (4\text{-}45)$$

Here it is understood that the slip velocity is tangent to the gas–solid interface. Maxwell's result was restricted to single component systems for which l represents the mean free path and for which $l \ll l_\gamma$. Thus Maxwell produced an expression for the slip velocity for viscous flow. Since this velocity will always be small compared to the average velocity, it is traditionally ignored and the "no slip" boundary condition illustrated by Eq. (4-44) is used in its place.

In diffusive transport processes, velocities tend to be small and the slip condition *cannot be ignored*. The mean free path arguments of Maxwell were extended to binary systems by Kramers and Kistemaker [19] in 1943 and generalized to N-component systems by Jackson [18] in 1977. Jackson's expression for the slip velocity is given by

$$\mathbf{v} = -\sum_{A=1}^{A=N} x_A \mathbf{u}_A \sqrt{M_A} \Big/ \sum_{A=1}^{A=N} x_A \sqrt{M_A}\,, \qquad \text{at } A_{\gamma\kappa} \qquad (4\text{-}46)$$

a result that is consistent with Eq. (4-44) rather than Eq. (4-45) for single component systems. From a practical point of view this is quite acceptable since the slip velocity given by Eq. (4-45) is of no importance for the case under consideration ($l_A \ll l_\gamma$) while the slip velocity given by Eq. (4-46) is crucial to the understanding of mass transport in porous catalysts.

If a slip in the mass average velocity occurs at the $\gamma-\kappa$ interface, it seems plausible that this would be comparable to the solid matrix itself moving. This means that the velocity $\mathbf{u}_0$ in Eq. (4-43) should be replaced by the slip velocity given by Eq. (4-46). This is in the nature of a plausible intuitive hypothesis and we express this idea as

Hypothesis I

$$\langle \mathbf{v} \rangle^\gamma = -\frac{\mathbf{K}_\gamma}{\mu} \cdot \nabla \langle p \rangle^\gamma - \frac{\sum\limits_{A=1}^{A=N} \langle x_A \rangle^\gamma \langle \mathbf{u}_A \rangle^\gamma \sqrt{M_A}}{\sum\limits_{A=1}^{A=N} \langle x_A \rangle^\gamma \sqrt{M_A}} \qquad (4\text{-}47)$$

Here we have replaced the values of $\mathbf{u}_A$ and x_A at the $\gamma-\kappa$ interface with the intrinsic phase average values with the thought that these are the appropriate quantities to be used in a volume-averaged transport equation.

Motivation for Eq. (4-47) is available both on theoretical grounds and on the basis of the following experimental results:

1. For single component systems, Eq. (4-47) reduces to Darcy's law.
2. For $\langle \mathbf{v} \rangle^\gamma \gg \langle x_A \rangle^\gamma \langle \mathbf{u}_A \rangle^\gamma$ we see that Eq. (4-47) reduces to Darcy's law.
3. For $\nabla \langle p \rangle^\gamma = 0$ and binary systems, a bit of algebra will show that Eq. (4-47) reduces to the Graham relation given by Eq. (4-17).

With some algebraic manipulation we can use Eqs. (4-41) and (4-47) to obtain

$$\sum_{A=1}^{A=N} \left\{ \varepsilon_\gamma \mu \mathbf{B} \cdot \mathbf{K}_\gamma^{-1} \cdot \langle \omega_A \rangle^\gamma \langle \mathbf{v}_A \rangle^\gamma + \frac{1}{\mathscr{V}} \int_{A_{\gamma\kappa}} \mathbf{n}_{\gamma\kappa} \cdot \boldsymbol{\tau}_A \, dA \right.$$
$$\left. + \frac{\varepsilon_\gamma \mu \mathbf{B} \cdot \mathbf{K}_\gamma^{-1} \cdot \langle x_A \rangle^\gamma \langle \mathbf{u}_A \rangle^\gamma \sqrt{M_A}}{\sum\limits_{B=1}^{B=N} \langle x_B \rangle^\gamma \sqrt{M_B}} \right\} = 0 \qquad (4\text{-}48)$$

in which $\langle \mathbf{v} \rangle^\gamma$ has been replaced with $\sum_{A=1}^{A=N} \langle \omega_A \rangle^\gamma \langle \mathbf{v}_A \rangle^\gamma$.

While this result provides a constraint on the *sum* of all the species stresses, we need an expression for the *individual* species stress if we are to obtain a form of Eq. (4-36) that can be used to determine the individual molar fluxes. Some guidance is available from the form that Eq. (4-36) must take for the case of a uniform mixture (no concentration gradients) flowing in a porous medium. The flow of a uniform mixture is described in terms of the constraints

$$\langle \mathbf{v}_A \rangle^\gamma = \langle \mathbf{v}_B \rangle^\gamma = \cdots \langle \mathbf{v}_N \rangle^\gamma, \qquad \mathbf{\nabla} \langle x_A \rangle^\gamma = \mathbf{\nabla} \langle x_B \rangle^\gamma = \cdots \mathbf{\nabla} \langle x_N \rangle^\gamma = 0$$

$$(4\text{-}49)$$

and under these circumstances Eq. (4-36) reduces to

$$0 = \varepsilon_\gamma \mathbf{B} \cdot \mathbf{\nabla} \langle p_A \rangle^\gamma + \frac{1}{\mathcal{V}} \int_{A_{\gamma\kappa}} \mathbf{n}_{\gamma\kappa} \cdot \boldsymbol{\tau}_A \, dA \qquad (4\text{-}50)$$

In the absence of any concentration gradients we have

$$\mathbf{\nabla} \langle p_a \rangle^\gamma = \langle x_A \rangle^\gamma \mathbf{\nabla} \langle p \rangle^\gamma \qquad (4\text{-}51)$$

and Eq. (4-50) can be expressed as

$$\frac{1}{\mathcal{V}} \int_{A_{\gamma\kappa}} \mathbf{n}_{\gamma\kappa} \cdot \boldsymbol{\tau}_A \, dA = \varepsilon_\gamma \langle x_A \rangle^\gamma \mathbf{B} \cdot \mathbf{\nabla} \langle p \rangle^\gamma \qquad (4\text{-}52)$$

When the conditions indicated by Eq. (4-49) are imposed on Eq. (4-47) we have

$$\langle \mathbf{v}_A \rangle^\gamma = -\frac{\mathbf{K}_\gamma}{\mu} \cdot \mathbf{\nabla} \langle p \rangle^\gamma \qquad (4\text{-}53)$$

and this result can be used in Eq. (4-52) to obtain

$$\frac{1}{\mathcal{V}} \int_{A_{\gamma\kappa}} \mathbf{n}_{\gamma\kappa} \cdot \boldsymbol{\tau}_A \, dA = -\varepsilon_\gamma \mu \langle x_A \rangle^\gamma \mathbf{B} \cdot \mathbf{K}_\gamma^{-1} \cdot \langle \mathbf{v}_A \rangle^\gamma \qquad (4\text{-}54)$$

One must keep in mind that $\boldsymbol{\tau}_A$ must reduce to this form *only* when the constraints indicated by Eq. (4-49) are valid, and that the general constraint given by Eq. (4-48) must be satisfied under *all* conditions. It is in the nature of an hypothesis to assume that Eq. (4-48) is uniquely satisfied by the following expression

Hypothesis II

$$\frac{1}{\mathcal{V}} \int_{A_{\gamma\kappa}} \mathbf{n}_{\gamma\kappa} \cdot \boldsymbol{\tau}_A \, dA = -\varepsilon_\gamma \mu \mathbf{B} \cdot \mathbf{K}_\gamma^{-1} \cdot \langle x_A \rangle^\gamma \langle \mathbf{v} \rangle^\gamma$$

$$-\frac{\varepsilon_\gamma \mu \mathbf{B} \cdot \mathbf{K}_\gamma^{-1} \cdot \langle x_A \rangle^\gamma \langle \mathbf{u}_A \rangle^\gamma \sqrt{M_A}}{\displaystyle\sum_{B=1}^{B=N} \langle x_B \rangle^\gamma \sqrt{M_B}} \qquad (4\text{-}55)$$

When this result is used in Eq. (4-36) it provides N independent equations relating the molar fluxes to the partial pressure gradients. These flux relations are to be used in Eq. (4-16) to predict concentrations and reaction rates. To be useful, Eq. (4-55) should be expressed in terms of molar fluxes and this form given by

$$-\frac{1}{\mathcal{V}} \int_{A_{\gamma\kappa}} \mathbf{n}_{\gamma\kappa} \cdot \boldsymbol{\tau}_A \, dA = -\varepsilon_\gamma \mu \mathbf{B} \cdot \mathbf{K}_\mu^{-1}$$

$$\cdot \left\{ \langle x_A \rangle^\gamma \frac{\displaystyle\sum_{B=1}^{B=N} M_B \langle \mathbf{N}_B \rangle^\gamma}{\displaystyle\sum_{B=1}^{B=N} M_B \langle c_B \rangle^\gamma} + \frac{\sqrt{M_A} \langle \mathbf{N}_A \rangle^\gamma}{\displaystyle\sum_{B=1}^{B=N} \langle c_B \rangle^\gamma \sqrt{M_B}} \right.$$

$$\left. - \frac{\langle c_A \rangle^\gamma \sqrt{M_A} \displaystyle\sum_{B=1}^{B=N} M_B \langle \mathbf{N}_B \rangle^\gamma}{\left(\displaystyle\sum_{B=1}^{B=N} M_B \langle c_B \rangle^\gamma\right)\left(\displaystyle\sum_{B=1}^{B=N} \langle c_B \rangle^\gamma \sqrt{M_B}\right)} \right\} \qquad (4\text{-}56)$$

This result is now ready for use in Eq. (4-36).

Knudsen Flow

One must keep in mind that the development that led to Eq. (4-56) was based on the mean free path being small compared to the characteristic pore diameter, i.e. $l_A \ll l_\gamma$. The Knudsen or free molecule flow regime is illustrated in Fig. 4-2 and is characterized by $l_A \gg l_\gamma$. We begin our discussion of the momentum transfer problem with a result first obtained by Maxwell [28] and referred to as the *maxwellian effusive* stream [21, page 61]. In an infinite region the

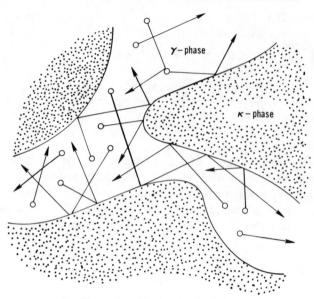

Figure 4-2   Free molecule flow.

number of molecules crossing a unit area *from one side* is given by

$$\left\{ \begin{array}{l} \text{one-sided flux} \\ \text{of species } A \end{array} \right\} = \tfrac{1}{4} n_A C_A \qquad (4\text{-}57)$$

Here n_A is the number density of species A and C_A represents the mean speed given by

$$C_A = \sqrt{\frac{8\mathscr{R}T}{\pi M_A}} \qquad (4\text{-}58)$$

Maxwell obtained Eq. (4-57) by treating the molecules as elastic spheres; however, the general form of Eq. (4-57) should be valid for more complex models of molecular interaction. The result given by Eq. (4-57) is also applicable to the flux of molecules striking a plane wall bounding a semi-infinite gas at equilibrium, and we will assume that the *form* of Eq. (4-57) is applicable to the gas in a bounded region such as that illustrated in Fig. 4-2. The mass flux of molecules striking the γ–κ interface illustrated in Fig. 4-2 is assumed to be of the form

$$\left\{ \begin{array}{l} \text{one-sided mass} \\ \text{flux of species } A \end{array} \right\} = \alpha c_A M_A C_A , \qquad \text{at } A_{\gamma\kappa} \qquad (4\text{-}59)$$

where α is a coefficient on the order of unity. Our interpretation of the species concentration c_A in Eq. (4-59) is that it is an average concentration in a region that is small compared to the size of the averaging volume, thus in terms of the method of volume averaging c_A represents a point concentration.

In order to determine the rate of momentum transfer to the κ-phase, we assume that the molecules that *leave* the $\gamma-\kappa$ interface have no preferred direction and thus carry no *net* momentum back into the gas phase. This means that the surface is treated as a *diffuse emitter*. Under these circumstances the momentum flux is determined solely by the rate of momentum transfer *from* the gas *to* the solid surface. To be more specific, the term $\mathbf{n}_{\gamma\kappa} \cdot \boldsymbol{\tau}_A$ in Eq. (4-36) represents the tangential momentum flux, thus we use Eq. (4-59) to obtain

$$\mathbf{n}_{\gamma\kappa} \cdot \boldsymbol{\tau}_A = -\boldsymbol{\lambda}(\boldsymbol{\lambda} \cdot \mathbf{v}_A)\alpha c_A M_A C_A = -\boldsymbol{\lambda}\boldsymbol{\lambda} \cdot \alpha \mathbf{N}_A M_A C_A , \qquad \text{at } A_{\gamma\kappa}$$

$$(4\text{-}60)$$

Here $\boldsymbol{\lambda}$ is a unit vector parallel to the vector given by $\mathbf{v}_A - (\mathbf{n}_{\gamma\kappa} \cdot \mathbf{v}_A)\mathbf{n}_{\gamma\kappa}$ and is thus tangent to the $\gamma-\kappa$ interface. A mapping exists between $\mathbf{N}_A$ and $\langle \mathbf{N}_A \rangle^\gamma$ so that we can write

$$\mathbf{N}_A = \mathbf{G} \cdot \langle \mathbf{N}_A \rangle^\gamma \qquad (4\text{-}61)$$

and the integral in Eq. (4-36) takes the form

$$\frac{1}{\mathcal{V}} \int_{A_{\gamma\kappa}} \mathbf{n}_{\gamma\kappa} \cdot \boldsymbol{\tau}_A s A = -\frac{1}{\mathcal{V}} \int_{A_{\gamma\kappa}} \boldsymbol{\lambda}\boldsymbol{\lambda} \cdot \mathbf{G} \cdot \alpha \langle \mathbf{N}_A \rangle^\gamma M_A C_A \, dA \qquad (4\text{-}62)$$

As usual we will neglect variations of $\langle \mathbf{N}_A \rangle^\gamma$ within the averaging volume as expressed in Eq. (4-62) as

$$\frac{1}{\mathcal{V}} \int_{A_{\gamma\kappa}} \mathbf{n}_{\gamma\kappa} \cdot \boldsymbol{\tau}_A \, dA = -\frac{1}{\mathcal{V}} \int_{A_{\gamma\kappa}} \boldsymbol{\lambda}\boldsymbol{\lambda} \cdot \mathbf{G}\alpha \, dA \cdot \langle \mathbf{N}_A \rangle^\gamma M_A C_A \qquad (4\text{-}63)$$

Use of the definition

$$\frac{1}{\mathcal{V}} \int_{A_{\gamma\kappa}} (\boldsymbol{\lambda}\boldsymbol{\lambda} \cdot \mathbf{G}\alpha) \, dA = \varepsilon_\gamma a_{\gamma\kappa} \mathbf{H} \qquad (4\text{-}64)$$

leads to

$$\frac{1}{\mathcal{V}} \int_{A_{\gamma\kappa}} \mathbf{n}_{\gamma\kappa} \cdot \boldsymbol{\tau}_A \, dA = -\varepsilon_\gamma a_{\gamma\kappa} \mathbf{H} \cdot \langle \mathbf{N}_A \rangle^\gamma M_A C_A \qquad (4\text{-}65)$$

Here one should keep in mind that $\mathbf{H}$ is order one, and the physics can be clarified somewhat by use of Eq. (4-58) to obtain

$$\frac{1}{\mathcal{V}} \int_{A_{\gamma\kappa}} \mathbf{n}_{\gamma\kappa} \cdot \boldsymbol{\tau}_A \, dA = -\varepsilon_\gamma a_{\gamma\kappa} \mathbf{H} \cdot \langle \mathbf{N}_A \rangle^\gamma \sqrt{8\mathcal{R}TM_A/\pi} \qquad (4\text{-}66)$$

This represents the Knudsen flow analog of the momentum flux given by Eq. (4-56), and the simplicity is a welcome relief from the complexity encountered in our study of bulk flow and diffusion.

A crucial step in the analysis leading to Eq. (4-66) is the mapping given by Eq. (4-61). In the Knudsen flow regime, illustrated in Fig. 4-2, we expect the tensor $\mathbf{G}$ to depend only on the geometry and *not* on the process. Such a mapping also exists for the bulk flow regime and is an important part of the theoretical derivation of Darcy's law [54, Sec. 4]; however, in the transition from Knudsen flow to bulk flow $\mathbf{G}$ changes from order one at the γ–κ interface to zero and is thus process dependent. The point here is that $\mathbf{H}$ should depend only on the geometry for the Knudsen flow regime and be of order one; however, as the transition from Knudsen flow to bulk flow takes place the general form of Eq. (4-66) breaks down and it must be replaced by Eq. (4-56).

The simplicity of Eq. (4-66) encourages its use in Eq. (4-36) and a search for special forms. This leads to

$$0 = -\varepsilon_\gamma \mathbf{B} \cdot \nabla \langle p_A \rangle^\gamma - \mathcal{R}T \left\{ \varepsilon_\gamma a_{\gamma\kappa} \mathbf{H} \sqrt{\frac{8M_A}{\pi \mathcal{R}T}} \cdot \langle \mathbf{N}_A \rangle^\gamma \right.$$

$$\left. - \varepsilon_\gamma \sum_{B=1}^{B=N} \frac{\langle x_A \rangle^\gamma \langle \mathbf{N}_B \rangle^\gamma - \langle x_B \rangle^\gamma \langle \mathbf{N}_A \rangle^\gamma}{\mathcal{D}_{AB}} \right\} \qquad (4\text{-}67)$$

Here we have replaced p/c with $\mathcal{R}T$ and a bit more algebraic manipulation leads to

$$0 = -\left\{ \frac{\mathbf{A}}{a_{\gamma\kappa}} \sqrt{\frac{\pi \mathcal{R}T}{8M_A}} \right\} \cdot \nabla \langle c_A \rangle^\gamma$$

$$- \left\{ \langle \mathbf{N}_A \rangle^\gamma - \frac{\mathbf{H}^{-1}}{a_{\gamma\kappa}} \sqrt{\frac{\pi \mathcal{R}T}{8M_A}} \cdot \sum_{B=1}^{B=N} \frac{\langle x_A \rangle^\gamma \langle \mathbf{N}_B \rangle^\gamma - \langle x_B \rangle^\gamma \langle \mathbf{N}_A \rangle^\gamma}{\mathcal{D}_{AB}} \right\}$$

$$(4\text{-}68)$$

Here the second order tensor **A** is given by

$$\mathbf{A} = \mathbf{H}^{-1} \cdot \mathbf{B} \tag{4-69}$$

and we define the Knudsen effective diffusivity tensor as

$$\mathbf{D}_{\text{eff}}^{K} = \frac{\mathbf{A}}{a_{\gamma\kappa}} \sqrt{\frac{\pi \mathscr{R} T}{8 M_A}} \tag{4-70}$$

This result was given earlier as Eq. (3-7) and one should keep in mind that **A** is order one. The last term in Eq. (4-68) originates from the *diffusion momentum source* given originally in Eq. (1-14). Since it represents the contribution from molecule–molecule collisions it should not be included for the type of process illustrated in Fig. 4-2. Nevertheless we will continue to retain this term and examine the case of an isotropic process. Under these circumstances Eq. (4-67) can be expressed as

$$0 = -B \nabla \langle p_A \rangle^{\gamma} - \mathscr{R} T \left\{ a_{\gamma\kappa} H \sqrt{\frac{8 M_A}{\pi \mathscr{R} T}} \langle \mathbf{N}_A \rangle^{\gamma} \right.$$

$$\left. - \sum_{B=1}^{B=N} \frac{\langle x_A \rangle^{\gamma} \langle \mathbf{N}_B \rangle^{\gamma} - \langle x_B \rangle^{\gamma} \langle \mathbf{N}_A \rangle^{\gamma}}{\mathscr{D}_{AB}} \right\} \tag{4-71}$$

For either the case of equimolar counterdiffusion or the dilute solution approximation, we can express this result as

$$\langle \mathbf{N}_A \rangle^{\gamma} \left(\frac{1}{D_{\text{eff}}^{K}} + \frac{1}{D_{\text{eff}}} \right) = -\nabla \langle c_A \rangle^{\gamma} \tag{4-72}$$

The total effective diffusivity can be defined as

$$\frac{1}{\mathbb{D}_{\text{eff}}} = \frac{1}{D_{\text{eff}}^{K}} + \frac{1}{D_{\text{eff}}} \tag{4-73}$$

so that Eq. (4-72) takes the form

$$\langle \mathbf{N}_A \rangle^{\gamma} = -\mathbb{D}_{\text{eff}} \nabla \langle c_A \rangle^{\gamma} \tag{4-74}$$

In these results we see a more rational justification for use of the Bosanquet interpolation formula than that given by Eqs. (4-1) through (4-5). Although Pollard and Present [35] indicate that Eq. (4-73) gives excellent agreement with an exact solution for dilute gas transport in a capillary tube, one must still wonder about its validity. The development leading to Eq. (4-73) hinges largely on the validity of Eqs. (4-60) and the idea that **G** in Eq. (4-61) depends on the geometry of the porous media and not on the process. For many bulk diffusion processes this would still appear to be true since species velocity profiles tend to be "flat" in the absence of significant pressure gradients. Arguments have been put forth elsewhere [58] that the species velocity profile in the Stefan diffusion tube shown in Fig. 1-3 is flat and this idea is illustrated for a porous medium in Fig. 4-3. For conditions such as those illustrated in Fig. 4-3 the arguments leading to Eq. (4-66) would seem to be satisfactory; however, if significant pressure gradients exist one would expect species velocity profiles of the type illustrated in Fig. 4-4.

In order to treat the full range of phenomena illustrated in Figs. 4-2 through 4-4 an interpolation rule is needed that makes use of both Eqs. (4-56) and (4-66). It will be convenient to represent Eq. (4-56) as

$$\frac{1}{\mathcal{V}} \int_{A_{\gamma\kappa}} \mathbf{n}_{\gamma\kappa} \cdot \boldsymbol{\tau}_A \, dA = \Omega_1 \,, \qquad l_A \ll l_\gamma \qquad (4\text{-}75)$$

and Eq. (4-66) as

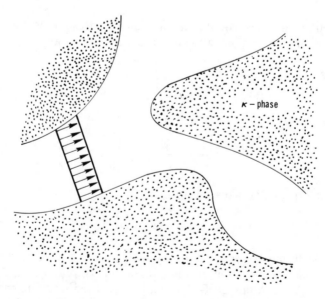

Figure 4-3 Species velocity profile for bulk diffusion.

$$\frac{1}{\mathcal{V}} \int_{A_{\gamma\kappa}} \mathbf{n}_{\gamma\kappa} \cdot \boldsymbol{\tau}_A \, dA = \Omega_2 \, , \qquad l_A \gg l_\gamma \qquad (4\text{-}76)$$

The proposed interpolation rule is given by

$$\frac{1}{\mathcal{V}} \int_{A_{\gamma\kappa}} \mathbf{n}_{\gamma\kappa} \cdot \boldsymbol{\tau}_A \, dA = \Omega_1 (1 + l_A a_{\gamma\kappa})^{-1} + \Omega_2 \left(\frac{l_A a_{\gamma\kappa}}{1 + l_A a_{\gamma\kappa}} \right) \qquad (4\text{-}77)$$

Here we have been explicit about our choice of the characteristic length for the γ-phase as $a_{\gamma\kappa}^{-1}$. This result is to be used in Eq. (4-36) in order to specify the N fluxes in terms of an appropriate set of boundary conditions. The complexity of Eq. (4-56) indicates that extensive numerical studies are required to elucidate the structure of the equation set comprised of Eqs. (4-16) and (4-36) which we list here as

$$\frac{\partial \langle c_A \rangle^\gamma}{\partial t} + \boldsymbol{\nabla} \cdot \langle \mathbf{N}_A \rangle^\gamma = \left(\frac{a_{\gamma\kappa}}{\varepsilon_\gamma} \right) \langle \tilde{R}_A \rangle_{\gamma\kappa} \qquad (4\text{-}78)$$

$$0 = -\varepsilon_\gamma \mathbf{B} \cdot \boldsymbol{\nabla} \langle c_A \rangle^\gamma + \frac{1}{\mathscr{R}T} \left[\Omega_1 (1 + l_A a_{\gamma\kappa})^{-1} + \Omega_2 \left(\frac{l_A a_{\gamma\kappa}}{1 + l_A a_{\gamma\kappa}} \right) \right]$$

$$+ \varepsilon_\gamma \sum_{B=1}^{B=N} \frac{\langle x_A \rangle^\gamma \langle \mathbf{N}_B \rangle^\gamma - \langle x_B \rangle^\gamma \langle \mathbf{N}_A \rangle^\gamma}{\mathscr{D}_{AB}} \qquad (4\text{-}79)$$

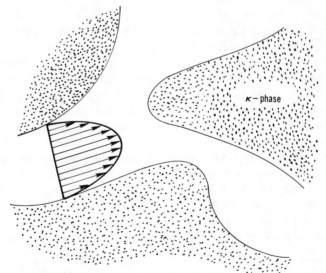

Figure 4-4   Species velocity profile for viscous flow.

In the preceding two sections we were able to both derive the volume-averaged transport equations and develop closure schemes that allowed for the direct theoretical determination of the effective diffusivity. The problem considered in this section is considerably more difficult and we have been forced to propose plausible constitutive equations, i.e. Eqs. (4-21) and (4-61). In addition, the development required the two plausible hypotheses given by Eqs. (4-47) and (4-55) along with the intuitive interpolation rule given by Eq. (4-77). Direct comparison between theory and experiment in the absence of adjustable parameters is not yet possible, but it remains as a worthwhile objective.

5. TWO-PHASE REACTORS

In the preceding three sections we have seen how the method of volume averaging can be used to analyze problems of diffusion and reaction in heterogeneous systems. For the simple systems studied in Secs. 2 and 3 we were able to derive the governing differential equations and predict the coefficients that appeared in these equations. The process studied in Sec. 4 was more complex and the method of volume averaging served as a guide to clarify the physics of the process. Plausible hypotheses were put forth and constitutive equations were proposed, both of which need to be tested by experiment. This same type of situation is encountered in the derivation of the design equations for two-phase reactors. The need for proposing constitutive equations, rather than developing closure schemes, often results from the fact that the flow is turbulent and the volume-averaged equations must also be time-averaged.

A two-phase reactor is illustrated in Fig. 5-1 and two averaging volumes are indicated. The spherical averaging volume must be considered as the preferred geometrical configuration; however, the cross-sectional slab illustrated in Fig. 5-1 is commonly used to derive what are sometimes known as *one-dimensional transport equations.†* It is important to keep in mind that the coefficients that appear in the equations that are derived from the two different averaging volumes are likely to take on different values. The classic example of this phenomenon has been discussed by Schlünder [39] in regard to heat transfer and dispersion phenomena in packed beds. Using

---

†The use of both one-dimensional and two-dimensional transport equations in the design of fixed bed reactors is discussed in Chapter 4 by Froment.

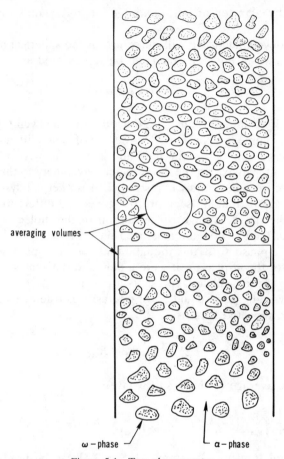

averaging volumes

ω – phase a – phase

Figure 5-1   Two-phase reactor.

relatively simple arguments, Schlünder was able to show that numerous investigators produced film heat transfer coefficients that were several orders of magnitude too low because they failed to understand the implications associated with the use of the cross-sectional slab as an averaging volume. Misinterpretation of experimental data led to correlations of the form [20]

$$\mathrm{Nu} \simeq 0.03\, \mathrm{Re}^{1.3} \tag{5-1}$$

for Reynolds numbers ranging from 100 to 10^{-1}. Since the asymptotic value of Nu for low Reynolds numbers is about 3.8 according to Sorensen and Stewart [43], the correlation given by Eq. (5-1) is incorrect by a factor of *one thousand* for a Reynolds number of 10^{-1}.

In this development we will make use of the spherical averaging volume shown in Fig. 5-1 which can be represented as

$$\mathcal{V} = V_\alpha(t) + V_\omega(t) \tag{5-2}$$

Note that $V_\alpha(t)$ and $V_\omega(t)$ may be functions of time even though the averaging volume is a constant. This degree of generality means that we are considering both fluid–fluid and fluid–solid systems and it is a trivial matter to extend the development given here to three-phase systems. Since our analysis is restricted to isothermal systems, the *design equations* that we seek are the governing differential equations for the phase average concentration of the molecular species involved in the reaction and mass transfer process. We will consider only the α-phase with the thought that the treatment would be identical for the ω-phase if it were a fluid and the analysis has been completed if the ω-phase represents a porous catalyst.

The governing point equation of interest was listed earlier as Eq. (1-71) and repeated here as

$$\frac{\partial c_A}{\partial t} + \mathbf{\nabla} \cdot \mathbf{N}_A = R_A \tag{5-3}$$

The phase average of this result is given by

$$\frac{1}{\mathcal{V}} \int_{V_\alpha(t)} \left(\frac{\partial c_A}{\partial t} \right) dV + \frac{1}{\mathcal{V}} \int_{V_\alpha(t)} \mathbf{\nabla} \cdot \mathbf{N}_A \, dV = \langle R_A \rangle \tag{5-4}$$

The general transport theorem [51, Sec. 3.4] can be used to write

$$\frac{d}{dt} \int_{V_\alpha(t)} c_A \, dV = \int_{V_\alpha(t)} \left(\frac{\partial c_A}{\partial t} \right) dV + \int_{A_{\alpha\omega}} c_A \mathbf{w} \cdot \mathbf{n}_{\alpha\omega} \, dA \tag{5-5}$$

where $\mathbf{w}$ represents the velocity of the α–ω interface. In our previous studies the phase interfaces were fixed in space; however, if we wish to deal with fluid–fluid reactors we must consider the movement of the interface. We can use Eq. (5-5) in Eq. (5-4) along with the averaging theorem to obtain

$$\frac{d}{dt} \left\{ \frac{1}{\mathcal{V}} \int_{V_\alpha(t)} c_A \, dV \right\} - \frac{1}{\mathcal{V}} \int_{A_{\alpha\omega}} c_A \mathbf{w} \cdot \mathbf{n}_{\alpha\omega} \, dA + \mathbf{\nabla} \cdot \langle \mathbf{N}_A \rangle$$

$$+ \frac{1}{\mathcal{V}} \int_{A_{\alpha\omega}} \mathbf{n}_{\alpha\omega} \cdot \mathbf{N}_A \, dA = \langle R_A \rangle \tag{5-6}$$

Since the phase averaged concentration $\langle c_A \rangle$ is *associated with* a fixed point in space we have

$$\frac{d}{dt}\left\{ \frac{1}{\mathscr{V}} \int_{V_\alpha(t)} c_A \, dV \right\} = \frac{\partial \langle c_A \rangle}{\partial t} \tag{5-7}$$

and we can use this result along with

$$N_A = c_A v_A \tag{5-8}$$

in order to express Eq. (5-6) as

$$\frac{\partial \langle c_A \rangle}{\partial t} + \nabla \cdot \langle N_A \rangle + \frac{1}{\mathscr{V}} \int_{A_{\alpha\omega}} c_A(v_A - w) \cdot n_{\alpha\omega} \, dA = \langle R_A \rangle \tag{5-9}$$

In this result the first term represents the obvious accumulation of species A, while the second term represents the convective transport. The third term represents the interfacial transport of species A and the right hand side represents the rate of production owing to chemical reaction. The interfacial flux term naturally couples the α-phase transport equation with the ω-phase. If the latter is a non-porous catalyst the coupling is quite simple; however, if the ω-phase is a fluid or a porous catalyst in which reaction takes place, the coupling is complex.

It is traditional to decompose the *convective* molar flux according to

$$N_A = c_A v_A = c_A v^* + c_A u_A^* \tag{5-10}$$

in which v^* is the molar average velocity and u_A^* is the molar diffusion velocity. In order to make reasonable approximation concerning the convective transport term in Eq. (5-9), we need a precise physical interpretation of $\nabla \cdot \langle N_A \rangle$. To obtain this, we use the divergence theorem to write

$$\langle \nabla \cdot N_A \rangle = \frac{1}{\mathscr{V}} \int_{A_{\alpha e}} n_\alpha \cdot N_A \, dA + \frac{1}{\mathscr{V}} \int_{A_{\alpha\omega}} n_{\alpha\omega} \cdot N_A \, dA \tag{5-11}$$

When this is compared with the averaging theorem we find

$$\nabla \cdot \langle N_A \rangle = \frac{1}{\mathscr{V}} \int_{A_{\alpha e}} n_\alpha \cdot N_A \, dA \tag{5-12}$$

and it is clear (mathematically and intuitively) that $\nabla \cdot \langle \mathbf{N}_A \rangle$ represents the net flux at entrances and exists. With this in mind, we express Eq. (5-10) as

$$\mathbf{N}_A = c_A \mathbf{v} - c_A \sum_{B=1}^{B=N} \omega_B \mathbf{u}_B^* + c_A \mathbf{u}_A^* \tag{5-13}$$

where $\mathbf{v}$ is the mass average velocity. In most reactor design problems convective effects at entrances and exists are large compared to diffusive effects. Under these circumstances, and given that

$$\sum_{B=1}^{B=N} x_A \mathbf{u}_B^* = 0 \tag{5-14}$$

one is inclined to write Eq. (5-13) as

$$\mathbf{N}_A = c_A \mathbf{v} + c_A \mathbf{u}_A^*, \qquad \text{at } A_{\alpha e} \tag{5-15}$$

Making use of the normal decompositions, this result leads to

$$\langle \mathbf{N}_A \rangle = \varepsilon_\alpha \langle c_A \rangle^\alpha \langle \mathbf{v} \rangle^\alpha + \langle \tilde{c}_A \tilde{\mathbf{v}} \rangle + \langle \mathbf{J}_A^* \rangle \tag{5-16}$$

The first term on the right hand side of this result represents the volume-averaged convective transport expressed in terms of the mass average velocity, the second term represents the dispersive transport, and the third term represents the molar diffusive flux. For most reactor design problems, these terms are arranged in order of descending importance. Use of Eq. (5-16) and

$$\langle c_A \rangle = \varepsilon_A \langle c_A \rangle^\alpha \tag{5-17}$$

in Eq. (5-9) yields

$$\varepsilon_\alpha \frac{\partial \langle c_A \rangle^\alpha}{\partial t} + \langle c_a \rangle^\alpha \frac{\partial \varepsilon_\alpha}{\partial t} + \nabla \cdot (\varepsilon_\alpha \langle c_a \rangle^\alpha \langle \mathbf{v} \rangle^\alpha)$$

$$+ \nabla \cdot \langle \tilde{c}_A \tilde{\mathbf{v}} \rangle + \nabla \cdot \langle \mathbf{J}_A^* \rangle + \frac{1}{\mathcal{V}} \int_{A_{\alpha \omega}} c_A (\mathbf{v}_A - \mathbf{w}) \cdot \mathbf{n}_{\alpha \omega} \, dA = \varepsilon_\alpha \langle R_A \rangle^\alpha$$

$$\tag{5-18}$$

This result should be thought of as a design equation for the two-phase reactor shown in Fig. 5-1; however, it cannot be used until the following problems have been resolved:

I. The volume fraction ε_α must be known. If the ω-phase is a rigid solid, one can determine ε_ω using simple experimental methods. If the ω-phase is a fluid, the problem is considerably more complex and "holdup correlations" are generally utilized. Strictly speaking, the volume fraction is determined by the laws of mechanics, but this approach is usually ignored because of the complexity of the problem. For fluid–fluid systems ε_α will depend on time and Eq. (5-18) must be time-averaged.

II. The velocity field must be determined in order to evaluate the convective transport. Often this is accomplished by assuming that the velocity profile is flat and $\varepsilon_\alpha \langle \mathbf{v} \rangle^\alpha$ can be replaced by the superficial velocity. That this can cause difficulties has been clearly demonstrated by Schlünder [39].

III. The dispersive term $\langle \tilde{c}_A \tilde{\mathbf{v}} \rangle$ must be modelled to produce reliable longitudinal and lateral dispersive fluxes. For steady laminar flow a closure scheme has been developed [10] that allows one to express the dispersive and diffusive flux as

$$\langle \tilde{c}_A \tilde{\mathbf{v}} \rangle + \langle \mathbf{J}_A^* \rangle = -\varepsilon_\alpha \mathbf{D}^* \cdot \mathbf{\nabla} \langle c_A \rangle^\alpha \qquad (5\text{-}19)$$

Here $\mathbf{D}^*$ is referred to as the total dispersivity tensor, and the theory provides for direct calculation of the components of $\mathbf{D}^*$ using spatially periodic models of porous media. The comparison between theory and experiment is shown in Fig. 5-2 and there we see that the calculated values [14] are in reasonably good agreement with experimental values for moderate particle Peclet numbers. Other theoretical studies [34] indicate that dispersion coefficients for pulsed systems are time-dependent for $t\mathscr{D}/d_p^2 \leqslant 1$ and this translates to a bed-length dependence for the experiments illustrated in Fig. 5-2. At high Peclet numbers, the bed length required to obtain asymptotic values of D_{xx}^* is excessively long and all of the high Peclet number data shown in Fig. 5-2 were obtained within the *dispersion entrance·region*. Because of this, they are predictably low relative to both the theory and the extrapolation of low Peclet number data.

IV. For completeness the diffusive flux in Eq. (5-18) needs to be represented as

$$\langle \mathbf{J}_A^* \rangle = -\varepsilon_A \mathscr{D}_{\text{eff}} \mathbf{\nabla} \langle c_A \rangle^\alpha \qquad (5\text{-}20)$$

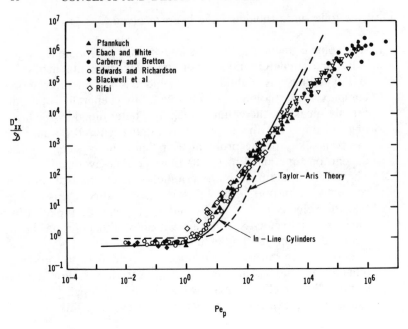

Figure 5-2 Theory and experiment for longitudinal dispersion coefficients in packed beds.

in which virtually any reasonable estimate of $\mathscr{D}_{eff}$ will be satisfactory. Because of the dominance of convective and dispersive transport.

V. The interfacial flux term in Eq. (5-18) represents a crucial term in the design of multiphase reactors and it is generally modelled according to

$$\frac{1}{\mathscr{V}} \int_{A_{\alpha\omega}} c_A(\mathbf{v}_A - \mathbf{w}) \cdot \mathbf{n}_{\alpha\omega} \, dA = a_{\alpha\omega} K(\langle c_A \rangle^\alpha - \langle c_A \rangle_{\alpha\omega})$$

(5-21)

Here one must take care to recognize that the mass transfer coefficient K will be influenced by the type of diffusion process that is occurring at the $\alpha-\omega$ interface. Often this is a high concentration, multicomponent diffusion process and the results obtained from the dilute solution diffusion of one component through a stagnant film should be used with caution.

VI. Homogeneous reaction rate expressions are relatively easy to determine experimentally, thus the term $\langle R_A \rangle^\gamma$ in Eq. (5-18) represents the least troublesome obstacle en route to a complete closure.

For a packed bed catalytic reactor operating at a sufficiently high Reynolds number, or a fluid–fluid reactor, the result given by Eq. (5-18) must be time-averaged. It is suggested elsewhere [57] that when $\bar{\varepsilon}_\alpha$ is independent of time the time-averaged version of Eq. (5-18) can be expressed as

$$\bar{\varepsilon}_\alpha \frac{\partial \langle \bar{c}_A \rangle^\alpha}{\partial t} + \nabla \cdot (\bar{\varepsilon}_\alpha \langle \bar{c}_A \rangle^\alpha \langle \bar{\mathbf{v}} \rangle^\alpha) =$$

$$\nabla \cdot (\bar{\mathbf{D}}^{(T)} \cdot \nabla \langle \bar{c}_A \rangle^\alpha) - \bar{a}_{\alpha\omega} \bar{K}^{(T)} (\langle \bar{c}_A \rangle^\alpha - \langle \bar{c}_A \rangle_{\alpha\omega}) + \bar{\varepsilon}_\alpha \langle \bar{R}_A \rangle^\alpha + \bar{\Phi}$$

$$(5\text{-}22)$$

Here the dispersive transport is defined by

$$\bar{\mathbf{D}}^{(T)} \cdot \nabla \langle \bar{c}_A \rangle^\alpha = \bar{\mathbf{D}}^* \cdot \nabla \langle \bar{c}_A \rangle^\alpha + \overline{\mathbf{D}^{*\prime} \cdot \nabla \langle c_A \rangle^{\alpha'}} - \bar{\varepsilon}_\alpha \overline{\langle c_A \rangle^{\alpha'} \langle \mathbf{v} \rangle^{\alpha'}}$$

$$- \langle \bar{c}_A \rangle^\alpha \overline{\varepsilon_\alpha' \langle \mathbf{v} \rangle^{\alpha'}} - \langle \bar{\mathbf{v}} \rangle^\alpha \overline{\varepsilon_\alpha' \langle c_A \rangle^{\alpha'}} \qquad (5\text{-}23)$$

and the interfacial flux is given explicitly as

$$\bar{a}_{\alpha\omega} \bar{K}^{(T)} (\langle \bar{c}_A \rangle^\alpha - \langle \bar{c}_A \rangle_{\alpha\omega}) = \bar{a}_{\alpha\omega} \bar{K} (\langle \bar{c}_A \rangle^\alpha - \langle \bar{c}_A \rangle_{\alpha\omega})$$

$$+ \bar{a}_{\alpha\omega} \overline{K' (\langle c_A \rangle^{\alpha'} - \langle c_A \rangle'_{\alpha\omega})}$$

$$+ \bar{K} \overline{a'_{\alpha\omega} (\langle c_A \rangle^{\alpha'} - \langle c_A \rangle'_{\alpha\omega})}$$

$$+ \overline{a'_{\alpha\omega} K' (\langle \bar{c}_A \rangle^\alpha - \langle \bar{c}_A \rangle_{\alpha\omega})} \qquad (5\text{-}24)$$

In both of these representations the third order correlations that arise naturally have been neglected. While these definitions yield a traditional form, as indicated by Eq. (5-22), there is a contradiction between Eqs. (5-23) and (5-24). For the representation given by Eq. (5-23) to be valid we require that

$$\overline{\varepsilon_\alpha' \langle \mathbf{v} \rangle^{\alpha'}} \sim 0, \qquad \langle c_A \rangle^{\alpha'} \sim \nabla \langle \bar{c}_A \rangle^\alpha \qquad (5\text{-}25)$$

The first of these requires that the correlation $\overline{\varepsilon_\alpha' \langle \mathbf{v} \rangle^{\alpha'}}$ provides a negligible contribution to the turbulent transport, and at this time there is no information to confirm this hypothesis. The second relation suggested in Eq. (5-25) is derivable [59] and contradicts the form of Eq. (5-24) since Eq. (5-25) would suggest that the time-averaged form of the interfacial flux gives rise to a convective-like transport mechanism. One always hopes that the temporal correlations make only small contributions and can therefore be neglected. This is, of course, the general expectation for the source term in Eq. (5-22) which takes the form

$$\bar{\Phi} = -\frac{\partial}{\partial t}(\overline{\varepsilon_\alpha' \langle c_A \rangle^{\alpha'}}) + \overline{\varepsilon_\alpha' \langle R_A \rangle^{\alpha'}} \qquad (5\text{-}26)$$

Assessing the importance of these terms either experimentally or theoretically would appear to be very difficult, thus there is a tendency to think of them as unimportant.

CONCLUSIONS

In this paper we have outlined a path of analysis that originates with the continuum axioms for the mass and momentum of multicomponent systems. Coupled with the method of volume averaging, these axioms have been used to analyze several multiphase mass transport processes. For the simplest of these, closure schemes are presented that allow for direct theoretical determination of the coefficients that appear in the volume-averaged transport equations. For more complex processes, methods of closure have not yet been developed and the method of volume averaging serves primarily as a tool to accurately account for the relevant physics. In this latter context, a design equation for two-phase reactors has been derived and discussed.

ACKNOWLEDGEMENT

This work was supported by NSF Grant CPE 8116528. Thanks are due to Roy Jackson of Princeton University for stimulating conversations on the subject of diffusion in porous catalysts. The closure schemes used in Secs. 2 and 3 were developed largely by Guillermo H. Crapiste of Planta Piloto de Ingeniería Quimíca (Bahía Blanca, Argentina) during a visit to Davis in March 1983 and while the author was a visiting lecturer at the Universidad del Sur in September 1983. The corrections and suggestions supplied by graduate students at the Universidad del Litoral (INTEC) and at the Universidad del Sur (PLAPIQUI) are greatly appreciated. Discussions with Christian Moyne of the Laboratoire d'Energétique et de Mécanique Théorique et Appliquée (Nancy, France), while the author was a Fulbright Research Scholar at the Université de Bordeaux in 1985, were instrumental in the development of Sec. 4 and are greatly appreciated.

NOMENCLATURE

Roman Letters

$\mathcal{A}_A(t)$ surface area of the species A body (m^2)

$\mathcal{A}_{\eta e}$ area of entrances and exits of the η-phase (m^2)

a_v area per unit volume (m^{-1})

$A_{\gamma\zeta}$ area of the $\eta-\zeta$ interface (m^2)

$A_{\gamma e}$ area of the entrances and exists of the η-phase contained within an averaging volume (m^2)

$\mathbf{b}_A$ body force per unit mass acting on species A (m/s^2)

$\mathbf{b}$ mass average body force (m/s^2)

c_A species A concentration $(moles/m^3)$

c total molar concentration $(moles/cm^3)$

$\langle c_A \rangle$ spatial average concentration or phase average concentration for the η-phase $(moles/m^3)$

$\langle c_A \rangle^\eta$ intrinsic phase average concentration for the η-phase $(moles/m^3)$

$\{c_A\}$ overall intrinsic phase average $(moles/m^3)$

C speed of sound (m/s)

C_A mean molecular speed for species A (m/s) and undefined average concentration for species A $(moles/m^3)$

$\hat{c}_A$ surface concentration $(moles/m^3)$

$\tilde{c}_A$ $c_A - \langle c_A \rangle^\eta$, concentration deviation $(moles/m^3)$

c_A^0 constant concentration $(moles/m^3)$

c_σ intrinsic phase average concentration in the micropore phase $(moles/m^3)$

c_β point concentration in the macropore phase $(moles/m^3)$

$\mathcal{D}_{AB}$ binary diffusivity (m^2/s)

$\mathbf{D}_{eff}$ effective diffusivity tensor (m^2/s)

$\mathcal{D}_{eff}$ undefined effective diffusivity (m^2/s)

$D_{A,eff}^K$ species A diffusivity relative to the "dust" species (m^2/s)

$\mathbf{D}_{\text{eff}}^K$ species A effective Knudsen diffusivity tensor (m^2/s)

$\mathscr{D}_\sigma$ effective diffusivity in the micropore phase (m^2/s)

$\mathscr{D}_\beta$ molecular diffusivity in the macropore phase (m^2/s)

$\mathbb{D}_{\text{eff}}$ combined bulk and Knudsen diffusivity (m^2/s)

d_p effective particle diameter (m)

$\mathbf{f}, \mathbf{g}$ vector fields that map $\nabla\langle c_A\rangle^\eta$ onto $\tilde{c}_A$

$\mathbf{g}$ gravity vector (m/s^2)

$\mathbf{I}$ unit tensor

$\mathbf{J}_A^*$ molar diffusive flux (moles/m s)

k_1 adsorption rate coefficient (m/s)

k_2 desorption rate coefficient (s^{-1})

$\hat{k}$ surface reaction rate coefficient (s^{-1})

k general reaction rate coefficient

$\mathbf{K}_\gamma$ modified Darcy's law permeability (m^2)

K equilibrium coefficient (m), and mass transfer coefficent (m/s)

L macroscopic length scale (m)

$L_\mathscr{D}$ diffusion length (m)

L_μ viscous length (m)

L_ρ inertial length (m)

L_ε characteristic length for variations in the void fraction (m)

l_A mean free path for species A (m)

l_γ characteristic length for the γ-phase (m)

$\mathbf{l}_i$ lattice vectors (m)

M Mach number

M_α atomic weight of α-type atoms (kg/kg moles)

M_A molecular weight of species A (kg/kg moles)

m_A mass of species A (kg)

n_A number density of species A (m^{-3})

$\mathbf{n}_{\eta\zeta}$ unit normal vector pointing from the η-phase into the ζ-phase

$\mathbf{N}_A$ molar flux of species A $(moles/m^2 s)$

$N_{\alpha A}$ number of α-type atoms in species A

p total pressure (N/m^2)

p_A partial pressure of species A (N/m^2)

$\mathbf{P}_{AB}$ force per unit volume exerted by species B on species A (N/m^3)

r_A mass rate of production per unit volume of species A owing to chemical reaction $(kg/m^3 s)$

R_A molar rate of production per unit volume of species A owing to chemical reaction $(moles/m^2 s)$

$\hat{R}_A$ molar rate of production per unit area of species A owing to chemical reaction $(moles/m^2 s)$

$\mathscr{R}_A$ mass rate of production of species A in a species A body owing to chemical reaction (kg/s)

r_0 radius of an av~r~~in~ ~~lume (m)

$\mathscr{R}$ universal gas constant (N m/kg moles °K)

Re vL_μ/ν, Reynolds numbers

$\mathbf{r}$ position vector (m)

s scalar field that maps $\langle c_A \rangle^\eta$ onto $\tilde{c}_A$

T temperature (°K)

$\mathbf{T}$ total stress tensor (N/m^2)

$\mathbf{T}_A$ species A stress tensor (N/m^2)

$\mathbf{t}_A$ species A stress vector (N/m^2)

t time (s)

t^* characteristic time (s)

$\mathbf{u}_A$ mass diffusion velocity (m/s)

$\mathbf{u}_A^*$ molar diffusion velocity (m/s)

V volume (m^3)

$\mathbf{v}_A$ species A velocity (m/s)

$\bar{\mathbf{v}}_A$ species A velocity resulting from chemical reaction (m/s)

$\mathbf{v}$ mass average velocity (m/s)

$\mathscr{V}_A(t)$ volume of a species A body (m^3)

$\mathscr{V}$ averaging volume (m^3)

$\mathscr{V}_\sigma$ averaging volume for the σ-phase (m^3)

x_A mole fraction of species A

Greek Letters

ε_η volume fraction of the η-phase

κ $\varepsilon_\gamma \mathscr{D}_\sigma/\mathscr{D}_\beta$, ratio of diffusivities

μ viscosity (N s/m^2)

ρ_A species A density (kg/m^3)

ρ total density (kg/m^3)

$\boldsymbol{\tau}_A$ species A viscous stress tensor (N/m^2)

$\boldsymbol{\tau}$ total viscous stress tensor (N/m^2)

ν kinematic viscosity (m^2/s)

ϕ $L\sqrt{a_v k/\mathscr{D}}$, Thiele modulus

ϕ_σ $l_\sigma\sqrt{k/\mathscr{D}_\sigma}$ in which $k = a_{\gamma\kappa}(k_1\hat{k})/\varepsilon_\gamma(\hat{k} + k_2)$, micropore Thiele modulus

ω_A mass fraction of species A

Ω_1 momentum flux per unit volume for bulk flow and diffusion (N/m^3)

Ω_2 momentum flux per unit volume for Knudsen diffusion (N/m^3)

REFERENCES

1. Anderson, T. B., and Jackson, R. (1967). A fluid mechanical description of fluidized beds, *Ind. Eng. Chem. Fundam.*, **6**, 527–538.
2. Aris, R. (1962). *Vectors, Tensors, and the Basic Equations of Fluid Mechanics*, Prentice-Hall, Inc., Englewood Cliffs, N.J.
3. Aris, R. (1975). *The Mathematical Theory of Diffusion and Reaction in Permeable Catalysts*. Vol. 1. *The Theory of the Steady State*. Vol. 2. *Questions of Uniqueness, Stability, and Transient Behavior*, Clarendon Press, Oxford.
4. Astarita, G. (1975). *An Introduction to Non-Linear Continuum Thermodynamics*, Societa Editrice di Chimica, Milano.
5. Balder, J. R., and Petersen, E. E. (1968). Application of the single pellet reactor for direct mass transfer studies, *J. Catalysis*, **11**, 195–210.
6. Bosanquet, C. H. (1944). British TA Report BR-507, September 27.
7. Bird, R. B., Stewart, W. E., and Lightfoot, E. N. (1960). *Transport Phenomena*, John Wiley & Sons, Inc., New York.
8. Birkhoff, G. (1960). *Hydrodynamics: A Study in Logic, Fact, and Similitude*, Princeton University Press, Princeton, N.J.
9. Carberry, J. J. (1962). The micro–macro effectiveness factor for the reversible catalytic reaction, *AIChE Journal*, **8**, 557–558.
10. Carbonell, R. G., and Whitaker, S. (1983). Dispersion in pulsed systems. Part II. Theoretical developments for passive dispersion in porous media, *Chem. Eng. Sci.*, **38**, 1795–1802.
11. Carbonell, R. G., and Whitaker, S. (1984). Heat and mass transport in porous media, in *Mechanics of Fluids in Porous Media* (Eds. J. Bear and Y. Corapcioglu), Martinus Nijhoff, Brussels.
12. Chapman, S., and Cowling, T. G. (1970). *The Mathematical Theory of Nonuniform Gases*, 3rd ed., Cambridge University Press, Cambridge.
13. Currie, J. A. (1960). Gaseous diffusion in porous media: Non-steady state method, *Brit. J. Appl. Physics*, **11**, 314–320.
14. Eidsath, A. B., Carbonell, R. G., Whitaker, S., and Herrmann, L. R. (1983). Dispersion in pulsed systems. Part III. Comparison between theory and experiments for packed beds, *Chem. Eng. Sci.*, **38**, 1803–1816.
15. Graham, T. (1833). On the law of diffusion of gases, *Phil. Mag.*, **2**, 175, 269, 351.
16. Gray, W. G. (1975). A derivation of the equations for multiphase transport, *Chem. Eng. Sci.*, **30**, 229–233.
17. Hayes, H. W., and Sarma, P. N. (1973). A model for the application of gas chromatography to measurements of diffusion in bidispersed structured catalysts, *AIChE Journal*, **19**, 1043–1046.
18. Jackson, R. (1977). *Transport in Porous Catalysts*, Elsevier Publishing Co., New York.
19. Kramers, H. A., and Kistemaker, J. (1943). On the slip of a diffusing gas mixture along a wall, *Physica*, **10**, 699.
20. Kunii, D., and Levenspiel, O. (1969). *Fluidization Engineering*, John Wiley & Sons, Inc., New York.
21. Kennard, E. H. (1938). *Kinetic Theory of Gases*, McGraw-Hill Book Co., New York.
22. Luss, D. (1977). Steady-state and dynamic behavior of a single catalyst pellet, in *Chemical Reactor Theory: A Review* (Eds. L. Lapidus and N. R. Amundson), Prentice-Hall, Inc., Englewood Cliffs, N.J.
23. Mason, E. A., Evans, R. B. III, and Watson, G. M. (1961). Gaseous diffusion in porous media at uniform pressure, *J. Chem. Phys.*, **35**, 2076.
24. Mason, E. A., Evans, R. B. III, and Watson, G. M. (1962). Gaseous diffusion in porous media. II. Effect of pressure gradients, *J. Chem. Phys.*, **36**, 1894.
25. Mason, E. A., Evans, R. B. III, and Watson, G. M. (1963). Gaseous diffusion in porous media. III. Thermal transpiration, *J. Chem. Phys.*, **38**, 1808.

26. Maxwell, J. C. (1860). On the process of diffusion of two or more kinds of moving particles among one another, *Phil. Mag.*, **20**, 21.
27. Maxwell, J. C. (1879). On stresses in rarefied gases arising from inequalities of temperature, *Phil. Trans. Roy. Soc.*, **170**, 231.
28. Maxwell, J. C. (1867). On the dynamical theory of gases, *Phil. Trans. Roy. Soc.*, **157**, 49.
29. Marle, C. M. (1967). Ecoulements monophasiques en milieu poreux, Rev. Inst. Francais du Petrole **22** (10) 1471–1509.
30. Mee, A. J. (1964). *Physical Chemistry*, Aldine Pub. Co., Chicago.
31. Nozad, I., Carbonell, R. G., and Whitaker, S. (1985). Heat conduction in multiphase systems. Part I. Theory and experiments for two-phase systems, *Chem. Eng. Sci.* **40**, 843–855.
32. Nozad, I. (1983). An experimental and theoretical study of heat conduction in two- and three-phase systems, Ph.D. Thesis, Department of Chemical Engineering, University of California at Davis.
33. Otani, S., and Smith, J. M. (1966). Effectiveness of large catalyst pellets—An experimental study, *J. Catalysis*, **5**, 332.
34. Paine, M. A., Carbonell, R. G., and Whitaker, S. (1983). Dispersion and adsorption in pulsed systems. Part I. Heterogeneous reaction and reversible adsorption in capillary tubes, *Chem. Eng. Sci.*, **38**, 1781–1793.
35. Pollard, W. G., and Present, R. D. (1948). On gaseous self-diffusion in long capillary tubes, *Physical Rev.*, **73**, 762–774.
36. Ruckerstein, E., Vaidyanathan, A. S., and Youngquist, G. R. (1971). Sorption by solids with bidisperse pore structures, *Chem. Eng. Sci.*, **26**, 1305–1318.
37. Ryan, D., Carbonell, R. G., and Whitaker, S. (1981). A theory of diffusion and reaction in porous media, *AIChE Symposium Series # 202*, Vol. 77, 46–62.
38. Scatterfield, C. M. (1970). *Mass Transfer in Heterogeneous Catalysis*, MIT Press, Cambridge, MA.
39. Schlunder, E. U. (1978). Transport phenomena in packed bed reactors, in *Chemical Reaction Engineering Reviews–Houston* (Eds. D. Luss and V. W. Weekman, Jr.), American Chemical Society, Washington, DC.
40. Slattery, J. C. (1967). Flow of viscoelastic fluids through porous media, *AIChE Journal*, **13**, 1066–1071.
41. Slattery, J. C. (1981). *Momentum, Energy, and Mass Transfer in Continua*, R. E. Krieger Pub. Co., Melbourne, FL.
42. Smith, J. M. (1981). *Chemical Engineering Kinetics*, 3rd ed., McGraw-Hill Book Co., New York.
43. Sorensen, J. P., and Stewart, W. E. (1974). Computation of forced convection in slow flow through ducts of packed beds. Part III. Heat and mass transfer in a simple cubic array of spheres, *Chem. Eng. Sci.*, **29**, 827–832.
44. Strieder, W., and Aris, R. (1973). *Variational Methods Applied to Problems of Diffusion and Reaction*, Springer-Verlag, New York.
45. Truesdell, C., and Toupin, R. (1960). The classical field theories, in *Handbuch der Physik*, Vol. III, Part 1 (Ed. S. Flugge), Springer-Verlag, New York.
46. Truesdell, C. (1969). *Rational Thermodynamics*, McGraw-Hill Book Co., New York.
47. Truesdell, C. (1968). *Essays in the History of Mechanics*, Springer-Verlag, New York.
48. Tsing, S. C.-A. (1982). A theoretical and computational analysis of radiant energy transport in porous media, MS Thesis, Department of Chemical Engineering, University of California at Davis.
49. Wakao, N., and Smith, J. M. (1962). Diffusion in catalyst pellets, *Chem. Eng. Sci.*, **17**, 825–834.
50. Whitaker, S. (1982). Laws of continuum physics for single-phase, single-component systems, in *Handbook of Multiphase Systems* (Ed. G. Hetstroni), Hemisphere Pub. Co., New York.

51. Whitaker, S. (1981). *Introduction to Fluid Mechanics*, R. E. Krieger Pub. Co., Melbourne, FL.
52. Whitaker, S. (1982). *Fundamental Principles of Heat Transfer*, R. E. Krieger Pub. Co., Melbourne, FL.
53. Whitaker, S. (1967). Diffusion and dispersion in porous media, *AIChE Journal*, **13**, 420–427.
54. Whitaker, S. (1969). Advances in the theory of fluid motion in porous media, *I & EC*, **61**, 14–28.
55. Whitaker, S. (1982). Diffusion and reaction in a micropore–macropore model of a porous medium, *Lat. Am. J. Chem. Eng. Appl. Chem.*, **13**, 143–183.
56. Whitaker, S. (1981). Heat and mass transfer in granular porous media, in *Advances in Drying*, Vol. 1 (Ed. A. S. Mujumdar), Hemisphere Pub. Co., New York.
57. Whitaker, S. (1973). The transport equations for multiphase systems, *Chem. Engr. Sci.*, **28**, 139–147.
58. Whitaker, S. (1967). Velocity profile in the Stefan diffusion tube, *Ind. Eng. Chem. Fundam.*, **6**, 476.
59. Whitaker, S., and Berker, A. (1978). A two-dimensional model of second order rapid reactions in turbulent tubular reactors, *Chem. Eng. Sci.*, **33**, 889.

Chapter 2

Effectiveness Factor Calculations

JUAN CARLOS GOTTIFREDI, ELIO EMILIO GONZO and
OSCAR QUIROGA
Instituto de Investigaciones para la Industria Química (INIQUI), UNSa.-CONICET, Buenos Aires 177-4400 Salta, Argentina

Abstract

In this review the first part is devoted to the analysis of the asymptotic behavior of the effectiveness factor for the single reaction case. The development of resulting equations is made under general conditions so that final results are valid for any kind of kinetic expression under isothermal or non-isothermal conditions. The influence of pellet shape as well as a non-uniform activity distribution is considered. Next, existing rational expressions to match the asymptotes are analyzed in order to show the advantages and disadvantages of each of them.

The results obtained for the case of a single reaction are extended to the case of a system formed by a pair of coupled, parallel reactions. Again the asymptotic behavior of the effectiveness factor for each reaction is developed and then rational expressions are used to match the asymptotes.

In the last part it is shown how to apply these rational expressions to take into account the effect of interphase resistances and to deduce general criteria for neglecting intraparticle and interphase resistances. It is also indicated how the falsification of kinetic parameters by transport limitations can be predicted with the appropriate use of these rational expressions.

INTRODUCTION

Almost all the active surface of a catalyst pellet is internal. The reaction that takes place within the pellet produces internal concentration and temperature gradients which, eventually, can be large enough to cause a significant variation in the rate of reaction inside the pellet. Moreover, the transport of heat and mass can strongly affect the fluid surrounding the catalyst particle producing important temperature and concentration differences between the bulk and the fluid in contact with the external pellet surface. Thus, in the more general case, both intraparticle and interphase transport phenomena are involved.

The general theory of chemical reactor analysis and design of heterogeneous catalytic systems relies on the effectiveness factor concept. Instead of considering a two phase chemical reactor, the actual system is modelled on the basis of an homogeneous reactor in which the rate of reaction is corrected by multiplying the value, calculated at bulk conditions, by the effectiveness factor, η. The net activity of each catalyst pellet is then considered in terms of η given by:

$$\eta = \frac{\text{observed (average) rate of reaction}}{\text{rate of reaction at bulk conditions}}$$

Because there is a continuous variation in concentration and temperature with the position inside the pellet, differential conservation equations are required to determine concentration and temperature profiles. These are used with the intrinsic rate of reaction to obtain the average rate of reaction by integration over the pellet.

After the pioneering works of Thiele [39], Zeldovich [48], Wagner [41], Wheeler [46] and Weisz [43] a great number of contributions were published treating new aspects of the effect of diffusion on the observed rate of reaction and its role in modifying the activity and selectivity of porous catalysts. Relevant conclusions on this subject have been concisely set forth in the texts by Satterfield and Sherwood [36], Petersen [32], Satterfield [37] and more recently by Aris [5], Carberry [9] and Butt [8].

Effectiveness factor estimation is of great concern in heterogeneous catalytic reactor analysis and design. The properties of the solution and its asymptotic behavior, as a function of relevant kinetic and diffusional parameters, are well established and thoroughly reviewed by Aris [5]. Analytical solutions for estimating η are only possible for particular cases in which the governing differential conservation equation is linear. However, actual cases involve a non-linear dependence of the kinetic expression on concentration and temperature.

Catalytic chemical reactor design requires repeated effectiveness factor calculations for different sets of parameters. This need is especially acute in the simulation of packed-bed reactors where local values of η are affected by bulk concentration and temperature changes produced along the reactor. As a consequence, η has to be evaluated at each grid-point except in the case of isothermal first order irreversible kinetic behavior. The importance of this calculation has been recently emphasized by Froment and Bishoff [13].

Classical numerical methods are quite time consuming and in some circumstances can become non-convergent. Recent experience in this field has shown that orthogonal collocation techniques, as described by Villadsen and Michelsen [40], seem to be the most adequate procedure to solve the resulting non-linear ordinary differential equation. However, as shown by Villadsen and Michelsen [40], the procedure can fail when there are large gradients in concentration and temperature inside the pellet. The problem can be overcome by taking some care, but this normally leads to an increase in computational time. The accuracy of the method depends upon the correct choice of the number of collocation points. As a rule the procedure is accurate with 8 to 10 interior points, thus reducing the original problem to the solution of a set non-linear algebraic equations whose number coincides with the number of internal collocation points. Our own experience has shown that the set can be solved by the Newton–Raphson method if an adequate initial guess of the concentration profile is supplied as starting point for the search of the final solution

The vast literature on this subject contains very few attempts to develop fast and sufficiently accurate approximations of the effectiveness factor. One possible alternative is the one-point collocation method of Stewart and Villadsen [38] that requires the solution of an algebraic non-linear equation and the well-known asymptotic solution for large Thiele moduli. A large inherent error exists in this crude approximation around the point where the collocation solution crosses the asymptote. Similar difficulties were observed by Paterson and Creswell [31], Karanth et al. [23] and Ramachandran et al. [34].

The first attempt to predict the effectiveness factor for arbitrary kinetic expressions and pellet geometry with a unique approximate algebraic expression was probably due to Petersen [32] who suggested that the analytical expression for a first order irreversible reaction in a slab isothermal pellet be used as the approximate equation for estimating the effectiveness factor. However, the Thiele modulus was replaced by a "normalized" one to fit the asymptotic behavior of η for large Thiele moduli. The same concept was advanced almost simultaneously by Aris [2, 3] and Bishoff [6].

It should be noted, however, that this approach implies a unique dependence of η on this normalized Thiele modulus. Unfortunately, it is well known that, even for the simplest case, the effect of pellet geometry is not accounted for by this simple representation. It is interesting to mention that the choice of the analytical expression for a slab pellet as a basis was surely due to its simplicity. From a practical point of view, in order to compensate errors, the corresponding expression for a cylindrical pellet would have had been a better choice.

After these first attempts, a number of empirical expressions were proposed to estimate effectiveness factor values under non-isothermal conditions. Liu [27] proposed some empirical formulas which can only be applied to first and second order non-isothermal reactions. Though explicit the method is not accurate and cannot deal with systems having steady-state multiplicity. Much more accurate is the empirical approximation of Jouven and Aris [22] which is only valid for first order non-isothermal reactions. The variational methods of Aris and Streider [4], Rester and Aris [35], as well as Jouven and Aris [22], have not found many applications since they are difficult to apply. More recently Rakadhyaksha et al. [33] presented an empirical formula aimed at predicting effectiveness factor values for large values of Thiele moduli for the case of Langmuir–Hinshelwood non-isothermal kinetics.

None of the forementioned methods, except the case of the one collocation point method, can be easily extended to rate expressions not covered in the corresponding original work. Moreover, the effects of pellet geometry, non-uniform catalytic activity, and mass and thermal diffusivity dependence on concentration and temperature, are difficult to incorporate in the traditional methods.

Churchill [12], based on a previous work (Churchill and Usagi [11]), suggested a new expression for the effectiveness factor that fits the asymptotic behavior for small and large values of the Thiele moduli. However, the effectiveness factor is only a function of the normalized Thiele modulus, thus it cannot account for the effect of pellet geometry, the reaction rate expression and eventually a non-uniform activity distribution inside the catalyst pellet. Nevertheless, it is the first rational attempt to match the asymptotic behavior of the effectiveness factor with an analytically continuous expression. Moreover, it will be shown later that this expression almost coincides with the analytical expression for first order reaction in a cylindrical pellet. Churchill [12] has also shown the importance of this kind of expression when effectiveness factor calculations involving interphase resistances are required. More recent-

[19, 20] presented an approximate expression for the effectiveness factor based on a rational procedure to match asymptotic expressions deduced for small and large values of the Thiele modulus.
for small and large values of the Thiele modulus.

This review is specifically designed to discuss the present status of the existing rational approximations and their capability to predict effectiveness factors under a great variety of circumstances usually encountered when performing chemical reactor design calculations. It will be clearly shown that the methods based on a rational approximation do not have the disadvantages of the collocation procedures and, at most, the solution of a unique non-linear algebraic equation is needed to solve for the unknowns. The essential feature of this method is the ability to match well-known asymptotic expressions for the effectiveness factor.

Another great concern in catalytic experimental work is to determine whether data are free from all transport influences before they are processed to yield kinetic parameters along with the activity and selectivity of the catalysts. The problem will be re-analyzed showing that all the existing criteria, based on special cases, can be reduced to only two theoretical expressions. One is for intraparticle diffusional phenomena and the other is for interphase phenomena, and each of them applies independently of the other. These theoretical criteria can also be applied to reactor design calculations since the set of kinetic parameters should be known in order to simulate the reactor performance.

In this review, the first part is devoted to the analysis of the asymptotic behavior of the effectiveness factor for the single reaction case. The development of resulting equations is made under general conditions so that the final results are valid for any type of kinetic expression under isothermal or non-isothermal conditions. The influence of pellet shape as well as a non-uniform activity distribution is considered. Next, existing rational expressions, used to match the asymptotes, are analyzed in order to show the advantages and disadvantages of each of them. The results obtained for the case of a single reaction are extended to the case of a kinetic system formed by a pair of coupled parallel reactions. Again, the asymptotic behavior of η for each reaction is determined first and then rational expressions are used to match the asymptotes.

In the final part of this review, it is shown how to apply these rational expressions to take into account the effect of interphase resistances and to deduce general criteria for neglecting intraparticle and interphase resistances.

A. ASYMPTOTIC EFFECTIVENESS FACTOR BEHAVIOR

A.1.   Single Reaction

Let us consider the case of a single reaction taking place in a porous catalyst pellet. The conservation of species A at steady-state conditions is most conveniently described by a second order ordinary differential equation which accounts for the balance established between diffusion and reaction. It will be assumed first that interphase resistances can be neglected and afterwards it will be shown how they can be incorporated into a calculation scheme once the final expression for the effectiveness factor is determined. The dimensionless mass balance for the reacting species A can be written as:

$$x^s \frac{d}{dx}\left(x^s \mathscr{D}(C)\, \frac{dC}{dx}\right) = \phi^2 f(x) R(C) \qquad (1)$$

where $s = 0, 1, 2$ stands for the geometrical shape of the particle (plane, cylindrical and spherical pellets respectively) and ϕ represents the Thiele modulus defined by:

$$\phi^2 = (r_s L^2 / D_s C_s') \qquad (2)$$

Here L represents the characteristic length of the particle, C' the dimensional concentration of species A, D the effective diffusivity of component A, r the dimensional rate of reaction and x the dimensionless coordinate normalized with respect to L. The subscript s denotes the values of the corresponding variable evaluated at the outer surface of the pellet. In Eq. (1) we have used $\mathscr{D}(C)$ and $R(C)$ to represent the dimensionless diffusivity and rate of reaction respectively normalized in terms of their corresponding surface values (D_s and r_s):

$$R(C) = (r/r_s) ; \qquad \mathscr{D}(C) = (D/D_s) \qquad (3a, b)$$

The quantity $f(x)$ is a normalized catalytic activity distribution function defined in such a way that:

$$\int_0^1 [(s + 1)f(x)x^s]\, dx = 1 \qquad (4)$$

In writing Eq. (1) it is assumed that an explicit relation exists between the dimensionless concentration of any other reactant or product taking place in the single reaction and the temperature with the dimensionless concentration of component $A(C)$. Such relations are straightforward when effective diffusivities and thermal conductivities are assumed to be independent of temperature and concentration. In such a case, as shown by Gottifredi et al. [19], one can write

$$C_i = \Gamma_i(C - 1) + 1$$

$$T = \beta(1 - C) + 1$$

(5a, b)

where:

$$\Gamma_i = -\left(\frac{DC'}{D_iC_i'}\right)_s \nu_i$$

$$\beta = \left(\frac{DC'}{\kappa T'}\right)_s (-\Delta H)$$

(6a, b)

Here κ represents the effective thermal conductivity of the pellet, ΔH the heat of reaction, T' the dimensional absolute temperature, and ν_i the stoichiometric coefficient. Appropriate boundary conditions required to solve Eq. (1) are

$$C = 1, \quad x = 1; \quad dC/dx = 0, \quad x = 0 \qquad (7a, b)$$

Under realistic situations Eq. (1) does not have an analytical solution since the term $R(C)$ is normally a non-linear function of C. Fortunately, from a chemical engineering point of view, one is not interested in solving Eq. (1) but rather in estimating the effectiveness factor given by

$$\eta = (s + 1) \int_0^1 [R(C)f(x)x^s] \, dx \qquad (8)$$

for the case of negligible interphase diffusional resistance. By integrating Eq. (1) once, with the boundary condition given by Eq. (7b), it can be easily shown that:

$$\eta = \frac{(s + 1)}{\phi^2} \frac{dC}{dx}\bigg|_{x=1} \qquad (9)$$

Both Eqs. (8) and (9) can be used to obtain expressions for η. However, since $C(x)$ is not known η cannot be estimated. Neverthe-

less, Eq. (1) can be solved approximately by perturbation procedures when $\phi \ll 1$ or when $\phi \gg 1$.

Perturbation Solution for Small Values of ϕ

When $\phi^2 \ll 1$ one notices that Eq. (1) itself suggests the following
series as an approximate solution:

$$C \simeq 1 + \phi^2 G_1(x) + \phi^4 G_2(x) + O(\phi^6) \tag{10}$$

One can also expand $\mathscr{D}(C)$ and $R(C)$ in Taylor series to obtain

$$R(C) \simeq 1 + \phi^2 R'(1)G_1 + \phi^4(R'(1)G_2 + R''(1)G_1^2/2) + O(\phi^6) \tag{11}$$

along with

$$\mathscr{D}(C) \simeq 1 + \phi^2 \mathscr{D}'(1)G_1 + O(\phi^6) \tag{12}$$

By replacing C, $R(C)$ and $\mathscr{D}(C)$ as given by Eqs. (10), (11), and
(12) in Eq. (1) and collecting terms of equal powers of ϕ, the
following set of linear ordinary uncoupled equations is found:

$$\frac{d}{dx}(x^s G_1') = x^s f(x) \tag{13}$$

$$\frac{d}{dx}(x^s G_2') = x^s f(x) R'(1) G_1 - \mathscr{D}'(1)\frac{d}{dx}(G_1 G_1') \tag{14}$$

Here $R'(1)$ and $\mathscr{D}'(1)$ denote first derivatives and $R''(1)$ the second
derivative with respect to concentration, all which are evaluated at
$C = 1$. One must be careful to note that G_1' and G_2' represent first
derivatives with respect to x. Taking into account boundary conditions (7a, b), Eqs. (13) and (14) must be solved subject to

$$G_1(1) = G_2(1) = 0; \qquad G_1'(0) = G_2'(0) = 0 \tag{15a, b}$$

Solutions to Eqs. (13) and (14) need no comment since they can
be found with standard analytical methods. By substituting Eq. (11)
into Eq. (8) the asymptotic expression for η is easily found:

$$\eta \simeq 1 - \sigma_1^* \phi^2 + \sigma_2^* \phi^4 + O(\phi^6) \tag{16}$$

where:

$$\sigma_1^* = -\int_0^1 [(s+1)x^2 f(x)R'(1)G_1]\, dx \qquad (17)$$

and

$$\sigma_2^* = (s+1)\int_0^1 x^s f(x)[R'(1)G_2 + R''(1)G_1^2/2]\, dx \qquad (18)$$

Taking into account Eqs. (13) and (14) with auxiliary conditions (15a, b) it is possible to show that:

$$\sigma_1^* = \alpha R'(1) = -G_2'(1)(s+1) \qquad (19)$$

where:

$$\alpha = (s+1)\int_0^1 x^{-s}\left[\int_0^x x^s f(x)\, dx\right]^2 dx \qquad (20)$$

Equation (16) is a valid asymptotic expression for the effectiveness factor when the Thiele modulus (ϕ) is small ($\phi \ll 1$). Equations (19) and (20) clearly show that σ_1^* is the product of two parameters. One is a function of the geometrical shape of the particle and the catalytic activity distribution, while the other depends on the form of the kinetic expression. Moreover, it is clearly seen that σ_1^* will be positive or negative depending upon the sign of $R'(1)$. Table I presents, in tabulated form, a number of practical cases in which the original dimensional kinetic expression is given together with $R(C)$ and $R'(1)$. In addition, Table II gives values of α for a number of cases.

Table I   Relation Between Dimensional (r) and Dimensionless Rates of Reaction (R) and $R'(1)$.

Dimensional Expression r	Dimensional Expression R	$R'(1)$
kC'^m	C^m	m
$\dfrac{kC_A'}{(1 + K_A C_A' + K_I C_I')}$	$(1+K)C/(1+KC)$	$1/(1+K)$
$kC_A'^m C_B'^m$	$C^m(1+(C-1)\Gamma_B)^n$	$m + n\Gamma_B$
$k_0 \exp(-E/RT')C'^m$	$C^m \exp\left(\dfrac{\beta\gamma(1-C)}{1+\beta(1-C)}\right)$	$m - \gamma\beta$

Table II   α Values as a Function of s and $f(x)$.

Activity Distribution	$f(x)$	$s=0$	$s=1$	$s=2$
Constant	1	1/3	1/8	1/15
Lineal	a_1x	1/5	1/12	1/21
Parabolic	a_2x^2	1/7	1/16	1/27
To fulfill	a_1	2	3/2	4/3
Eq. (4)	a_2	3	2	5/3

It should be noticed that the influence of concentration (or temperature) on the effective diffusivity only affects σ_2^* although there could be some influence on $R'(1)$ if other relations, different than those given by Eqs. (5a, b), linking C with C_i and T were used.

Perturbation Solution for Large Values of ϕ

When $\phi^2 \to \infty$ it is most convenient to introduce the following stretched coordinate, as suggested by Petersen [32]:

$$\xi = \phi(1-x) \tag{21}$$

By performing the change of the independent variable, Eq. (1) yields:

$$\frac{d}{d\xi}\left(\mathscr{D}(C)\frac{dC}{d\xi}\right) - \frac{s}{\phi}\left(1-\frac{\xi}{\phi}\right)^{-1}\mathscr{D}(C)\frac{dC}{d\xi} = R(C)f(\xi) \tag{22}$$

subject to the following boundary conditions:

$$C = 1 \qquad \xi = 0$$

$$C \to 0 \qquad \xi \to \infty$$

The approximate solution to Eq. (22) can be represented by the following series:

$$C = H_0 + (1/\phi)H_1 + O(\phi^{-2}) \tag{23}$$

Once again $R(C)$, $\mathscr{D}(C)$ and $f(\xi)$ can be expanded in Taylor series on the basis that $\phi \gg 1$. Thus:

$$R(C) = R(H_0) + \phi^{-1}R'(H_0)H_1 + \cdots \tag{24}$$

$$\mathscr{D}(C) = \mathscr{D}(H_0) + \phi^{-1}\mathscr{D}'(H_0)H_1 + \cdots \tag{25}$$

$$f(1 - \xi/\phi) = f(1) + \phi^{-1}f'(1)\xi + \cdots \tag{26}$$

By inserting C, given by Eq. (23) and Eqs. (24)–(26), into Eq. (22) and collecting terms of equal powers of ϕ the following set of ordinary differential equations is found:

$$\frac{d}{d\xi}\left(\mathscr{D}(H_0)\frac{dH_0}{d\xi}\right) = f(1)R(H_0) \tag{27}$$

$$\frac{d}{d\xi}\left(\mathscr{D}'(H_0)H_1\frac{dH_0}{d\xi} + \mathscr{D}(H_0)\frac{dH_1}{d\xi}\right) - s\mathscr{D}(H_0)\frac{dH_0}{d\xi}$$

$$= f(1)R'(H_0)H_1 - R(H_0)f'(1)\xi \tag{28}$$

Once again we note that a prime denotes the first derivative with respect to the original independent variable.

At first sight, the solution to Eq. (28) seems to be a rather difficult task. However, following the elegant procedure of Wedel and Luss [42], it is possible to deduce the asymptotic expression for the effectiveness factor based on Eq. (9). By defining:

$$p = \mathscr{D}(H_0)\left(\frac{dH_0}{d\xi}\right) \tag{29}$$

Eq. (27) can be rewritten in terms of H_0 as the independent variable. This leads to

$$\frac{d}{dH_0}(p^2) = 2f(1)\mathscr{D}(H_0)R(H_0) \tag{30}$$

Also, a new variable, p^*, is defined:

$$p^* = -\left[2f(1)\int_0^1 \mathscr{D}(H_0)R(H_0)\,dH_0\right]^{1/2} \tag{31}$$

Here it was assumed that $p^* \to 0$ as $H_0(C) \to 0$. Now Eq. (28) can be rewritten in terms of H_0 in the following manner:

$$\frac{p}{\mathscr{D}(H_0)}\left[\frac{d}{dH_0}\mathscr{D}'H_1p^*/\mathscr{D} + p^*\frac{dH_1}{dH_0}\right] = sp^* + f(1)R'H_1 - Rf'(1)\xi \tag{32}$$

Integration by parts allows us to express Eq. (32) as

$$p^{*2}\frac{dH_1}{dH_0}\Big]_0^1 - \int_0^1 p^*\left(\frac{dp}{dH_0}\right)\left(\frac{dH_1}{dH_0}\right)dH_0 = -\frac{\mathcal{D}'p^{*2}}{\mathcal{D}}H_1\Big]_0^1$$

$$+\int_0^1 \left(\frac{\mathcal{D}'}{\mathcal{D}}\right)p^*\left(\frac{dp^*}{dH_0}\right)H_1\,dH_0 + \int_0^1 f(1)\mathcal{D}R'H_1\,dH_0$$

$$-\int_0^1 \mathcal{D}Rf'(1)\xi\,dH_0 + \int_0^1 s\mathcal{D}p^*\,dH_0 \qquad (33)$$

However, by integrating once more the integral on the left hand side of Eq. (33), using Eq. (30) and the fact that H_1 must vanish when $H_0 = 1$ as well when $H_0 = 0$, Eq. (33) can be greatly simplified to obtain:

$$p^{*2}\frac{dH_1}{dH_0}\Big]_0^1 = -\int_0^1 \frac{d}{dH_0}(f(1)\mathcal{D}R)H_1\,dH_0 + \int_0^1 \frac{d}{dH_0}(f(1)\mathcal{D}R)$$

$$H_1\,dH_0 - \int_0^1 (f'(1)/2f(1))\frac{d}{dH_0}(p^{*2})\xi\,dH_0 + \int_0^1 s\mathcal{D}p^*\,dH_0$$

$$(34)$$

Finally, by integrating by parts the third term in the right hand side of Eq. (34), it is possible to solve for $(dH_1/d\xi)$ at $\xi = 0$. Thus, with Eq. (9) and Eqs. (30) and (34) it is possible to write the asymptotic expression for the effectiveness factor for large values of ϕ as:

$$\eta = \rho_1/\phi + \rho_2/\phi^2 + O(\phi^{-3}) \qquad (35)$$

where:

$$\rho_1 = (s+1)\left\{2\int_0^1 f(1)\mathcal{D}(H_0)R(H_0)\,dH_0\right\}^{1/2} \qquad (36)$$

$$\rho_2 = \frac{(s+1)^2[s+f'(1)/2f(1)]}{\rho_1}\int_0^1 \mathcal{D}(H_0)p^*(H_0)\,dH_0 \qquad (37)$$

It should be noticed that in the case of slab pellets ($s = 0$) with uniform activity distribution ($f'(1) = 0$), ρ_2 vanishes leaving Eq. (35) with only the first term. In some cases ρ_1 and ρ_2 can be evaluated anaytically with the help of Eq. (31). However it is always possible to calculate ρ_1 and ρ_2 by very simple numerical procedures such as Radau or Gauss quadrature methods.

Rational Approximations for the Effectiveness Factor
In the previous paragraphs it was shown that is possible to find asymptotic expressions for η either when $\phi \ll 1$ or when $\phi \gg 1$. These expressions can be rewritten in the following form:

$$\eta = 1/\hat{\phi} + \hat{\rho}_2/\hat{\phi}^2 + O(\hat{\phi}^{-3}) \tag{38}$$

when $\hat{\phi} \gg 1$, where the normalized Thiele modulus was introduced:

$$\hat{\phi} = (\phi/\rho_1) \tag{39}$$

and

$$\eta = 1 - \sigma_1\hat{\phi}^2 + \sigma_2\hat{\phi}^4 + O(\hat{\phi}^6) \tag{40}$$

when $\hat{\phi} \ll 1$.

Equations (38) and (40) contain most of the information available from the rate expression about the limiting values of η. The challenge is to find a rational expression which would be capable of reproducing Eqs. (38) and (40) when $\hat{\phi} \gg 1$ and $\hat{\phi} \ll 1$ respectively.

As mentioned before, the first attempt was due to Petersen [32] who simply proposed the analytical expression for η obtained in a slab pellet for first order isothermal reaction, but for the case in which ϕ was replaced by $\hat{\phi}$, i.e.,

$$\eta = \tanh(\hat{\phi})/\hat{\phi} \tag{41}$$

It must be stressed that this expression predicts η only as a function of $\hat{\phi}$. However, by expanding Eq. (41) for large and small values of $\hat{\phi}$ it yields:

$$\eta = (1/\hat{\phi}) - (2/\hat{\phi}) \exp(-2\hat{\phi}) + \cdots \tag{42}$$

and

$$\eta = 1 - 1/3\,\hat{\phi}^2 + O(\hat{\phi}^4) \tag{43}$$

respectively. By comparing Eqs. (42) and (43) with Eqs. (38) and

(40) respectively, it is concluded that Eq. (41) only reproduces the asymptotic behavior of η up to terms of order $(1/\hat{\phi})$ when $\hat{\phi} \gg 1$ and of order $\hat{\phi}^0$ when $\hat{\phi} \ll 1$. Nevertheless, this could be a good approximation in the case of a slab pellet with uniform activity distribution, since in such a case it was shown that $\hat{\rho}_2 = 0$. Later, results produced with Eq. (41) will be compared with exact values of η. However, it is well known that results given by Eq. (41) produce deviations as large as 15% from the exact results even for first order irreversible reactions in a spherical pellet. Moreover, Eq. (41) is not able to account for the effect of non-uniform activity within the pellet.

Churchill [12] proposed the following rational expression:

$$\eta = (1 + \hat{\phi}^2)^{-1/2} \tag{44}$$

which after being expanded for large and small values of $\hat{\phi}$ gives:

$$\eta = (1/\hat{\phi}) - 0.5\hat{\phi}^{-3} + \cdots \tag{45}$$

and

$$\eta = 1 - 0.5\hat{\phi}^2 + O(\hat{\phi}^4) \tag{46}$$

A comparison between Eqs. (45) and (46) with Eqs. (38) and (40) respectively would lead to the same conclusions as those stated for Petersen's expression (Eq. (41)). However it should be mentioned that σ_1 is exactly equal to 0.5 for the case of first order irreversible kinetics and a cylindrical pellet. In fact, when results produced with Eq. (44) are compared with exact values of η for first order irreversible reactions in a cylindrical pellet of uniform activity $(f(x) = 1)$ maximum deviations are always below 4%. Thus, although Eq. (44) has the same disadvantages as Eq. (41), it can be used on a more general basis since it is well known that results obtained for the case of a cylindrical pellet are located between those for spherical and slab pellets. In fact, if Eq. (44) had been used instead of Eq. (41), the maximum deviations for the classical case of first order irreversible reactions would have been less than 7%. This is shown in Fig. 1 where the quantity

$$PD = 100(\eta - \eta_N)/\eta_N \tag{47}$$

is plotted as a function of $\hat{\phi}$ for first order irreversible reactions

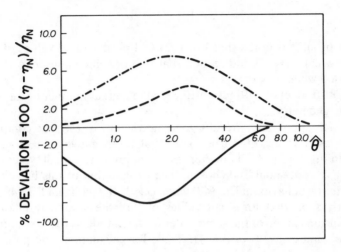

Figure 1   Comparison between approximate Eq. (44) and exact values of the first order irreversible reactions under isothermal conditions (——— slab, – – – – cylindrical and – · – · – spherical pellets).

under isothermal conditions for spherical, cylindrical and slab pellets. Here η is the value of η given by Eq. (44) and η_N is the exact value.

The next logical step leads to the two-parameter rational expression:

$$\eta = (1 + a^2\hat{\phi}^2)^{1/2}/(1 + a\hat{\phi}^2) \tag{48}$$

which coincides with Eq. (44) in the particular case when $a = 1$. By expanding Eq. (48) for large and small values of $\hat{\phi}$ it yields:

$$\eta = (1/\hat{\phi}) - a^{-2}(a - 0.5)\hat{\phi}^{-3} + \cdots \tag{49}$$

and

$$\eta = 1 + \left(\frac{1}{2}a^2 - a\right)\hat{\phi}^2 + O(\hat{\phi}^4) \tag{50}$$

By comparing Eq. (50) with Eq. (40) the expression for the unknown coefficient "a" is found:

$$a - \frac{1}{2}a^2 = \sigma_1 = \sigma_1^*\rho_1^2 \tag{51}$$

Thus:

$$a = 1 \pm \sqrt{1 - 2\sigma_1} \qquad (52)$$

First of all it is noted that Eq. (52) will produce real values of "a" only when $\sigma_1 \leqslant 0.5$ and second that two possible values of "a" can be used when $\sigma_1 \leqslant 0.5$.

Gottifredi et al. [19] tested Eq. (48) in a great number of cases for plane geometry.

When approximate results generated with Eq. (48) are compared with exact values of η it is found that the negative root should be taken in Eq. (52). However as σ_1 approaches small values the positive root should be chosen. The explanation of this behavior is due to the behavior of Eq. (48) when $\hat{\phi} \gg 1$. In fact, from Eq. (49) it is clearly seen that for small values of "a" the second term of the expansion can be of the same order of magnitude as the first even for large values of $\hat{\phi}$ (i.e. $\hat{\phi} \sim 4$). Thus, for calculational purposes the following criterion can be chosen:

$$a = 1 - \sqrt{1 - 2\sigma_1} \quad \text{if} \quad 1 - 2\sigma_1 \leqslant 0.75 \qquad (53a)$$

and

$$a = 1 + \sqrt{1 - 2\sigma_1} \quad \text{if} \quad 1 - 2\sigma_1 > 0.75 \qquad (53b)$$

It should be also noticed that in accordance with Eq. (49), the results given by Eq. (48) will approach the asymptote $(1/\hat{\phi})$ from above or from below depending on the value of the parameter "a". Since it is well known that exact results approach the asymptote from below, it should be expected that, in some circumstance, Eq. (48) will not be able to follow the true behavior of $\eta(\hat{\phi})$.

The other problem related to the case in which $2\sigma_1 > 1$ cannot be overcome with Eq. (48) itself. The only possible alternative would be to take $a = 1$ in an arbitrary form so that Eq. (48) will simply reduce to Churchill's expression (44). In order to better understand this limitation, it will be interesting to analyze the case of an mth order irreversible reaction under isothermal conditions. From Eqs. (18) and (36), and assuming $f(x) = 1$ and $\mathscr{D}(C) = 1$, it can be shown that:

$$2\sigma_1 = 4 \frac{(s + 1)}{(s + 3)} \frac{m}{(m + 1)} \qquad (54)$$

Thus $2\sigma_1 = 1$ implies $m = 3, 1, (5/7)$ for a slab, cylindrical and spherical pellet respectively. Moreover the situation can be even worse when the effect of a non-uniform activity distribution is considered. It can then be concluded that Eq. (48) has a limited range of applicability although better than Eq. (44). Nevertheless, Gottifredi et al. [19] tested the results produced by Eq. (48) for a number of kinetic expressions in slab pellets with uniform activity. The observed deviations were very small in most cases analyzed (mth order irreversible reactions, (m, p)th order irreversible and reversible kinetic expressions) although, as will be shown below, large deviations were observed in the case of Langmuir–Hinshelwood kinetics. Also, Gonzo and Gottifredi [15] tested Eq. (48) for the case of an (m, p)th order irreversible reaction under non-isothermal conditions. Both endothermic and exothermic reactions were considered. There is generally good agreement between approximate and exact results for the case of endothermic reactions which have also been tested with Langmuir–Hinshelwood kinetics. However, Eq. (48) is not able to reproduce the multiple solution region in the case of exothermic reactions. Nevertheless a practical situation was chosen to test the validity of Eq. (48). From the experimental results of Maymó and Smith [29], the effectiveness factor was calculated with Eq. (48) and compared with experimental data as well as with exact numerical predictions. These results are presented in Table III. It can be clearly seen that for most practical situations Eq. (48) gives very good estimates of η. Moreover Eq. (48) can be used to establish the region where $\eta > 1$ should be expected. In many practical cases isothermal conditions are assumed in order to simplify the calculations. With Eq. (48) there is no actual need to assume isothermal conditions since the calculation of η is as simple as for isothermal conditions as it is for non-isothermal conditions.

Table III Comparison Between Approximate (A), Experimental (Exp.) and Numerical (N) Results. Maymo and Smith [26].

θ	2.67	2.47	2.44	2.00	2.17	3.14	4.53
η(Exp.)	0.67	0.87	0.93	0.96	1.10	0.80	0.60
η_A	0.78	0.85	0.89	0.95	1.07	0.84	0.65
η_N	0.78	0.87	0.94	1.01	1.42	0.89	0.94

Wedel and Luss [42] presented a rather different rational expression. They assumed, as a fitting expression, the ratio between two polynomials:

$$\eta = \frac{1 + b_1\hat{\phi} + b_2\hat{\phi}^2}{1 + b_3\hat{\phi} + b_4\hat{\phi}^2 + b_5\hat{\phi}^3} \tag{55}$$

By expanding Eq. (55) for large and small values of $\hat{\phi}$ it is found:

$$\eta = \frac{b_2}{b_5}\hat{\phi}^{-1} + \left(\frac{b_1}{b_5} - \frac{b_4 b_2}{b_5^2}\right)\hat{\phi}^{-2} + O(\hat{\phi}^{-3}) \qquad (56)$$

and

$$\eta = 1 + (b_1 - b_3)\hat{\phi} + (b_2 - b_1 b_3 - b_4 - b_3^2)\hat{\phi}^2$$
$$+ (2b_3 b_4 - b_5 - b_1 b_4 - b_2 b_3 - b_3^3 + b_1 b_3^2)\hat{\phi}^3 + O(\hat{\phi}^4) \qquad (57)$$

A comparison with Eqs. (38) and (40) requires:

$$b_1 = b_3; \qquad b_2 = b_5; \qquad b_4 - b_2 = \sigma_1$$
$$b_2 = (b_1 - \sigma_1)/(1 + \hat{\rho}_2) \qquad (58a, b, c, d)$$

while from the condition that the coefficient of the term multiplying $\hat{\phi}^{-3}$ should vanish, it is found:

$$b_1 = \frac{\sigma_2}{(1 - (1 + \hat{\rho}_2)\sigma_1)} \qquad (59)$$

Thus Eq. (55) reproduces the asymptotic behavior of η up to terms of order $\hat{\phi}^{-2}$ and to terms of order $\hat{\phi}^2$ when $\hat{\phi} \gg 1$ and $\hat{\phi} \ll 1$ respectively. As can be seen, the unknowns are very easily determined by the linear uncoupled system given by Eqs. (58a, b, c) and (59).

Wedel and Luss [42] tested Eq. (55) for a great variety of cases using a spherical catalyst pellet with uniform activity. They have shown that Eq. (55), with auxiliary conditions (58a, b, c, d) and (59), produces extremely accurate results, provided $\sigma_1 > 1$. This implies an apparent reaction order greater than one. When σ_1 is small, b_1, b_2, and b_5 become very small and since the coefficient multiplying $\hat{\phi}^{-3}$ in Eq. (56) is of the order b_5^{-2} large deviations arise as $\hat{\phi}$ becomes greater than one. To avoid this problem, Wedel and Luss [42] forced Eq. (55) to fit the asymptotic expression (38) at $\hat{\phi} = 1$. Thus, instead of using condition (59) they assumed:

$$(1 + b_1 + b_2) = (1 + \hat{\rho}_2)(1 + b_3 + b_4 + b_5) \qquad (60)$$

and Eqs. (58a, b, c, d) were used in conjunction with Eq. (60) to determine b_1. Since $(1 + \hat{\rho}_2)$ is an extremely good estimate of η at $\hat{\phi} = 1$, when $\sigma_1 < 1$, they succeeded in producing extremely accurate values of η. Maximum deviations were always below 7% for Langmuir–Hinshelwood kinetic expressions under isothermal conditions. Even for non-isothermal (exothermic) reactions predicted η values were in general sufficiently accurate. Of course, Eq. (55) is not able to account for the multiple solution behavior of the effectiveness factor.

Gottifredi et al. [20], in an attempt to overcome the limitation given by the condition $2\sigma_1 > 1$ in Eq. (52), proposed a more complete rational expression in terms of $\hat{\phi}^2$:

$$\eta = \frac{(1 + a^2\hat{\phi}^2)^{1/2}}{(1 + a\hat{\phi}^2)} + \frac{d_1\hat{\phi}^2}{(1 + d_2\hat{\phi}^2)^2} \tag{61}$$

After being expanded for large and small values of $\hat{\phi}$ Eq. (61) yields:

$$\eta = \hat{\phi}^{-1} + \frac{d_1}{d_2^2}\hat{\phi}^{-2} + O(\hat{\phi}^{-3}) \tag{62}$$

and

$$\eta = 1 - (a - 0.5a^2 - d_1)\hat{\phi}^2 + \left(a^2 - \frac{1}{2}a^3 - \frac{1}{8}a^4 - 2d_1d_2\right)\hat{\phi}^4 + O(\hat{\phi}^6) \tag{63}$$

By comparing Eqs. (62) and (63) with Eqs. (38) and (40) respectively the following set of non-linear algebraic equations is established to determine the three unknowns:

$$d_1 = \hat{\rho}_2 d_2^2$$

$$a - 1/2a^2 - d_1 = \sigma_1 \tag{64a, b, c}$$

$$a^2 - 1/2a^3 - 3/8a^4 - 2d_1d_2 = \sigma_2$$

After eliminating d_1 and d_2 the quantity "a" can be determined from:

$$\left\{\left(1 - \frac{1}{8}a^2 - \frac{1}{2}a\right)a^2 - \sigma_2\right\}^2 = \frac{4}{\hat{\rho}_2}\left\{\left(1 - \frac{1}{2}a\right)a - \sigma_1\right\}^3 \tag{65}$$

Since $\hat{\rho}_2 < 0$, the initial guess for "a" should be such that the term in brackets on the right hand side of Eq. (65) be negative. In most cases tested by Gottifredi et al. [19] the parameter "a" ranged between 1.1 to 4 approximately.

Equation (61) produces extremely accurate values of η as should be expected since Eq. (61) reproduces very well the asymptotic behavior of η for large as well for small values of $\hat{\phi}$. Gottifredi et al. [20] have shown that maximum deviations are below 2% for most kinetic expressions, under isothermal conditions. Nevertheless the main disadvantage of Eq. (61) is the non-linearity of Eq. (65) which must be used to find the unknown parameter "a" as a function of σ_1, σ_2, and $\hat{\rho}_2$. It should be stressed that Eq. (61) reduces to Eq. (48) when $\hat{\rho}_2 = 0$ (slab pellets with uniform activity).

Gonzo et al. [17] proposed the following rational expression to overcome the inconvenience caused by Eq. (65):

$$\eta = (\hat{\phi}^2 + \exp(-a\hat{\phi}^2))^{-1/2} + \hat{\rho}_2(\hat{\phi}^2 + \exp(-d\hat{\phi}^2))^{-2}\hat{\phi}^2 \quad (66)$$

which after being expanded for large and small values of $\hat{\phi}$ yields:

$$\eta = \hat{\phi}^{-1} + \hat{\rho}_2\hat{\phi}^{-2} + \frac{1}{2}\hat{\phi}^{-3}\exp(-a\hat{\phi}^2) + \cdots \quad (67)$$

and

$$\eta = 1 - \left(\frac{1}{2}(1 - a) - \hat{\rho}_2\right)\hat{\phi}^2 + \left(\frac{3}{8}(1 - a)^2 - \frac{a^2}{4} - 2\hat{\rho}_2(1 - d)\right)\hat{\phi}^4$$

$$+ O(\hat{\phi}^6) \quad (68)$$

By comparing Eqs. (67) and (68) with Eqs. (38) and (40) respectively conditions for the unknowns are found:

$$\frac{1}{2}(1 - a) - \hat{\rho}_2 = \sigma_1$$

$$\frac{3}{8}(1 - a)^2 - \frac{1}{4}a^2 - 2\hat{\rho}_2(1 - d) = \sigma_2$$

$$(69a, b)$$

The rational expression given by Eq. (66) has many advantages when compared with previous expressions. It reproduces the asymptotic behavior of η up to terms of order $\hat{\phi}^{-2}$ and $\hat{\phi}^4$ when $\hat{\phi} \gg 1$ and $\hat{\phi} \ll 1$ respectively. The term of order $\hat{\phi}^{-3}$ is damped out by an exponential term except when $a \ll 1$. Equation (66) reduces to Churchill's expression when $a = d = \hat{\rho}_2 = 0$. In some cases d could become negative and when this situation is encountered calculations are performed with $d = 0$ which implies a violation of condition (69b).

Results obtained with Eq. (66) were compared with exact values of η obtained either by analytical or by numerical techniques. Figure 2 presents a comparison for the case of spherical pellets with uniform activity distribution ($f(x) = 1$), constant diffusivity coefficient and irreversible reactions of zero, one half, first and second order, all under isothermal conditions. It becomes quite clear that Eq. (66) is able to predict η values in very close agreement with the corresponding exact values for the entire range of Thiele modulus. The maximum deviation is 15% for the most unfavorable situation since the analytical solution presents a discontinuity at $\hat{\phi} = 1$ for a zeroth order reaction. However, Fig. 2 shows that the region where the deviations are significant is very narrow. Nevertheless, any of the other expressions present a deviation larger than 20% at $\hat{\phi} = 1$. Fortunately, the zeroth order reaction does not exist in practice. Thus a Langmuir–Hinshelwood kinetic expression of the type:

$$R(C) = (1 + K)C/(1 + KC) \qquad (70)$$

was also considered for the worst situation, which is the slab pellet with uniform activity distribution. When K becomes very large $R'(1) \ll 1$ and so $\sigma_1 \ll 1$. Since $\hat{\rho}_2 = 0$ the method proposed by Wedel and Luss [42] is not applicable. When the criterion given by the Eq.

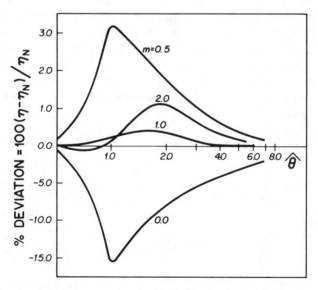

Figure 2   Deviation between the rational (η) and exact (η_N) as function of Thiele modulus for isothermal mth order reactions. Spherical geometry.

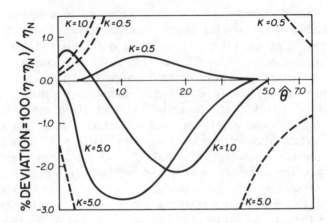

Figure 3   Deviation between rational and exact η values for Langmuir–Hinshelwood kinetics. Plane geometry. (—— Eq. (66); – – – – Eq. (48)).

(60) was applied, b_4 vanished and $b_2 = b_5$ were negative. If Eq. (59) is applied $b_1 \simeq \sigma_1$ and again b_2 will become negative. For that reason Fig. 3 presents a comparison between deviations obtained with Eq. (66) and with (48) since in this case Eq. (61) reduces to Eq. (48). It should be noticed that condition (69a) gives a unique value of "a" instead of the dual condition solved by the criterion (53a, b). Figure 3 clearly shows that Eq. (66) is able to fit much better than Eq. (48) the behavior of η for all values of $\hat{\phi}$ and K. The results given by Eq. (48) (broken lines) for $K \geqslant 5$ are not included since the actual values are located outside the scale of the figure. Maximum deviation for $K = 100$ is less than 5% for $\hat{\phi} \sim 1$, when Eq. (66) is used.

A.2.   Coupled Parallel Reactions

The case of a set of coupled parallel reactions was recently analyzed by Gonzo et al. [16]. The following reaction scheme was assumed:

$$A + \nu_1 B \rightarrow \text{products}$$

$$A + \nu_2 C \rightarrow \text{products}$$

Although the analysis here will be restricted to the case of a constant diffusion coefficient and uniform activity distribution, it could be easily extended.

As in the case of a single reaction, the dimensionless mass balances for each species are most conveniently written as:

$$x^{-s}\frac{d}{dx}\left(x^s\frac{dA}{dx}\right) = \phi^2(R_1(A, B, C) + \lambda^2 R_2(A, B, C))$$

$$x^{-s}\frac{d}{dx}\left(x^s\frac{dB}{dx}\right) = \gamma_B\phi^2 R_1(A, B, C) \qquad (71a, b, c)$$

$$x^{-s}\frac{d}{dx}\left(x^s\frac{dC}{dx}\right) = \gamma_C\lambda^2\phi^2 R_2(A, B, C)$$

where A, B, and C represent the dimensionless concentration of species A, B, and C respectively normalized with respect to their respective surface values. As before, the Thiele modulus is defined as:

$$\phi = (r_{1s}L^2/D_A C'_{As}) ; \qquad \lambda^2 = (r_{2s}/r_{1s}) \qquad (72a, b)$$

while R_1 and R_2 denote the dimensionless rates of reaction for the first and second reaction respectively and γ_B and γ_C are given by

$$\gamma_B = \left(\frac{D_A C'_{As}}{D_B C'_{Bs}}\right)\nu_1 ; \qquad \gamma_C = \left(\frac{D_A C'_{As}}{D_C C'_{Cs}}\right)\nu_2 \qquad (73a, b)$$

Since the reaction order is arbitrary it can always be chosen in such a way that

$$\lambda < 1 \qquad (74)$$

There is no actual need to solve the set of three differential equations since the dimensionless concentrations of the three components are easily related by:

$$A = (1/\gamma_B)B - (1/\gamma_C)C = 1 - 1/\gamma_B - 1/\gamma_C = \Gamma \qquad (75)$$

Before replacing A as function of B and C it should be noted that it is assumed that the concentration of A will never vanish. Otherwise the concentration of the excess component should be replaced by the other two. By defining the effectiveness factor of the two reactions as:

$$\eta_1 = \int_0^1 [(s + 1)x^s R_1] \, dx = \frac{(s + 1)}{\gamma_B\phi^2} \frac{dB}{dx}\bigg|_{x=1} \qquad (76)$$

$$\eta_2 = \int_0^1 [(s+1)x^s R_2]\, dx = \frac{(s+1)}{\gamma_C \lambda^2 \phi^2} \frac{dC}{dx}\bigg|_{x=1} \qquad (77)$$

The resulting pair of differential equations must be solved subject to the following boundary conditions:

$$A = B = 1 \quad \text{at } x = 1\,; \qquad \frac{dA}{dx} = \frac{dB}{dx} = 0 \quad \text{at } x = 0$$

The same procedure as that used for a single reaction will be developed here.

Perturbation Solution for Small Values of ϕ
When $\phi^2 \ll 1$ the following series can be used as an approximate solutions to Eqs. (71b, c):

$$B = 1 + G_1(x)\phi^2 + \phi^4 G_2(x) + \cdots$$
$$C = 1 + J_1(x)\phi^2 + \phi^4 J_2(x) + \cdots \qquad (78a, b)$$

By expanding $R_1(B, C)$ and $R_2(B, C)$ in a Taylor series we obtain

$$R_1 \simeq 1 + [R'_{1B}G_1 + R'_{1C}J_1]\phi^2 + \left[\frac{1}{2}R''_{1B}G_1^2 + R'_{1BC}G_1 J_1\right.$$
$$\left. + \frac{1}{2}R''_{1C}J_1^2 + R'_{1B}G_2 + R'_{1C}J_2\right]\phi^4 + \cdots \qquad (79)$$

$$R_2 \simeq 1 + [R'_{2B}G_1 + R'_{2C}J_1]\phi^2 + \left[\frac{1}{2}R''_{2B}G_1^2 + R''_{2BC}G_1 J_1\right.$$
$$\left. + \frac{1}{2}R''_{2C}J_1^2 + R'_{2B}G_2 + R'_{2C}J_2\right]\phi^4 + \cdots \qquad (80)$$

where the primes and double primes denote first and second derivatives with respect to either B or C or both, according to the subscripts. In all cases these derivatives are evaluated at $B = C = 1$.

By replacing Eqs. (78a, b), (79) and (80) into Eqs. (71b, c) and then collecting terms of equal powers of ϕ we obtain

$$\frac{d}{dx}\left[x^s \frac{dG_1}{dx}\right] = x^s \gamma_B \qquad (81)$$

$$\frac{d}{dx}\left[x^s \frac{dH_1}{dx}\right] = \lambda^2 x^s \gamma_C \qquad (82)$$

$$\frac{d}{dx}\left[x^s\frac{dG_2}{dx}\right] = x^s[R'_{1B}G_1 + R'_{1C}J_1]\gamma_B \tag{83}$$

$$\frac{d}{dx}\left[x^s\frac{dH_2}{dx}\right] = \lambda^2 x^s[R'_{2B}G_1 + R'_{2C}J_1]\gamma_C \tag{84}$$

subject to:

$$G_1(1) = J(1) = G_2(1) = J_2(1) = 0$$

and

$$G'_1(0) = G'_2(0) = J'_1(0) = J'_2(0) = 0$$

so that:

$$J_1 = \lambda^2 G_1\gamma_C/\gamma_B = w^2 G_1 \tag{85}$$

$$J_2 = gG_2 = w^2[R'_{2B} + w^2R'_{2C}]G_2/[R'_{1B} + w^2R'_{1C}] \tag{86}$$

To find the asymptotic expressions for η_1 and η_2 one has to solve Eqs. (81) and (83) which are similar to those deduced for the case of a single reaction (see Eqs. (13) and (14)). It is possible to show that:

$$\eta_1 = 1 - \sigma^*_{11}\phi^2 + \sigma^*_{21}\phi^4 + O(\phi^6) \tag{87}$$

$$\eta_2 = 1 - \sigma^*_{12}\phi^2 + \sigma^*_{22}\phi^4 + O(\phi^6) \tag{88}$$

where:

$$\sigma^*_{11} = \frac{\gamma_B}{(s+1)(s+3)}[R'_{1B} + w^2R'_{1C}] \tag{89}$$

$$\sigma^*_{12} = \frac{\gamma_B}{(s+1)(s+3)}[R'_{2B} + w^2R'_{2C}] \tag{90}$$

$$\sigma^*_{21} = (s+1)\int_0^1 x^s\left\{\left[\frac{1}{2}R''_{1B} + w^2R''_{1BC} + \frac{1}{2}w^4R''_{1C}\right]G_1^2\right.$$

$$\left. + [R'_{1B} + gR'_{1C}]G_2\right\}dx \tag{91}$$

$$\sigma_{22}^* = (s+1) \int_0^1 x^s \left\{ \left[\frac{1}{2} R_{2B}'' + w^2 R_{2BC}'' + \frac{1}{2} w^4 R_{2C}'' \right] G_1^2 \right.$$

$$\left. + [R_{2B}' + g R_{2C}'] G_2 \right\} dx \qquad (92)$$

where g can be found from Eq. (86).

Perturbation Solution for Large Values of ϕ

When $\phi^2 \gg 1$ it is further assumed that the product $(\lambda\phi)^2$ will also be very large. Otherwise one is faced with an almost uncoupled system. In other words the range of interest for λ should be $0.1 \ll \lambda \ll 1$. In order to solve this problem Cukierman and Lemcoff [10] assumed that a very simple relation will always exist between B and C. Such a relation was assumed as:

$$C = B^w \qquad (93)$$

with w already defined by Eq. (85):

$$w = (\gamma_C/\gamma_B)^{1/2} \lambda \qquad (94)$$

Such a relation is strictly valid for the particular case where $\gamma_C \gg 1$, $\gamma_B \gg 1$, and both kinetic expressions being first order with respect to component B and C. Nevertheless Eq. (93) was used by Cukierman and Lemcoff [10] with reasonable success to predict asymptotic values of η_1 and η_2. In fact, coming back to Eqs. (71b, c), they can now be written in the following form:

$$x^{-s} \frac{d}{dx} \left(x^s \frac{dB}{dx} \right) \simeq \gamma_B \phi^2 R_1(B, C(B)) \qquad (95)$$

$$x^{-s} \frac{d}{dx} \left(x^s \frac{dC}{dx} \right) \simeq \gamma_C \lambda^2 \phi^2 R_2(B(C), C) \qquad (96)$$

when ϕ^2 (and so $(\lambda\phi)^2$) $\gg 1$. However under these circumstances Eqs. (95) and (96) are subject to the same analysis carried out for the particular case of a single reaction. So:

$$\eta_1 = \rho_{11} \phi^{-1} + \rho_{21} \phi^{-2} + \cdots \qquad (97)$$

$$\eta_2 = \rho_{12} \phi^{-1} + \rho_{22} \phi^{-2} + \cdots \qquad (98)$$

where:

$$\rho_{11} = \left\{ \frac{2}{\gamma_B} \int_0^1 R_1(B, C(B)) \, dB \right\}^{1/2} (s + 1) \qquad (99)$$

$$\rho_{12} = \left\{ \frac{2}{\lambda^2 \gamma_C} \int_0^1 R_2(B(C), C) \, dC \right\}^{1/2} (s + 1) \qquad (100)$$

$$\rho_{21} = -\frac{(s + 1)^2 s}{\rho_{11}} \int_0^1 \left\{ \frac{2}{\gamma_B} \int_0^Z R_1(B, C(B)) \, dB \right\}^{1/2} dZ \qquad (101)$$

$$\rho_{22} = -\frac{(s + 1)^2 s}{\rho_{12}} \int_0^1 \left\{ \frac{2}{\gamma_C \lambda^2} \int_0^Z R_2(B(C), C) \, dc \right\}^{1/2} dZ \qquad (102)$$

These coefficients can be easily calculated by Radau or Gauss quadrature techniques. Once again we can resume the asymptotic behavior of η_1 and η_2 in following fashion:

$$\eta_1 = \hat{\phi}_1^{-1} + \hat{\rho}_{21}\hat{\phi}_1^{-2} + O(\hat{\phi}_1^{-3}) \qquad \text{when } \hat{\phi}_1 \gg 1$$

$$\eta_1 = 1 + \sigma_{11}\hat{\phi}_1^2 + \sigma_{21}\hat{\phi}_1^4 + O(\hat{\phi}_1^6) \qquad \text{when } \hat{\phi}_1 \ll 1$$

(103a, b)

and

$$\eta_2 = \hat{\phi}_2^{-1} + \hat{\rho}_{22}\hat{\phi}_2^{-2} + O(\hat{\phi}_2^{-3}) \qquad \text{when } \hat{\phi}_2 \gg 1$$

$$\eta_2 = 1 + \sigma_{12}\hat{\phi}_2^2 + \sigma_{22}\hat{\phi}_2^4 + O(\hat{\phi}_2^6) \qquad \text{when } \hat{\phi}_2 \ll 1$$

(104a, b)

where:

$$\hat{\phi}_1 = (\phi_1/\rho_{11}) ; \qquad \hat{\phi}_2 = (\phi_2/\rho_{12}) \qquad (105a, b)$$

and

$$\sigma_{11} = \sigma_{11}^* \rho_{11}^2 ; \qquad \sigma_{21} = \sigma_{21}^* \rho_{11}^4 ; \qquad \hat{\rho}_{21} = \rho_{21}/\rho_{11}^2$$

$$\sigma_{21} = \sigma_{21}^* \rho_{12}^2 ; \qquad \sigma_{22} = \sigma_{22}^* \rho_{12}^4 ; \qquad \hat{\rho}_{22} = \rho_{22}/\rho_{12}^2$$

(106a, b, c, d, e, f)

All rational expressions outlined above can be now used to predict η_1 and η_2 for all ranges of $\hat{\phi}_1$ and $\hat{\phi}_2$ values.

Table IV.

h_1	m = 1				n = 1				p = 1				q = 1			
	$\gamma_B=2; \gamma_C=2; \lambda=0.5$				$\gamma_B=10; \gamma_C=2; \lambda=0.5$				$\gamma_B=2; \gamma_C=2; \gamma=0.8$				$\gamma_B=10; \gamma_C=2; \lambda=0.8$			
	η_1	η_{1N}	η_2	η_{2N}	η_1	η_{1N}	η_2	η_{2N}	η_1	η_{1N}	η_2	η_{2N}	η_1	η_{1N}	η_2	η_{2N}
0.1	0.9883	0.9893	0.9950	0.9942	0.9642	0.9642	0.9964	0.9943	0.9881	0.9881	0.9904	0.9904	0.9630	0.9631	0.9915	0.9905
0.5	0.7997	0.8031	0.8944	0.8857	0.5624	0.5646	0.9195	0.9066	0.7834	0.7868	0.8185	0.8221	0.5558	0.5576	0.8355	0.8393
1	0.5425	0.5510	0.7071	0.7063	0.3052	0.3051	0.7601	0.7696	0.5235	0.5310	0.5771	0.5873	0.3010	0.3003	0.6053	0.6271
2	0.2986	0.3033	0.4472	0.4642	0.1530	0.1523	0.5049	0.5384	0.2875	0.2912	0.3311	0.3393	0.1510	0.1499	0.3554	0.3744
4	0.1523	0.1528	0.2425	0.2587	0.0763	0.0759	0.2807	0.3034	0.1469	0.1473	0.1723	0.1754	0.0754	0.0749	0.1867	0.1940
8	0.0764	0.0763	0.1240	0.1327	0.0381	0.0389	0.1447	0.1544	0.0738	0.0738	0.0870	0.0882	0.0377	0.0383	0.0946	0.0973

Gonzo et al. [16] tested Eq. (48) as rational expressions for η_1 and η_2 in the case of plane geometry for the following kinetic expressions:

$$R_1 = A^p B^m$$
$$R_2 = A^q C^n$$
(107a, b)

under isothermal conditions. In those cases where the parameter "a" becomes imaginary, Churchill's expression was used. Results are presented in Table IV for the case $m = n = p = q = 1$ and for different values of γ_B, γ_C, and λ. Again the subscript N is used to denote values of η obtained by the collocation method (see Gonzo et al. [16]). It is clearly seen that the approximate rational expression (48) produces fairly accurate results since maximum deviations are always below 5%. In Table V the same situation is presented, but m is now set equal to 2 instead of one. Again the approximate results are in fair agreement with numerical calculations although it is clearly seen that asymptotic behavior of η_2, as predicted by our analysis, presents a systematic deviation in the coefficient ρ_{12}. The same trends are observed in Table VI where the effect of reaction order on component A is investigated. Nevertheless, it can be concluded that under the worst conditions maximum deviation between the approximate and the numerical results is smaller than 15%.

Although this is the first attempt to predict effectiveness factors for a pair of coupled parallel reactions, there is a strong need to further investigate the asymptotic behavior of η_1 and η_2.

The authors are studying the behavior of rational expressions with other kinetic expressions, especially those of the Langmuir–Hinshelwood type. When these kinds of kinetic expressions are

Table V.

	$m = 2;$	$n = 1;$	$p = 1;$	$q = 1$			
	$\gamma_B = 2;\ \gamma_C = 2;\ \lambda = 0.5$				$\gamma_B = 2;\ \gamma_C = 2;\ \lambda = 0.8$		
h_1	η_1	η_{1N}	η_2	η_{2N}	η_1	η_{1N}	η_2	η_{2N}
0.1	0.9828	0.9830	0.9950	0.9943	0.9817	0.9817	0.9904	0.9904
0.5	0.7270	0.7317	0.8944	0.8919	0.7159	0.7185	0.8185	0.8272
1	0.4646	0.4709	0.7071	0.7266	0.4545	0.4563	0.5771	0.6001
2	0.2524	0.2536	0.4472	0.4880	0.2466	0.2449	0.3311	0.3509
4	0.1291	0.1280	0.2425	0.2727	0.1261	0.1241	0.1723	0.1815
8	0.0649	0.0644	0.1240	0.1394	0.0634	0.0626	0.0870	0.0911

Table VI.

h_1	$m = 1$; $n = 1$; $p = 2$, $q = 1$ $\gamma_B = 2$; $\gamma_C = 2$; $\lambda = 0.5$				$m = 2$; $n = 1$; $p = 1$; $q = 2$ $\gamma_B = 10$; $\gamma_C = 2$; $\lambda = 0.8$			
	η_1	η_{1N}	η_2	η_{2N}	η_1	η_{1N}	η_2	η_{2N}
0.1	0.9853	0.9853	0.9950	0.9943	0.9344	0.9349	0.9869	0.9856
0.5	0.7454	0.7551	0.8944	0.8899	0.4526	0.4606	0.7747	0.7952
1	0.4714	0.4932	0.7071	0.7210	0.2431	0.2457	0.5224	0.5712
2	0.2485	0.2646	0.4472	0.4829	0.1238	0.1236	0.2929	0.3354
4	0.1249	0.1324	0.2425	0.2705	0.0622	0.0617	0.1514	0.1746
8	0.0625	0.0660	0.1240	0.1385	0.0311	0.0309	0.0764	0.0877

used it is increasingly difficult to obtain reliable numerical values of η_1 and η_2 as will be shown in future studies.

A.3. Concluding Remarks

All rational approximations reviewed here give accurate results with minimal computation efforts as long as multiplicity is absent. It was shown that Petersen's and Churchill's expressions are equivalent although it was clearly shown that Eq. (44) produces intermediate rather than extreme results.

Equation (48) produces much better results than any of the two expressions mentioned above since it is able to take into account reaction order and in some cases the effect of pellet shape. Nevertheless it can only be used in a limited range of parameters such that $2\sigma_1 \leqslant 1$, although it has been tested with great success in a great number of cases including non-isothermal systems.

Equation (55), as proposed by Wedel and Luss [42], produces extremely accurate results for the case of spherical and cylindrical pellets, since it is essential that $\rho_2 \neq 0$. Wedel and Luss [42] have tested their expression for quite a number of cases with very good success. It was shown that Eq. (55) is very easy to apply since the unknowns are easily determined. The only disadvantage of this expression is the fact that, in some cases, one has to force the entire expression to fit the value of η at $\hat{\phi} = 1$. This is given by the derived asymptotic expression for $\hat{\phi} \gg 1$ and that cannot be fully applied to plane geometry.

A much more complete rational expression is that given by Eq. (61) although one of the unknowns has to be found by solving a non-linear algebraic equation. However, it was shown that this expression can be applied to quite a number of situations with almost the same accuracy. It reduces to Eq. (48) for the case of plane geometry with uniform activity distribution. Finally it was shown that most of the disadvantages of Eq. (61) will be removed if Eq. (66) is used. It has the same, or better, accuracy than Eq. (61). The two unknowns are given by linear expressions and the asymptotic behavior at large $\hat{\phi}$ values is much better than any of the previous expressions. It was also shown how to extend the use of rational expressions to the case of a pair of coupled parallel reactions under isothermal conditions.

Much needs to be done in the case of coupled parallel reactions to reduce the effort required for effectiveness factor calculations. However this first attempt provides a foundation for subsequent studies of more complex systems.

B. INTERPHASE HEAT AND MASS TRANSFER EFFECTS

As shown above, it is always possible to relate concentrations of any component and the temperature field with the key component concentration especially in those cases where mass and thermal diffusivity can be assumed to be independent of concentration and temperature.

However when significant concentration and temperature differences between the surface of the pellet and the bulk of the fluid arise, the effectiveness factor as calculated in section A, must be corrected in order to express the main reaction rates in terms of their corresponding bulk values. For simplicity let us show the procedure for the case of a single reaction. By defining:

$$\eta_0 = \eta(r_s/r_0) \tag{108}$$

as the effectiveness factor calculated in terms of the reaction rate evaluated at bulk conditions (r_0). Since now r_s is not known, mass and heat balances at the outer pellet surface must be taken into account:

$$\left(\frac{C'_{i0}}{C'_{is}} - 1\right) B_{imi} = \frac{dC}{dx}\bigg|_{x=1} \tag{109}$$

$$\left(\frac{T'_0}{T'_s} - 1\right) B_{ie} = \frac{dT}{dx}\bigg|_{x=1} \tag{110}$$

where B_{imi} and B_{ie} represent Biot numbers for mass and heat transport respectively and the subscript "0" is used to denote bulk value. Thus:

$$B_{imi} = (k_{gi}L/Di) ; \qquad B_{ie} = (h_e L/\kappa) \qquad (111a, b)$$

Here k_{gi} and h_e denote film mass and heat transport coefficients. From the Thiele modulus definition:

$$\phi_0^2 = \phi^2(r_0/r_s)(C_s'/C_0') \qquad (112)$$

By combining Eqs. (109) and (110) with Eqs. (5a, b) and Eq. (9) and using Eq. (112) the following relations are obtained:

$$\frac{C_{is}'}{C_{i0}'} = 1 - \frac{\phi_0^2 \eta_0 \Gamma_{i0}}{B_{imi}(s+1)} \qquad (113)$$

$$\frac{T_s'}{T_0'} = 1 + \frac{\phi_0^2 \eta_0 \beta_0}{B_{ie}(s+1)} \qquad (114)$$

It should be noticed that Eq. (113) is also valid for the key component with $\Gamma_{i0} = 1$. A trial and error procedure can be now used to estimate η_0. By setting the values of B_{imi}, B_{ie}, β_0, Γ_{i0} and the dimensionless activation energy (Arrhenius number) η can be calculated first with some adequate rational expression by assuming negligible interphase resistance ($\eta = \eta_0$ and $\phi = \phi_0$). With this approximation a correction is made with Eqs. (113) and (114) and the resulting values can be used to correct η_0 with Eq. (108) and ϕ_0 with Eq. (112). The procedure is repeated until two successive calculations with Eqs. (113) and (114) indicate the desired convergence is achieved. In a chemical reactor calculation the procedure can be accelerated by using values of ϕ_0 and η_0 calculated at the previous grid point as initial guesses.

The advantage of this procedure is due to the rational expression which allows a rapid and accurate calculation of η for each trial. Convergence is achieved very quickly even with strongly exothermic reactions, as was shown by Gonzo and Gottifredi [15]. Moreover, under these conditions, the interphase heat transport limitation plays an important role and cannot be neglected since large temperature differences between pellet surface and the bulk of the fluid can build up.

This increase in temperature at the outer surface produces in turn a dramatic increase of ϕ so that the multiple solution region of the η

vs ϕ relation is finally avoided by the system. This supports the importance of rational expressions in handling non-isothermal systems.

Finally it should be noted that the final error on η_0 will be of the same order of magnitude of η since the trial and error procedure does not increase the inherent deviation produced by the rational expression.

C. GENERAL DIAGNOSTIC CRITERIA FOR TRANSPORT LIMITATIONS

Criteria to establish whether or not mass and heat transfer resistances can be neglected are of prime interest for catalytic reactor design and in experimental studies dealing with heterogeneous catalytic systems.

Some useful experimental criteria have been presented in the past. The main advantages of them has been pointed out by Madon and Boudart [28]. For instance the criteria presented by Koros and Nowak [25] can be used to establish mass and heat transport limitations with supported as well as with unsupported catalysts (Gonzo and Boudart [14] and Boudart et al. [7]). On the other hand, theoretical criteria can be used whenever some knowledge of the kinetic behavior of the system is known. This is just the situation in chemical reactor simulation where the chemical kinetics must be known a priori.

Theoretical criteria have been extensively reviewed since the first contribution of Weisz and Prater [44] which is strictly valid for first order irreversible reactions. However, all of them deal with some particular case e.g. isothermal mth order irreversible reactions (Hudgins [21], Weisz and Hicks [45]) or just the effect of temperature gradients inside a catalyst (Anderson [1]). Excellent summaries of these particular case existing criteria were presented by Froment and Bishoff [13], Butt [8] and very recently by Madon and Boudart [28].

The purpose of this section is to show that there is no need to assume any chemical kinetic expression to deduce useful criteria. Only two general criteria will be deduced; one for interphase and the other for intraparticle mass and heat transport effects. Each of them applies independently of the other as shown recently by Gonzo and Gottifredi [18].

The general criterion used to establish the absence of intraparticle transport limitations is always written in the following form:

$$|1 - \eta| \leq 0.05 \tag{115}$$

where 0.05 is an arbitrary figure. However the essential idea is that to fulfill conditions given by Eq. (115) ϕ^2 should be small enough so that Eq. (16) is a very good approximation for η. Thus by combining Eq. (115) and Eq. (16) it yields:

$$|\sigma_1^* \phi^2| = |R'(1)\alpha\phi^2| \leq 0.05 \tag{116}$$

The effect of the kinetic expression is concentrated in $R'(1)$ and ϕ^2 while the effect of geometry and activity distribution is included in the parameter α. Equation (116) can be regarded as the explict criterion for intraparticle transport limitation which encompasses all previous cases deduced on the basis of a particular situation, as will be shown below.

Along the same line of reasoning, when interphase phenomena are to be considered, the criterion to establish negligible effects can be written as:

$$\left|1 - \frac{\eta_0}{\eta}\right| \leq 0.05 \tag{117}$$

However from Eqs. (108) and (113) we have

$$\frac{\eta_0}{\eta} = \frac{r_s}{r_0} = F(C^*) \tag{118}$$

where:

$$C^* = \frac{C_s'}{C_0'} = 1 - \frac{\eta_0 \phi_0^2}{(s+1)B_{im}} \tag{119}$$

To accomplish the condition given by Eq. (115) it will be necessary that:

$$\varepsilon = \frac{\eta_0 \phi_0^2}{(s+1)B_{im}} \ll 1 \tag{120}$$

By expanding Eq. (118) in terms of ε:

$$F(C^*) = 1 + F'(1)\varepsilon + O(\varepsilon^2) \tag{121}$$

where $F'(1)$ denotes first derivative with respect to C^* evaluated at

$C^* = 1$. Substituting Eq. (121) into Eq. (117) yields:

$$|F'(1)\varepsilon| \leq 0.05 \qquad (122)$$

It should be stressed that Eqs. (122) and (116) were deduced without invoking any particular assumption except that the single reaction case was analyzed. These two expressions can be regarded as the general explicit *theoretical* criteria to establish the absence of transport limitations. Each of them applies independent of the other.

As examples of applications, some particular cases will be considered. Let us first analyze the case of a reversible reaction. The dimensional rate of reaction is given by:

$$r = k\left(C'^m C_B'^m - \frac{1}{K} C_C'^P C_D'^q \right) \qquad (123)$$

where K denotes the thermodynamic equilibrium constant and m, n, p, and q the reaction orders for the key component and species B, C, and D respectively. Assuming that the effect of temperature on K can be neglected, the criterion given by Eq. (116) will take this particular form:

$$\left(\frac{r_{0b} L^2}{D C_s'} \right) \leq \frac{(0.05/\alpha)}{m + n\Gamma_B - \gamma\beta - \dfrac{1}{K'}(p\Gamma_C + q\Gamma_D)} \qquad (124)$$

where:

$$K' = K[C'^m C_B'^m / C_C'^P C_D'^q]_s \qquad (125)$$

and

$$\gamma = (E/RT_s') \qquad (126)$$

Here we note that Γ and β are defined by Eqs. 6a, b, E' denotes activation energy for the forward specific rate constant (k), and R the universal gas constant.

The example was chosen to show that Eq. (124) is an extension of all previous findings deduced for particular cases. In fact Anderson's [1] expression is obtained when $K' \to \infty$ (irreversible kinetics) and $|\gamma\beta| \gg (m + n\Gamma_B)$. On the other hand, Kubota and Yamanaka's [26] criterion is restored when $K' \to \infty$ and $|m - \gamma\beta| \gg n\Gamma_B$. Mears' [30] criterion is also a particular case of Eq. (124). Gonzo and Gottifredi [18] illustrated the application of the general expression given by Eq. (117) for Langmuir–Hinshelwood kinetic expressions with the same simplicity.

As second example let us analyze the case of an irreversible (m, n)th order reaction but with the aim of establishing the particular conditions where interphase resistances are negligible. $F(C^*)$ becomes:

$$F(C^*) = C^{*m}\{1 - \chi_B(1 - C^*)\}^B \exp\left(\frac{\gamma_0 X(1 - C^*)}{1 + X(1 - C^*)}\right) \quad (127)$$

where:

$$\chi_B = \Gamma_{B0}(B_{imB}/B_{im}) ; \qquad X = \beta_0(B_{im}/B_{ie}) \quad (128)$$

Application of criterion (122) yields:

$$\left(\frac{r_{0b}L^2}{DC_0'}\right) \leq \frac{(s + 1)0.05B_{im}}{(m + \chi_B n - X\gamma_0)} \quad (129)$$

which is also an extension of Mears' [30] results deduced for the particular case of an irreversible mth order reaction.

It is worth mentioning that Gonzo and Gottifredi [18] compared the general criterion given by Eq. (116) for the particular case of the experimental results of Wu and Nobe [47] who measured η values for NO reduction with ammonia on cylindrical pellets, with good success. They have also shown that Eq. (122) when applied to Kehoe and Butt's [24] results for benzene hydrogenation predicts negligible interphase resistances when temperature differencies are below 2 K.

D. CONCLUSIONS

Throughout this contribution a methodology to construct rational expressions for effectiveness factor was reviewed. Some applications were shown to demonstrate the ability of these expressions to predict η values under a variety of situations. New rational expressions can be developed in the future which can predict even better effectiveness factor behavior over the entire range of ϕ values.

Finally it was shown how these expressions can be used to take into account interphase mass and heat transfer resistances and the theoretical criteria to predict whether or not these mass and heat transfer effects can be safely neglected.

The falsification of kinetic parameters by transport limitations can also be predicted with the appropriate use of rational expressions.

REFERENCES

1. Anderson, J. B. (1963). A criterion for isothermal behaviour of a catalyst pellet, *Chem. Eng. Sci.*, **18**, 147.
2. Aris, R. (1965a). A normalization for the Thiele modulus, *Ind. Eng. Chem. Fundam.*, **4**, 227.
3. Aris, R. (1965b). A normalization for the Thiele modulus, *Ind. Eng. Chem. Fundam.*, **4**, 487.
4. Aris, R., and Strieder, W. (1971). Variational bounds for problems in diffusion and reaction, *J. Inst. Math. its Appl.*, **8**, 328.
5. Aris, R. (1975). *The Mathematical Theory of Diffusion and Reaction in Permeable Catalysts*, Claredon, Oxford.
6. Bishoff, K. B. (1965). Effectiveness factors for general reaction rate forms, *AIChE J.*, **11**, 351.
7. Boudart, M., Aldag, A., Benson, J. E., Dougharty, N. A., and Harkin, G. C. (1966). On the specific activity of platinum catalysts, *J. Catal.*, **6**, 92.
8. Butt, J. B. (1980). *Reaction Kinetics and Reactor Design*, Prentice-Hall, Englewood Cliff, New Jersey.
9. Carberry, J. J. (1980). *Chemical and Catalytic Reaction Engineering*, McGraw-Hill, New York.
10. Cukierman, A., and Lemcoff, N. (1979). Selectivity in complex catalytic reactions, *Lat. Am. J. Chem. Eng. Appl. Chem.*, **9**, 57.
11. Churchill, S. W., and Usagi, R. (1972). A general expression for the correlation of rate of transfer and other phenomena, *AIChE J.*, **18**, 1121.
12. Churchill, S. W. (1977). A generalized expression for the effectiveness factor of porous catalyst pellets, *AIChE J.*, **23**, 208.
13. Froment, G., and Bishoff, K. (1980). *Chemical Reactor Analysis and Design*, Wiley, New York.
14. Gonzo, E. E., and Boudart, M. (1978). Catalytic hydrogenation of cyclohexene: 3 Gas-phase and liquid phase reaction on supported palladium, *J. Catal.*, **52**, 462.
15. Gonzo, E. E., and Gottifredi, J. C. (1982). Non isothermal effectiveness factor—Estimation with simple analytical expressions, *Lat. Am. J. Heat Mass Transfer*, **6**, 113.
16. Gonzo, E. E., Gottifredi, J. C., and Lemcoff, N. (1983a). Effectiveness factor and selectivity estimation for a parallel reaction system, *Chem. Eng. Sci.*, **38**, 849.
17. Gonzo, E. E., Gottifredi, J. C., and Romero, L. C. (1983b). On the rational approximation of the effectiveness factor for simple and parallel reactions, *33rd Canadian Chemical Engineering Conference 1983, Toronto, Ontario, Canada*, Vol. 2, 598.
18. Gonzo, E. E., and Gottifredi, J. C. (1983). General diagnostic criteria for transport limitation in porous solid chemical reactions, *J. Catal.*, **83**, 25.
19. Gottifredi, J. C., Gonzo, E. E., and Quiroga, O. D. (1981a). Isothermal effectiveness factor I. Analytical expression for simple reaction with arbitrary kinetics. Slab geometry, *Chem. Eng. Sci.*, **36**, 705.
20. Gottifredi, J. C., Gonzo, E. E., and Quiroga, O. D. (1981b). Isothermal effectiveness factor II. Analytical expression for single reaction with arbitrary kinetics. Geometry and activity distribution, *Chem. Eng. Sci.*, **36**, 713.
21. Hudgins, R. R. (1968). A general criterion for absence of diffusion control in an isothermal catalyst pellet, *Chem. Eng. Sci.*, **23**, 93.
22. Jouven, F. G., and Aris, R. (1972). A method of representing the non isothermal effectiveness factor for fixed bed calculations, *AIChE. J.*, **18**, 402.

132   CONCEPTS AND DESIGN OF CHEMICAL REACTORS

23. Karanth, N. G., Koh, H. P., and Hughes, R. (1974). Analysis of non-isothermal catalytic reactions. Limiting effect of non-key component, *Chem. Eng. Sci.*, **29**, 451.
24. Kehoe, J. P., and Butt, J. (1972). Interactions of inter and intraphase gradients in a diffusion limited catalytic reaction, *AIChE J.*, **18**, 347.
25. Koros, R., and Nowak, E. (1967). A diagnostic test of the kinetic regime in a packed bed reactor, *Chem. Eng. Sci.*, **22**, 470.
26. Kubota, H., and Yamanaka, Y. (1969). Remarks on approximate estimation of catalyst effectiveness factor, *J. Chem. Eng. Jpn.*, **2**, 238.
27. Liu, S.L. (1970). The influence of intraparticle diffusion in fixed bed catalytic reactors, *AIChE J.*, **16**, 742.
28. Madon, R., and Boudart, M. (1982). An experimental criterion for the absence of artifacts in the measurement of rates of heterogeneous catalytic reactions, *Ind. Eng. Chem. Fundam.*, **21**, 438.
29. Maymo, J. A., and Smith, J. M. (1966). Catalytic oxidation of hydrogen-intrapellet heat and mass transfer, *AIChE J.*, **12**, 845.
30. Mears, D. E. (1971). Test for transport limitation in experimental catalytic reactors, *Ind. Eng. Chem. Proc. Des. Dev.*, **10**, 541.
31. Paterson, W. R., and Creswell, D. L. (1971). Simple method for the calculation of effectiveness factors, *Chem. Eng. Sci.*, **26**, 605.
32. Peterse, E. E. (1965). *Chemical Reaction Analysis*, Prentice-Hall, Englewood Cliffs, New Jersey.
33. Rajadhyaksha, R. A., Vasudeva, K., and Doraiswamy, L. (1976). Effectiveness factors in Langmuir–Hinshelwood and general order kinetics, *J. Catal.*, **41**, 61.
34. Ramachandran, P. A., Kam, E. K. T., and Hughes, R. (1976). A simple method for the calculation of effectiveness factors for complex reactions, *Chem. Eng. Sci.*, **31**, 244.
35. Rester, S., and Aris, R. (1972). Theory of diffusion and reaction VIII: Variational bounds on the effectiveness factor, *Chem. Eng. Sci.*, **27**, 347.
36. Satterfield, C. N., and Sherwood, T. K. (1963). *The Rate of Diffusion in Catalysis*, Addison-Wesley, Reading, Mass.
37. Satterfield, C. N. (1970). *Mass Transfer in Heterogeneous Catalysis*, M.I.T. Press, Cambridge, Mass.
38. Stewart, W. E., and Villadsen, J. (1969). Graphical calculation of multiple steady states and effectiveness factors for porous catalysts, *AIChE J.*, **15**, 28.
39. Thiele, E. W. (1939). Relation between catalytic activity and size of particle, *Ind. Eng. Chem.*, **31**, 916.
40. Villadsen, J., and Michelsen, M. L. (1978). *Solution of Differential Equation Model by Polynomial Approximation*, Prentice-Hall, Englewood Cliffs, New Jersey.
41. Wagner, C. (1943). The interaction of streaming diffusion and chemical reaction on heterogeneous catalysts, *Z. Physick. Chem.*, **193**, 1.
42. Wedel, S., and Luss, D. (1980). A rational approximation of the effectiveness factor, *Chem. Eng. Commun.*, **7**, 145.
43. Weisz, P. B. (1957). Diffusivity of porous particles I. Measurements and significance for internal reaction velocities, *Z. Physik. Chem.*, **11**, 1.
44. Weisz, P. B., and Prater, C. D. (1954). Interpretation of measurements in experimental catalysis, *Advan. Catal.*, **6**, 143.
45. Weisz, P. B., and Hicks, J. S. (1967). Behaviors of porous catalyst particles in view of internal mass and heat diffusion effects, *Chem. Eng. Sci.*, **17**, 265.
46. Wheeler, A. (1955). Reaction rates and selectivity in catalyst pores, in *Catalysis*, Vol. 2 (Ed. P. H. Emmett), Reinhold, New York.
47. Wu, S. C., and Nobe, K. (1977). Reduction of nitric oxide with ammonia on vanadium pentoxide, *Ind. Eng. Chem. Prod. Res. Dev.*, **16**, 136.
48. Zeldovich, Y. B. (1939). On the theory of reactions on powders and porous substances, *Acta Phys. Chem. URSS*, **10**, 583.

Chapter 3

Noncatalytic Gas–Solid Reactions with Applications to Char Combustion and Gas Desulfurization

GEORGE GAVALAS

Department of Chemical Engineering California Institute of Technology, Pasadena, California 91125

Abstract

This is a review of recent work in noncatalytic gas–solid reactions emphasizing structural changes accompanying reaction. After a first section on general thermodynamic relations, section two treats the shrinking-core model including volume changes. The third section discusses the sulfur dioxide–calcium oxide reaction including the effect of pore blockage. The last four sections deal with the combustion and gasification of porous carbonaceous particles. After an introduction in section four, sections five and six analyze the progress of reaction using a random capillary model to describe intraparticle pore growth and diffusion. Section seven treats heat and mass transfer in the gas film surrounding the reaction particles.

INTRODUCTION

Gas–solid reactions, with the solid being a reactant rather than a catalyst, are widely encountered in various industrial applications. Iron ore reduction and sulfide roasting are prominent examples from the metallurgical industry. Coal combustion and gasification are widely applied for power generation and for production of fuel gas and synthesis gas. These two applications almost always necessitate the removal of sulfur oxides or hydrogen sulfide, which can be carried out by reaction with metal oxides. More exotic applications to the production of electronic materials, specialty metals and other specialty materials are small scale but growing operations.

133

The physicochemical principles of gas–solid reactions are as varied as those of heterogeneous catalysis. However in gas–solid reactions the purely chemical steps are coupled with diffusional steps even more intimately than in heterogeneous catalysis. In fact, in many cases the rates of chemical reactions and solid-state diffusion are essentially inseparable. The coupling between chemical reaction and solid state diffusion is often complicated by the phenomena of nucleation of the product phase and the complex morphology of the two phase region. Such phenomena have received limited attention, perhaps because of the lack of unifying principles.

In addition to diffusion in the solid phase and nucleation of new phases, diffusion through porous reactants or products plays a significant role in shaping the overall rate of reaction. Diffusion in a porous reactant or product and reaction at the reactant–product interface are central concepts in systematic mathematical modeling and have been widely adopted in the chemical engineering and related literature. This literature has undergone significant rejuvenation in the last few years by the introduction of a previously neglected complication, the changes in the porous structure engendered by the reactions. Such changes either expand the porous structure, as in char gasification and combustion, or contract it as in the reaction of metal oxides with sulfur oxides.

The purpose of this paper is to review in detail a few recent contributions in gas–solid reactions where structural changes play a prominent role. The first section is of a somewhat more general character, however. It deals with the role of thermodynamics which is sometimes not adequately emphasized. The second section is an elementary one. It discusses the shrinking core model, including the effect of volume changes. This model is used as an element in a more general problem concerning a porous reactant treated in the third section. The treatment is based on the well known grain model, although capillary models have also been used for this purpose. The problem is illustrated with the reaction between sulfur dioxide and calcium oxide, important in fluid bed coal combustion.

Sections four to seven treat the combustion and gasification of carbonaceous particles. After the short introductory section four, the gasification of char under chemically controlled conditions is examined in section five and under more general conditions in section six. These two sections emphasize the evolution of pore surface area with conversion. A recently developed random capillary model provides an effective tool for this purpose.

Intraparticle phenomena are coupled with heat and mass transfer and homogeneous reaction in the surrounding gas phase, however. These processes are examined in section seven dealing specifically with char combustion. The general problem is very complicated and requires extensive numerical treatment. Section seven limits discussion to a simplified but useful version of the problem which allows considerable progress by analytical means.

The recent advances included in this review emphasize changes in the porous structure accompanying and influencing chemical reaction. The assumption, however, is made that reaction takes place on a mathematically well defined surface. As further progress is made it will surely be found that this idealization is not always accurate and may obscure important processes such as nucleation of solid product phases in the case of desulfurization or micropore enlargement in the case of char gasification. After five to ten years the time may become ripe for a review of such more esoteric phenomena.

THERMODYNAMICS

Thermodynamic considerations play a crucial role in gas–solid reactions because of the several possible solid phases that can be encountered in the progress of the reaction. The length scale and physical structure of such solid phases influence reaction and diffusion rates and, hence, determine the rate of the overall reaction. In the thermodynamic analysis of such systems it must be kept in mind that published free energies and heats of formation are frequently uncertain by several kcal/gmol. In this section we shall illustrate thermodynamic analysis in two industrially important reactions, hydrogen sulfide–zinc oxide and carbon monoxide–ferric oxide.

Hydrogen Sulfide–Zinc Oxide Reaction

The removal of hydrogen sulfide from fuel gas and synthesis gas is a very large volume operation. Gas produced by the steam reforming of coal or heavy petroleum fractions must be cleaned of hydrogen sulfide and other sulfur impurities before being brought in contact with catalysts for the synthesis of methanol, ammonia and other chemicals. Fuel gas from coal, also known as low Btu gas, destined for combustion in conventional or combined-cycle power cycles must be purified of sulfur compounds to prevent air pollution. Fuel cells with molten salt electrolytes operating at high temperatures offer exciting new avenues to small or large scale power generation. These fuel cells, however, require very clean fuel gas, with less than one part per million of sulfur compounds.

Removal of hydrogen sulfide from synthesis gas is currently carried out by absorption in methanol or in an aqueous solution of ethanolamines, at essentially room temperature. The same type of purification can also be applied to fuel gas. However, cooling the gasifier product from 800°C, or higher, to room temperature in order to purify it, then reheating it to the combustion temperature entails significant loss of thermodynamic efficiency. A great deal of effort is, therefore, devoted to high-temperature (600–900°C) hydrogen sulfide removal by solid sorbents, chiefly metal oxides.

In this section we will use the reaction between hydrogen sulfide and zinc oxide to illustrate the importance of thermodynamic considerations. Although several oxides have been and are currently being investigated, iron oxide and zinc oxide alone or in combination with other components are the most promising. Iron oxide is suitable for purification to about 100 parts per million (ppm) of hydrogen sulfide while zinc oxide could, in principle, provide purification to as low as one ppm. Both iron sulfide and zinc sulfide can be regenerated to the oxide form by reaction with air or air–steam mixtures with the evolution of sulfur dioxide and elemental sulfur. The latter product is marketable but sulfur dioxide must be disposed of after reaction with lime, or converted to elemental sulfur by reaction with coke.

The removal of hydrogen sulfide by zinc oxide involves the following reactions:

sulfidation (removal)

$$H_2S + ZnO = H_2O + ZnS \qquad (1)$$

regeneration

$$ZnS + \tfrac{3}{2}O_2 = ZnO + SO_2 \qquad (2)$$

side reactions

$$ZnO + H_2 = Zn(\ell) + H_2O \qquad (3)$$

$$ZnO + SO_2 + \tfrac{1}{2}O_2 = ZnSO_4 \qquad (4)$$

Table 1 below lists the free energies of formation for various compounds and the equilibrium constants for four reactions.

Because of the regenerative scheme of reactions, a combination of two interconnected fluidized beds, one for sulfidation (reaction 1), the other for regeneration (reaction 2) would be most convenient. However the desired high degree of purification dictates the use of two or more fixed beds in parallel, alternating between sulfidation and regeneration.

We start the thermodynamic analysis with the sulfidation part of the cycle. Assuming very fast rates for the reaction and diffusional processes, the concentration profiles along the sorbent bed would have the form shown in Fig. 1. In the section of the bed to the left of the plane AA', the sorbent is fully sulfided and the ratio p_{H_2S}/p_{H_2O} is equal to its feed value $(p_{H_2S}/p_{H_2O})_0$. At the right of AA' the sorbent consists of pure ZnO and the ratio p_{H_2S}/p_{H_2O} has the equilibrium value K_1^{-1}. It is understood that the gas phase contains H_2, CO, CO_2, N_2, in addition to H_2S and H_2O.

Because of finite reaction and diffusion rates, the actual concentration profiles involve gradual transition between the two levels. The step change AA' is replaced by the gradual transition $BB'B''$. The slower the rate processes, the smaller the slope of the reaction front $BB'B''$. As the reaction progresses, the reaction front $BB'B''$ moves to the right as a wave of constant shape and constant velocity determined by the concentration of fresh sorbent and by the feed rate of H_2S.

Breakthrough occurs when the reaction front reaches the reaction exit. More precisely, breakthrough is defined as the time at which

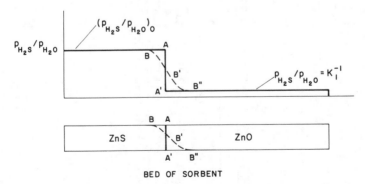

Figure 1   Schematic of concentration profiles during sulfidation of ZnO in a packed bed.

Table 1 Free Energies and Equilibrium Constants for the ZnO–H_2S System.

| T(K) | $\Delta_f G_i^0$(kcal/gmol) | | | | | | K_1 | | | |
	H_2O	H_2S	ZnO	ZnS	$ZnSO_4$	SO_2	1	2	3	4
800	−48.65	−12.19	−63.92	−45.02	−162.70	−72.57	62.7×10^3	1.5×10^{25}	6.7×10^{-5}	1.5×10^7
900	−47.35	−11.03	−61.35	−42.56	−151.80	−70.82	18×10^3	5.8×10^{21}	4.0×10^{-4}	6.1×10^4
1000	−46.04	−9.85	−58.80	−40.12	−141.11	−69.07	7.31×10^3	1.5×10^{19}	1.6×10^{-3}	7.9×10^2
1100	−44.71	−8.67	−56.26	−37.69	−130.43	−67.33	2.96×10^3	1.2×10^{17}	5.1×10^{-3}	2.3×10^1

Sources: JANAF Thermochemical Tables [9], Barin and Knacke [3].

the concentration of H_2S exceeds a specified level. Shortly before breakthrough, sulfidation would be interrupted and regeneration would be initiated by flow in the reverse direction.

To get some actual numbers we take $T = 900°K$, $p_{H_2O} = 0.1$ atm and total pressure 1 atm. The equilibrium value of H_2S is $p_{H_2S} = p_{H_2O}K^{-1} = 5.6 \times 10^{-6}$ atm which is equivalent to 5.6 ppm. This is the lowest level that can be achieved under these conditions. Thermodynamic data on solids are somewhat inaccurate and an error of ± 2 kcal in $\Delta G°$ would not be surprising. A change of ± 2 kcal at $900°K$ would change K_1 and the equilibrium level of H_2S by a factor of 3.

We next examine the effect of side reaction (3) during the sulfidation stage. Elemental zinc melts at $793°K$ and boils at $1180°K$. According to the equilibrium of reaction (3), liquid zinc would be present if

$$\frac{p_{H_2O}}{p_{H_2}} < K_3$$

The values of K_3 suggest that as long as the gas contains some water vapor, liquid zinc would not appear. However, we must also examine the equilibrium

$$ZnO + H_2 = Zn(g) + H_2O \qquad (3')$$

The values of the equilibrium constant K_3' at 800, 900, 1000, 1100°C are 2.1×10^{-7}, 9.3×10^{-6}, 1.8×10^{-4}, 2.1×10^{-3}. The equilibrium vapor pressure of Zn is given by

$$p_{Zn} = K_3' \frac{p_{H_2}}{p_{H_2O}} p_0$$

where $p_0 = 1$ atm. Assuming $p_{H_2} = 0.25$ atm, $p_{H_2O} = 0.1$ atm we find $p_{Zn} = 5.25 \times 10^{-7}$, 2.3×10^{-5}, 4.5×10^{-4}, 5.2×10^{-3} atm at 800, 900, 1000, 1100°K, respectively. The loss of zinc would become significant above 900 or 1000°K. Side reaction (3') sets an upper limit on the operating temperature during sulfidation.

The thermodynamics of regeneration also impose operational constraints. The extremely large values of K_2 guarantee that ZnS will be completely converted, assuming adequate kinetics. However, the relatively large values of the equilibrium constant for reaction

(4) suggest that the conversion might first lead to $ZnSO_4$ which would subsequently decompose to ZnO, since the feed contains no SO_2. As a matter of fact, during regeneration the sorbent bed would consist of three sections, the first containing ZnO, the second $ZnSO_4$ and the last ZnS. The two interfaces would move to the right with constant velocity. At the interface between ZnO and $ZnSO_4$, equilibrium provides the relationship

$$K_4 \frac{p_{SO_2}}{p_0} \left(\frac{p_{O_2}}{p_0} \right)^{1/2} = 1$$

where $p_0 = 1$ atm. In this relationship p_{O_2} is the oxygen pressure in the feed. Taking $p_{O_2} = 0.2$ atm we calculate for 800, 900, 1000, 1100°K: $p_{SO_2} = 1.5 \times 10^{-7}$, 3.7×10^{-5}, 2.8×10^{-3}, 9.7×10^{-2} atm. From the standpoint of further processing and disposal of SO_2, it is important that p_{SO_2} is as high as possible. Thus, the regeneration temperature should be as high as possible, at least 1100°K.

Reduction of Iron Ore

Production of iron by reduction of iron ore is the highest volume metallurgical process. It is commonly carried out in a packed bed of iron ore and coke, known as the blast furnace. Less commonly used are various "direct reduction' processes, in which iron ore is contacted directly with a reducing gas, natural gas or fuel gas from a coal gasifier.

The blast furnace reduction of coal ore involves the following reactions

$$3Fe_2O_3(\text{hematite}) + CO = 2Fe_3O_4(\text{magnetite}) + CO_2 \qquad (5)$$

$$Fe_3O_4 + CO = 3FeO(\text{Wüstite}) + CO_2 \qquad (6)$$

$$FeO + CO = Fe + CO_2 \qquad (7)$$

$$C + CO_2 = 2CO \qquad (8)$$

A phase diagram for iron oxides incorporating the equilibrium of reactions (5)–(7) is shown in Fig. 2. In a detailed model of the blast furnace Rao [11, 12] assumed that reactions (5)–(7) were locally at equilibrium so that the overall rate was determined by reaction (8).

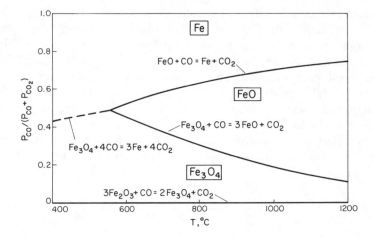

Figure 2   Thermodynamic diagram of iron oxides vs. temperature and gas composition.

KINETICS AND INTRAPARTICLE DIFFUSION—GENERAL

Let us consider a single gas-solid reaction

$$A(\text{gas}) + \nu_S S(\text{solid}) = \nu_B B(\text{gas}) + \nu_P P(\text{solid}) \tag{9}$$

The first fundamental problem that has to be addressed, in principle, is to identify the reaction mechanism. The actual detailed chemical steps and intermediates are seldom known, but since solids S and P involve the same cation, if ionic, the reaction involves change from one anion to another. In most cases S and P have different crystal structures, therefore the reaction involves a phase transformation. Attar and Sahly [2] have suggested the following sequence of steps: (i) formation of a certain amount of solid with the stoichiometry of P dissolved in the crystal structure of S, (ii) nucleation in this metastable solid to form small domains of P with the stable crystal structure of P; this nucleation may occur at random locations or at specific defect sites, (iii) reaction (9) progressing at the surface of the nuclei leads to growth of phase P. These steps obviously require diffusion of gases A, B as well as diffusion and rearrangement of cationic and ionic species.

The steps outlined above take place in an interfacial layer between the two phases A and B. The thickness and geometry of this reaction layer depends on the rates of solid state diffusion and on the porous structure of the reactant solid S. In many cases the overall reaction rate could be set proportional to the interfacial surface, ignoring the detailed phenomena of reaction, nucleation and solid state diffusion. The following sections treat a few of the most useful models of gas–solid reactions always based on the phenomenological description of the local reaction rate as proportional to the local interfacial area.

We start by considering reaction (9) in an isothermal spherical particle. The equations of change may be written provisionally in the general form

$$\frac{\partial}{\partial t}\left(\varepsilon c_A\right) = \frac{1}{r^2}\frac{\partial}{\partial r}\left(\mathcal{D}_A r^2 \frac{\partial c_A}{\partial r}\right) - f(c_A, \varepsilon_S) \tag{10}$$

$$\frac{\partial \varepsilon_S}{\partial t} = -\nu_S V_S f(c_A, \varepsilon_S) \tag{11}$$

$$t = 0: \quad c_A = 0, \quad \varepsilon_S = \varepsilon_{S0} \tag{12}$$

$$r = 0: \quad \partial c_A/\partial r = 0 \tag{13}$$

$$r = a: \quad \mathcal{D}_A \frac{\partial c_A}{\partial r} = k_m(c_{A0} - c_A) \tag{14}$$

Although general in appearance, these equations incorporate several assumptions:

(i) The reaction rate depends only on reactants A, S. This assumption is not fundamental in nature; dependence on the products B, P can be accommodated simply by adding similar equations for B, P. Dependence on the concentration of the solid S is assumed to be fully described by the local volume fraction ε_S of S. This assumption will be further clarified in the special cases discussed later on.

(ii) Equation (11) is applicable when the particle radius a remains constant, or when the radius diminishes because all products are gaseous, e.g. in gasification and combustion of coal or char. The equation requires modification in cases where the particles swell or contract due to the difference between the molar volumes of S and P.

(iii) Diffusion can be characterized by an effective diffusion coefficient $\mathcal{D}_A$. This coefficient is generally a function of c_A and also depends on the porous structure of the solid which

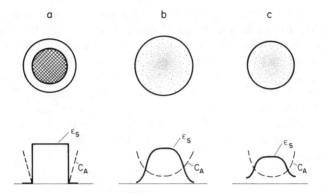

Figure 3 Schematic of gas and solid concentration profiles: (a) nonporous reactant, porous product (b) porous reactant, constant particle size (c) porous reactant, gaseous products only, decreasing particle size.

according to assumption (i) is a function of ε_S. In many applications the dependence on c_A is neglected on the basis of equimolar counterdiffusion or diffusion in an excess of inert.

At pressures near atmospheric the gas to solid concentration ratio $c_{A0}V_S/\varepsilon_S$ is very small on the order of 0.001, and the accumulation term $\partial(\varepsilon c_A)/\partial t$ in Eq. (10) can be neglected yielding

$$\frac{1}{r^2}\frac{\partial}{\partial r}\left(\mathscr{D}_A r^2 \frac{\partial c_A}{\partial r}\right) = f(c_A, \varepsilon_S) \qquad (10')$$

Equation $(10')$ states that the concentration profile of the gas is at steady state with respect to the concentration profile of the solid, and forms the usual starting point of gas–solid reaction models.

Before further progress can be made it is necessary to specify the form of $\mathscr{D}_A$ and f as functions of the volume fraction ε_S of the solid reactant. Proceeding in full generality is not practical. Instead, it is convenient to consider certain special cases which have found wide applicability. These are illustrated schematically in Fig. 3 and analyzed in the following sections.

NONPOROUS REACTANT—THE SHRINKING-CORE MODEL

This limiting case is illustrated by diagram (a) in Fig. 3. It involves nonporous solid reactant and porous solid product. With the reactant being nonporous, reaction is strongly limited by the slow solid-state diffusion and takes place in a thin layer. This layer is idealized as a sharp interface between the core of unreacted S and the shell of product P. Under these conditions an effective reaction rate per unit interfacial area can be defined. This will be taken as first order in the concentration of the gaseous reactant. This very simple model, known as the shrinking-core model has been employed extensively, even for cases of porous reactants.

On the basis of sharp reaction zone, Eqs. (10′) and (11)–(14) are replaced by

$$\frac{1}{r^2} \frac{\partial}{\partial r} \left(r^2 \frac{\partial c_A}{\partial r} \right) = 0 \tag{15}$$

$$t = 0: \qquad r_2 = a . \tag{16}$$

$$r = r_2: \qquad \mathscr{D}_A \frac{\partial c_A}{\partial r} = k_i c_A \tag{17}$$

$$r = r_2: \qquad \frac{c_S}{\nu_S} \frac{dr_2}{dt} = -\mathscr{D}_A \frac{\partial c_A}{\partial r} \tag{18}$$

$$r = a(t): \qquad \mathscr{D}_A \frac{\partial c_A}{\partial r} = k_m (c_{A0} - c_A) \tag{19}$$

where r_2 is the radius of unreacted core and $a(t)$ is the radius of the particle. If the volume ratio between equivalent stoichiometric amounts of the product P (including its voids) to reactant S is γ, $a(t)$ is related to r_2 by

$$a^3(t) - r_2^3(t) = \gamma(a_0^3 - r_2^3(t)) \tag{20}$$

where a_0 is the initial particle radius. Equation (19) relates the consumption of solid at the reaction surface with the supply of reactant gas. Equations (15)–(20) can be written in the dimensionless form

$$\frac{1}{x^2} \frac{\partial}{\partial x} \left(x^2 \frac{\partial \psi_A}{\partial x} \right) = 0 \tag{21}$$

$$\tau = 0: \qquad x_2 = 1 \tag{22}$$

$$x = x_2: \qquad \frac{\partial \psi_A}{\partial x} = q \psi_A \tag{23}$$

$$x = x_2: \qquad \frac{dx_2}{d\tau} = -\frac{\partial \psi_A}{\partial x} \tag{24}$$

$$x = \alpha(\tau): \qquad \frac{\partial \psi_A}{\partial x} = \beta(1 - \psi_A) \tag{25}$$

$$\alpha^3(\tau) = \gamma + (1 - \gamma)x_2^3(\tau) \tag{26}$$

where

$$x = \frac{r}{a_0}, \qquad \tau = \frac{t}{t_c}, \qquad \psi_A = \frac{c_A}{c_{A0}}, \qquad q = \frac{a_0 k_i}{\mathscr{D}_A} \tag{27}$$

$$t_c = \frac{a_0^2 c_S}{\mathscr{D}_A \nu_S c_{A0}}, \qquad \beta = \frac{k_m a_0}{\mathscr{D}_A}, \qquad \alpha(\tau) = \frac{a(t)}{a_0} \tag{28}$$

Solution of Eq. (21) with boundary conditions (23) and (25) yields

$$\psi_A = \frac{\dfrac{1}{qx_2^2} + \dfrac{1}{x_2} - \dfrac{1}{x}}{\dfrac{1}{qx_2^2} + \dfrac{1}{x_2} + \dfrac{1}{\beta\alpha^2} - \dfrac{1}{\alpha}} \tag{29}$$

which is inserted in (24) to provide

$$\frac{dx_2}{d\tau} = -\frac{1}{x_2^2\left(\dfrac{1}{\beta\alpha^2} + \dfrac{1}{qx_2^2} + \dfrac{1}{x_2} - \dfrac{1}{\alpha}\right)} \tag{30}$$

Integrating Eq. (30) in conjunction with Eqs. (22), (26) yields

$$\frac{1 - \alpha}{\beta(1 - \gamma)} + \frac{1}{q}(1 - x_2) + \tfrac{1}{2}(1 - x_2^2) - \frac{1 - \alpha^2}{2(1 - \gamma)} = \tau \tag{31}$$

The time for complete conversion is obtained from (31) by setting $x_2 = 0$, $\alpha = \gamma^{1/3}$:

$$\tau_f = \left[\frac{1}{2} - \frac{1 - \gamma^{2/3}}{2(1 - \gamma)}\right] + \frac{1}{\beta}\frac{1 - \gamma^{1/3}}{1 - \gamma} + \frac{1}{q} \tag{32}$$

The three terms in this expression represent resistances by film diffusion, pore diffusion in the product layer and chemical reaction.

When film diffusion controls, $\beta \ll 1$

$$\tau_f \cong \frac{1-\gamma^{1/3}}{1-\gamma} \frac{1}{\beta}, \qquad t_f = \frac{1-\gamma^{1/3}}{1-\gamma} \frac{a_0 c_S}{k_m \nu_S c_{A0}} \qquad (33)$$

When diffusion in the product layer controls $\beta \gg 1$, $q \gg 1$ and

$$\tau_f = \frac{1}{2} - \frac{1}{2} \frac{1-\gamma^{2/3}}{1-\gamma}, \qquad t_f = \left(\frac{1}{2} - \frac{1}{2} \frac{1-\gamma^{2/3}}{1-\gamma} \right) \frac{a_0^2 c_S}{\mathscr{D}_A \nu_S c_{A0}}$$

$$(34)$$

When reaction controls $q \ll 1$ and

$$\tau_f = \frac{1}{q}, \qquad t_f = \frac{c_S a_0}{\nu_S c_{A0} k_i} \qquad (35)$$

The dependence of t_f on a_0 can be used to determine the rate controlling step from experimental data. This dependence on a_0 is the same, save for a proportionality constant, as that given in the standard textbooks, e.g. Froment and Bischoff [5], for the case of constant particle size, $\gamma = 1$.

POROUS REACTANT: SULFUR DIOXIDE–CALCIUM OXIDE REACTION

For porous reactants, the analysis is considerably more complex because the reaction zone has considerable thickness and includes both reactant and product solids. The porosity of this zone varies with local conversion because of the different molar volumes of S and P. This broad class of reactions will be illustrated with the industrially important reaction between sulfur oxides and calcium oxide.

Removal of sulfur dioxide from combustion gases, known as flue gas desulfurization, is currently being practiced by conducting the combustion effluent gases through a slurry of an alkaline hydroxide, chiefly $Ca(OH)_2$, at essentially ambient temperatures. The high cost associated with equipment maintenance and slurry disposal has stimulated intensive research and development in alternative ways of burning coal and removing sulfur oxides from the combustion gases. One of the most promising developments in this area is fluidized combustion of coal. In this combustion process, coal is burned at relatively low temperatures (800–1000°C) in a fluidized bed of calcined limestone (CaO) or dolomite (CaO–MgO). The mass of coal constitutes only about 2% of the mass of the solids. Sulfur dioxide evolving from the oxidation of sulfur in coal is removed according to the reaction

$$CaO + SO_2 + \tfrac{1}{2}O_2 = CaSO_4 \qquad (36)$$

The chief characteristics of this system are (i) the reactant is porous, therefore, reaction takes place throughout the particle, (ii) concentration gradients develop as a result of pore diffusion resistance, (iii) the local reaction rate is controlled by diffusion through a product layer developing on the pore surface, (iv) the molar volumes of $CaSO_4$ and CaO are in the ratio of about $3:1$, hence the pore volume fraction decreases with conversion. This phenomenon can cause complete pore blockage.

An analysis of this problem has been given by Georgakis et al. [8] extending the grain model of Sohn and Szekely [13] to take into account the variation of porosity with reaction. Figure 4 defines the geometric parameters of the model. For a single grain, a_0, $a(t)$, $r_2(t)$ are the initial grain radius, the radius of the grain at time t and

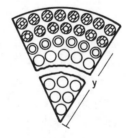

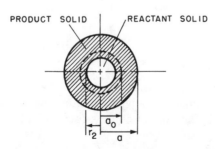

Figure 4   Schematic of grain model; upper section shows a slice of a particle, lower section shows a single grain.

the radius of the unreacted core at time t. The three radii are related by Eq. (20) above where γ is the ratio of molar volumes of $CaSO_4$ and CaO.

The dimensionless equations for diffusion and reaction within the grain are

$$\frac{1}{x^2} \frac{\partial}{\partial x} \left(x^2 \frac{\partial \psi_A}{\partial x} \right) = 0 \tag{37}$$

$$x = x_2: \qquad \frac{\partial \psi_A}{\partial x} = q \psi_A \tag{38}$$

$$x = \alpha: \qquad \psi_A = \omega_A \tag{39}$$

$$x = x_2: \qquad \frac{dx_2}{d\tau} = -\frac{\partial \psi_A}{\partial x} \tag{40}$$

where

$$\psi_A = \frac{C_A(r, y, t)}{C_{A0}}, \qquad \omega_A = \frac{C_A(a, y, t)}{C_{A0}} \tag{41}$$

The dimensionless quantity ω_A is the concentration in the pore volume surrounding the grains and is a function of time and the radial distance within the particle. Integrating (37)–(39) we obtain

$$\psi_A = \frac{\dfrac{1}{qx_2^2} + \dfrac{1}{x_2} - \dfrac{1}{x}}{\dfrac{1}{qx_2^2} + \dfrac{1}{x_2} - \dfrac{1}{\alpha}} \tag{42}$$

which is inserted in Eq. (40) to yield

$$\frac{dx_2}{d\tau} = -\frac{\omega_A(\xi, \tau)}{\dfrac{1}{q} + x_2 - \dfrac{x_2^2}{\alpha}} \tag{43}$$

$$x_2(0) = 1 \tag{44}$$

where α, x_2 are coupled as before by Eq. (26).

Equation (43) must be solved simultaneously with the diffusion equation for ω_A. The dimensionless form of the latter equation is

$$\frac{1}{\xi^2} \frac{\partial}{\partial \xi} \left(\xi^2 \delta_e \frac{\partial \omega_A}{\partial \xi} \right) = \lambda^2 \left(\frac{1}{qx_2^2} + \frac{1}{x_2} - \frac{1}{\alpha} \right)^{-1} \omega_A \tag{45}$$

$$\xi = 0: \qquad \frac{\partial \omega_A}{\partial \xi} = 0 \qquad (46)$$

$$\xi = 1: \qquad \delta_e \frac{\partial \omega_A}{\partial \xi} = \beta'(1 - \omega_A) \qquad (47)$$

where

$$\xi = \frac{y}{A}, \qquad \delta_e = \frac{\mathscr{D}_e(\varepsilon)}{\mathscr{D}_{e0}} \qquad (48)$$

$$\beta' = \frac{k_m A}{\mathscr{D}_{e0}}, \qquad \lambda^2 = 3(1 - \varepsilon_0) \frac{\mathscr{D}_A}{\mathscr{D}_{e0}} \frac{A^2}{a_0^2} \qquad (49)$$

In these equations A is the particle radius, assumed to be constant and $\mathscr{D}_e$ is the effective diffusion coefficient in the particle assumed to be a function of pore volume fraction ε. The value of $\mathscr{D}_e$ at the initial pore volume fraction is denoted by $\mathscr{D}_{e0}$. The pore volume fraction is easily related to x_2 or α:

$$\varepsilon = 1 - (1 - \varepsilon_0)\alpha^3 = 1 - (1 - \varepsilon_0)[\gamma - (\gamma - 1)x_2^3] \qquad (50)$$

For bulk diffusion one can often use the empirical relation

$$\mathscr{D}_e(\varepsilon) = K\varepsilon^n \qquad (51)$$

The coupled Eqs. (43)–(47), (26) can be solved only numerically. In the case of negligible resistance to film diffusion ($\beta' \gg 1$), $\omega_A(1, \tau) = 1$ and Eq. (43) becomes for $\xi = 1$:

$$\frac{dx_2}{d\tau} = -\frac{1}{\dfrac{1}{q} + x_2 - \dfrac{x_2^2}{\alpha}} \qquad (52)$$

This is integrated in conjunction with Eq. (26) and the initial condition $x_2(0) = 1$ to obtain

$$\tau = \tfrac{1}{2}(1 - x_2^2) - \frac{1}{2(\gamma - 1)}(\alpha^2 - 1) + \frac{1}{q}(1 - x_2) \qquad (53)$$

which can be solved simultaneously with Eq. (26) to provide x_2 and α as functions of τ at $\xi = 1$.

Examination of Eq. (50) shows that the pore volume fraction ε at the pellet surface $\xi = 1$ will vanish during reaction if ε_0 and γ satisfy

$$\gamma(1 - \varepsilon_0) \geqslant 1 \tag{54}$$

The values of α, x_2 at the instant when ε vanishes are found from Eq. (50) as

$$\alpha = \left(\frac{1}{1 - \varepsilon_0}\right)^{1/3}, \qquad x_2 = \left[\frac{\gamma}{\gamma - 1} - \frac{1}{(\gamma - 1)(1 - \varepsilon_0)}\right]^{1/3} \tag{55}$$

The time at which ε vanishes at the pellet surface can be found simply by inserting Eq. (55) into Eq. (53). At that instant, diffusion to the pellet becomes completely obstructed and reaction stops. Thus, if $\gamma(1 - \varepsilon_0) < 1$, complete conversion of the reactant solid will be achieved, given sufficient time, while if $\gamma(1 - \varepsilon_0) > 1$ reaction will cease and conversion will freeze.

Numerical solution of Eqs. (43)–(47) was carried out by Georgakis et al. [8] for reaction (36) with SO_2 being the limiting gaseous reactant, A. They used the experimental parameters $\gamma = 3$, $\varepsilon_0 = 0.54$, for which $\gamma(1 - \varepsilon_0) > 1$ predicting incomplete conversion of CaO. Figure 5 compares calculated and experimental results for three pellet sizes. The agreement is generally good and establishes

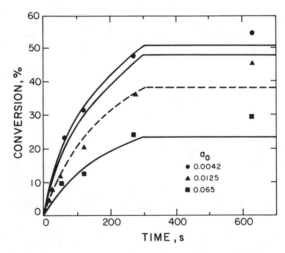

Figure 5   Grain model calculations and experimental data on CaO conversion (from Georgakis et al. [8]).

the ability of this model to describe the interesting feature of incomplete conversion.

The grain model is one of two popular geometric models of pore structure that have been proposed for gas–solid reactions. The other is the capillary model. Which of the two models provides a better description of pore structure depends on the particular solid material. Pellets produced by compaction of small particles would be more naturally described by the grain model. Materials produced by drying or calcination would generally be described more accurately by capillary models. Bhatia and Perlmutter [4] have analyzed gas–solid reactions such as that between CaO and SO_2, using a random capillary model and took into account changes in porosity due to molar volume differences. Their numerical results are in good agreement with those obtained with the grain model, despite the very different geometries.

POROUS REACTANT: CHAR GASIFICATION AND COMBUSTION—GENERAL

The principal features of gasification or combustion of carbonaceous particles are the enlargement of pores and gradual decrease of particle size due to reaction. Depending on particle size and temperature we may distinguish four regimes of reaction.

Regime (i): control by reaction: particles burn at constant size and diminishing density

Regime (ii): control by reaction and pore diffusion: particles burn at diminishing size and density

Regime (iii): control by reaction, pore diffusion and film diffusion: reaction occurs in a thin surface layer, particles burn at constant density and diminishing size

Regime (iv): control by film diffusion: burning behavior as in (iii).

Figure 6 classifies various combustion and steam gasification processes according to the reaction regime they follow. Because of the much lower reaction rates of steam gasification, the transition between the various regimes is displaced towards higher temperature and larger particle size. It should be pointed out that a particle may start reacting in a particular regime, e.g. (iii), but after some size reduction enter a lower regime, e.g. (ii).

As seen in Fig. 6, regime (i) where control is by chemical reaction alone is not encountered in actual combustion or gasification proces-

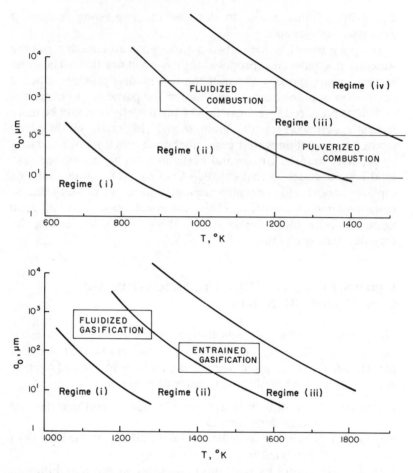

Figure 6   Regimes of combustion (upper diagram) and gasification (lower diagram).

ses. Yet, this regime is often pertinent to laboratory investigations and forms the basis for the theoretical analysis of regimes (ii) and (iii). Consequently, our analysis will start with regime (i) and then proceed to the other regimes.

The analysis throughout this section will treat the particle temperature as a known parameter. It will also assume that only one reaction takes place, that between solid carbon and gaseous oxygen. The reaction between carbon and carbon dioxide and between carbon monoxide and oxygen will be neglected. With these restrictions, the analysis does not provide a complete solution of the particle combustion problem. In the solution of the complete problem, the particle temperature is one of the unknowns which must be determined simultaneously with the particle radius by consideration of the equations for energy and chemical species conservation in the gas film. A simplified version of this more general problem will be considered in the last section.

CHAR GASIFICATION AND COMBUSTION AT CHEMICALLY CONTROLLED RATES

In regime (i) reaction is taking place uniformly throughout the internal pore surface area. However, how is the "internal pore surface area" defined? Unlike porous alumina or silica, carbonaceous materials such as coal, char and petroleum coke possess a complex porous structure that requires careful consideration. The pores of these materials are customarily classified according to pore radius into micropores $(R < 6\,\text{Å})$, transitional pores $(6\,\text{Å} < R < 150\,\text{Å})$ and macropores $(150\,\text{Å} < R < 5000\,\text{Å})$. Macropores serve as channels for transport of reactant and product gases. Transitional pores contribute to transport as well as to chemical reactions while micropores contribute to chemical reactions alone. Micropores form what is known as the molecular sieve structure of carbons. Diffusion in these pores is very slow and "activated". The diffusion coefficient depends very strongly on the size of the diffusing molecule. At the temperatures of experimental interest the micropores are penetrated only very slightly by the reactant gas, oxygen or steam, hence only a small fraction of the micropore surface participates in the reaction. The intrinsic reaction rate should, consequently, be defined in terms of the surface area of transitional pores and macropores. The intrinsic rate constant would then incorporate the rate-enhancing effect of surface roughness due to micropores. The reaction rate per unit volume will be expressed as $Sf_i(c, T)$ where S is the surface area of transitional pores and macropores per unit particle volume and $f_i(c, T)$, the intrinsic rate, is a function of temperature and one or more gaseous concentrations.

We now introduce an assumption which is almost always employed in the modeling of char reactions. Assumption 1: As reaction proceeds and the internal pore surface recedes, the intrinsic reaction rate remains constant. This assumption could, for example, be violated due to changes in the catalytic activity per unit pore surface area or due to penetration of micropores at low temperatures. A consequence of the assumption is the existence of master curves that we now proceed to describe.

Assuming that the density of the solid phase remains constant, the pore volume fraction ε and the fractional conversion X of a char sample are related by

$$X = \frac{\varepsilon - \varepsilon_0}{1 - \varepsilon_0} \tag{56}$$

Consider now a char sample reacting in an atmosphere of fixed composition and at temperatures sufficiently low for reaction to be solely under chemical control (regime (i)). The changes in ε, X are given by

$$\frac{d\varepsilon}{dt} = bf_i(c, T)S(X) \tag{57}$$

$$\frac{dX}{dt} = \frac{b}{1 - \varepsilon_0} f_i(c, T)S(X) \tag{58}$$

where b is a constant involving reaction stoichiometry and density of the solid. The surface area per unit volume, S, is clearly a function of conversion.

Integration of this equation yields

$$F(X) = \int_0^X \frac{dx}{S(x)} = \frac{b}{1 - \varepsilon_0} f_i(c, T)t \tag{59}$$

The time $t_{0.5}$ required to obtain a fractional conversion of 0.5 is given by

$$F(0.5) = \frac{b}{1 - \varepsilon_0} f_i(c, T)t_{0.5} \tag{60}$$

Dividing Eq. (59) by Eq. (60) yields

$$\frac{F(X)}{F(0.5)} = \frac{t}{t_{0.5}}$$

which defines X as a function of $t/t_{0.5}$ alone, independent of temperature and composition of the gas phase:

$$X = F_1\left(\frac{t}{t_{0.5}}\right) \tag{61}$$

Equation (58) can be evaluated at any particular conversion level, e.g.

$$\left(\frac{dX}{dt}\right)_{t_{0.5}} = \frac{b}{1 - \varepsilon_0} S(0.5) f_i(c, T) \tag{62}$$

Dividing Eq. (58) by Eq. (62) provides a relation independent of temperature and pressure:

$$\frac{(dX/dt)}{(dX/dt)_{t=0.5}} = \frac{S(X)}{S(0.5)} \tag{63}$$

Another useful way of presenting the reaction rate is obtained by combining Eqs. (58) and (60):

$$t_{0.5} \frac{dX}{dt} = F(0.5)S(X) \tag{64}$$

Equations (61), (63), (64) are all direct consequences of assumption 1 via Eq. (58). By plotting conversion and reactivity versus conversion or dimensionless time $t/t_{0.5}$, data corresponding to different temperatures and reactant pressures should all collapse to a single "master curve" as a consequence of assumption 1. Figure 7 shows conversion data of Tseng and Edgar [16] in the temperature

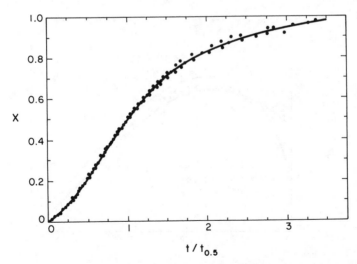

Figure 7   Master curve for the oxidation of a lignite char in the temperature range 400–475°C (from Tseng and Edgar [16]).

range 400–475°C. The data points cluster very close to a single master curve. By contrast, Fig. 8 shows the conversions measured at 350–375°C deviating from the high temperature curve. The deviation of the 350–375°C conversions is most likely due to the fact that at sufficiently low temperatures the reactant gas penetrates the microporous structure rendering assumption 1 untenable. Another interesting feature of Fig. 8 is the maximum in the rate-conversion curve. This maximum is usually observed under conditions of chemical control for chars of initial pore volume fraction below about 0.35. The curve becomes monotonically declining at higher temperatures as pore diffusion becomes important.

The preceding general conclusions are independent of the particular char under study. The specific form of the master curves, however, depends on the char considered, in other words each char possesses its own characteristic master curves.

To develop a quantitative description of surface area and reaction rate as functions of conversion it is necessary to adopt some geometric model of the porous structure. Models of porous materials are generally of two types, periodic and random. Random models are simpler to apply and more in keeping with the actual physical structure of porous chars. Among random pore models, two are particularly useful because of their simplicity and generality: the random capillary model and random sphere model. In the first

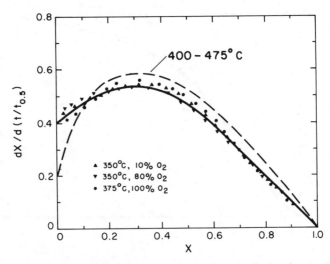

Figure 8   Rate vs. conversion data from the oxidiation of a lignite char at temperatures above and below 400°C (from Tseng and Edgar [16]).

the void phase consists of the interior of cylindrical capillaries located in space in a completely random fashion. The solid phase consists of the space outside the capillaries. In the random spheres model, the solid phase consists of the space occupied by spheres while the void phase consists of the space outside the spheres. We will only discuss the random capillary model.

Random Capillary Model

This model of pore structure is constructed by distributing the axes of the infinitely long capillaries completely randomly, according to a Poisson process [6]. As shown in Fig. 9, the density of capillaries is defined by a parameter λ such that $\lambda \, dS$ is the mean number of capillary axes intersections with a surface element of area dS, arbitrarily located in space. According to the properties of the Poisson process, the probability that a finite plane surface of area S is not intersected by any capillary is $\exp(-\lambda S)$. For a closed surface of area S this probability is $\exp(-\lambda S/2)$ because λS is the expected number of intersections, but these correspond to $\lambda S/2$ axes.

Suppose first that all capillaries have the same radius R. Then the probability that an arbitrary point P lies in the solid phase is equal to the probability that no capillary axis is closer than a distance R from P, which is to say that no capillary axis intersects a sphere of radius R centered at P (see Fig. 9). This probability is $\exp(-2\pi\lambda R^2)$. However, the same probability is equal to $1 - \varepsilon$, hence

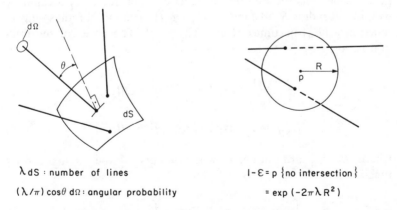

$\lambda \, dS$: number of lines

$(\lambda/\pi)\cos\theta \, d\Omega$: angular probability

$1 - \varepsilon = p\,\{$no intersection$\}$

$= \exp(-2\pi\lambda R^2)$

Figure 9   Definition of probabilities in random capillary model.

$$\varepsilon = 1 - \exp\left(-2\pi\lambda R^2\right) \tag{65}$$

For low values of the product λR^2 this expression can be approximated by

$$\varepsilon \cong 2\pi\lambda R^2$$

which is also the volume of capillaries per unit total volume, in the absence of overlap. The complete expression (65) includes the effect of overlap.

A distribution of capillary sizes is described by a probability density function $\lambda(R)$ such that $\lambda(R)\, dR\, dS$ is the mean number of intersections of a surface element of area dS with capillaries with radii in the range $[R, R + dR]$. The void volume fraction is now given by

$$\varepsilon = 1 - \exp\left[-2\pi \int_{R_*}^{R^*} \lambda(R)R^2\, dR\right] \tag{66}$$

where R_*, R^* are the smallest and largest capillary radii in the porous material.

The random capillary model yields simple expressions for the number of capillary intersections per unit volume, the mean length of a capillary between intersections and other related quantities [6]. For our purposes it suffices to derive the relationship between conversion and reaction rate or, equivalently, between pore volume fraction and surface area.

Let $q(t)$ be the reaction distance, i.e., the distance by which the pore surface has receded due to reaction. The radius of a capillary with initial radius R_0 at time t is $R_0 + q(t)$. The initial radius density function $\lambda_0(R_0)$ becomes $\lambda(R) = \lambda_0(R - q)$. The pore volume fraction is given by

$$\frac{1 - \varepsilon(q)}{1 - \varepsilon_0} = \exp\left[-2\pi(B_0 q^2 + 2B_1 q)\right] \tag{67}$$

$$X(q) = 1 - \exp\left[-2\pi(B_0 q^2 + 2B_1 q)\right] \tag{68}$$

where B_0, B_1 are the first two moments of the radius density function

$$B_0 = \int_{R_*}^{R^*} \lambda_0(R)\, dR \,, \qquad B_1 = \int_{R_*}^{R^*} R\lambda_0(R)\, dR \tag{69}$$

Finally, the surface area per unit total volume is given by

$$S(q) = \frac{d\varepsilon(q)}{dq} = 4\pi[1 - \varepsilon(q)](B_0 q + B_1) \tag{70}$$

All quantities of interest have been expressed in terms of a single variable $q(t)$ which characterizes the extent of reaction. The derivative of q is proportional to the intrinsic reaction rate,

$$\frac{dq}{dt} = bf_i(c, T) \tag{71}$$

The parameters B_0, B_1 characterizing the initial probability density of capillary radius can be computed from porosimetry data [6]. Eliminating q between (68), (70) we obtain a relationship between surface area and conversion:

$$S(X) = 4\pi(1 - \varepsilon_0)(1 - X)\left[B_1^2 - \frac{B_0}{2\pi} \ln(1 - X) \right]^{1/2} \tag{72}$$

$$S(0) = 4\pi(1 - \varepsilon_0)B_1 \tag{73}$$

We can further calculate

$$\frac{(dX/dt)}{(dX/dt)_{t=0.5}} = \frac{S(X)}{S(0.5)} = \frac{2(1 - X)\left[1 + \dfrac{B_0}{2\pi B_1^2} \ln(1 - X) \right]^{1/2}}{\left(1 + \dfrac{B_0}{2\pi B_1^2} \ln 2 \right)^{1/2}} \tag{74}$$

Equation (74) provides the functional form for one of the master curves discussed previously. The master curve depends on the initial pore size distribution via a single parameter B_0/B_1^2. This can be determined by curve-fitting the data. The parameter B_1 can be independently determined from a measurement of the initial surface area.

Solving Eq. (68) for q in terms of X we obtain

$$q(X) = -\frac{B_1}{B_0} + \left[\frac{B_1^2}{B_0^2} + \frac{1}{2\pi B_0} \ln\left(\frac{1}{1 - X} \right) \right]^{1/2} \tag{75}$$

$$q(X = 0.5) = -\frac{B_1}{B_0} + \left[\frac{B_1^2}{B_0^2} + \frac{1}{2\pi B_0} \ln 2 \right]^{1/2} \tag{76}$$

Since q is proportional to t, Eq. (71), we also have

$$\frac{q(X)}{q(X=0.5)} = \frac{t}{t_{0.5}} = \frac{-1+\left[1+\dfrac{B_0}{2\pi B_1^2}\ln\left(\dfrac{1}{1-X}\right)\right]^{1/2}}{-1+\left[1+\dfrac{B_0}{2\pi B_1^2}\ln 2\right]^{1/2}} \qquad (77)$$

which defines the master curve between the conversion X and the dimensionless time $t/t_{0.5}$, again in terms of the single parameter B_0/B_1^2.

Figures 10 and 11 compare calculated and experimental conversions and reaction rates. The parameters B_0/B_1^2 and dq/dt needed for the calculations were determined from the experimental data. The agreement between model calculations and experimental data is very good at conversions below 0.7. At higher conversions the calculated values exceed the experimental values. This discrepancy may be due to inhomogeneity in the material, to the presence of mineral matter and to other experimental features not accounted by the model.

The applicability of the model is somewhat restricted by the following curious feature. Equation (68) implies that complete conversion ($X = 1$) requires infinite q, i.e. infinite time. The reason behind this physically unrealistic feature is that a completely random

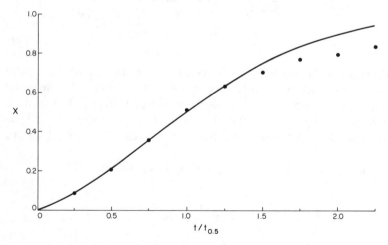

Figure 10   Conversions calculated by the random capillary model vs. experimental conversions for char oxidation (from Gavalas [6]).

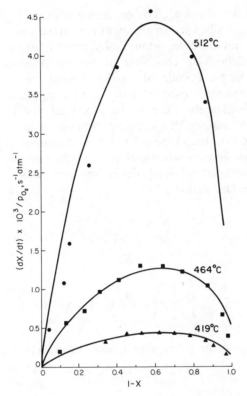

Figure 11   Reaction rate vs. conversion in char oxidation; results of the random capillary model compared with experimental data (from Gavalas [6]).

porous structure (Poisson) must allow, with very small but finite probability, the occurrence of large regions of space occupied completely by the solid phase. This difficulty can be resolved if we take into account that the particle radius decreases by the same velocity associated with pore growth, i.e.

$$\frac{da}{dt} = -bf_i(c, T) \tag{78}$$

$$a = a_0 - q \tag{79}$$

Conversion is now given by the more precise expression

$$X(q) = 1 - \left(1 - \frac{q}{a_0}\right)^3 \exp\left[-2\pi(B_0 q^2 + 2B_1 q)\right] \tag{80}$$

giving $X = 1$ when $q = a_0$. The difference between expressions (68) and (80) is actually very small, except at conversions close to one.

A second and more important modification must be introduced to resolve the difficulty encountered at high conversions. As reaction progresses and the cylindrical capillaries widen, at some point the solid phase becomes disconnected, i.e. disintegrates to small particles. This disintegration cannot be described exactly. One simple approach is to assume that disintegration occurs at some particular conversion X^* (in the range 0.7 to 0.9). At disintegration, the solid is assumed to become a collection of small, nonporous, spherical particles of uniform radius a^*, defined by the correspondence in the volume to surface ratio:

$$\frac{a^*}{3} = \frac{1 - \varepsilon(X^*)}{S(X^*)} \tag{81}$$

or using Eq. (72)

$$a^* = \frac{3}{4\pi} \left[B_1^2 - \frac{B_0}{2\pi} \ln(1 - X^*) \right]^{-1/2} \tag{82}$$

From that point on, conversion follows the reaction of nonporous spheres.

CHAR GASIFICATION AND COMBUSTION IN REGIMES (ii), (iii)

As defined earlier, regime (iii) is characterized by slow pore diffusion. The overall reaction rate, of course, depends on diffusion as well as reaction rate because the two processes are not in series. To be definite, we will consider combustion described by the reaction

$$C + \tfrac{1}{2}O_2 = CO \tag{83}$$

In regimes (ii) and (iii) conversion is a function of radial position in the particle, as well as time, characterized by the reaction distance $q(x, t)$. Diffusion in the pores will be described by an effective diffusion coefficient given by the expression

$$\mathcal{D}_e(q) = \frac{1}{\zeta} \int \varepsilon(R, q) \mathcal{D}(R) \, dR \tag{84}$$

where $\varepsilon(R, q)\,dR$ is the pore volume fraction of capillaries with radius in $[R, R + dR]$. The pore volume fraction changes with pore enlargement, therefore is a function of q. $\mathscr{D}(R)$ is the diffusion coefficient for a single capillary, which can be often approximated by

$$\frac{1}{\mathscr{D}(R)} = \frac{1}{\mathscr{D}_b} + \frac{1}{\mathscr{D}_k} \tag{85}$$

where $\mathscr{D}_b$, $\mathscr{D}_k$ are the bulk and Knudsen diffusion coefficients. Finally ζ is a tortuosity factor, in the range 2–3.

Utilizing an effective diffusion coefficient contains the implicit assumption that the length scale associated with concentration gradients is much larger than the length scale associated with the geometry of the porous solid, which could be taken as the length of capillary segments between intersections. This assumption is not always justifieid but is, nevertheless, retained in the absence of alternatives to this macroscopic characterization.

With this preparation we can now describe combustion in terms of equations for the oxygen concentration $c(r, t)$ and the reaction distance $q(r, t)$:

$$\frac{1}{r^2} \frac{\partial}{\partial r} \left(r^2 \mathscr{D}_e(q) \frac{\partial c}{\partial r} \right) = b' f_{ci}(c, T) S(q) \tag{86}$$

$$\frac{\partial q}{\partial t} = \frac{1}{\rho_c} f_{ci}(c, T) \tag{87}$$

$$t = 0: \qquad q = 0 \tag{88}$$

$$r = 0: \qquad \partial c / \partial r = 0 \tag{89}$$

$$r = r^*(t): \qquad \mathscr{D}_e \frac{\partial c}{\partial r} = k_m(c_0 - c) \tag{90}$$

$$t < t^*: \qquad r^*(t) = a_0, \qquad q(r^*, t) < q^* \tag{91a}$$

$$t > t^*: \qquad r^*(t) < a_0, \qquad q(r^*, t) = q^* \tag{91b}$$

The rate f_{ci} is expressed in terms of mass of char per unit surface area, b' is a stoichiometric coefficient for oxygen ($b' = 1/24$) and r^* is the instantaneous particle radius. During an initial period $t < t^*$, the particle radius decreases according to $dr^* / dt = -f_{ci}/\rho_c$. This velocity is very low, hence the particle radius can be considered constant, as stated in Eq. (91a). When the conversion at the surface reaches the critical value X^*, the solid phase disintegrates and the particle surface starts receding at a velocity $dr^*/dt \gg f_{ci}/b'$. The value of q corresponding to X^* is denoted by q^*.

Equations (86)–(91) apply to regime (ii) as well as (iii) and must be solved numerically. A series of numerical calculations have been presented by Gavalas [7]. In regime (iii), characterized by strong limitations due to pore diffusion, reaction takes place in a thin shell at the particle surface. Neglecting curvature we can rewrite Eq. (86) for a semi-infinite region with x measuring the distance from the original position of the boundary.

$$\frac{\partial}{\partial x}\left(\mathscr{D}_e(q)\,\frac{\partial c}{\partial x}\right) = b'f_{ci}(c,\,T)S(q) \tag{92}$$

while boundary conditions (89) become

$$x \to \infty: \quad c \to 0, \quad q \to 0 \tag{93}$$

and (90) becomes

$$x = x^*(t): \quad \mathscr{D}_e\,\frac{\partial c}{\partial x} = -k_m(c_0 - c) \tag{94}$$

where x^* is the instantaneous position of the surface with respect to its initial position.

Condition (91) describing the change in the particle size can be rewritten in terms of x^*. However, in view of the thin reaction layer, a pseudosteady-state is quickly established characterized by a constant boundary velocity, $dx^*/dt = v \gg f_{ci}/b'$. Under these conditions c and q depend on a single variable, the position with respect to the instantaneous boundary, $z = x - x^*(t)$. Equations (86)–(90) now become

$$\frac{d}{dz}\left(\mathscr{D}_e(q)\,\frac{dc}{dz}\right) = b'f_{ci}(c,\,T)S(q) \tag{95}$$

$$\frac{dq}{dz} = -\frac{1}{\rho_c v}\,f_{ci}(c,\,T) \tag{96}$$

$$z = 0: \quad \mathscr{D}_e\,\frac{dc}{dz} = -k_m(c_0 - c) \tag{97}$$

$$z = 0: \qquad q = q^* \tag{98}$$

$$z \to \infty: \qquad c \to 0, \qquad q \to 0 \tag{99}$$

Equations (95)–(99) can be solved exactly by a suitable change of variables as detailed in Gavalas [7]. The solution can be written in terms of the boundary velocity v as

$$v = \frac{1}{\rho_c b'^{1/2}} \frac{I^{1/2}}{J^{1/2}} \tag{100}$$

where

$$I = \int_0^{c^*} f_{ci}(c, T) \, dc \tag{101}$$

$$c^* = c_0 - \frac{b' v \rho_c}{k_m} (\varepsilon^* - \varepsilon_0) \tag{102}$$

$$J = \int_0^{q^*} \frac{\varepsilon(q) - \varepsilon_0}{\mathscr{D}_e(q)} \, dq \tag{103}$$

The combustion rate per unit external surface area is given by

$$f_{ce} = \rho_c v (1 - \varepsilon_0) \tag{104}$$

When the resistance to film diffusion can be neglected, $c^* = c_0$ and (100) assumes the simple form

$$f_{ce} = \frac{1 - \varepsilon_0}{b'^{1/2}} \frac{I^{1/2}}{J^{1/2}} \tag{105}$$

where now

$$I = \int_0^{c_0} f_{ci}(c, T) \, dc \tag{106}$$

This rate is the product of a chemical term $I^{1/2}$ and a physical term $J^{-1/2}$. When the reaction rate is described by power-law kinetics, $f_{ci} = k_i c^m$, the rate f_{ce} assumes the expression

$$f_{ce} = (1 - \varepsilon_0) \left[\frac{k_i}{b'(m-1)J} \right]^{1/2} c_0^{(m-1)/2} \tag{107}$$

which exhibits the dependence on temperature and concentration encountered in heterogeneous catalytic reactions at large values of the Thiele modulus.

For the case when the resistance due to film diffusion is not negligible, deriving an expression for f_{ce} requires elimination of c^* between Eqs. (100)–(102). This can be carried out explicitly only for zero and first order kinetics. In the case of the first order kinetics the result is

$$f_{ce} = \frac{c_0}{\dfrac{b'}{k_m} + \dfrac{1}{1 - \varepsilon_0} \left(\dfrac{2b'J}{k_i} \right)^{1/2}} \tag{108}$$

which displays the customary combination of film diffusion with pore diffusion and chemical reaction.

Figures 12 and 13 show some numerical results based on Eqs. (103) through (107). Figure 12 plots the quantity $J^{-1/2}$ which is proportional to the reaction rate f_{ce} as a function of total initial porosity for a char possessing uniformly sized capillaries. The rate varies slowly with porosity in the range of low porosities. Above $\varepsilon_0 = 0.10$, however, the rate increases rapidly with increasing porosity. It is also observed that the rate is a very weak function of pore radius, represented by the dimensionless parameter $\mathscr{D}_k / \mathscr{D}_b$. Variation of this parameter by a factor of 80 induces a variation in the rate by a factor of 2 only. This insensitivity to increasing the pore radius

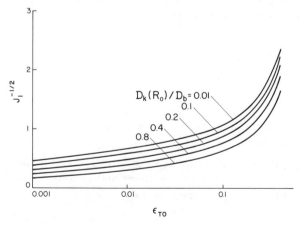

Figure 12   The dependence of relative rate on initial porosity for a char initially with uniformly sized pores (from Gavalas [7]).

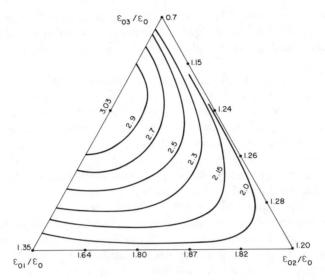

Figure 13   The dependence of relative rate on initial pore size distribution (from Gavalas [7]).

is due to the cancellation of two opposing effects, increasing the diffusion coefficient and decreasing the surface area.

Figure 13 shows numerical results for an initial pore size distribution consisting of three discrete sizes ($R_{01} = 30$, $R_{02} = 500$, $R_{03} = 5000$ Å). The relative rate is plotted as a function of the initial pore volume fractions of the three capillary sizes, keeping the total pore volume fixed. The diagram uses customary triangular coordinates and each curve is a locus of fixed relative reaction rate. With the exception of narrow regions near the sides $\varepsilon_{01} = 0$, $\varepsilon_{03} = 0$, the relative rate varies only between 2 and 3. The results indicate that as long as each size range contains nonnegligible pore volume, the relative rate is quite insensitive to the pore size distribution.

CHAR COMBUSTION: COUPLING BETWEEN PARTICLE AND GAS FILM

In the previous section char combustion was analyzed with emphasis on intraparticle reaction, diffusion and pore growth. The temperature was treated as a known parameter. In this section particle radius along with particle temperature will be treated as time dependent quantities to be determined from appropriate conservation equations. The solution of these equations is required to calculate the time for complete conversion and to understand the important and intriguing phenomena of particle ignition and extinction.

In a complete study of particle combustion, the conservation equations must describe processes occurring within the particle and in the surrounding gas phase. The former include reaction, pore growth and heat and mass transfer. Two heterogeneous reactions must be taken into account, the oxygen–carbon and the carbon dioxide–carbon. The processes occurring in the surrounding gas include homogeneous oxidation of carbon monoxide and heat and mass transfer from the particle surface to a position further away where temperature and composition are known.

Simultaneous solution of the species and energy equations for the gas phase and the equations describing diffusion, reaction and pore growth in the particle has only been recently carried out [14]. The detailed numerical calculations involved are beyond the scope of this review. It is useful, however, to treat the simpler limiting case wherein the particle burns in regime (iii) or (iv), i.e. at constant density in a shrinking core fashion. Under these conditions intraparticle phenomena need not be considered explicitly. They are combined with the intrinsic kinetics to effective rates concentrated on the external particle surface. The rates of the surface reactions are employed as boundary conditions to the species and energy equations for the gas phase. This problem has a long history culminating with the analyses of Amundson and co-workers [15]. A review on the subject has been presented recently by Alfano and Arce [1].

In this paper we shall present a relatively simple analysis due to Libby and Blake [10]. These authors have simplified the problem by assuming frozen or equilibrium conditions for the gas phase reactions and by using a number of approximations concerning the transport coefficients. These approximations allow the derivation of results with a minimum of numerical effort.

Like other investigators, Libby and Blake decompose the problem into a pseudosteady-state subproblem and a dynamic subproblem. The first subproblem, discussed in the first section below, assumes that the gas phase temperature and composition are at steady state with respect to the instantaneous particle radius and particle temperature. This is nearly always a sound simplification, for the characteristic time for transport in the gas film is much shorter than the characteristic combustion time. In the second and final section we discuss the dynamic problem involving the variation of particle radius and temperature. The solution of the dynamic problem draws on the results of the pseudosteady-state problem.

The Pseudosteady-State Problem

The assumptions made to simplify the mathematical treatment are
  (i) All diffusivities (mass and thermal) are equal $\mathscr{D}_i = \mathscr{D} = \lambda/\rho c_p$.
  (ii) The product $\rho \mathscr{D}$ is independent of temperature.
  (iii) No homogeneous reactions take place in the gas phase around the particle, i.e. the reactions are "frozen".
Assumptions (i) and (ii) do not alter the qualitative features of the problem, although they affect the numerical results. Assumption (iii) is considered adequate for small particles and high temperatures, e.g. under conditions of pulverized combustion. The opposite to assumption (iii), that of very fast homogeneous oxidation of CO, has also been investigated by Libby and Blake [10]. The general case of finite rates has been treated in the aforementioned studies of Amundson and his co-workers.

The analysis starts with the total mass balance

$$\rho v r^2 = m_p a^2 \tag{109}$$

where ρ, v are gas density and radial velocity and m_p is the mass flux at the particle surface $r = a$.

Let Y_1, Y_2, Y_3, Y_4 be the mass fractions of O_2, CO_2, CO and N_2. In view of the assumption of frozen reactions these mass fractions satisfy the conservation equations

$$\frac{d}{d\xi}(\xi^2 m_i) = 0 \tag{110}$$

$$\xi = 1: \qquad Y_i = Y_{ip} \tag{111}$$

$$\xi \to \infty: \qquad Y_i = Y_{i\infty} \tag{112}$$

where $\xi = r/a$ and m_i, the flux of species i, is given by

$$m_i = \rho v Y_i - \eta \frac{dY_i}{dr} \tag{113}$$

and

$$\eta = \rho \mathscr{D} \equiv \text{constant} \tag{114}$$

Equations (110)–(112) are routinely integrated to

$$Y_i = \frac{(Y_{i\infty} - Y_{ip})\, e^{-B/\xi} + (Y_{ip} - Y_{i\infty}\, e^{-B})}{1 - e^{-B}} \qquad (115)$$

where

$$B = \frac{m_p a}{\eta} \qquad (116)$$

is a mass-loss parameter that must be determined along with the other unknowns Y_{1p}, Y_{2p}, Y_{3p}.

We denote by f_{ce1}, f_{ce2} the rates of the two surface reactions $C + \frac{1}{2}O_2 = CO$, $C + CO_2 = 2CO$ in mass of carbon per unit external surface area and time. Clearly

$$m_p = f_{ce1} + f_{ce2} \qquad (117)$$

We also introduce the dimensionless quantities

$$B_1 = \frac{f_{ce1} a}{\eta}, \qquad B_2 = \frac{f_{ce2} a}{\eta} \qquad (118)$$

so that

$$B = B_1 + B_2 \qquad (119)$$

The boundary conditions for O_2, CO_2, CO at the particle surface $\xi = 1$ are determined by stoichiometry as follows:

$$m_{1p} = -\frac{16}{12}\, f_{ce1}, \qquad m_{2p} = -\frac{44}{12}\, f_{ce2}, \qquad m_{3p} = \frac{28}{12}\, f_{ce1} + \frac{56}{12}\, f_{ce2} \qquad (120)$$

or using (113), (114), (118)

$$\left(BY_1 - \frac{dY_1}{d\xi} \right)_{\xi=1} = -\frac{4}{3}\, B_1 \qquad (121)$$

$$\left(BY_2 - \frac{dY_2}{d\xi} \right)_{\xi=1} = -\frac{11}{3}\, B_2 \qquad (122)$$

$$\left(BY_3 - \frac{dY_3}{d\xi} \right)_{\xi=1} = \frac{7}{3}\, B_1 + \frac{14}{3}\, B_2 \qquad (123)$$

Introducing (115) into (121)–(123) we find the surface mass fractions

$$Y_{1p} = e^{-B}Y_{1\infty} - \frac{4B_1}{3B} \tag{124}$$

$$Y_{2p} = e^{-B}Y_{2\infty} - \frac{11B_2}{3B} \tag{125}$$

$$Y_{3p} = e^{-B}Y_{3\infty} + \frac{7B_1 + 14B_2}{3B} \tag{126}$$

At this point, expressions for the reaction rates must be introduced. It is simplest to use first order kinetics, although other rate forms can be accommodated at the cost of somewhat more tedious expressions.

$$f_{ce1} = k_1\, e^{-E_1/RT_p}c_{1p} \tag{127}$$

$$f_{ce2} = k_2\, e^{-E_2/RT_p}c_{2p} \tag{128}$$

where c_{1p}, c_{2p} are the concentrations of O_2, CO_2 at the surface. Each of these concentrations is a function of all three mass fractions Y_{1p}, Y_{2p}, Y_{3p}; namely

$$c_{ip} = \frac{p}{RT_p}\,\frac{Y_{ip}/M_i}{\displaystyle\sum_{i=1}^{4} Y_{ip}/M_i} \tag{129}$$

where M_i are the corresponding molecular weights. Thus Eqs. (118), (119), (127)–(129) may be combined to yield

$$B_1 = G_1(B_1, B_2; T_p, Y_\infty) \tag{130}$$

$$B_2 = G_2(B_1, B_2; T_p, Y_\infty) \tag{131}$$

where G_1, G_2 are functions defined implicitly.

Equations (130), (131) can be solved numerically for B_1 and B_2. The problem can be simplified considerably by employing the approximation

$$\sum_{i=1}^{4} Y_{ip}/M_i \cong \sum_{i=1}^{4} Y_{i\infty}/M_i = \frac{1}{M_\infty} \tag{132}$$

so that B_1, B_2 are now given by

$$B_1 = A_1 Y_{1p}, \qquad B_2 = A_2 Y_{2p} \tag{133}$$

where A_1, A_2 are functions of T_p and a. Introducing (133) into (124) and (125) provides

$$Y_{1p} = \frac{Y_{1\infty} e^{-B}}{1 + \dfrac{A_1}{2B}(1 - e^{-B})}, \qquad Y_{2p} = \frac{Y_{2\infty} e^{-B}}{1 + \dfrac{A_2}{B}(1 - e^{-B})} \tag{134}$$

which is introduced back into (133), (119) to yield an equation in the single unknown B

$$\frac{A_1 Y_{1\infty} e^{-B}}{1 + \dfrac{A_1}{2B}(1 - e^{-B})} + \frac{A_2 Y_{2\infty} e^{-B}}{1 + \dfrac{A_2}{B}(1 - e^{-B})} = B \tag{135}$$

This equation is solved numerically for B to give the rate of carbon loss as a function of particle radius and particle temperature.

The Dynamic Response Problem

The mass-loss parameter B provides the rate of change of the particle radius

$$\frac{d}{dt}\left(\rho_c \frac{4\pi}{3} a^3\right) = -4\pi a^2 m_p \tag{136}$$

An additional equation is needed to obtain a determined problem in the two unknowns a, T_p. The energy equation will serve this role.
   An energy balance on the particle yields

$$\frac{d}{dt}\left(\frac{4\pi}{3} a^3 \rho_c \mathbf{U}_c\right) = -4\pi a^2 \left(\sum_{i=1}^{4} m_{ip} \mathbf{H}_{ip} + q_r + q_c\right) \tag{137}$$

where $\mathbf{U}_c$ is the specific internal energy, $\mathbf{H}_{ip}$ are the specific species enthalpies at the surface and q_r, q_c are radiative and conductive fluxes

$$q_r = \sigma \varepsilon'(T_p^4 - T_w^4) \tag{138}$$

$$q_c = -\lambda\left(\frac{dT}{dr}\right)_{r=a} \tag{139}$$

where T, T_p, T_w are the temperature of gas, particle, and wall.

The various terms in (137) must be expressed in terms of the basic unknowns a, T_p. We start with the derivative at the left side

$$\frac{d}{dt}\left(\frac{4\pi}{3}\,a^3\rho_c\mathbf{U}_c\right) = 4\pi a^2\,\frac{da}{dt}\,\rho_c\mathbf{U}_c + \frac{4\pi}{3}\,a^3\rho_c\,\frac{d\mathbf{U}_c}{dt} \qquad (140)$$

Internal energy and enthalpy are nearly equal for the solid. Thus assuming constant heat capacity and using the reference $\mathbf{H}_c(T_\infty) = 0$ we obtain

$$(\mathbf{U}_c)_p \cong (\mathbf{H}_c)_p = c_{pc}(T_p - T_\infty)$$

$$\frac{d\mathbf{U}}{dt} \cong \frac{d\mathbf{H}_c}{dt} = \frac{d\mathbf{H}_c}{dT}\frac{dT}{dt} = c_{pc}\frac{dT}{dt}$$

so that

$$\frac{d}{dt}\left(\frac{4\pi}{3}\,a^3\rho_c\mathbf{U}_c\right) = 4\pi a^2\left[-\rho_c c_{pc}(T_p - T_\infty)\,\frac{\eta}{a}\,B + \frac{a\rho_c c_{pc}}{3}\,\frac{dT}{dt}\right]$$

$$(141)$$

The sum of the first two terms at the right side of Eq. (137) can be evaluated by considering the energy equation in the gas phase:

$$\frac{d}{dr}\left[r^2\left(-\lambda\,\frac{dT}{dr} + \sum m_i\mathbf{H}_i\right)\right] = 0 \qquad (142)$$

Using Eq. (113) for the mass fluxes we have

$$\sum m_i\mathbf{H}_i = \rho v \sum Y_i\mathbf{H}_i - \eta \sum \mathbf{H}_i\,\frac{dY_i}{dr} \qquad (143)$$

On the other hand

$$\frac{d}{dr}\sum Y_i\mathbf{H}_i = \sum \mathbf{H}_i\,\frac{dY_i}{dr} + \sum Y_i\,\frac{d\mathbf{H}_i}{dr} \qquad (144)$$

The expressions are simplified by assuming that the heat capacities of all species are equal to each other and temperature independent, so that

$$\frac{d\mathbf{H}_i}{dr} = \frac{d\mathbf{H}_i}{dT}\frac{dT}{dr} = c_p\frac{dT}{dr} \qquad (145)$$

Combining (143)–(145) and using the assumption of unity Lewis number, $\lambda = \rho c_p \mathcal{D} = \eta c_p$, we have

$$-\lambda \frac{dT}{dr} + \sum m_i \mathbf{H}_i = \rho v \mathbf{H} - \eta \frac{d\mathbf{H}}{dr} \tag{146}$$

where

$$\mathbf{H} = \sum Y_i \mathbf{H}_i \tag{147}$$

Thus the energy Eq. (142) becomes

$$\frac{d}{dr} \left[r^2 \left(\rho v \mathbf{H} - \eta \frac{d\mathbf{H}}{dr} \right) \right] = 0 \tag{148}$$

which by analogy to (110)–(115) accepts the solution

$$\mathbf{H} = \frac{(\mathbf{H}_\infty - \mathbf{H}_p) e^{-B/\xi} + (\mathbf{H}_p - \mathbf{H}_\infty e^{-B})}{1 - e^{-B}} \tag{149}$$

At the particle surface we have

$$\left(-\lambda \frac{dT}{dr} + \sum m_i \mathbf{H}_i \right)_p = m_p \mathbf{H}_p - \eta \left(\frac{d\mathbf{H}}{dr} \right)_p = \frac{\eta B (\mathbf{H}_p - e^{-B} \mathbf{H}_\infty)}{a(1 - e^{-B})} \tag{150}$$

Introducing (141) and (150) into (137) we finally obtain

$$\frac{dT}{dt} = \frac{3\eta B}{a^2} (T_p - T_\infty) - \frac{3\eta B}{\rho_c c_{pc} a^2} \frac{\mathbf{H}_p - e^{-B} \mathbf{H}_\infty}{1 - e^{-B}} - \frac{3\sigma \varepsilon'}{\rho_c c_{pc} a} (T_p^4 - T_w^4) \tag{151}$$

This equation for the particle temperature must be solved simultaneously with Eq. (136), keeping in mind that $\mathbf{H}_p$ is a function of T_p, Y_{ip}, therefore, a function of T_p, a. Likewise, B is a function of T_p, a.

The paper of Libby and Blake [10] must be consulted for further details and numerical results. Figure 14 reproduces a sample of their calculations showing the evolution of particle temperature and radius for hot particles injected in a cold atmosphere, in terms of the dimensionless time $\tau = \eta t / \rho_c a_0^2$. The figure shows that when the

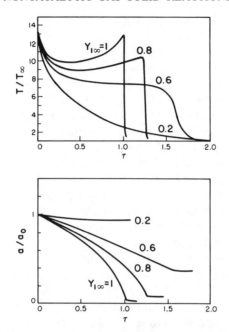

Figure 14 Particle temperature and radius vs. time for a hot particle in a cold oxidizing atmosphere; $a_0 = 50$ μm, $p = 1$ atm (from Libby and Blake [19]).

oxygen mass fraction is less than 0.6, the reaction is quenched at some intermediate level of conversion. With pure oxygen, conversion proceeds essentially to completion. Much of the interest in such theoretical analyses of particle combustion is to delineate the phenomena of ignition and quenching, or extinction, and to predict the time required for complete conversion as functions of operating parameters such as particle size, oxygen concentration and ambient temperature.

NOTATION

A particle radius (grain model)

A_1, A_2 functions by Eq. (133)

a particle or grain radius

a_0 initial particle or grain radius

B_0, B_1, B_2 moments of capillary size distribution

B, B_1, B_2 dimensionless mass loss parameters

b, b' stoichiometric coefficients
c concentration
c_p heat capacity
F, F_1, F_2 denote functions
f reaction rate
f_i, f_{ci} intrinsic surface reaction rate
f reaction rate per unit external particle area
$\mathbf{H}$ specific enthalpy for mixture
$\mathbf{H}_i$ specific enthalpy of species i
I defined by Eqs. (101) or (106)
J defined by Eq. (103)
k_i, k_1, k_2 rate constants for surface reactions
k_m film mass transfer coefficient
M_i molecular weight of species i
m_i mass flux of species i
m_p mass flux at particle surface
q dimensionless reaction parameter, Eq. (27)
$q, q(t)$ reaction distance, pp. 158
q_c, q_r conductive and radiative heat fluxes
R capillary radius
R_*, R^* minimum and maximum capillary radius
r radial position
r_2 radius of unreacted core
S pore surface area per unit volume
T temperature
T_w wall temperature
t time
t_c characteristic time
t_f time for complete conversion
$t_{0.5}$ time for fraction conversion equal to 0.5
$\mathbf{U}_c$ specific internal energy of carbonaceous particle
V molar volume
v velocity of receding particle boundary $(= dx^*/dt)$
v gas velocity, pp. 169
X conversion
x dimensionless radial position, pp. 144–145; position, p. 164
x_2 dimensionless radius of unreacted core
y radial position in a particle
Y_i mass fraction of species i
z distance from receding particle surface

Greek Symbols

α dimensionless particle radius

β, β' dimensionless mass transfer parameters, Eqs. (28), (49)

γ volume ratio of reactant and product solids

$\mathscr{D}$ diffusion coefficient

$\mathscr{D}_b$ bulk diffusion coefficient

$\mathscr{D}_k$ Knudsen diffusion coefficient

$\mathscr{D}_e$ effective diffusion coefficient

$\mathscr{D}_{e0}$ effective diffusion coefficient at initial porosity

δ_e dimensionless effective diffusion coefficient

ε pore volume fraction

ε_0 initial pore volume fraction

ε' emissivity

ζ tortuosity factor

$\xi = \dfrac{y}{A}$ dimensionless radial position

λ probability density parameter in capillary model

λ thermal conductivity

$\lambda(R), \lambda_0(R)$ probability density of capillary radius

ν stoichiometric coefficient

η defined by Eq. (114)

ρ gas density

ρ_c density of char particle

σ Stefan–Boltzmann constant

τ dimensionless time

τ_f dimensionless time for complete conversion

ψ dimensionless concentration of reactant gas

Subscripts

A refers to gaseous reactant

c refers to properties of carbon particle

p refers to surface of carbon particle

S refers to solid material

∞ refers to free stream conditions

Subscripts

$*$ refers to conditions of pore collapse

REFERENCES

1. Alfano, O. M., and Arce, P. E. (1982). Non-catalytic gas–solid reactions and their application to combustion and gasification. A review, *Lat. Am. J. Heat Mass Transf.*, **6**, 191.
2. Attar, A., and Shaly, W. (1982). Reduction of iron pyrite with molecular hydrogen—The initial stages, presented at the 75th Annual A.I.Ch.E. Meeting in Los Angeles, Nov. 13–17.
3. Barin, I., and Knacke, O. (1974). *Thermochemical Properties of Inorganic Substances*, Springer-Verlag, Berlin.
4. Bhatia, S. K., and Perlmutter, D. D. (1983). Unified treatment of structural effects in fluid solid reactions, *A.I.Ch.E. J.*, **29**, 281.
5. Froment, G. F., and Bischoff, K. B. (1979). *Chemical Reactor Analysis and Design*, Wiley, New York.
6. Gavalas, G. R. (1980). A random capillary model with application to char gasification at chemically controlled rates, *A.I.Ch.E. J.*, **26**, 577.
7. Gavalas, G. R. (1981). Analysis of char combustion including the effect of pore enlargement, *Comb. Sci. Tech.*, **24**, 197.
8. Georgakis, C., Chang, C. W., and Szekely, J. (1979). A changing grain size model for gas–solid reactions, *Chem. Eng. Sci.*, **34**, 1072.
9. JANAF Thermochemical Tables, NSRDS-NBS 37, 2nd ed. (1971).
10. Libby, P. A., and Blake, T. R. (1979). Theoretical study of burning carbon particles, *Comb. Flame*, **36**, 139.
11. Rao, Y. K., (1974). A physico-chemical model for reactions between particulate solids occurring through gaseous intermediates—I. Reduction of hematite by carbon, *Chem. Eng. Sci.*, **9**, 1435.
12. Rao, Y. K., and Chuang, Y. K. (1974). A physico-chemical model for reactions between particulate solids occurring through gaseous intermediates—II. General solutions, *Chem. Eng. Sci.*, **9**, 1933.
13. Sohn, H. Y., and Szekely, J. (1972). A structural model for gas–solid reactions with a moving boundary, *Chem. Eng. Sci.*, **27**, 763.
14. Sotirchos, S. V., and Amundson, N. R. (1983). Dynamic behavior of a porous char particle burning in an oxygen containing environment, Parts 1, 2. *A.I.Ch.E. J.*, **30**, 537.
15. Sundaresan, S. and Amundson, N. R. (1980). Diffusion and reaction in a stagnant boundary layer about a carbon particle, *I&EC Fundamentals*, **19**, 344.
16. Tseng, H. P., and Edgar, T. F. (1984). Identification of the combustion behavior of lignite char between 35– and 900°C, *Fuel*, **63**, 385.

Chapter 4

Fixed Bed Reactor Design

GILBERT F. FROMENT

Laboratorium voor Petrochemische Techniek, Rijksuniversiteit, Gent, Belgium

Abstract

This article provides a methodical approach to fixed bed, catalytic reactor design. It treats a series of increasingly complex models and carefully identifies the limitations of each one. The concept of a close connection between mathematical models and experimental results is stressed throughout, and the dominant characteristics of reactors of industrial importance are discussed. This article provides both a guide to the designer in terms of current methodology and a guide to the researchers in terms of a series of challenging unresolved problems.

I. INTRODUCTION

A successfully designed reactor achieves a given conversion with optimal selectivity by operating in a certain temperature and pressure range, with minimum hazard and downtime and maximum profit. This review deals with a modern approach to this objective. It does not consider the mechanical aspects involved in the design but concentrates instead on the sizing and on the selection of optimal operating conditions. This is generally done by means of information—hopefully quantitative—obtained at a smaller scale in a reactor which is not necessarily conformal, neither in geometry nor in operating principle. Before the definitive design can be undertaken, several major issues have to be settled. Decisions concerning the type of reactor and its configuration or about the desirability of recycling have to be made at an early stage, to avoid getting involved into costly detailed studies of inadequate diversions. Only after the number of alternatives has been reduced to

179

two or three will the definitive design be tackled, making use of all available quantitative information.

A number of qualitative design considerations will be briefly mentioned first. They mainly aim at shaping the reactor and defining its mode of operation.

II. QUALITATIVE DESIGN CONSIDERATIONS

For its simplicity of construction and operation, an adiabatic reactor is the first possibility to be considered. When the heat of reaction is pronounced, a single adiabatic bed is not always possible: with endothermic reactions the temperature drop may extinguish the reaction and with exothermic reactions the temperature rise may be detrimental to the selectivity or deactivate the catalyst, either directly through sintering or indirectly through coke formation. Staging is required, to limit the ΔT per bed. With endothermic processes like catalytic reforming intermediate heating is provided by furnaces. With exothermic processes the intermediate cooling is achieved by means of multitubular heat exchangers, internal or external, or by injection of cold reactant or by a combination of both means.

With exothermic equilibrium reactions like SO_2-oxidation, water–gas shift, ammonia- and methanol synthesis, high temperatures do not necessarily favor the conversion, because of equilibrium limitations. To achieve high conversions with a reasonable amount of catalyst requires a progressive lowering of the temperature as conversion increases. The $x–T$ relation which is optimal in this respect can be rigorously determined, but this is part of the quantitative design stage already. At the stage dealt with here it can only be said that a reactor with continuous heat exchange is more appropriate for reaching this objective than the multibed adiabatic reactor with its saw-tooth $x–T$ trajectory. Two common types of reactors with continuous heat exchange are: the multitubular reactor, in which the catalyst is packed inside rather narrow tubes and the fixed bed reactor with an internal heat exchanger whose tubes are distributed over the packed bed of catalyst. With strongly exothermic reactions the multitubar reactor is preferred, to avoid runaway, which would lead to a near-adiabatic temperature rise. Special heat transfer fluids may be necessary. In less demanding cases, cooling of the reaction mixture by means of the feed may be adequate. When the heat of reaction suffices to preheat the feed to the desired inlet temperature the operation is called autothermal. Many important industrial processes are carried out in this way. With endothermic reactions, like steam reforming for hydrogen or synthesis gas production, the reaction mixture has to be heated as fast as possible to the highest allowable temperature level to maximize the production. Steam reforming tubes are placed inside a furnace.

Another concern is the limitation of the pressure drop over the reactor, particularly when a stream has to be recycled and compression is an important cost item. For a given flow rate, catalyst volume and dimensions, lowering the pressure drop implies reducing the bed depth. With axial flow this would lead to excessive bed diameters. To avoid this, radial flow reactors were introduced, e.g. in catalytic reforming and ammonia-synthesis. Care should be taken to have a bed depth sufficient to avoid bypassing of the catalyst, however. Increasing the particle size also reduces the pressure drop, of course, but diffusional limitations inside the particle may then develop, while a high ratio d_p/d_t favors the bypassing of the fluid along the wall.

III. VARIOUS APPROACHES TO QUANTITATIVE DESIGN

Once the reactor type and its mode of operation have been selected, the sizing of the reactor and the optimization of the operating conditions can be undertaken.

The design of a commercial-size reactor is generally based upon information obtained on smaller scale equipment. There are various ways that can be followed in this scaling up.

A. Scaling up by Similarity

Complete similarity has frequently been shown to be impossible between two non-isothermal and non-isobaric reactors. Himmelblau and Bischoff [49] provide an example that strikingly illustrates the unrealistic conditions which would have to be satisfied to achieve this. The scaling up will, therefore, always involve some compromise and some uncertainties. Certain cases are more favorable than others: operating a single tube pilot plant to test the scaling up of a multitubular reactor is a feasible task, although it does not solve the important problem of equalizing the heat transfer over all the tubes of the bundle. On the other hand, true adiabatic operation is practically impossible in small pilots.

B. Progressive Extrapolation

The awareness that complete similarity is impossible has frequently led to development procedures involving several steps, with progressively increasing size of the equipment.

It is general practice to proceed from the bench scale to the commercial scale in a gradual way, involving at least two intermediate scales: the pilot and the demonstration unit. It should be borne in mind that even large pilots cannot be expected to behave exactly like the ultimate commercial reactor. The pilot is no guarantee for the successful scale-up if it is not used for producing more fundamental information about the process. The pilot may not even be the most adequate type of equipment for the generation of such information. It is unlikely that a single equipment would be adequate for studying widely different phenomena like chemical kinetics, heat transfer, diffusional limitations and dispersion, among others. Useful additional information on hydrodynamics may be collected in mock up.

C. The Fundamental Modeling Approach

It is not an exaggeration to say that to date no reactor is being developed any more without a certain degree of modeling. What the present text is trying to advocate is an approach which dissects the process into its basic phenomena, considers each of them separately by means of literature information or appropriate experimentation

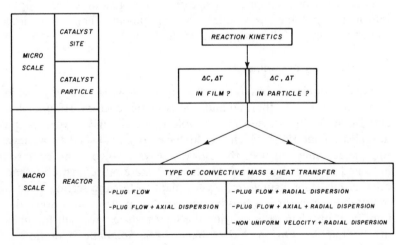

Figure 1   Aspects to be considered in the modeling of a fixed bed catalytic reactor.

and reconstructs the mosaic, accounting for the interaction between the various phenomena. In designing or simulating a catalytic reactor a number of problems have to be addressed first. These problems and their interrelation are shown in Fig. 1. In the following sections the aspects given in Fig. 1 will be discussed in detail and quantitatively expressed.

IV. ASPECTS INVOLVED IN THE FUNDAMENTAL MODELING

A. Micro-Scale Design Aspects at the Active Site Level. Chemical Kinetics

Reliable reactor design is based in the first place on accurate kinetic equations. Too often these equations are simplified to such an extent that they no longer reflect essential features of the reaction and are only valid for extremely narrow operating conditions. Frequently first order kinetics are used, purely for reasons of mathematical convenience. When the interaction with the catalyst is explicitly accounted for, Hougen–Watson-type rate equations are obtained. These typically contain denominators in which terms related to the various adsorbed species appear. If there is only one rate determining step per overall reaction, as is frequently observed for well balanced catalysts, the complexity of the equations is reasonably limited. But since each reaction of a complex network may have a rate determining step, a large set of "rival" models results. Accurate experimental results, covering a wide range of operating variables and powerful optimization routines are required to discriminate between these [32, 33].

With complex reactions an overall rate equation for the disappearance of the feed yields insufficient information. In general the selectivity with respect to a certain product is at least as important as the overall converison. What is required to investigate the effect of operating variables on this selectivity is a set of rate equations, describing the formation of at least the major reaction products and the formation and disappearance of intermediates. This set can only be derived when the structure of the reaction network itself is known. The elucidation of the latter requires a detailed identification of the reaction products and intermediates, backed by physicochemical insight in possible reaction mechanisms and the role of the catalyst [71]. This is by no means an easy task and the depth to which it has to be carried through should be continuously confronted with its pay-off. No general answer can be given to this question, but a sensitivity analysis of the process and its objectives with respect to the accuracy of the kinetics provides important indications.

Selectivity vs conversion plots provide useful tools for getting insight into a reaction network: primary products exhibit a finite initial selectivity, whereas secondary products show a zero initial selectivity. It is of importance also to determine the number of independent reactions. The rank of the matrix of stoichiometric coefficients is a widely used tool for this. Sequential design of experiments for optimal discrimination and optimal parameter estimation have been shown to improve the efficiency of the task described above, although there are practically no applications so far to multiresponse problems, which is an area in which one could expect them to contribute most [39]. Generally, lumping is necessary to limit the complexity. Several criteria for rigorous lumping have been proposed; but more often common sense has guided the investigator in this matter. A well documented example is the catalytic cracking of gas oil, presented by Weekman [86] and Jacob· et al. [50]. Other examples were dealt with by Marin and Froment in catalytic reforming [62] and Steyns and Froment in hydrocracking [73].

An additional complication is the deactivation of the catalyst by poisoning, coking or sintering. Too often the activity has been correlated with respect to time, instead of the true deactivating agent like coke or poisons. Reviews on deactivation by poisoning were presented by Hegedus and McCabe [48] and by Butt [10, 11], on deactivation by coking by Froment [34, 35, 37] and on sintering by Delmon and Grange [16].

B.   Micro Scale Design Aspects at the Particle Level

B.1.   Interfacial Gradients
When the rate of mass and heat transfer from the bulk fluid to the catalyst particle is intrinsically lower than that of the chemical reaction, concentration and temperature gradients may develop around the particle. The flux of heat and mass from the bulk fluid at (C, T) to a particle in the catalyst bed may be written, for a single reaction with key component A:

$$k_g(C - C_s^s) = \frac{d_p}{6} r_A(C_s^s, T_s^s)\rho_s \qquad \text{(B.1-1)}$$

$$h_f(T_s^s - T) = \frac{d_p}{6}(-\Delta H)r_A(C_s^s, T_s^s)\rho_s \qquad \text{(B.1-2)}$$

Many correlations for k_g and h_f have been published. They were recently reviewed by Schlünder [69]. Frequently used correlations are:

for k_g [67]:

$$\epsilon \, Sh = 0.357 \, Re^{0.641} \, Sc^{0.33} \qquad 3 \leqslant Re \leqslant 200 \qquad \text{(B.1-3)}$$

for h_f (Whitaker [83]):

$$\epsilon \, Nu = (1 - \epsilon)\left[0.5\left(\frac{Re}{1 - \epsilon}\right)^{0.5} + 0.2\left(\frac{Re}{1 - \epsilon}\right)^{0.66}\right]Pr^{0.33} \qquad \text{(B.1-4)}$$

More recent work of Balakrishnan and Pei [4] dissects h_f into various contributions. Mears [63, 64] proposed the following criteria, to be satisfied when the rate of the global phenomenon is to deviate less than 5% from the intrinsic rate of the reaction, considered to be irreversible and of order n:

when only interfacial concentration gradients occur:

$$\frac{r_A \rho_s d_p}{2 k_g C} < \frac{0.15}{n} \qquad \text{(B.1-5)}$$

when only interfacial temperature gradients occur:

$$\frac{(-\Delta H)r_A \rho_s d_p}{2 h_f T} < 0.15 \frac{R'T}{E} \qquad \text{(B.1-6)}$$

The most likely is the temperature gradient.

Interfacial gradients may cause, theoretically at least, multiplicity of steady states. A priori criteria for testing multiplicity arising from interfacial gradients were reviewed by Luss [59]. For exothermic reactions multiplicity is guaranteed for given bulk conditions (C, T) when:

$$\frac{\beta_f \gamma}{1 + \beta_f} \leqslant f(\beta_f, n) \qquad \text{(B.1-7)}$$

For orders $0 \leqslant n \leqslant 2$ the function $f(\beta_f, n)$ is bounded between 1 (for

$n = 0$ and $\beta_f = 0$) and 5.82 (for $n = 2$ and $\beta_f = 0$). For large β_f the function value asymptotically approaches 4, for all n. For $n = 1$ the function $f(\beta_f, n)$ equals 4 for all β_f. When $\beta_f \gamma / 1 + \beta_f$ exceeds the value of $f(\beta_f, n)$, multiplicity is possible for some values of the dimensionless group [78]

$$\alpha = \frac{V_p k(T) \rho_s C^{n-1}}{S_p k_g \beta_f^{n-1}}.$$

Recently, Tsotsis et al. [76] presented criteria for Hougen–Watson-type kinetics. With such rate equations multiplicity would even become possible for isothermal situations ($\beta_f = 0$) and with endothermic reactions.

B.2. Interfacial and Intraparticle Gradients
The particle is generally considered as a continuum so that heat and mass transfer through it are described in terms of Fick- and Fourrier-type laws. For a particle surrounded by bulk fluid at (C, T):

$$\frac{D_e}{\xi^2}\frac{d}{d\xi}\left(\xi^2 \frac{dC_s}{d\xi}\right) - r_A(C_s, T_s)\rho_s = 0 \qquad \text{(B.2-1)}$$

$$\frac{\lambda_e}{\xi^2}\frac{d}{d\xi}\left(\xi^2 \frac{dT_s}{d\xi}\right) + r_A(C_s, T_s)\rho_s(-\Delta H) = 0 \qquad \text{(B.2-2)}$$

with boundary conditions:

$$\text{at } \xi = 0 \qquad \frac{dC_s}{d\xi} = \frac{dT_s}{d\xi} = 0$$

$$\xi = \frac{d_p}{2} \qquad k_g(C_s^s - C) = -D_e \frac{dC_s}{d\xi} \qquad \text{(B.2-3)}$$

$$h_f(T_s^s - T) = -\lambda_e \frac{dT_s}{d\xi}$$

The structure of the network of pores inside the particle only weakly enters into the model through the effective diffusivity, D_e. For the case that only molecular diffusivity has to be considered e.g.

$$D_e = \frac{\epsilon_s}{\tau} D_m \qquad \text{(B.2-4)}$$

The tortuosity τ enters into this formula because the pores are not oriented along the normal from the surface to the center. The effective diffusivity, D_e, can be measured by means of the steady state Wicke–Kallenbach technique [84] or by the transient gas chromatographic technique that also accounts for the effect of the dead end pores [77]. Pore structure models have been presented for calculating D_e or τ [81, 24]. The recently proposed stochastic model of Beeckman and Froment [5] leads to a value of $\tau = 4$ for a chromia–alumina catalyst used in butene dehydrogenation, close to the experimental value of 5. No such models are available for λ_e, except for that of Butt [9], based upon the simple pore model of Wakao and Smith [81], but temperature gradients inside the particle are generally negligible for catalytic reactions. When only interfacial temperature and intraparticle concentration gradients occur, only (B.2-1) has to be considered, of course, with the boundary conditions:

$$\text{at } \xi = 0 \qquad \frac{dC_s}{d\xi} = 0$$

$$\text{at } \xi = \frac{d_p}{2} \qquad -D_e \frac{dC}{d\xi} = r_A(C_s^s, T_s^s)\rho_s \qquad \text{(B.2-5)}$$

$$h_f(T_s^s - T) = -\lambda_e \frac{dT_s}{d\xi}$$

For a first order reaction inside an isothermal particle an analytical solution is possible. It can be cast in the following form:

$$\eta = \frac{1}{\phi} \frac{(3\phi)\coth(3\phi) - 1}{3\phi} \qquad \text{(B.2-6)}$$

where η, the effectiveness factor is the ratio of the actual rate to that which would be experienced in the absence of diffusional limitations. In this expression for the effectiveness factor

$$\phi = \frac{V_p}{S_p}\sqrt{\frac{k\rho_s}{D_e}}$$

is the Thiele modulus, which is valid with sufficient accuracy for any shape of the particle. It varies with temperature. In the above treatment r_A is evaluated at surface conditions (C_s^s, T_s^s). Sometimes it is referred to bulk conditions (C, T) to include the external concentration gradient into the η-value. The analytical expression for the isothermal effectiveness factor for first order reactions becomes:

$$\eta_G = \frac{3\,\text{Sh}}{\phi^2}\left[\frac{\phi\cosh\phi - \sinh\phi}{\phi\cosh\phi + (\text{Sh} - 1)\sinh\phi}\right] \qquad \text{(B.2-7)}$$

where $\text{Sh} = (k_g\delta/D_e)$ and δ is the film thickness.

For orders different from one, or Hougen–Watson-type rate equations, η and η_G cannot be obtained in an analytical way. This is why approximate solutions for the relation η vs ϕ have been worked out [3, 7, 66], based upon the introduction of a generalized modulus:

$$\phi = \frac{V_p r_A(C_s^s, T_s^s)\rho_s}{\sqrt{2}S_p}\left[\int_{C_{s_0}}^{C_s^s} D_e r_A(C_s^s, T_s^s)\rho_s\, dC\right]^{1/2} \qquad \text{(B.2-8)}$$

This modulus contains both the temperature and the concentration so that η varies through the reactor, even when it is isothermal. The use of the generalized modulus is limited to a single reaction. For complex reactions (B.2-1) has to be numerically integrated. More empirical relations for η vs ϕ were derived by Gottifredi et al. [41] for an isothermal particle and any power law kinetic expression with n positive, but also for Hougen–Watson expressions. The relation is written:

$$\eta = a\frac{\sqrt{r + \phi^2}}{s + \phi^2}$$

with

$$\phi = \frac{V_p}{S_p}\sqrt{\frac{k\rho_s}{D_e}(C_s^s)^{m-1}}$$

while a, s, and r are specific for the kinetic equation, but always algebraically related to its parameters. Wedel and Luss [87] presented an approximate solution for any power law and simple Hougen–Watson rate equation, valid also for non-isothermal situations, but it may involve the numerical computation of two integrals.

Again a number of a priori criteria are available for evaluating, on the basis of observables, whether or not it is necessary to account for intraparticle gradients.

The Weisz–Prater criterion [82] is well known. In order to have a value of $\eta > 0.95$, so that intraparticle concentration gradients can be neglected in an isothermal particle, the following inequality has to be satisfied:

$$\frac{r_A \rho_s d_p^2}{4 D_e C_s^s} < \begin{array}{l} 6.0 \quad \text{for zero order reactions} \\ 0.6 \quad \text{for first order} \\ 0.3 \quad \text{for second order} \end{array} \qquad \text{(B.2-9)}$$

in which r_A is the observed rate.

For the absence of combined intraparticle and interfacial gradients ($\eta = 1 \pm 0.05$) Mears [63, 64] proposed:

$$\frac{r_A \rho_s d_p^2}{4 D_e C} < \frac{1 + 0.33 \dfrac{\gamma(-\Delta H) r_A \rho_s d_p}{2 h_f T}}{|n - \beta_p \gamma| \left(1 + 0.33 n \dfrac{r_A \rho_s d_p}{2 k_g C}\right)} \qquad \text{(B.2-10)}$$

with β_p and γ evaluated at bulk gas conditions, C and T.

Recently Madon and Boudart [60] further developed the Koros–Novak experimental criterion [56] for testing the absence of internal and external gradients.

For a certain range of parameter values, more than one steady state profile would be possible for given geometry and operating conditions. Van den Bosch and Luss [78] derived the following criteria for uniqueness when a first order irreversible reaction occurs in a particle with concentration and temperature gradients or in a particle having internal concentration gradients and interfacial concentration and temperature gradients:

$$\frac{\beta \gamma}{1 + \beta} < 4 \qquad \text{(B.2-11)}$$

where $\beta = \beta_p$ in the first case and $\beta = \epsilon_s \beta_f$ in the second case. When this condition is violated multiplicity may exist for a certain range of values of ϕ in the first case and of the Sherwood number $\epsilon_s = (k_g/D_e)(V_p/S_p)$ in the second case. Compared with the situation whereby only interfacial gradients occur, the presence of internal concentration gradients decreases the range in parameter space for which multiplicity is possible.

Recently, increased attention has been given to more detailed models of catalyst particles. Jensen and Ray [51] developed a model for supported metal catalysts in which the behavior of the individual crystallites is described by separate continuity and energy equations. These are—be it weakly—coupled to the particle equations, which are still of the continuum type. This "pebbly surface model" generates, for appropriate conditions, multiple steady states and oscillatory behavior, even at large Lewis numbers. It shows the same trends as the experimental data of Keil and Wicke [52] on CO-oxidation on Pt/alumina: decreasing amplitudes and periods of oscillations as the gas temperature increases, in contrast with continuum model simulation. At the higher temperature of 246°C the amplitudes of the conversion and temperature oscillations is less than 0.2% and 0.3°C respectively, so that the question arises whether such a model is required for simulating normal operating practice, except when informaton is required as to the true crystallite temperature, e.g. in sintering studies.

Beeckman and Froment [5] modeled the catalyst particle in a truly heterogeneous way: the pore network structure was carried along up to the final equations for the fluxes. Clearly, the equations are much more complicated than those of the continuum model. Such a heterogeneous model will be a valuable tool for situations in which the prediction of the surface flux yields insufficient information. This is the case when site coverage and pore blockage by coke and metals deposition lead to catalyst deactivation, as in resid hydrodesulfurization e.g. With the multilayered catalysts presently used in automobile mufflers, local information on the interaction between the pore structure, reaction and transport phenomena is also of importance.

C.   Macro-Scale Design Aspects. Reactor Level

C.1.   Heat and Mass Transfer in Packed Beds

In reactor design heat and mass are generally considered to be transported by plug flow only. It is thereby assumed that there are no gradients of concentration in a cross-section and that, in non-adiabatic reactors, the temperature gradient is localized in a thin film near the wall. The heat transfer from the bed to the wall is characterized by a coefficient, α_i, which is correlated as if it were purely convective.

Deviations from plug flow are caused by molecular and turbulent diffusion in the fluid phase—usually of little influence in industrial conditions—and by the presence of the packing itself. The latter generates a mass transfer whose flux is described in terms of a Fickian type equation, in analogy with molecular and turbulent diffusion. The three mechanisms are then lumped and the proportionality factor between the flux and the gradient is called the effective diffusivity or dispersion coefficient. The flux resulting from this mechanism is superimposed upon that caused by plug flow. Because of the forced flow direction and the radial non-homogeneity of the packing, the bed is not isotropic for the dispersion, so that a distinction is made between axial and radial components. Both the axial and the radial effective diffusivity are correlated in the form: Peclet- vs Reynolds-number. The Peclet number for effective axial diffusion, Pe_{ma}, is of the order of 1 to 2 in the turbulent flow regime [38]. The Peclet number for effective radial diffusion is of the order of 8 to 10. The ratio of particle to tube diameter also enters into this correlation, as exemplified by the Fahien and Smith equation [23]. The effective diffusion of mass evidently leads to a supplementary type of convective heat transfer. But still other mechanisms contribute to heat transfer, like conduction and radiation in both the fluid and the solid. Any mechanism distinct from the convection by plug flow is lumped into an effective conduction, whose flux is superimposed upon that caused by plug flow. The advantage of this lumping is clear: it avoids the need for solving a number of equations, some highly non-linear, describing all the different heat transfer contributions in each increment used in the reactor integration. The above approach simply leads to additional conduction-like terms in the energy equation, reflecting axial and radial components of this flux and characterized by effective conductivities, λ_{ea} and λ_{er}. The radial non-homogeneity of the void fraction and the flow conditions close to the wall, which differ from those in the core and which are insufficiently known to date, have necessitated the introduction of a second parameter characterizing heat transfer in radial direction: the wall heat transfer coefficient, α_w—not to be confused with α_i mentioned above, which is essentially different. Experimental correlations for λ_{ea}, λ_{er} and α_w versus Reynolds are available [38, 53]. Also, models have been developed for these parameters [54, 88, 90, 91]. They contain static and dynamic contributions, the former independent of flow, the latter arising from the dispersion. This concept of heat and mass transfer in packed beds, with its lumping of the non-plug flow mechanisms into effective diffusion and conduction, leads to so-called pseudo-homogeneous models, which do not, explicitly at least, distinguish between gas. and solid phase conditions. Under reacting conditions in particular, gradients of concentration and/of temperature may occur around and inside the catalyst particles, as discussed under B.1 and 2. Accounting for these gradients leads to a category of "heterogeneous" models. If the effective conduction concept is to be used consistently in these models, the effective diffusivity also has to distinguish between contributions in respectively the gas and the solid phase.

The mathematical description of heat and mass transfer according to the views presented is in terms of differential equations and yields so called continuum models. The continuity and energy equations cannot be solved analytically, except in unrealistic cases. Their numerical integration is not always easy, particularly for two-point boundary value problems. For mathematical convenience, cell models were introduced [2, 14, 65, 1]. These consider the bed to consist of a one- or two-dimensional array of cells in which the gas is completely mixed. The cell-dimensions have to be chosen to match observations on heat and mass transfer, which leads to a depth in the axial direction of one particle diameter and a width in the radial direction of approximately $0.8 d_p$. In the steady state the model leads to a set of algebraic equations with initial-value boundary conditions only. Most of the models used so far considered heat transfer to occur through the gas phase only, a simplification that may lead to important errors [29]. Kunii and Furusawa [55] developed an elaborate model including the various mechanisms of heat transfer which, in the continuum model, are lumped into the effective conductivity. The cell contains the front half of a particle on a row i, the back half of a particle on row $i + 1$ and voids in between and next to these particles. In a recent paper Sundaresan et al. [74] introduced a time lag to avoid the infinite speed of propagation associated with a sequence of completely mixed cells.

It may be hard to justify the application of a continuum model to reactors with ratios d_t/d_p of the order of 5, sometimes encountered in highly exothermic processes. At this time it is an illusion also to believe that our present day cell models are the answer to this problem. What is required in the first place is more information on the hydrodynamics of such cases. Meanwhile, the models which will be given here are of the continuum type.

C.2. Pressure Drop in Packed Beds
The pressure drop equation does not explicitly account for the real flow pattern and is usually written:

$$- \frac{dp_t}{dz} = \frac{f \rho_g u_s^2}{d_p} \qquad \text{(C.2-1)}$$

where f is an Ergun-type friction factor, related to the void fraction

Table 1 Classification of Models for Fixed Bed Tubular Reactors.

	Pseudo-homogeneous $T = T_s; C = C_s$		Heterogeneous $T \neq T_s; C \neq C_s$	
One-dimensional	A.I	Basic, ideal	B.I	+ interfacial gradients
	A.II	+ axial dispersion	B.II	+ intraparticle gradients
Two-dimensional	A.III	+ radial dispersion	B.III	+ radial dispersion

and the Reynolds number in the following way:

$$f = \frac{1 - \epsilon}{\epsilon^3}\left(a + b\frac{1 - \epsilon}{Re}\right)$$

Values for a reported in the literature are: 1.75 [22], $(Re/1 - \epsilon) <$ 500; 1.24 [43], $1000 < (Re/1 - \epsilon) < 5000$; 1.75 [75], $0.1 < (Re/1 - \epsilon) < 10.000$ and for b: 150, 368, and $4.2\,Re^{5/6}$ respectively.

The Reynolds number in these expressions contains a shape factor, to generalize the results. For hollow rings, however, specific correlations have to be used, e.g. those published by Brauer [8] and by Reichelt and Blasz [68].

C.3. Model Equations
The aspects mentioned in Fig. 1 and discussed under C.1 led to the following classification, introduced by Froment [29, 31] and widely accepted since.

In the following paragraph the models given in Table 1 will be expressed in mathematical terms, their features discussed and their application illustrated. The model equations are written for the steady state, for a single reaction and for constant density, so as not to obscure their main features.

The Basic Pseudo-Homogeneous Model (A.I.)

$$u_s\frac{dC}{dz} + r_A\rho_B = 0 \tag{C.3-1}$$

$$u_s\rho_g c_p\frac{dT}{dz} + 4\frac{\alpha_i}{d_t}(T - T_w) - (-\Delta H)r_A\rho_B = 0 \tag{C.3-2}$$

$$-\frac{dp_t}{dz} = \frac{f\rho_g u_s^2}{d_p} \tag{C.3-3}$$

with initial conditions:

at $z = 0$ $C = C_0$, $T = T_0$,   and $p_t = p_{t_0}$

In (C.3-2) T_w is the internal wall temperature and α_i the heat transfer coefficient between the bed and the internal wall. Several purely convective type correlations have been proposed for α_i e.g. by Leva [58], Verschoor and Schuit [79], Maeda [61]. De Wasch and Froment [20], on the other hand, also accounted for a contribution by conduction, so that:

$$\frac{\alpha_i d_p}{\lambda_g} = \frac{\alpha_i^0 d_p}{\lambda_g} + 0.033\left(\frac{c_p \mu}{\lambda_g}\right)\left(\frac{d_p G}{\mu}\right) \qquad 30 < \mathrm{Re} < 1000$$

$$(\text{C.3-4})$$

where

$$\alpha_i^0 = \frac{2.44\lambda_{er}^0}{d_t^{4/3}}$$

and λ_{er}^0 is calculated from a correlation to be presented further in the section on the two-dimensional pseudo-homogeneous model.

This is the model which has been—and still is—generally used for the majority of reactor simulations, in the form shown above or in its adiabatic version or in variants like e.g. when the bed contains a heat exchanger, to preheat the feed by means of the reacting gases, or for radial flow.

Axial Dispersion Model (A.II)

$$(D_{ea})_s \frac{d^2C}{dz^2} - u_s \frac{dC}{dz} - r_A \rho_B = 0 \qquad (\text{C.3-5})$$

$$\lambda_{ea} \frac{d^2T}{dz^2} - u_s \rho_p c_g \frac{dT}{dz} - 4\frac{\alpha_i}{d_t}(T - T_w) + (-\Delta H)r_A \rho_B = 0 \qquad (\text{C.3-6})$$

and the pressure drop equation (C.3-3).
The boundary conditions are:

$$u_s(C_0 - C) = (D_{ea})_s \frac{dC}{dz} \qquad \text{at } z = 0$$

$$u_s \rho_g c_p (T_0 - T) = -\lambda_{ea} \frac{dT}{dz}$$

$$\frac{dC}{dz} = \frac{dT}{dz} = 0 \qquad \text{at } z = L$$

These boundary conditions, which lead to a two-point boundary value problem, have been extensively discussed [85]. Notice also that $(D_{ea})_s$ is based upon the superficial velocity u_s, whereas in the correlations Pe_{ma} vs Re, the $Pe_{ma} = u_i d_p / (D_{ea})_i$ is based upon the interstitial velocity. Evidently, $(D_{ea})_s = \epsilon (D_{ea})_i$. For gases, Pe_{ma} lies between 1 and 2. The parameter λ_{ea} can be derived from $Pe_{ha} = u_s \rho_g c_p d_p / \lambda_{ea}$ which is close to 2, although Votruba [80] found values close to 0.5. Young and Finlayson [92] derived criteria for neglecting axial dispersion effects. For situations in which the rate is monotonically decreasing with bed length (isothermal operation, adiabatic operation of an endothermic reaction, exothermic reaction with excessive cooling) the criteria are set up for inlet conditions. Axial dispersion effects are negligible at the inlet when:

$$\frac{r_{A_0} \rho_B d_p}{u_s C_0} \ll Pe_{ma} \tag{C.3-7}$$

and

$$\frac{(-\Delta H) r_{A_0} \rho_B d_p}{(T_0 - T_w) u_s \rho_g c_p} \ll Pe_{ha} \tag{C.3-8}$$

In industrial reactors the flow is so high that these conditions are always satisfied.

When the rate of reaction is a maximum at some intermediate position, because of a hot spot e.g., the above criteria become inadequate. Evidently, the steepness of the temperature peak or the concentration gradient determines whether or not axial dispersion has to be accounted for, but general criteria are not available. Again according to Young and Finlayson [92] axial dispersion is negligible when:

$$\max \left| \frac{dx}{d\left(\frac{z}{d_p}\right)} \right| \ll Pe_{ma} \tag{C.3-9}$$

$$\max \left| \frac{dT'}{d\left(\frac{z}{d_p}\right)} \right| \ll Pe_{ha} \tag{C.3-10}$$

The gradients can be approximated from a simulation based upon the one-dimensional pseudo-homogeneous model with plug flow. To not satisfy (C.3-10) takes very steep temperature gradients, which would probably have to be avoided for reasons of selectivity, anyway. Consequently, industrial situations which would require simulation by means of the axial dispersion model are rather unlikely. In catalytic mufflers, however, in which pronounced hot spots are observed, some model more elaborate than the basic one may be required. Whether the axial dispersion model is sufficiently accurate for this purpose is rather doubtful.

The structure of the model equations introduces the possibility of multiple steady states. These would only become possible for strongly exothermic processes with a very high activation energy and considerable influence of axial mixing. For first order reactions multiplicity would arise when:

$$\beta_r \gamma > \frac{4\gamma}{\gamma - 4} \qquad (\text{C.3-11})$$

provided Pe_{ma} is smaller than a certain value and the first Damköhler number, $k(T_0)L/u_i$, is comprised between two values. Multiplicity would only be possible for very shallow beds [47]. Whether this predicted multiplicity corresponds to reality or is simply a mathematical artefact is not clear yet. Certainly, multiplicity has been generated in laboratory fixed beds, but the tendency is now to explain this in terms of phenomena occurring at the site level.

Two-Dimensional Pseudo-Homogeneous Model (A.III)

$$(D_{er})_s \left(\frac{\delta^2 C}{\delta r^2} + \frac{1}{r}\frac{\delta C}{\delta r} \right) - u_s \frac{\delta C}{\delta z} - r_A \rho_B = 0 \qquad (\text{C.3-12})$$

$$\lambda_{er} \left(\frac{\delta^2 T}{\delta r^2} + \frac{1}{r}\frac{\delta T}{\delta r} \right) - u_s \rho_g c_p \frac{\delta T}{\delta z} + (-\Delta H) r_A \rho_B = 0 \qquad (\text{C.3-13})$$

with boundary conditions:

at $z = 0$ $\quad C = C_0$; $\quad T = T_0$

at $r = 0$ $\quad \dfrac{\delta T}{\delta r} = 0$

at $r = 0$ $\quad$ and $\quad r = R \quad \dfrac{\delta T}{\delta r} = 0$ $\qquad (\text{C.3-14})$

at $r = R$ $\quad \dfrac{\delta T}{\delta r} = -\dfrac{\alpha_w}{\lambda_{er}}(T - T_w)$

Notice that axial dispersion terms have been neglected in this equation, for the reasons already mentioned. Notice also that two parameters are involved in the description of radial heat transfer, λ_{er} and α_w. The latter, which is essentially different from α_i used in the one-dimensional model, is introduced to compensate for the lack of knowledge of the velocity profile near the wall, so that D_{er} cannot be ascribed point values, but is given a uniform value up to the immediate vicinity of the wall. From an extensive set of experimental data De Wasch and Froment [20] derived the following correlation for α_w:

$$\alpha_w = \alpha_w^0 + \frac{0.0115 d_t}{d_p} \, \text{Re} \qquad (\text{C.3-15})$$

The numerical integration of this set of non-linear second order partial differential equations has been carried out by the Crank–Nicolson technique [42] or by orthogonal collocation [25].

Yet, for situations which are very demanding in computational efforts, like those involving transients, it may still be of interest to use a structure requiring less computation, but containing nevertheless certain features of the two-dimensional model. An "equivalent" one-dimensional model has been developed that aims at reproducing the radially averaged concentrations and temperatures generated by the two-dimensional model. By matching temperature profiles in the absence of reaction Froment [26] derived the following relation from which the parameter of the equivalent one-dimensional model can be calculated:

$$\frac{1}{\alpha_i'} = \frac{1}{\alpha_w} + \frac{d_t}{8\lambda_{er}} \qquad (\text{C.3-16})$$

Slightly different, but analogous relations were derived by Crider and Foss [13], Hlavacek [47] and Finlayson [25]. A comparison of both model predictions was made by Froment for o-xylene oxidation [27]. For severe, but still realistic operating conditions, the deviations between both models become important. And, of course, the values in the axis may still significantly differ from the radially averaged values, so that a more detailed model is required. Examples of simulations based upon the two-dimensional pseudo-homogeneous model have been presented by Froment [27], Hawthorn et al. [45], Carberry and White [12] and Garcia-Ochoa et al. [40] among others. These simulations reveal that the radial profiles are not very sensitive to Pe_{mr}, but that, for severe conditions, λ_{er} and λ_w have to be known with great accuracy.

Lerou and Froment [57] also introduced a radial variation in velocity into the model. For severe reaction conditions the predicted conversion and temperature profiles were very different from those simulated with uniform velocity. More information on the flow velocity distribution in a cross section perpendicular to flow is required to progress further in the design of reactors for processes with a pronounced heat effect.

Clearly, considerable sophistication is required in the modeling of severe situations. There are cases where there is no longer justification for not using a two-dimensional model in simulating a reactor. Further improvements can be expected from the introduction of a more realistic velocity profile. Also, for low ratios of d_t/d_p ratios the model may not be sufficiently adequate. To decide whether or not a two-dimensional model is required Mears [64] proposed to evaluate the following group in the hot spot predicted by the one-dimensional simulation:

$$\frac{|\Delta H| r_A(T) \rho_B d_t^2 E}{4 \lambda_{er} R' T_w^2} \left\{ 1 + \left[\frac{8}{\dfrac{\alpha_w d_p}{\lambda_{er}}} \left(\frac{d_p}{d_t} \right) \right] \right\} \qquad \text{(C.3-17)}$$

If the value of the group exceeds 0.4 then the simulation should be extended to the pseudo-homogeneous two-dimensional model.

Plug Flow Heterogeneous Model with Interfacial Gradients (B.I)
For the gas phase the continuity and energy equations are:

$$-u_s \frac{dC}{dz} = k_g a_v (C - C_s^s) \qquad \text{(C.3-18)}$$

$$u_s \rho_g c_p \frac{dT}{dz} + 4 \frac{\alpha_i}{d_t} (T - T_w) = h_f a_v (T_s^s - T) \qquad \text{(C.3-19)}$$

and for the solid:

$$k_g a_v (C - C_s^s) = r_A(C_s^s, T_s^s) \rho_B \qquad \text{(C.3-20)}$$

$$h_f a_v (T_s^s - T) = (-\Delta H) r_A(C_s^s, T_s^s) \rho_B \qquad \text{(C.3-21)}$$

with initial conditions at $z = 0$, $T = T_0$, and $C = C_0$. Since $a_v = 6(1 - \epsilon)/d_p$ (C.3-20) and (C.3-21) are nothing but (B.1-1) and

(B.1-2), but written more explicitly for the solid phase and not for a particle only.

The algebraic equations (C.3-20) and (C.3-21) are first iteratively solved, yielding the values of C_s^s and T_s^s entering in the right hand side of the differential equations for the fluid. These are then solved numerically by a Runge–Kutta routine. Notice that this model does not explicitly contain any axial coupling between the solid particles: heat and mass are transferred in the axial direction through the gas phase only. Eigenberger [21] has extended the model to account for this. To do so requires a reconsideration of h_f, which is really a conglomerate reflecting several mechanisms, in the sense of Bala-krishnan and Pei [4].

Interfacial gradients are generally not of concern in industrial reactors, since the flow velocity is high: ΔT of less than 3°C were simulated in ammonia-, methanol- and phthalic anhydride synthesis [38], while in steam reforming of natural gas De Deken et al. [15] simulated a ΔT of 4°C and less.

Heterogeneous Model with Plug Flow, Interfacial and Intraparticle Gradients (B.II)

The model equations are now for the fluid phase (C.3-18)–(C.3-19) and for the solid phase (B.2-1) and (B.2-2). When only a ΔT occurs over the film surrounding the particle and only a concentration gradient inside the particle, the equations are:

Gas phase:

$$u_s \frac{dC}{dz} = D_e a_v \frac{dC}{d\xi}\bigg|_{\xi=d_p/2} \qquad \text{(C.3-22)}$$

$$u_s \rho_g c_p \frac{dT}{dz} + 4\frac{\alpha_i}{d_t}(T - T_w) = h_f a_v(T_s^s - T) \qquad \text{(C.3-23)}$$

Solid phase:

$$\frac{D_e}{\xi^2} \frac{d}{d\xi}\left(\xi^2 \frac{dC_s}{d\xi}\right) - r_A(C_s, T_s)\rho_s = 0 \qquad \text{(C.3-24)}$$

with B.C. for the solid:

at $\xi = \dfrac{d_p}{2}$ $\qquad -D_e \dfrac{dC}{d\xi} = r_A(C_s^s, T_s^s)\rho_s$

$$h_f(T_s^s - T) = -(-\Delta H)D_e \frac{dC_s}{d\xi} \qquad \text{(C.3-25)}$$

at $\xi = 0$ $\qquad \dfrac{dC_s}{d\xi} = 0$

The use of the effectiveness factor concept permits the reduction of the equations for the solid phase to:

$$k_a a_v (C - C_s^s) = \eta r_A(C_s^s, T_s^s)\rho_B \tag{C.3-26}$$

$$h_f a_v (T_s^s - T) = \eta(-\Delta H)r_a(C_s^s, T_s^s)\rho_B \tag{C.3-27}$$

whereby there is only a benefit if η can be obtained by means of one of the formulae given in B.2. The system now consists of (C.3-18), (C.3-19), and (C.3-26)–(C.3-27). If, in addition, η is expressed in terms of bulk conditions (C, T) the system finally reduces to:

$$u_s \frac{dC}{dz} + \eta_G r_A(C, T)\rho_B = 0 \tag{C.3-28}$$

$$u_s \rho_g c_p \frac{dT}{dz} + 4\frac{\alpha_i}{d_t}(T - T_w) - \eta_G(-\Delta H)r_A(C, T)\rho_B = 0$$

$$\tag{C.3-29}$$

This is a set of equations having the same structure as the basic pseudo-homogeneous model.

Two-Dimensional Heterogeneous Model (B.III)

This is a rather recent development which is expected to gain more attention as heat transfer in packed beds and flow velocity distribution are analyzed and modeled in more detail. The model equations can be written [19]:

Gas phase:

$$(D_{er})_s\left(\frac{\delta^2 C}{\delta r^2} + \frac{1}{r}\frac{\delta C}{\delta r}\right) - u_s\frac{\delta C}{\delta z} = k_g a_v(C - C_s^s) \tag{C.3-30}$$

$$\lambda_{er}^f\left(\frac{\delta^2 T}{\delta r^2} + \frac{1}{r}\frac{\delta T}{\delta r}\right) - u_s\rho_g c_p\frac{\delta T}{\delta z} = h_f a_v(T - T_s^s) \tag{C.3-31}$$

Solid phase:

$$k_g a_v(C - C_s^s) = \eta\rho_B r_A \tag{C.3-32}$$

$$h_f a_v(T_s^s - T) = \eta(-\Delta H)\rho_B r_A + \lambda_{er}^s\left(\frac{\delta^2 T}{\delta r^2} + \frac{1}{r}\frac{\delta T}{\delta r}\right) \tag{C.3-33}$$

with boundary conditions:

$$C = C_0 \qquad\qquad\qquad \text{at } z = 0, \qquad \text{all } r$$

$$T = T_0$$

$$\frac{\delta C}{\delta r} = 0 \qquad\qquad\qquad \text{at } r = 0, \qquad \text{all } z$$

$$\frac{\delta T}{\delta r} = \frac{\delta T_s}{\delta r} = 0$$

$$\text{(C.3-34)}$$

$$\frac{\delta C}{\delta r} = 0$$

$$\alpha_w^f(T_w - T) = \lambda_{er}^f \frac{\delta T}{\delta r} \qquad \text{at } r = R, \qquad \text{all } z$$

$$\alpha_w^s(T_w - T_s) = \lambda_{er}^s \frac{\delta T_s}{\delta r}$$

The effectiveness factor η was used in (C.3-32) and (C.3-33) to abbreviate the equations, but for complex reactions it may have to be obtained through integration of a set of second order differential equations, as mentioned already. Notice that both λ_{er} and α_w have been split into solid phase and gas phase contributions. The second term in the right hand side of (C.3-33) expresses that heat transfer in the radial direction involves both the solid and the gas phase. There still is no explicit coupling in the axial direction. The two phase character of the model is also reflected in the boundary condition for heat transfer at the wall. De Wasch and Froment [20] and Dixon and Cresswell [17] derived relations between λ_{er}^f, λ_{er}^s, α_w^f and α_w^s and the parameters of the pseudo-homogeneous two-dimensional model, λ_{er} and α_w, but further refinement in the experimental data is required.

V. EXAMPLES OF APPLICATION

By now many examples of the simulation of fixed bed reactors used in commercial processes have been published, practically all based upon the one-dimensional pseudo-homogeneous model. Runaway, autothermal behavior, optimal temperature profiles, minimum amount of catalyst, radial rather than axial flow, the influence of inerts are but a sample of the aspects which have been investigated through this approach, in far greater depth and detail than previously. There are few examples of simulations based upon more complex models, however, although the pseudo-homogeneous two-dimensional model has become fairly standard when very exothermic processes have to be dealt with. Hatcher et al. [44] illustrated a case in which a heterogeneous one-dimensional model with interfacial gradients was required, namely the reoxidation of a nickel-catalyst in the secondary reformer of an ammonia synthesis plant. Dumez and Froment [18] dealt with a very complex case in which interfacial gradients, intraparticle gradients and rapid catalyst deactivation by coke formation were accounted for: the dehydrogenation of 1-butene into butadiene on a chromia–alumina catalyst.

Recently, De Deken and Froment [15] extensively simulated the primary steam reforming of natural gas on a nickel–alumina catalyst, accounting for interfacial and intraparticle gradients and using both one- and two-dimensional models.

Two independent rate equations:

$$CH_4 + H_2O \rightleftharpoons CO + 3H_2$$

$$CH_4 + 2H_2O \rightleftharpoons CO_2 + 4H_2$$

and, therefore, two continuity equations were required to describe the composition of the reacting mixture at any position in the reactor. Rate equations of the Hougen–Watson type were derived, one for the total methane-conversion, one for the production of CO. The set of continuity and energy equations was written:

$$F^\circ_{CH_4} \frac{dx_{CH_4}}{dz} = \Omega \eta_{CH_4} r_{CH_4} \rho_B \qquad (V.1)$$

$$F^\circ_{CH_4} \frac{dx_{CO}}{dz} = \Omega \eta_{CO} r_{CO} \rho_B \qquad (V.2)$$

$$\frac{dT}{dz} = \frac{U \pi d_{t_i}}{\Omega c_p G}(T_{we} - T) + \frac{h_f a_v}{c_p G}(T - T_s) \qquad (V.3)$$

$$h_f a_v(T - T_s) = -\rho_B \sum_{j=1}^{2} \eta_j r_j(-\Delta H_j) \qquad (V.4)$$

with the appropriate boundary conditions and the overall heat transfer coefficient calculated from:

$$\frac{1}{U} = \frac{d_t}{2\lambda_{ss}} \ln\left(\frac{d_{t_e}}{d_t}\right) + \frac{1}{\alpha_w} + \frac{d_t}{8\lambda_{er}} \qquad (V.5)$$

The model is an "equivalent" one-dimensional model that attempts at matching the radially averaged values of the two-dimensional model, with α_w and λ_{er} calculated according to the correlations of De Wasch and Froment [20]. The effectiveness factors were based upon the generalized modulus concept, which could be used here with sufficient accuracy because a simple relation was found to exist between the two conversions. The tortuosity factor entering into D_e was determined by means of the gas chromatographic method. A value of 4.7 was obtained. The pressure drop was calculated using Ergun's equation [22], with d_p defined according to Brauer [8] and Reichelt and Blasz [68], to account for the hollow-ring shaped catalyst particle. The simulation results for typical operating conditions are represented in Figs. 2 and 3. The discontinuities in the effectiveness factors shown after 6 m result from different catalyst sizes. The solid temperature T_s is not shown since the interfacial ΔT amounts to 4°C only up to a distance of 6 m, to further decrease to 2°C and even less. The methane conversions simulated without consideration of interfacial ΔT differs by less than 0.8% absolute in any position from the value given in Fig. 2.

The exit composition is in excellent agreement with industrial results. The effectiveness factors are very low, so that only a very thin layer of the catalyst is actually involved in the reaction.

De Deken and Froment also used a two-dimensional model which does not include interfacial gradients but does consider interparticle

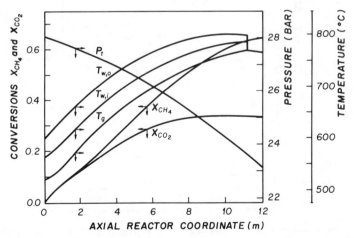

Figure 2   Steam reforming of natural gas. Conversion, temperature and pressure profiles. (From De Deken et al. [15]).

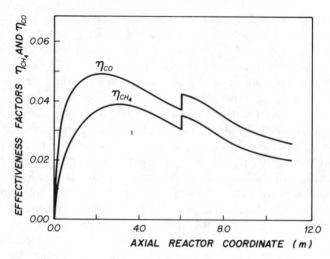

Figure 3   Steam reforming of natural gas. Effectiveness factor profiles versus reactor
length. (From De Deken et al. [15]).

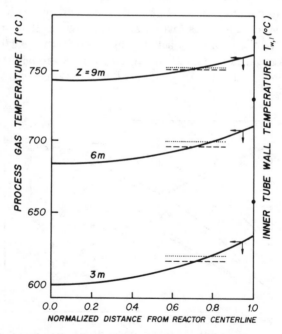

Figure 4   Steam reforming of natural gas. Radial process gas temperature profiles at
several bed depths; $----$ averaged radial; $\cdots\cdots$ predicted by equivalent one-
dimensional model; $\cdot$ inner tube wall temperature. (From De Deken et al. [15]).

gradients. The model equations are those of (C.3-12) and (C.3-13), with the rate multiplied by the corresponding effectiveness factor, however. Figure 4 shows radial temperature profiles at various bed depths simulated for the conditions of Figs. 2 and 3 and for an identical external tubeskin temperature. The temperature difference between the bed centerline and the wall amounts to 33°C, which is not excessive. Also the methane conversion at the wall never differed more than 2% absolute from that in the centerline. This explains why radially averaged temperatures could be accurately simulated by means of the above one-dimensional model with "equivalent" heat transfer parameters. This conclusion should not be generalized to cases with a much more pronounced heat effect, as illustrated by Froment in the simulation of o-xylene oxidation [27, 28].

VI. CONCLUSION

The fundamental models presented above combine the hydro-dynamic and heat transfer phenomena with the chemical reaction by means of a set of mathematical equations of varying complexity. With the present day computational facilities and numerical techniques the solution of these equations is no longer an obstacle to the application of these models to the real problems and configurations encountered in industrial operation.

The equations contain a number of parameters, some of which are "effective" in the sense that they lump effects of phenomena which would otherwise have to be expressed by means of additional equations, to be solved simultaneously in the reactor simulation. For reasons of convenience, correlations for these parameters were generally obtained from purely physical experiments, possibly without sufficient testing under conditions similar to those encountered in the presence of reaction. Under reaction conditions these parameters may yet have to be improved.

The paper presents an escalation of model sophistication and criteria are given for the selection of an appropriate model. Which model is "appropriate" depends in the first place on the reactor, the operating conditions and the reactions, but also on the required accuracy and on the availability and precision of the required fundamental information. Care should be taken not to introduce features into the simulation which may only reflect mathematical artefacts, resulting from the structure of the model equations. A confrontation with reliable experimental data, gathered over a sufficiently wide range of operating conditions is paramount—but not always possible. Some of the models presented here are a bare minimum for a realistic description of the reactor operation, others may seem to be too refined and of academic interest only. It is safe to predict, however, that what may seem to be too refined today may well be the minimum level of modeling tomorrow, particularly for severe conditions. Models will certainly be further refined, but realism should be a guide in this development.

ACKNOWLEDGMENT

A version of this review was presented as a Plenary Lecture at the Pachec 83 Conference—Seoul, Korea, May 8–11, 1983.

NOTATION

a_v external particle surface area per unit reactor volume, m_p^2/m_r^3

C gas phase concentration, $Kmol/m_g^3$

C_0 inlet concentration of component A, $Kmol/m_g^3$

C_s concentration inside catalyst, $Kmol/m_g^3$

C_s^s surface concentration, $Kmol/m_g^3$

c_p specific heat of gas, $kJ/kg\,K$

d_p particle diameter, m_p

d_t internal tube diameter, m_r

d_{t_e} external tube diameter, m_r

D_e effective diffusivity inside catalyst, $m_g^3/m_p s$

D_m molecular diffusivity, m_g^2/s

$(D_{ea})_s$ effective diffusivity in axial direction based upon superficial flow velocity, $m_g^3/m_r s$

$(D_{er})_s$ effective diffusivity in radial direction based upon superficial flow velocity, $m_g^3/m_r s$

f friction factor

$F_{CH_4}^o$ molar feed rate of methane, $Kmol/hr$

G superficial mass flow velocity, $kg/m_r^2 s$

h_f gas to solid heat transfer coefficient in a fixed bed, $kJ/m_p^2 s K$

$(-\Delta H)$ heat of reaction, $kJ/Kmol$

k reaction rate coefficients; for first order e.g., $m_g^3/kg\,cat\,s$

k_g gas to solid mass transfer coefficient in a fixed bed, $m_g^3/m_p^2 s$

n order of reaction

Nu   Nusselt number, $h_f d_t/\lambda_g$

p_t total pressure, kg/ms^2

Pe_{ma} Peclet number for axial effective diffusion in fixed bed, $d_p u_i/(D_{ea})_i$

Pe_{ha} id. for axial effective heat transfer, $u_s \rho_g c_p d_p/\lambda_{ea}$

Pr   Prandtl number, $c_p \mu/\lambda_g$

r radial position, m

r_A rate of reaction of A, Kmol/kg cat s

R tube radius, m

R' gas constant, kJ/Kmol K

Sh   Sherwood number, $k_g d_t/D_m$

S_p surface area of a particle, m^2

T gas temperature, K

T' dimensionless gas temperature, $T - T_w/T_0 - T_w$

T_0 feed temperature of gas, K

T_s solid temperature, K

T_s^s solid surface temperature, K

T_w wall temperature, K

u_i interstitial flow velocity, m_r/s

u_s superficial flow velocity, $m_g^3/m_r^2 s$

V_p particle volume, m_p^3

x conversion

z axial distance in reactor, m

Greek Letters

α_i bed to wall heat transfer coefficient in one-dimensional model, $kJ/m^2 sK$

α_w wall heat transfer coefficient in two-dimensional model, $kJ/m^2 sK$

β_r dimensionless adiabatic temperature rise in a reactor $(-\Delta H)C_0/\epsilon_g c_p T_0$

β_f dimensionless adiabatic temperature rise in a film surrounding a particle, $k_g(-\Delta H)C/h_f T$

β_p dimensionless adiabatic temperature rise in a particle, $D_e C_s^s(-\Delta H)/\lambda_e T_s^s$

γ dimensionless activation energy, E/RT

ϵ void fraction of bed, m_g^3/m_r^3

ϵ_s void fraction of particle, m_g^3/m_p^3

η catalyst effectiveness factor based upon surface conditions C^s_s, T^s_s.

η_G catalyst effectiveness factor based upon bulk gas conditions, C, T

λ_e effective thermal conductivity inside a particle, kJ/msK

λ_g gas conductivity, kJ/msK

λ_{ea} effective thermal conductivity in axial direction, kJ/msK

λ_{er} effective thermal conductivity in radial direction, kJ/msK

μ dynamic viscosity, kg/ms

ξ coordinate inside particle, oriented towards surface, m_p

ρ_B bulk density of bed, kgcat/m_r^3

ρ_g gas phase density, kg/m_g^3

ρ_s catalyst density, kgcat/m_p^3

τ tortuosity

ϕ Thiele modulus $(V_p/S_p)\sqrt{k\rho_s/D_e}$

REFERENCES

1. Amundson, N. R. (1970). Mathematical models of fixed bed reactors, Diskussinstagung Bunsen Gesellschaft und Dechema, Königstein, FRG, *Ber. Bunsen Ges.*, **74**.
2. Aris, R., and Amundson, N. R. (1957). Some remarks on longitudinal mixing or diffusion in fired beds, *AIChE J.*, **3**, 280.
3. Aris, R. (1965). A normalization for the Thiele modulus, *Ind. Eng. Chem. Fund.*, **4**, 227.
4. Balakrishnan, A. R., and Pei, P. C. T. (1979). Heat transfer in gas–solid packed bed systems. 1. A critical review. 2. The conduction mode. 3. Overall heat transfer rates in adiabatic beds, *Ind. Eng. Chem. Proc. Des. Dev.*, **18**, 31, 40, 47.
5. Beeckman, J., and Froment, G. F. (1982). Deactivation of catalysts by coke formation in the presence of internal diffusional limitation, *Ind. Eng. Chem. Fund.*, **21**, 243.
6. Bischoff, K. B. (1962). Axial thermal conductivity in packed beds, *Can. J. Chem. Eng.*, **40**, 161.
7. Bischoff, K. B. (1965). Effectiveness factors for general reaction rate forms, *AIChE J.*, **11**, 351.
8. Brauer, M. (1957). Druckverlust in Füllkörpersäulen bei Einphasen-Strömung, *Chem. Ing. Techn.*, **29**, 785.
9. Butt, J. B. (1965). Thermal conductivity of porous catalysts, *AIChE J.*, **11**, 106.
10. Butt, J. B. (1980). In *Catalyst Deactivation* (Eds. B. Delmon and G.F. Froment), Elsevier, Amsterdam.
11. Butt, J. B. (1986). Deactivation and regeneration of catalysts. Editor: in the present monograph.
12. Carberry, J. J., and White, D. (1969). Role of transport phenomena in catalytic reactor behavior, *Ing. Eng. Chem.*, **61**, 27.
13. Crider, J. E., and Foss, A. S. (1965). Effective wall heat tansfer coefficients and thermal resistances in mathematical models of packed beds, *AIChE J.*, **11**, 1012.

FIXED BED REACTOR DESIGN 209

14. Deans, H. A., and Lapidus, L. (1960). A computational model for predicting and correlating the behavior of fixed bed reactors. I. Derivation of model for nonreactive systems. II. Extension to chemically reactive systems, *AIChE J.*, **6**, 656, 663.
15. De Deken, J. C., Devos, E. F., and Froment, G. F. (1982). Stream reforming of natural gas. Instrinsic kinetics, diffusional influence and reactor design, *Proc. ISCRE7—A.C.S. Wash.*
16. Delmon, P., and Grange, P. (1980). Catalyst deactivation. Solid state chemical phenomena in aging and deactivation of catalysts (Eds. B. Delmon and G. F. Froment), Elsevier, Amsterdam.
17. Dixon, A. G., and Cresswell, D. L. (1979, 1982). Theoretical prediction of effective heat transfer parameters in packed beds. Reply to comments by Vortmeyer and Berninger on theoretical prediction..., *AIChE J.*, **25**, 663 (1979), **28**, 511 (1982).
18. Dumez, F. J., and Froment, G. F. (1976). Dehydrogenation of 1-B into butadiene. Kinetic catalyst coking and reactor design, *Ind. Eng. Chem. Proc. Des. Dev.* **15**, 291.
19. De Wasch, A. P., and Froment, G. F. (1971). A two-dimensional heterogeneous model for fixed bed catalytic reactors, *Chem. Eng. Sci.*, **26**, 629.
20. De Wasch, A. P., and Froment, G. F. (1972). Heat transfer in packed beds, *Chem. Eng. Sci.*, **27**, 567.
21. Eigenberger, G. (1972). On the dynamic behavior of the catalytic fixed-bed reactor in the region of multiple steady states. I. The influence of heat conduction in two-phase models. II. The influence of the boundary conditions in the catalyst phase, *Chem. Eng. Sci.*, **27**, 1909, 1917.
22. Ergun, S. (1952). Fluid flow through packed columns, *Chem. Eng. Progr.*, **48**(2), 89.
23. Fahien, R. W., and Smith, J. M. (1955). Mass transfer in packed beds, *AIChE J.*, **1**, 25.
24. Feng, C. F., and Stewart, W. E. (1973). Practical models for isothermal diffusion and flow of phases in porous solids, *Ing. Eng. Chem. Fund.*, **12**, 143.
25. Finlayson, B. A. (1971). Packed bed reactor analysis by orthogonal collocation, *Chem. Eng. Sci.*, **26**, 1081.
26. Froment, G. F. (1961). Design of fixed bed catalytic reactors based on effective transport models, *Chem. Eng. Sci.*, **7**, 29.
27. Froment, G. F. (1967). Fixed bed catalytic reactors. Current design status, *Ind. Eng. Chem.*, **59**(2), 18.
28. Froment, G. F. (1971). Some aspects of the design of fixed bed reactors for hydrocarbon oxidation, *Period. Polytechnica (Budapest)*, **15**, 219.
29. Froment, G. F. (1972). Analysis and design of fixed bed catalytic reactors. A state of the art review, *Proc. 1st. Int. Symp. Chem. React. Eng. Adv. Chem. Ser., A.C.S., Wash.*
30. Froment, G. F. (1972). Fixed bed reactors. Steady state conditions, *Proc. 5th Eur. Symp. Chem. React. Eng. Elsevier, Amsterdam.*
31. Froment, G. F. (1974). Fixed bed catalytic reactors. Technological and fundamental aspects, *Chem. Ing. Techn.*, **26**, 374.
32. Froment, G. F. (1975). Model discrimination and parameter estimation in heterogeneous catalysis, *AIChE J.*, **21**, 1041.
33. Froment, G. F. (1976). Kinetic analysis of reactions catalyzed by solids, chemical reaction engineering, *Proc. 4th Intl. Symp., Heidelberg, 1976, Dechema.*
34. Froment, G. F. (1976). Catalyst deactivation by coking, *Proc. 6th Int. Congr. Catalysis*, Vol. 1, p. 10, Chemical Society, London.
35. Froment, G. F. (1980). In *Catalyst Deactivation* (Eds. B. Delmon and G. F. Froment), Elsevier, Amsterdam.
36. Froment, G. F. (1980). Hot spots and runaway in fixed bed tubular reactors, p. 535, *Chemistry and Chemical Engineering of Catalytic Processes*, NATO Adv.

210   CONCEPTS AND DESIGN OF CHEMICAL REACTORS

Study Inst. Series (Eds. R. Prins and G. C. A. Schuit), Sijthoff and Noordhoff, Alphen/Rijn, Nederland.

37. Froment, G. F. (1982). A rigorous formulation of the effect of coke formation on catalyst activity, in *Progress in Catalyst Deactivation*, NATO Adv. Study Inst. Series E54 (Ed. J. L. Figueiredo), Nijhoff, the Hague.

38. Froment, G. F., and Bischoff, K. B. (1979). *Chemical Reactor Analysis and Design*, Wiley, New York.

39. Froment, G. F., and Hosten, L. H. (1981). In *Catalysis, Science and Technology*, Vol. 2 (Eds. J. R. Anderson and M. Boudart), Springer-Verlag, Berlin.

40. Garcia-Ochoa, F., Leuz, M., Emig, G., and Hofmann, H. (1981). Reaktionstechnische Untersüchung der Katalytischen Dehydrierung von Cyklohexanol zü Cyclohexanen, *Chem. Ztg.*, **105**, 349.

41. Gottifredi, J. C., Gonzo, E. E., and Quiroga, O. D. (1981). Isothermal effectiveness factor I. Isothermal effectiveness factor II, *Chem. Eng. Sci.*, **36**, 705, 713.

42. Grosjean, C. C., and Froment, G. F. (1962). Computation of axial and radial profiles of temperature and conversion in chemical reactors, *Meded. Kon. Vlaamse Akademie*, **XXIV**, nr. 1.

43. Handley, D., and Heggs, P. J. (1968). Momentum and heat transfer mechanisms in regular shaped packings, *Trans. Instn. Chem. Engrs.*, **46**, T251.

44. Hatcher, W. L., Viville, L., and Froment, G. F. (1978). The reoxidation of a nickel reforming catalyst. Kinetic analysis and fixed bed simulation, *Ind. Eng. Chem. Proc. Des. Dev.*, **17**, 491.

45. Hawthorn, R. D., Ackerman, G. H., and Nixon, A. C. (1968). A mathematical model of a packed bed heat exchanger reactor for dehydrogenation of methylcyclohexane. Comparison of predictions with experimental results, *AIChE J.*, **14**, 69.

46. Hennecke, F. W., and Schlünder, E. U. (1973). Wärmeübergang in Beheizten oder gekühlten Rohren mit Schüttungen aus Kugeln, Zylinder und Raschig-Ringen, *Chem. Eng. Techn.*, **45**, 277.

47. Hlavacek, V., and Hofmann, H. (1970). Steady state axial heat and mass transfer in tubular reactors. An analysis of the uniqueness of solution, *Chem. Eng. Sci.*, **25**, 173, 187.

48. Hegedus, L. L., and McCabe, R. W. (1980). In *Catalyst Deactivation* (Eds. B. Delmon and G. F. Froment), Elsevier, Amsterdam.

49. Himmelblau, D. M., and Bischoff, K. B. (1968). *Process Analysis and Simulation. Deterministic Systems*, Wiley, New York.

50. Jacob, S. M., Grass, B., Voltz, S. E., and Weeckman, V. W. (1976). A lumping and reaction scheme for catalytic cracking, *AIChE J.*, **22**, 701.

51. Jensen, K. F., and Ray, H. W. (1982). The bifurcation behavior of tubular reactors, *Chem. Eng. Sci.*, **37**, 199.

52. Keil, W., and Wicke, E. (1980). Uber die Kinetischen Instabilitäten bei der CO-Oxidation an Plantin-Katalysatoren, *Ber. Bensenges. Phys. Chem.*, **84**, 377.

53. Kulkarni, B. D., and Doraiswamy, L. K. (1980). Estimation of effective transport properties in packed bed reactors, *Catal. Rev. Sci. Eng.*, **22**(3), 431.

54. Kunii, D., and Smith, J. M. (1960). Heat transfer characteristics of porous rocks, *AIChE J.*, **6**, 71.

55. Kunii, D., and Furusawa, T. (1972). Prediction of hot-spot location in catalytic fixed bed reactors, *Chem. Eng. J.*, **4**, 268.

56. Koros, R. M., and Nowak, E. J. (1967). A diagnostic test of the kinetic regime in a packed bed reactor, *Chem. Eng. Sci.*, **22**, 470.

57. Lerou, J. J., and Froment, G. F. (1977). Velocity, temperature and conversion profiles in fixed bed catalytic reactors, *Chem. Eng. Sci.*, **32**, 853.

58. Leva, M. (1948). Cooling of gases through packed tubes, *Ind. Engng. Chem.*, **40**, 747.

59. Luss, D. (1976). Interactions between transport phenomena and chemical rate

processes, Chemical Reaction Engineering, *Proc. 4th Int./6th Eur. Symp.*, *Heidelberg*, p. 487, Dechema.
60. Madou, R. J., and Boudart, M. (1982). Experimental criterion for the absence of artifacts in the measurements of rates of heterogeneous catalytic reactions, *Ind. Eng. Chem. Fundam.*, **21**, 438.
61. Maeda, S. (1952). Heat transfer of granular catalysts. I. Theoretical equations and experiments for the heat transfer between the gas flowing through granular solids and the cylindrical wall, *Tech. Rep. Tohoku Univ.*, **16**, 1.
62. Marin, G. B., and Froment, G. F. (1982). Reforming of C_6-hydrocarbons on a Pt-Al_2O_3 catalyst, *Chem. Eng. Sc.*, **37**, 759.
63. Mears, D. E. (1971). Tests for transport limitations in experimental catalytic reactors, *Ind. Eng. Chem. Proc. Des. Dev.*, **10**, 541.
64. Mears, D. E. (1971). Diagnostic criteria for heat transport limitations in fixed bed reactors, *J. of Catal.*, **20**, 127.
65. Olbrich, W. E., Agnew, J. B., and Potter, O. E. (1966). Dispersion in packed beds and the cell model, *Trans. Instn. Chem. Engrs.*, **44**, T207.
66. Petersen, E. E. (1965). A general criterion for diffusion influenced chemical reactions in porous solids, *Chem. Eng. Sci.*, **20**, 587.
67. Petrovic, L. F., and Thodos, G. (1968). Mass transfer in the flow of gases. Through packed beds, *Ind. Eng. Chem. Fund.*, **7**, 274.
68. Reichelt, W., and Blasz, E. (1971). Strömungstechnische Untersuchungen an mit Raschig-Ringen gefullten Füllkörperzöhren und-säulen, *Chem. Ing. Techn.*, **43**, 949.
69. Schlunder, E. U. (1978). Transport phenomena in packed bed reactors, *Chem. React. Eng. Reviews—Houston, A.C.S. Symp. Ser.*, **72**.
70. Singer, E., and Wilhelm, R. E. H. (1950). Heat transfer in packed beds. Analytical solutions and design methods, *Chem. Eng. Progr.*, **46**, 343.
71. Smith, J. M. (1982). Thirty-five years of applied catalytic kinetics, *Ind. Eng. Chem. Fund.*, **21**, 327.
72. Spechia, V., Baldi, G., and Sicardi, S. (1980). Heat transfer in packed bed reactors with one-phase flow, *Chem. Eng. Commun.*, **4**, 361.
73. Steyns, M., and Froment, G. F. (1981). Hydroisomerization and hydrocracking. 3. Kinetic analysis of rate data for *n*-decane and *n*-dodecane, *Ind. Eng. Chem. Prod. Res. Dev.*, **20**, 660.
74. Sundaresan, S., Amundson, N. R., and Aris, R. (1980). Observations on fixed-bed dispersion models. The role of the interstitial fluid, *AIChE J.*, **26**, 529.
75. Tallmadge, J. A. (1970). Packed bed pressure drop. An extension to higher Reynolds numbers, *AIChE J.*, **16**, 1092.
76. Tsotsis, T. T., Haderi, A. E., and Schmitz, R. A. (1982). Exact uniqueness and multiplicity criteria for a class of lumped reaction systems, *Chem. Eng. Sci.*, **37**, 1235.
77. Van Deemter, J. J., Zuiderweg, F. J., and Klinkenberg, A. (1956). Longitudinal diffusion and resistance to mass transfer as causes of nonideality in chromatography, *Chem. Eng. Sci.*, **5**, 271.
78. Van den Bosch, B., and Luss, D. (1977). Uniqueness and multiplicity criteria for an *n*th order chemical reaction, *Chem. Eng. Sci.*, **32**, 203.
79. Verschoor, H., and Schuit, G. (1950). Heat transfer to fluids flowing through á bed of granular solids, *Appl. Sci. Res.*, **A2**, 97.
80. Vortruba, J., Hlavacek, V., and Marek, M. (1970). Packed bed axial thermal conductivity, *Chem. Eng. Sci.*, **27**, 1845.
81. Wakao, N., and Smith, J. M. (1962). Diffusion in catalyst pellets, *Chem. Eng. Sci.*, **17**, 825.
82. Weisz, P. B., and Prater, C. D. (1959). Interpretation of measurements in experimental catalysis, *Adv. Catal.*, **6**, 143.
83. Whitaker, S. (1972). Forced convection heat transfer correlations for flow in pipes, past flat plates, single spheres and for flow in packed beds and tube bundles, *AIChE J.*, **18**, 361.

84. Wicke, E., and Kallenbach, A. (1941). Surface diffusion of carbon dioxide in activated charcoals, *Koll. Zeitschrift*, **97**, 135.
85. Wicke, E. (1975). Boundary conditions at input of a solid bed reactor, *Chem. Eng. T.*, **47**, 547.
86. Weekman, V. W. (1979). Lumps, models and kinetics in practice, *AIChE Monograph Series*, **11**, Vol. 75 (1979).
87. Wedel, S., and Luss, D. (1980). A rational approximation of the effectiveness factor, *Chem. Eng. Commun.*, **7**, 245.
88. Yagi, S., and Kunii, D. (1957). Studies on effective thermal conductivities in packed beds, *AIChE J.*, **3**, 373.
89. Yagi, S., and Kunii, D. (1960). Studies on heat transfer near wall surface in packed beds, *AIChE J.*, **6**, 97.
90. Yagi, S., Kunii, D., and Wakao, N. (1960). Studies on axial effective thermal conductivity in packed beds, *AIChE J.*, **6**, 543.
91. Yagi, S., and Wakao, N. (1959). Heat and mass transfer from wall to fluid in packed beds, *AIChE J.*, **5**, 79.
92. Young, L. C., and Finlayson, B. A. (1973). Axial dispersion in nonisothermal packed bed chemical reactors, *Ind. Eng. Chem. Fund.*, **12**, 412.

Chapter 5

Modeling of Fluidized Bed Reactors

GILBERT F. FROMENT

Laboratorium voor Petrochemische Techniek, Rijksuniversiteit, Gent, Belgium

Abstract

In this article current models of transport in fluidized bed catalytic reactors are presented along with a brief discussion of the hydrodynamics of the bubble phase and emulsion phase. The relation between conversion and operating parameters is discussed in terms of current industrial operations in which catalyst deactivation is an integral part of the design for both dense beds and risers.

I. INTRODUCTION

The fluidization of solids is a complex phenomenon involving many parameters and its application requires an extensive background on the relation between observables and variables. Yet, in the present paper this background and the associated technological aspects will be supposed to be known, so that the focus is on the reactor modeling.

The main field of application of fluidization in the chemical industry is the catalytic cracking of gas oil for enhancing the gasoline production. The fluidized bed operation was introduced into this process to permit in situ regeneration of the catalyst, which rapidly deactivates because of coke deposition. Figure I.1 shows a conventional catalytic cracking unit, consisting of a reactor and regenerator, both operating in the so-called dense bed mode. Catalyst is continuously withdrawn from the reactor, flows to the regenerator

213

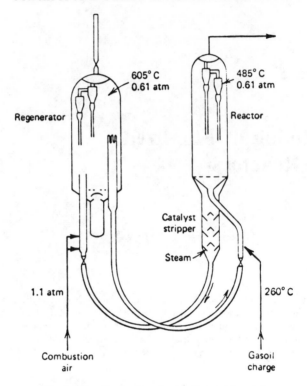

Figure I.1   Conventional catalytic cracking unit.

where the coke is burnt off and returns to the reactor, under the influence of differences in hydrostatic pressure heads.

In the late sixties synthetic zeolites were introduced in catalytic cracking. Their extremely high activity caused considerable over-cracking. This led to the development of riser cracking. Figure I.2 schematically represents a modern catalytic cracking unit with riser cracking. The former reactor vessel has shrunk considerably with respect to the original version and mainly serves for separating the reaction products from the catalyst.

Fluidization is also used for processes in which the catalyst deactivation is of no concern, but whose exothermicity is so high that the temperature control would become difficult in a fixed bed reactor. Examples of such processes are: acrylonitrile synthesis from propylene, ammonia and oxygen; ethylene dichloride synthesis from ethylene, hydrochloric acid and oxygen and phthalic anhydride synthesis from naphtalene and air.

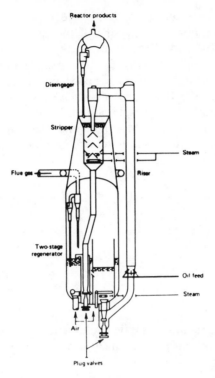

Figure I.2   Kellogg Orthoflow model *F* unit with riser cracker and two-stage regeneration.

II.   TWO-PHASE MODELS FOR FLUIDIZED BEDS

Visual observations of fluidized beds reveal the existence of bubbles rising through a more homogeneous mixture of solids and fluid. The amount of catalyst in the bubbles is very small. The bubbles are said to bypass the catalyst. It is known that under the conditions of industrial fluidization the amount of gas flowing through the bed in the form of bubbles is very important. The bubbles grow with bed height by the effect of the decreasing hydrostatic pressure, and also by coalescence and absorption of gas from the surrounding homogeneous phase, called the emulsion phase. As a result, there is a certain amount of interchange between the two phases.

From the above description it is not surprising that simple models with plug flow or complete mixing for the gas phase were not successful in simulating fluidized bed reactors. To explain experimental results falling below curves based upon complete mixing, when plotted in a conversion vs. space time diagram, requires a model that incorporates bypassing of the catalyst. These considerations led, at the end of the fifties, to the two-phase models. They are still the preferred basis for modeling today, although the fundamentals behind them have been considerably extended and documented. Figure II.1 represents such a two-phase model.

Early modeling efforts based upon this concept were published by Mathis and Watson [23], May [24] and Van Deemter [32]. The equations were written as follows by Froment and Bischoff [17]:

For the bubble phase:

$$f_b u_b \frac{dC_{Ab}}{dz} + k_I (C_{Ab} - C_{Ae}) + r_A \rho_b f_b = 0 \qquad (\text{II.1})$$

For the emulsion phase:

$$f_e u_e \frac{dC_{Ae}}{dz} - k_I (C_{Ab} - C_{Ae}) - f_e D_e \frac{d^2 C_{Ae}}{dz^2} + r_A \rho_e (1 - f_b) = 0 \qquad (\text{II.2})$$

where f_e is the fraction of the bed volume taken by the emulsion *gas* (not by the emulsion *phase*, which also comprises catalyst), f_b the

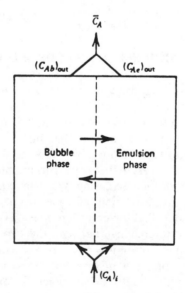

Figure II.1   Basic two-phase model for fluidized bed.

fraction of the bed volume taken by the bubbles and k_I is an interchange coefficient. Since the amount of catalyst in the bubble phase is so small, the reaction term in (II.1) may be neglected. In (II.2) u_e represents the velocity of the emulsion gas, on an interstitial basis. It is generally assumed—and it has also been experimentally verified, at least with small particles—that all the gas flowing through the bed in excess of that required for minimum fluidization is in the bubble phase. Therefore, $u_e = u_{mf}/\epsilon_{mf}$. The above assumption also determines f_b.

Both streams, with their respective conversion, are hypothetically mixed at the bed exit to yield the exit stream, with its mean conversion

$$u_s \bar{C}_A = f_b u_b (C_{Ab})_{\text{out}} + f_e u_e (C_{Ae})_{\text{out}} . \tag{II.3}$$

The violent motion of the catalyst in the emulsion is considered to cause a mixing in the emulsion gas. In the absence of gross mixing patterns this is expressed in terms of an effective diffusion mechanism, characterized by D_e. May [24] determined D_e from the residence time distribution of tagged catalyst particles. Knowing D_e, the interchange coefficient, k_I, was determined by fitting gas phase residence time distribution measurements by means of the model equations (II.1) and (II.2). Van Deemter [32] combined the results of gas residence time distribution measurements and steady state tracer experiments to obtain k_I and D_e.

The system (II.1) and (II.2) is integrated accounting for the boundary conditions:

$$z = 0 \text{ bubble phase } C_{Ab} = (C_A)_i$$

$$\text{emulsion} \quad -D_e \frac{dC_{Ae}}{dz} = u_e((C_A)_i - C_{Ae})$$

$$z = Z \quad \frac{dC_{Ae}}{dz} = 0 .$$

Notice also that it is assumed that the bed is isothermal. Figure II.2 summarizes calculations by May [24]. The two-phase model indeed yields conversions which have to be lower than those predicted by the complete mixing model. The higher k_I the closer plug flow is approached (curves 3 and 3′), but the more it matters to accurately know D_e.

In essence, the above model is still in use to date for the simulation and design of fluidized beds, be it eventually with

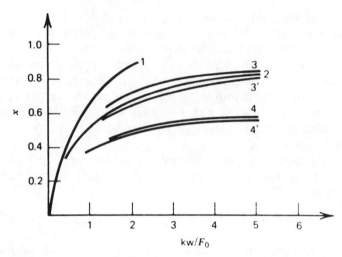

Figure II.2 Conversion in a fluidized bed reactor. Curve 1: single-phase model with plug flow ($k_I = \infty$, $D_e = 0$); curve 2: single-phase model with complete mixing ($k_I = \infty$, $D_e = \infty$); curve 3: two-phase model $k_I = 3$, $D_e = 0$; curve 3': two-phase model $k_I = 3$, $D_e = \infty$; curve 4: two-phase model $k_I = 1$, $D_e = 0$; curve 4': two-phase model $k_I = 1$, $D_e = \infty$ (after [24]).

refinements as to the role of the grid [2, 19, 11] and of the freeboard [43, 18, 7]. Also, the parameters of the model, in particular k_I, have been further developed.

III. A HYDRODYNAMIC INTERPRETATION OF THE INTERCHANGE COEFFICIENT, k_I

More recently great efforts were undertaken to predict the interchange coefficient from fundamentals. Hydrodynamic theory has enabled the calculation of the streamlines of gas and particles in the vicinity of a single bubble. The flow patterns strongly depend upon the relative velocity of the bubble with respect to the emulsion gas. For all practical situations the bubble rises faster than the emulsion gas, so that the relative movement of the emulsion gas is downward. Figure III.1.a shows gas streamlines, calculated by Murray [26] starting from the Davidson and Harrison theory [4] and assuming spherical bubbles.

Bubble gas is seen leaving the bubble in the upper part and entering it again in the lower part, without leaving a sphere which is surrounding the bubble, however. It is in this space between the two

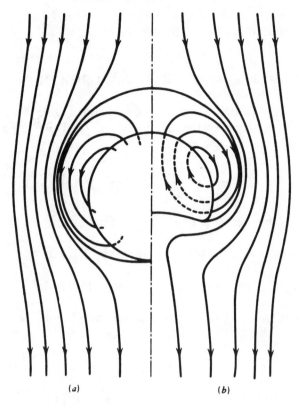

Figure III.1   Gas streamlines for a three dimensional bubble and $u_b/(u_{mf}/\epsilon_{mf}) = 1.5$. (a) according to Murray [26], (b) according to Partridge and Rowe [29].

concentric spheres, called the interchange zone, that the bubble gas comes into contact with the catalyst particles and is converted. The interchange zone decreases in importance as the relative velocity of the bubble, u_{br}, increases.

In reality the bubble is not spherical, but more like that shown in the right hand side of Fig. III.1. Partridge and Rowe [29] found experimentally that the wake occupies some 25% of the spherical bubble volume and that the ratio of the interchange zone volume to that of the bubble itself equals $\delta/(\delta - 1)$; where $\delta = u_b \epsilon_{mf}/u_{mf}$. What is of importance also is that the bubble grows, but only up to a certain height in the bed. The maximum size of the bubble is a function of the particle size distribution.

From this fundamental picture k_I has been derived in several ways. What is clear at this point is that k_I depends upon the bubble size. Therefore, prior to calculating k_l it is necessary to briefly mention correlations for the bubble diameter, d_b.

IV. BUBBLE VELOCITY, -SIZE AND -GROWTH

The following equation was derived by Darton et al. [3] for d_b in the zone $0 < z < z_m$ between the distributor and the height z_m at which the maximum size is reached:

$$d_b = \frac{0.54}{g^{1/5}} (u_s - u_{mf})^{2/5}(z + 4\sqrt{A_0})^{4/5} \qquad \text{(IV.1)}$$

where A_0 is the area of plate per orifice (m^2) and d_b is expressed in m.
For a porous plate distributor Werther [40] derived:

$$d_b = 0.00853[1 + 27.2(u_s - u_{mf})]^{1/3}(1 + 6.84z)^{1.21} \qquad \text{(IV.2)}$$

The bubble rise velocity among a swarm of bubbles is given, according to Werther [40] by the equation:

$$u_{br} = \Psi\sqrt{d_b g} \qquad \text{(IV.3)}$$

where $\Psi = 0.64$ when the bed diameter, d_t, <0.1 m

$\quad = 1.6d_t^{0.4}$ when $0.1 < d_t < 1.0$ m

$\quad = 1.6$ $\quad$ when $d_t > 1.0$ m

whereas $\Psi = 0.711$, for a single bubble, according to the well known Davis-formula.

V. DERIVATION OF k_l FROM HYDRODYNAMICS

Correlations for k_l were derived by Van Swaay and Zuiderweg [33] in their work on the Shell chlorine process. Using ozone decomposition on iron oxide deposited on sand or silica particles, in beds with diameters ranging from 0.1 to 0.6 m and heights ranging from 0.5 to 3.0 m, they derived the following correlation for the height of a transfer unit, in m:

$$\frac{u_s}{k_I} = \left(1.8 - \frac{1.06}{d_t^{1/4}}\right)\left(3.5 - \frac{2.5}{Z^{1.4}}\right).$$ (V.1)

The correlation reveals a dependence on bed diameter and bed height. A correlation by Mirreur and Bischoff [25] yields values which are in good agreement with these of (V.1).

Recently, Krishna [21] tried to confirm values and trends for k_I starting from the model Davidson [5] derived from hydrodynamic observations and theory. Davidson expresses the interaction between the bubble phase and the emulsion in terms of two contributions: cross flow and diffusion. The total flow rate of gas transferred, Q (m^3/s), is derived to be:

$$Q = 1.19\pi d_b^2 u_{mf} + 0.91\pi d_b^{7/4} D_{be}^{1/2} g^{1/4} \frac{\epsilon_{mf}}{1 + \epsilon_{mf}}$$ (V.2)

The first term in (V.2) is the flow term and dominates in the range of flow rates encountered in commercial practice. The interchange coefficient k_I is related to Q by:

$$Q = \frac{k_I}{f_b} \frac{\pi d_b^3}{6}$$ (V.3)

Since d_b varies with bed height, k_I also depends on Z. It also follows from (V.3) that k_I/f_b is independent of d_t, in contradiction with the Van Swaay–Zuiderweg correlation. One should keep in mind, however, that k_I derived from experiments is really an integral over the total bed height of point values, k_I.

$$\frac{k_I}{f_b} = \frac{1}{Z} \int_0^Z \frac{k_I}{f_b} dz$$ (V.4)

Krishna extensively confronted k_I-values calculated from (V.2) and (V.3) with experimental data of De Groot [6], De Vries et al. [9] and Van Swaay and Zuiderweg [33]. The results are shown in Fig. V.1.

It follows that:

$$\frac{k_I}{f_b} \sim (u_s - u_{mf})^{-0.4 \text{ to } -0.5}$$

$$\sim z^{-0.8 \text{ to } -1} \qquad z \leqslant z_m$$

$$\sim C_1 + C_2/z \qquad z > z_m$$

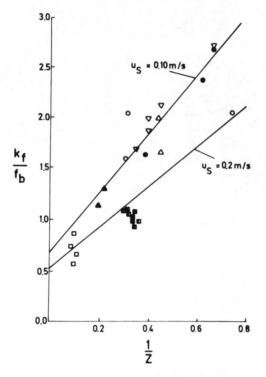

Figure V.1   Variation of k_I/f_b with bed height. Comparison of model predictions and experimental results.

where C_1 and C_2 are functions of the operating conditions. A couple of recommendations can be extracted from the above results for the experimental determination of k_I. First, k_I/f_b should be determined in beds with a sufficient diameter, d_t, to avoid slugging. The bed height should exceed z_m. For a d_t between 0.5 to 1 m the required height is of the order of 3 m. The flow velocity should exceed $5u_{mf}$.

Refinements are possible. Kunii and Levenspiel [22] distinguish between transfer from the bubble to the interchange zone and from the latter to the emulsion phase, with the two steps purely in series. The corresponding mass transfer coefficients, based upon unit bubble volume, are obtained respectively from:

$$(k_{bc})_b = 4.5\,\frac{u_{mf}}{d_b} + 5.85\left(\frac{D_{bc}^{1/2}g^{1/4}}{d_b^{5/4}}\right) \tag{V.5}$$

with $(k_{bc})_b$ in $m_g^3/m_b^3 s$ and from:

$$(k_{ce})_b = 6.78\left(\frac{\epsilon_{mf}D_{ce}u_b}{d_b^3}\right)^{1/2} \tag{V.6}$$

with the same units.

The coefficients may be combined by the rule of addition of resistances, since both steps are purely in series:

$$\frac{1}{(k_{be})_b} = \frac{1}{(k_{bc})_b} + \frac{1}{(k_{ce})_b}. \tag{V.7}$$

The coefficient $(k_{be})_b$ is related to k_I by the equation:

$$k_I = (k_{be})_b f_b. \tag{V.8}$$

A diagram of Pyle [30] shows differences between k_I calculated by the above and other models of the order of a factor 5 to 10.

VI.   THE EFFECTIVE DIFFUSIVITY IN THE EMULSION PHASE

As mentioned already, the mixing in the emulsion phase is the second parameter of the two-phase model. The mixing is generally expressed in terms of effective diffusion.

The correlation of Baird and Rice [1] for the effective diffusivity is:

$$D_e = 0.35(gu_s)^{1/3}d_t^{4/3} \tag{VI.1}$$

where D_e is expressed in m_r^2/s.

Mirreur and Bischoff [25] derived the following correlation from their experiments in small diameter beds:

$$\frac{u_s z}{D_e\left(\dfrac{d_t}{Z}\right)^{1.5}} = 0.33 \tag{VI.2}$$

valid over a wide range of u_s/u_{mf}.

De Vries et al. [9] measured a value of $1.5\,m^2/s$ in a commercial reactor. The value calculated from Baird and Rice's correlation would be $1.25\,m^2/s$, which is in good agreement. For this situation (VI.2) yields a value of $0.35\,m_r^2/s$ only.

VII.   THE RELATION BETWEEN CONVERSION AND OPERATING PARAMETERS. MODEL PREDICTIONS

Figures VII.1 and VII.2 represent conversions calculated by Krishna on the basis of the two-phase model equations (II.1) and (II.2) for varying parameter values.

The model equations still have the form (II.1) and (II.2). In reality they are applied over height increments, to account for the variation of d_b, k_1, f_b etc. along the bed.

For $u_s Z / f_e D_e < 3$ the effect of backmixing in the emulsion is important. This would be a situation encountered in shallow beds with Z/d_t of the order of 1. For $Z/d_t > 3$ the group $u_s Z / f_e D_e$ will usually exceed 5 and plug flow is fairly well approximated in the emulsion phase.

For $u_s > 5 u_{mf}$ almost all the gas enters the bubble phase. The rate coefficient k in Figs. VII.1 and VII.2 is that of a first order reaction and ρ_B is the catalyst bulk density, in kg cat/m^3 reactor.

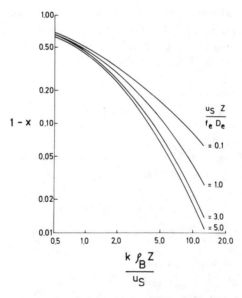

Figure VII.1   Fraction of unconverted gas in effluent vs. $k\rho_B \cdot Z/u_s$ at various $u_s Z / f_e D_e$ and constant $f_b = 1.0$ and $k_1 Z / u_s = 4$.

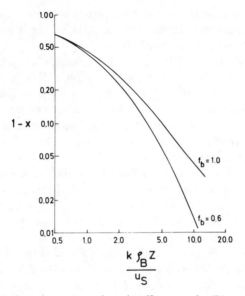

Figure VII.2   Fraction of unconverted gas in effluent vs. $k\rho_B Z/u_s$ at various f_b and $u_s Z/f_e D_e = 5$ and $k_1 Z/u_s = 5$.

VIII.   AN EXAMPLE OF COMMERCIAL APPLICATION: THE CATALYTIC CRACKING OF GAS OIL

1.   Mobil's Kinetic Models for Catalytic Cracking

In the early sixties Mobil developed a three-lump model for catalytic cracking [36]. The experimental results were reported in terms of conversion of the gas oil charge, the amount of gasoline produced and the amount of light gases and coke. The reaction model was represented by:

$$GO \xrightarrow{k_1} G \xrightarrow{k_2} C$$

$$\underset{k_3}{\big\lfloor\underline{}\big\rfloor\!\!\uparrow}$$

where GO represents gas oil, G the C_5-210°C gasoline and C the C_4^- and coke.

Each of the rate coefficients accounted for the deactivation of the catalyst through a deactivation function $\phi = \exp(-\alpha t_c)$ where t_c represents the residence time of the catalyst.

The kinetic studies were carried out in a moving bed bench scale reactor, to permit independent variation of the catalyst residence time and of the gas oil space time.

The order of the global rate of conversion of gas oil into both gasoline and "coke" was found to be two. This is a result of the wide range of crackability of the various types of molecules in the feed. The cracking of gasoline—a much narrower fraction—behaved as a first order reaction. Figure VIII.1 shows a selectivity plot for gasoline, revealing that the rates of the different reactions decay according to identical functions.

Clearly, a three lump model for a complex reaction like catalytic cracking of gas oil cannot be universal. Indeed, each time the feed stock was changed a set of rate coefficients k and deactivation constant, α, had to be determined, even for the same catalyst. Recycle oil and thermally coked stock from a delayed coker did not correlate well when the k and α were plotted versus the aromatics to naphthene ratio. Mobil then embarked into the development of a 10 lump model [20]. A proper set of lumps was to possess rate

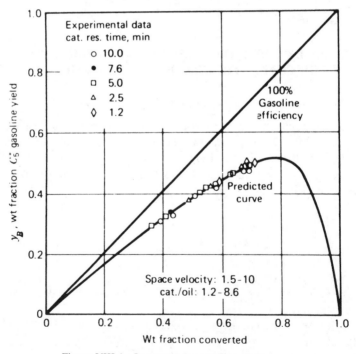

Figure VIII.1   Instantaneous gasoline yield curve.

constants invariant with the feed. The resulting set of lumps and the reactions between them are shown in Fig. VIII.2.

The continuity equation for the lumped species in a fluid bed catalytic cracker with plug flow for the gas phase can be written, provided the residence time of the gas phase is much smaller than that of the catalyst, so that the gas contacts catalyst of the same age and therefore activity:

$$\frac{d\mathbf{a}}{d\zeta} = \frac{1}{1 + K_{AR_h}AR_h} \frac{p_t \bar{M}}{R'T} \frac{\phi(t_c)}{SV} \mathbf{k'a} \qquad (\text{VIII.1})$$

where

  a: composition vector of reacting lumped species

  $\left(a_j = \dfrac{\text{moles } j}{\text{g gas}} \right)$

  k': matrix of rate coefficients, Arrhenius dependent
      $[\text{cm}^3/\text{gcat} \cdot \text{s}]$

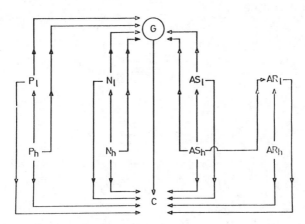

Figure VIII.2   10 Lump reaction network for catalytic cracking of gas oil.

P_l: wt% paraffinic molecules (221–343°C)
N_l: wt% naphthenic molecules (id.)
AR_l: wt% carbon atoms among aromatic rings (id.) } LFO
AS_l: wt% aromatic substituent groups (id.)

P_h: wt% paraffinic molecules (>343°C)
N_h: wt% naphthenic molecules (id.)
AR_h: wt% carbon atoms among aromatic rings } HFO
AS_h: wt% aromatic substituent groups (id.)

G: gasoline lump (C_5 – 221°C)
C: C-lump (C_1 to C_4 + coke)

K_{Ar_h}:   adsorption constant of aromatic rings in fraction $>343°C$ $(wt\%)^{-1}$

ζ: dimensionless axial position $= z/Z$

p_t: absolute pressure (atm)

R':   gas constant $(82.05 \text{ atm cm}^3/\text{gmole K})$

$\bar{M}$:   mean molecular weight of gas $= \dfrac{1}{\sum a_j(\zeta)}$

SV: weight space velocity $\left(\dfrac{\text{g feed oil} + \text{inerts}}{\text{g cat} \cdot \text{s}}\right)$

t_c: residence time of catalyst, min

$\phi(t_c)$:   deactivation function $= \dfrac{1}{p^n[1 + \beta(60t_c)^\gamma]}$

p: partial pressure of oil (atm)

The poisoning effect of basic nitrogen compounds was accounted for by a specific deactivation function multiplying the gasoline formation rate constants:

$$f(N) = \cfrac{1}{1 + \cfrac{K_N}{1000} \cfrac{N}{\text{cat/oil}} \theta} \qquad \text{(VIII.2)}$$

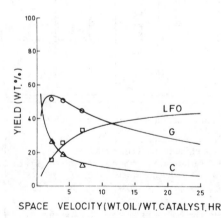

SPACE   VELOCITY(WT.OIL/WT.CATALYST.HR

Figure VIII.3   Comparison of experimental and computed cracking yields for a paraffinic charge stock as a function of weight hourly space velocity: $t_c = 5$ min; $T = 482°C$.

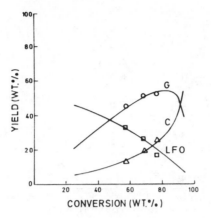

Figure VIII.4 Comparison of experimental and computed cracking yields for a paraffinic charge stock as a function of conversion. $t_c = 5$ min; $T = 482°C$.

where θ is the normalized catalyst residence time; N the mass of basic nitrogen (g) to which the catalyst has been exposed at a catalyst residence time t_c and K_N the adsorption coefficient of basic nitrogen. Correlations were derived for the detailed composition of the light ends fraction, C. The coke content of the catalyst, C_c (g coke/g cat) was obtained from the Voorhies equation:

$$C_c = \alpha t_c^n$$

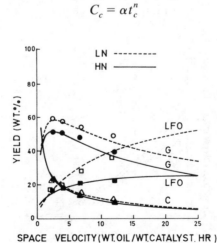

Figure VIII.5 Comparison of experimental and computed cracking yields for a heavy (HN) and a light naphthenic (LN) charge stock as a function of weight hourly space velocity. $t_c = 5$ min; $T = 482°C$.

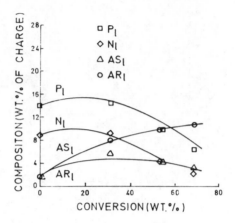

Fig. VIII.6 Comparison of experimental and computed LFO composition from a highly aromatic charge stock as a function of conversion.

with α a function of the feedstock composition. Over a dozen feedstocks, covering wide ranges of compositions, were investigated. Figures VIII. 3 to 7 compare model predictions with experimental results and illustrate the trends.

Performance maps were generated by the model and are presently used in all Mobil's commercial FCC-units to predict the most profitable mode of operation for each charge stock.

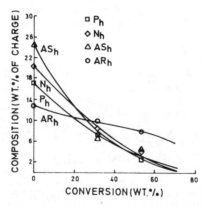

Fig. VIII.7 Comparison of experimental and computed heavy fuel oil (343°C+) composition from a highly aromatic charge stock as a function of conversion.

2. Modeling of Riser Cracking

Riser cracking is substantially different from the dense phase cracking modeled under Section II because the catalyst is transported, together with the vapor, in the vertical direction. Since the catalyst and vapor residence time are practically identical, the catalyst activity can no longer be considered as uniform in the reactor. Figure VIII.8 [34] schematically represents profiles of compositions of the vapor phase and catalyst properties in a riser cracker for 343°C+ gas oil.

As the conversion increases there is significant expansion and acceleration of both catalyst and vapor, while coke and poison deposit on the catalyst, affecting its activity and selectivity. Weekman [37] reported measurements of such profiles in a commercial Mobil riser cracker. The situation was also simulated using the 10 lump kinetic model discussed above. The reactor model was not disclosed, but it is likely that a plug flow model of the type given by Eq. (VIII.1) was used. Figures VIII.9 and VIII.10 show a comparison of measured and computed values.

Figure VIII.11 shows coke contents of the catalyst in a commercial Gulf riser cracker [31]. The question of whether or not catalyst and vapor phase have the same velocity is a matter of debate. Figure VIII.12 [42] shows information on the slip, $S = u_s/u_c$. Increasing the vapor velocity and decreasing the particle size reduces the slip.

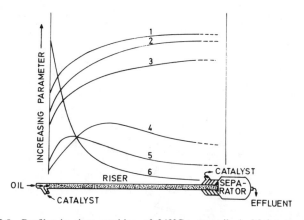

Fig. VIII.8 Profiles in riser cracking of 343°C+ gas oil. 1. Molar expansion. 2. Conversion of 343°C+ charge. 3. Coke, metals, nitrogen on catalyst. 4. C_5-220°C gasoline. 5. 220–343°C light fuel oil. 6. Catalyst activity decay.

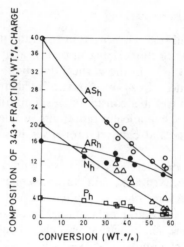

Figure VIII.9 Composition of 343°C+ fraction versus conversion in commercial riser cracking (after [37]).

Paraskos et al. [28] and Shah et al. [31] modeled riser cracking using the three lump model of Weekman and Nace [36] and the following reactor model, in which it is assumed that the gas oil vaporization is instantaneous, that the slip is negligible and that plug flow is valid for the vapor:

Gas oil conversion:

$$-\frac{F_{t_0}}{\Omega} \frac{dy_{GO}}{dz} = (k_1^0 + k_3^0)\rho_B \, e^{-\alpha\eta} \, \frac{\rho_g}{\rho_{go}} \, y_{GO}^2 \qquad \text{(VIII.3)}$$

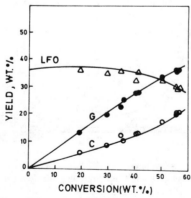

Fig. VIII.10 Yields versus conversion in commercial riser cracking (after [37]).

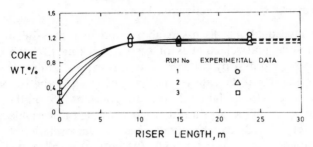

Figure VIII.11   Coke contents of the catalyst measured in a commercial Gulf
riser-cracker.

Gasoline formation:

$$\frac{F_{t_0}}{\Omega} \frac{dy_G}{dz} = \left(\nu k_1^0 \rho_B y_{GO}^2 - k_2^0 \rho_B y_G \right) e^{-\alpha\eta} \frac{\rho_g}{\rho_{GO}} \qquad (\text{VIII.4})$$

where $\eta = \dfrac{\Omega Z F_{\text{cat}}}{F_{t_0} F_{GO}}$.

It can be seen that the same deactivation function applies to the different reactions.

Whereas Paraskos et al. [28] only considered isothermal operation, Shah et al. [31], who dealt with a commercial unit, extended the above model to adiabatic operation.

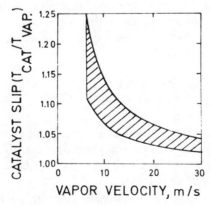

Fig. VIII.12   Slip versus vapor velocity in riser cracking (after [42]).

3. Remarks on the Deactivation Function

In the investigations discussed above and in others dealing more with kinetics and selectivities of gas oil cracking [27] the deactivation of the catalyst by coke deposition is expressed in terms of the time during which the catalyst has been exposed to the vapors. This idea goes back to a relation derived by Voorhies [35] on the basis of measurements made over a rather narrow range of variables. Later, Eberly et al. [10] experimentally showed this relation to be an oversimplification. Froment and Bischoff [15, 16] and Froment [12, 13, 14] developed kinetic and reactor models accounting in a more rigorous way for the reaction path leading to coke formation. Clearly, the coke content of a catalyst not only depends upon the time during which the catalyst was exposed to the vapor but also upon the composition of the latter. Therefore, the residence time of the catalyst is an insufficient yardstick of the coke content of the catalyst and, consequently, of its activity. The catalyst activity should be related to the true deactivating agent, which is the coke content. In kinetic studies this approach requires a separate rate equation for the coking and, consequently, measurements of the coke content of the catalyst as a function of time. In the simulation and design stage the coking rate equation has to be coupled with the continuity equations for the vapor phase components.

4. Modeling of the Regenerator

Regenerators are always operated with a fluidized dense bed, so that their modeling is based upon the two phase model of Section II, eventually with extensions. By way of example, work carried out by de Lasa and Grace [7] will be briefly discussed here.

Coke was considered to have the formula CH_2, so that stoichiometry of the reactions with oxygen was written, according to Weisz and Goodwin [38, 39]:

$$CH_2 + O_2 \longrightarrow CO + H_2O$$
$$CH_2 + 1.5O_2 \longrightarrow CO_2 + H_2O$$

with $r_c = k_c C_c y$

where k_c is the rate coefficient of carbon burning. The rate coefficient for the global oxygen consumption, k_{O_2}, has to account for the relative importance of each of the two reactions through the amounts of CO and CO_2 formed.

The emulsion phase was considered to be completely mixed. Bubbles erupt at the surface of the bed, ejecting particles into the freeboard. This is why this region is also accounted for. Small particles with a diameter such that their terminal velocity is lower than u_s are entrained to the cyclone and return to the bed (provided they are not too small). Particles with a d_p such that their terminal velocity exceeds u_s do not leave the freeboard and fall back into the dense bed. In the freeboard the gas and particles are in plug flow. There is no combustion anymore in the cyclone or standpipe.

The model equations can now be written:

For the bubble phase in the dense bed

Continuity equation for oxygen:

$$-u_b \frac{dy_b}{dz} = \frac{k_I}{f_b} (y_b - y_e). \qquad\qquad \text{(VIII.5)}$$

For the emulsion phase in the dense bed

Oxygen balance:

$$(F_A)_{mf}[(y_e)_{in} - (y_e)_{out}] + \frac{N_b Q \rho_A u_b}{Z M_A} \int_0^{\frac{Z}{u_b}} (y_b - y_e) d\left(\frac{z}{u_b}\right)$$
$$= k_{O_2} y_e C_c W \qquad\qquad \text{(VIII.6)}$$

Coke balance:

$$[(F_{cat} C_c)_{in} + (F_{cat} C_c)_{cycl} + (F_{cat} C_c)_Z] - [F_{cat} C_c + (F_{cat} C_c)_Z]$$
$$= k y_e C_c W . \qquad\qquad \text{(VIII.7)}$$

Notice that the two terms $(F_{cat} C_c)_Z$ in the left hand side of (VIII.7) are not identical, since their coke content has been modified by the excursion of the catalyst in the freeboard.

Freeboard region:
Continuity equation for coke on a catalyst particle with size $(d_p)_i$:

$$\pm (F_{cat})_i \frac{d(C_c)_i}{dz} = r_{c_i} \Omega (1 - \epsilon)_i \rho_s \qquad\qquad \text{(VIII.8)}$$

($-$ sign for d_p with $u_T < u_s$ and $+$ sign for d_p with $u_T > u_s$).

Oxygen balance:

$$-F_A \frac{dy_f}{dz} = k_{O_2}\left[\sum_i (1 - \epsilon)_i (C_c)_i\right] y_f \Omega \rho_s \qquad (VIII.9)$$

Energy equations:
Overall:

$$(F_{cat}c_{ps}T)_{in} + (F_A c_{pA}T)_{in} - (F_{cat}c_{ps}T)_{out} - (F_g c_{pg}T)_{out}$$
$$= \frac{\Delta H F_{cat}(C_c)_{in} x_c}{M_c} \qquad (VIII.10)$$

Freeboard:

$$\frac{d}{dz}\left(c_{ps_i}\sum_i (F_{cat})_i T_i + F_g c_{pg}T\right) = \frac{(-\Delta H)\Omega \rho_s}{M_c}\left[\sum_i r_{c_i}(1 - \epsilon)_i\right]$$

$$(VIII.11)$$

Notice that the summation in the left hand side of (VIII.11) contains positive terms for a catalyst particle with upward flow and negative terms for particles flowing downwards.

For catalyst particles with d_p whose terminal velocity is smaller than u_s the mass flow rate $(F_{cat})_i$ is obtained, according to George and Grace [18], from:

$$(F_{cat})_i = M_0 \cdot \Omega \qquad (VIII.12)$$

where

$$M_0 = \frac{1}{\xi} \rho_s (1 - \epsilon_{mf})\beta(d_b)(u_s - u_{mf}) \qquad (VIII.13)$$

with $\beta = 0.025 d_b - 0.014$ (with d_b in cm) for a cracking catalyst e.g. and where $\xi = 1$ if one of the basic hypotheses of the two-phase model is satisfied, namely that all the gas flowing through the bed in excess of that required for minimum fluidization is in the bubble phase. If it were not, $\xi > 1$.

For particles with diameters leading to a terminal velocity exceeding u_s:

$$(F_{cat})_i = C_{ij}u_{ij} = M_0 w_i w_j \Omega \qquad (VIII.14)$$

The absolute upward velocity, u_{sj}, is obtained from a numerical integration of the equation of motion for the particle, w_i is the weight fraction of particles with diameter $(d_p)_i$ and w_j the weight fraction of particles with ejection velocity u_{ij}. Results from the simulation of a commercial unit based upon this model are shown in Fig. VIII.13.

Figure VIII.13 shows the evolution of conversion in the dense bed with and without accounting for the reaction in the freeboard. The difference can be quite significant for shallow beds, not only for the coke conversion, but also for the temperature. When the coke content of the catalyst fed to the regenerator is higher than in the example, the difference decreases, however. In that case the re-generator operation is controlled by oxygen and the oxygen content of the gases is very low in the freeboard. It is also seen from Fig. VIII.13 that the coke is very rapidly converted in the particles with small diameter, which are frequently recycled through the freeboard. Below 150 μ this effect is somewhat reversed: the velocity of these particles is so high that the lower residence time per pass in the freeboard is now being felt. As already mentioned, another aspect that has to be accounted for in the development of more realistic models for dense bed is the effect of jets originating from the grid. Behie and Kehoe [2] used the model illustrated in Fig. VIII.14 to study this effect.

Grace and de Lasa [19] showed that a model of this type predicts significantly higher conversions than the classical two-phase model, since the resistance to transfer between grid jets and emulsion phase

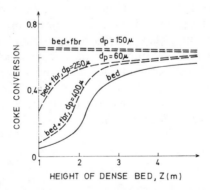

Fig. VIII.13   Influence of particle size and depth of dense bed on coke conversion with and without freeboard. Inlet coke content: 7.5×10^{-3} kg coke/kg cat; height of dense bed + freeboard: 10 m; air flow rate: 73.9 m^3/s; $T_0 = 474°C$; $F_{cat} = 506$ kg/s. Dense bed modeled as CSRT (after [7]).

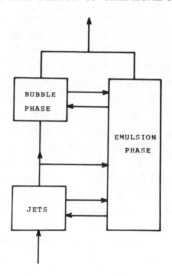

Figure VIII.14   Two-phase model accounting for jets originating from the grid.
(after [2]).

is much lower than that between the bubbles and the emulsion
phase.

IX. CONCLUSION

Fluidization is a very complex phenomenon and its modeling is still
in the development stage. Many models have been presented, some
of which focus on aspects which are of minor importance only. What
is required in the first place for further progress are more experimental data obtained in equipment of a scale which is sufficiently
representative. Direct observations of hydrodynamic nature would
be extremely useful. These would enable us to develop much more
realistic models, which will necessarily be complex. Their simulation
will require huge computer capacity.

NOTATION

$\quad$ **a** $\quad$ composition vector, moles/g gas
$\quad A_0 \quad$ area of plate per orifice in distributor, m^2
$\quad C_A \quad$ concentration of component A, kmol/m^3 gas

C_c coke content of catalyst, g coke/g cat

c_{pg} specific heat of gaseous mixture, kJ/kg K

c_{ps} specific heat of cracking catalyst, kJ/kg K

c_{pA} specific heat of air, kJ/kg K

d_b bubble diameter, m

d_t vessel diameter, m

D_{bc} molecular diffusivity of gas in interchange zone, $m_g^3/m_b s$

D_{be} molecular diffusivity of gas from bubble phase into emulsion phase, $m_g^3/m_b s$

D_{ce} molecular diffusivity of gas in interchange zone, $m_g^3/m_b s$

D_e effective diffusivity in emulsion phase, $m_r^3/m_r s$

f_b fraction of bed volume taken by bubble phase, m_b^3/m_r^3

f_e fraction of bed volume taken by emulsion gas, m_g^3/m_r^3

F_A molar feed rate of air, kmol/s

F_{cat} mass flow rate of catalyst, kg cat/s

F_g mass flow rate of gaseous mixture, kg/s

F_{GO} mass flow rate of gas oil feed, kg gas oil/s

F_{t_0} total mass flow rate at inlet, kg fluid/s

k rate coefficient, $m_g^3/$kg cat · s

k' rate coefficients in VIII.1, $cm^3/$g cat · s

k_c rate coefficient for coke combustion, kmol air/ kmol O_2 · s

k_{O_2} rate coefficient for oxygen consumption, kmol air/ kg coke · s

k_1^0, k_2^0, k_3^0 rate coefficients in VIII-4, $s^{-1}(kg\,cat)^{-1}\dfrac{(kg\,feed)^n}{(kg\,comp)^n}$

k_I interchange coefficient, $\dfrac{m_g^3}{m_r^3 \cdot s}$

$(k_{bc})_b$ mass transfer coefficient from bubble to interchange zone, based on bubble volume, $m_g^3/m_b^3 s$

$(k_{ce})_b$ mass transfer coefficient from interchange zone to emulsion, based on bubble volume, $m_g^3/m_b^3 s$

$(k_{be})_b$ mass transfer coefficient from bubble to emulsion, based on bubble volume, $m_g^3/m_b^3 s$

K_{ARh} adsorption constant of aromatic rings in 345°C + fraction

K_N adsorption constant of basic nitrogen lump

M molecular weight of vapor at height z, kg/kmol

M_c molecular weight of coke, kg/kmol

M_A molecular weight of air, kg/kmol

M_0 mass flux of particles entering freeboard, $kg/m^2 \cdot s$

n order of reaction

N mass of basic nitrogen to which catalyst has been exposed, g

N_b number of bubbles in the bed at any instant

p partial pressure of oil, atm

p_t total pressure, atm

Q cross flow, m^3/s

r_A rate of reaction of A, kmol/kg cat $\cdot$ s

r_c rate of coke combustion, kg coke/kg cat $\cdot$ s

R' gas constant (82.05 in VII.1), atm cm^3/gmol K

t_c residence time of catalyst, min

u_b bubble rising velocity, m_r/s or m_b/s

u_{br} bubble rising velocity relative to emulsion, $m_g^3/m_r^2 s$

u_c superficial catalyst velocity, m_{cat}^3/m_r^2 s

u_e emulsion gas flow velocity, interstitial, m_r/s

u_{mf} superficial minimum fluidization velocity, m_g^3/m_r^2 s

u_s superficial gas velocity, m_g^3/m_r^2 s

w_i weight fraction of particles with diameter d_{p_i}

w_j weight fraction of particles with ejection velocity u_{ij}

x_c coke conversion

y, y_b, y_e, y_f oxygen mol fraction in regenerator

y_G, y_{GO} weight fraction of gasoline or gas oil, kg/kg feed

z axial position in fluidized bed, m

z_m height at which bubbles reach maximum size, m

Z total height of fluidized bed, m

Greek Letters

ϵ void fraction in freeboard, m_g^3/m_e^3

ϵ_{mf} void fraction of bed at minimum fluidization, m_g^3/m_r^3

ν stoichiometric coefficient

ρ_A density of air, kg/m_g^3

ρ_b density of bubble phase, kg/m_b^3

ρ_e density of emulsion phase, kg/m_e^3

ρ_B catalyst bulk density, kg/m_r^3

ρ_g point density of inerts + HC vapors, kg/m_g^3

ρ_{g_0} id. at inlet, kg/m_g^3

ρ_s true catalyst density, kg/m^3 cat

θ normalized residence time

$$\eta \quad \text{flowing space time} = \frac{\Omega z}{F_{t_0}} \frac{F_{\text{cat}}}{F_{\text{GO}}}, \frac{m_r^3 \, \text{kg cat} \cdot \text{s}}{\text{kg feed} \cdot \text{kg}_{\text{GO}}}$$

$\zeta \quad \dfrac{z}{Z}$ dimensionless axial position

$\Omega \quad$ cross section of reactor, m_r^2

Subscripts

b   bubble

g   gas

r   reactor

REFERENCES

1. Baird, H. M. I., and Rice, R. G. (1975). Axial dispersion in large unbaffled columns, *Chem. Eng. J.*, **9**, 171.
2. Behie, L. A., and Kehoe, P. (1973). The grid region in a fluidized bed reactor, *A.I.Ch.E. J.*, **19**, 1070.
3. Darton, R. C., La Nauze, R. D., Davidson, J. F., and Harrison, D. (1977). Bubble growth due to coalescence in fluidized beds, *Trans. Inst. Chem. Engrs.*, **55**, 274.
4. Davidson, J. F., and Harrison, D. (1963). *Fluidized Particles*, Cambridge University Press.
5. Davidson, J. F. (1977). In *Chemical Reactor Theory*, Chapter 10 (Eds. L. Lapidus and N. R. Amundson), Prentice-Hall, Englewood Cliffs, N.J.
6. De Groot, J. H. (1967). *Proc. Symp. Fluidization, Eindhoven*, p. 348, Netherlands University Press.
7. De Lasa, H. I., and Grace, J. R. (1979). The influence of the freeboard region in a fluidized bed catalytic cracking regenerator, *A.I.Ch.E. J.*, **25**, 984.
8. De Lasa, H. I., Errazu, A., Porras, J., and Barreiro, E. (1981). Influence of the pneumatic transport line in the simulation of a fluidized bed catalytic cracking regenerator, *Lat. Am. J. Chem. Eng. Appl. Chem.*, **11**, 139.
9. De Vries, R. J., Van Swaay, W. P. M., Mantovani, C., and Meijkoop, A. (1972). *Proc. 5th European/2nd Int. Symp. Chem. React. Eng., Amsterdam 1972*, Elsevier, Amsterdam.
10. Eberly, P. E., Kimberlin, C. N., Muller, W. H., and Drushel, H. V. (1966). Coke formation on silica–alumina cracking catalysts, *Ind. Eng. Chem. Proc. Des. Dev.*, **5**, 193.
11. Errazu, A. F., de Lasa, H. I., and Sarti, F. (1979). A fluidized bed catalytic cracking regenerator model, *Can. J. Chem. Eng.*, **57**, 191.
12. Froment, G. F. (1976). *Proc. Int. Congress on Catalysis, London.*
13. Froment, G. F. (1980). In *Catalyst Deactivation* (Eds. B. Delmon and G. F. Froment), Elsevier, Amsterdam.
14. Froment, G. F. (1982). In *Progress in Catalyst Deactivation* (Ed. J. L. Figueiredo), M. Nijhoff, The Hague.
15. Froment, G. F., and Bischoff, K. B. (1961). Non-steady state behaviour of fixed bed catalytic reactors due to catalyst fouling, *Chem. Eng. Sci.*, **16**, 189.
16. Froment, G. F., and Bischoff, K. B. (1962). Kinetic data and product distributions from fixed bed catalytic reactors subject to catalyst fouling, *Chem. Eng. Sci.*, **17**, 105.
17. Froment, G. F., and Bischoff, K. B. (1979). *Chemical Reactor Analysis and Design*, John Wiley, New York.

18. Goerge, S. E., and Grace, J. R. (1978). Entrainment of particles from aggregative fluidized beds, $A.I.Ch.E. Symp. Ser.$, **74**, 176.
19. Grace, J. R., and de Lasa, H. I. (1978). Reaction near the grid in fluidized beds, $A.I.Ch.E. J.$, **24**, 364.
20. Jacob, S. M., Gross, B., Voltz, S. E., and Weekman, V. M. (1976). A lumping and reaction scheme for catalytic cracking, $A.I.Ch.E. J.$, **22**, 701.
21. Krishna, R. (1981). In NATO Advanced Study Institute Series, $Multiphase$ $Chemical Reactors$ (Eds. A. Rodriguez, J. M. Calo, and N. Swede), Sijthoff & Noordhoff, The Netherlands.
22. Kunii, D., and Levenspiel, O. (1969). $Fluidization Engineering$, John Wiley, New York.
23. Mathis, J.F., and Watson, C. C. (1956). Effect of fluidization on catalytic cumene dealkylation, $A.I.Ch.E. J.$, **2**, 518.
24. May, W. G. (1959). Fluidized-bed reactor studies, $Chem. Eng. Progr.$, **55**, 49.
25. Mirreur, J. P., and Bischoff, K. B. (1967). Mixing and contacting models for fluidized beds, $A.I.Ch.E. J.$, **13**, 839.
26. Murray, J. D. (1965). $J. Fluid. Mech.$, **21**, 465; **22**, 57.
27. Pachovsky, R. A., John, I. M., and Wojciechowski, B. N. (1973). Theoretical interpretation of gas oil selectivity data on X-sieve catalyst, $A.I.Ch.E. J.$, **19**, 802.
28. Paraskos, J. A., Shah, Y. T., McKinney, J. D., and Carr, N.L. (1976). A kinematic model for catalytic cracking in a transfer line reactor, $Ind. Eng. Chem.$ $Proc. Des. Dev.$, **15**, 165.
29. Partridge, B. A., and Rowe, P. N. (1966). Chemical reaction in a bubbling gas-fluidized bed, $Trans. Instn. Chem. Engrs.$, **44**, T335.
30. Pyle, D. L. (1972). Fluidized-bed reactors. Review, $Adv. Chem. Ser.$, **109**, 106 (A.C.S. Washington).
31. Shah, Y. T., Huling, G. P., Paraskos, J. A., and McKinney, J. D. (1977). A kinematic model for an adiabatic transfer line catalytic cracking reactor, $Ind.$ $Eng. Chem. Proc. Des. Dev.$, **16**, 89.
32. Van Deemter, J. J. (1961). Mixing and contacting in gas-solid fluidized beds, $Chem. Eng. Sci.$, **16**, 89.
33. Van Swaay, W. P. M., and Zuiderweg, F. J. (1972). $Proc. 5th Eur. /2nd Int.$ $Symp. Chem. React. Eng.$, $Amsterdam 1972$, Elsevier, Amsterdam.
34. Venuto, P. B., and Habib, E. T. (1978). Catalyst-feedstock-engineering interactions in fluid catalytic cracking, $Catal. Rev. Sci. Eng.$, **18**(1), 1.
35. Voorhies, A. (1945). C formation in catalytic cracking, $Ind. Eng. Chem.$, **37**, 318.
36. Weekman, V. W., and Nace, D. M. (1970). Kinetics of catalytic cracking selectivity in fixed, moving, and fluid bed reactors, $A.I.Ch.E. J.$, **16**, 397.
37. Weekman, V. W. (1979). Lumps, models, and kinetics in practice, $A.I.Ch.E.$ $Monogr. Series$, **11**, Vol. 75.
38. Weisz, P. B., and Goodwin, R. B. (1963). Combustion of carbonaceous deposits within porous catalyst particles. 1. Diffusion-controlled kinetics, $J. Catalysis$, **2**, 397.
39. Weisz, P. B., and Goodwin, R. B. (1966). Combustion of carbonaceous deposits within porous catalyst particles, 2. Intrinsic burning rate, $J. Catalysis$, **6**, 227; 3. CO_2/CO product ratio, $J. Catalysis$, **6**, 425.
40. Werther, J. (1977). Zur Problematik der Massstabsvergrössering von Wirbelschichtreaktoren, $Chem. Ing. Techn.$, **49**, 777.
41. Werther, J. (1978). Mathematische Modellierung von Wirbelschichtreaktoren, $Chem. Ing. Techn.$, **50**, 850.
42. Whittington, E. L., Murphy, J. R., and Lutz, L. H. (1972). $Prepr. Div. Petr.$ $Chem. Am. Chem. Soc.$, **17**(3), B-66.
43. Yates, J., and Rowe, P. N. (1977). A model for chemical reaction in the freeboard region above a fluidized bed, $Trans. Inst. Chem. Eng.$, **55**, 137.

Chapter 6

Recent Advances in Slurry Reactors

R. V. CHAUDHARI
National Chemical Laboratory, Pune, India

and

Y. T. SHAH
Chemical and Petroleum Engineering Department, University of Pittsburgh, Pittsburgh, PA 15261

Abstract

Slurry reactors are widely used in hydroprocessing, chemical, biochemical and fermentation industries. The design and scaleup of such reactors requires proper understanding of hydrodynamics, mixing and transport characteristics as a function of operating variables. This paper reviews recent advances on this subject. The paper also evaluates the modelling of a slurry reactor for a variety of simple and complex reactions. While the major emphasis is given to conventional, mechanically agitated and bubble column reactors, novel reactors such as the loop recycle slurry reactor, three-phase fluidized bed reactors and multistage column reactors are also considered. The review indicates that the future work should place more emphasis on the experimental study of the hydrodynamics, mixing and transport characteristics of real reactors processing reacting materials with complex physical properties. The reactor modelling should also be directed more towards real case studies.

1. INTRODUCTION

Slurry reactors have a wide range of applications in chemical processes and in particular catalytic processes (see Table 1). In this type of contactor, the solid catalyst is suspended either by mechanical or gas induced agitation. The slurry is often a continuous phase with dispersed gas bubbles. Two major types of slurry reactors, (a) mechanically agitated slurry reactors and (b) bubble column slurry reactors, are schematically shown in Fig. 1. Besides these, specially

243

Table 1 Important Applications of Slurry Reactors

System	Catalyst	Reactor Type	Pressure Atm.	Temperature °C	References
Hydrogenation					
(1) Soybean Oil	Cu-chromite or	Stirred Slurry	2.0	170–200	Koritala and Dutton [86]
	Supported Ni	Bubble Column Slurry	1.5–4.8	180–200	Schmidt [143]
		Bubble Column Slurry	1.0	185–215	Snyder et al. [156]
(2) Rapseseed Oil	Nickel	Stirred Slurry	0.3–10	140–220	Bern et al. [9]
(3) Glucose	Raney Ni	Stirred Slurry and Bubble Column	100–250	170–220	Brahme and Doraiswamy [17]
(4) 2,4-Dinitrotoluene	Pd-alumina	Stirred Slurry	1–5	25–70	Acres and Cooper [1]
(5) Butynediol	Pd–Zn–CaCO$_3$	Stirred Slurry	10–20	50–90	Chaudhari et al. [24]
(6) Carbon Monoxide (F–T Synthesis)	Ni–MgO	Bubble Column	8–30	200–320	Kolbel et al. [84]
(7) Chloronitro Compounds	Pd–C	Loop-Recycle	5–10	60–90	Malone [97]
(8) Adiponitrile	Pd-Supported	Bubble Column or Loop Recycle	1–20	50–120	Weissermel and Arpe [168]
Hydrodesulfurization of Oil	CoO–MoO$_2^-$––Al$_2$O$_3$	Fluidized bed Column	50–140	250–350	Mounce and Rubin [109]
Oxidation of SO$_2$	Activated-C	Stirred Slurry	1	25	Komiyama and Smith [85]

244

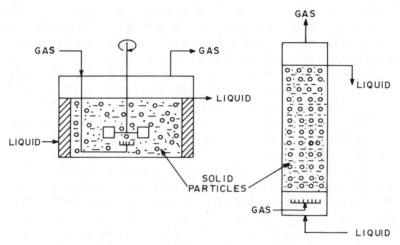

Figure 1   Schematic diagrams of three phase slurry reactors.

designed novel reactors like: (a) loop recycle slurry reactor; (b) jet reactor; (c) three-phase fluidized bed reactor; (d) multistage column reactor; (e) column reactor with multiple agitators; and (f) sieve tray column reactor have also been used in industrial practice. The subject of slurry reactors has been reviewed by Chaudhari and Ramachandran [25], Ramachandran and Chaudhari [132] and Shah [146]. Various aspects of mass transfer effects, complex kinetics, reactor models and hydrodynamics have been considered in detail in these reviews. However, following these reviews several new aspects, particularly on hydrodynamics and reactor modeling, have been reported. It is the purpose of this article to review the recent advances in slurry reactors. Although the major emphasis will be placed on mechanically agitated and bubble column slurry reactors, some aspects of novel reactors will also be briefly evaluated.

The well-known advantages of the slurry reactor over other types of three phase reactors (such as trickle bed or fixed bed upflow reactor) are: (1) intraparticle diffusional resistance is less compared to trickle or packed bubble column reactors due to the use of smaller catalyst particles; (2) better control of temperature can be achieved due to higher heat transfer efficiency and the heat capacity of slurries. The possibility of hot spot formation and temperature runaway is also minimized; (3) it is not necessary to shut down plants for catalyst replacement as it can be removed continuously or added while the plant is in operation; (4) the problem of partial wetting of catalyst (as observed in trickle bed reactors) is not encountered in slurry reactors, which leads to better utilization of the catalyst surface; and (5), under comparable conditions, the space time yield is often higher in slurry reactors.

In spite of the several advantages of slurry reactors, some technical difficulties are involved in the operation of these reactors such as: (1) separation of the catalyst and handling of the slurry is difficult; (2) the solids can produce erosion of the equipment; and (3) significant backmixing of the liquid phase does not allow operation in a plug flow manner.

The choice of a particular type of slurry reactor often depends on the process as well as the scale of operation. For example, for batch hydrogenation of nitroaromatics or manufacture of small volume chemicals, mechanically agitated slurry reactors are often used. In recent years however, these reactors have been replaced by loop recycle reactors. Buss engineering design of the loop slurry reactor is reported to be highly efficient [85] for the fast hydrogenation reactors particularly because of its high mass and heat transfer efficiency. When a high quality product is required, such as in the pharmaceutical industries, a mechanically agitated batch reactor is preferred. The well-known Fischer–Tropsch reaction is generally carried out in a bubble column slurry reactor, which is particularly suited for continuous operation. The hydrogenation of fatty acids and esters is carried out in a bubble or a multistage agitated slurry reactor. Thus, the reactor type is generally chosen to achieve a particular objective in the process. In order to understand the design principles of these reactors, hydrodynamic, mass transfer, mixing, as well as surface chemical reaction characteristics need to be well understood. The important design parameters are (1) flow regimes, (2) bubble size distribution, (3) gas, liquid and solid holdups, (4) gas–liquid and liquid–solid mass transfer coefficients, (5) gas–liquid, liquid–particle interfacial areas, (6) dispersion coefficients in gas, liquid and solid phases, (7) heat transfer characteristics and (8) reaction kinetics parameters. For the rational design and scaleup of reactors, it is important to have a thorough knowledge of how these parameters will vary with operating conditions and physical properties of the systems. It is also necessary to understand the effect of mass transport, mixing, exothermicity and complex kinetics on the performance of the reactors. While some of these aspects have been dealt with in earlier publications [25, 146], it is the purpose of this review to present the recent advances in the design and analysis of slurry reactors. In all equations presented in

this manuscript, the units to be used are those given in the nomenclature except where a specific set of units is given.

2. HYDRODYNAMICS AND FLOW REGIMES IN
SLURRY REACTORS

A slurry reactor generally contains gas and liquid phase reactants
with a solid phase catalyst. The overal performance of such a
reactor (conversion of reactants) depends upon gas–liquid and
liquid–particle mass transfer, surface chemical reaction and the
prevailing mixing characteristics. The mass transfer and mixing
characteristics are influenced by the hydrodynamical interaction
between the various phases which in turn depends upon the engineering design of the reactor. Important hydrodynamic parameters are also dependent on the flow regime prevailing within the
reactor.

2.1. Mechanically Agitated Reactors

2.1.1. Flow Regimes
The flow patterns in a mechanically agitated reactor depend largely
on the type of stirrer and its speed, and the range of gas and liquid

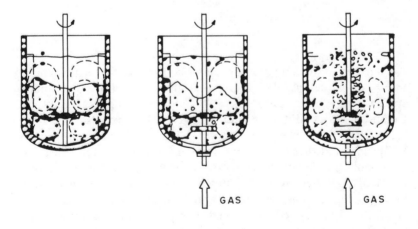

GAS   GAS

SLURRY   SLIGHTLY AERATED   AERATED SLURRY AT
                     SLURRY   FLOODING POINT

Figure 2   Schematic of flow patterns in mechanically agitated slurry reactor. (After
Wiedman et al. [170]).

velocities. Wiedmann et al. [170] studied the flow regimes and flooding characteristics of this type of reactor. A schematic of the flow patterns proposed by them is shown in Fig. 2. They found that at low gas velocity, the gas stream does not change the flow pattern significantly, and the bubbles and particles are essentially carried along by the circulation flow produced by the stirrer. With increasing gas flow rate, however, the liquid circulation gets more and more pronounced. Further increase of gas rate leads to the so-called flooding point at which the gas breaks through in the middle of the vessel without dispersing. Under such conditions, the radial flow from the stirrer declines drastically and the escaping gas determines the flow in the vessel. Also, in this situation, the particles tend to settle down at the bottom due to reduced effect of the stirrer. Further work on this aspect is necessary to characterize conditions under which flooding behavior occurs. Meaningful data on systems of practical interest are presently not available.

2.1.2. Power Consumption

It is essential to determine the power consumption of the stirrer in agitated slurry reactors, as this quantity is required in the prediction of parameters such as gas holdup, and mass transfer coefficients. For the gas–liquid systems Michel and Miller [104] proposed the following correlation to predict the power consumption:

$$P = 0.812 \left[\frac{P_0^2 N d_I^3}{Q_G^{0.56}} \right]^{0.45} \tag{1}$$

All units in above equation are described in the list of nomenclature. Loiseau et al. [94] found that their data for non-foaming systems agreed well with Eq. (1). Calderbank [20], Hassan and Robinson [50] and Luong and Volesky [95] have also proposed correlations for power consumption in gas–liquid systems. Nagata [112] suggested that power consumption for agitated slurries can be reasonably predicted from these correlations by the correction factor ρ_{sl}/ρ_L, where ρ_{sl} is the density of slurry. Power consumption for a gas–liquid–solid system has also been studied by Wiedmann et al. [170]. They studied the influence of gas velocity, solid loading, type of stirrer and position of the stirrer blades on power consumption; a plot of power number vs. Reynolds number for propeller and turbine type impellers proposed by them is shown in Fig. 3. They found that optimal conditions of the position of stirrer exist in the range of $0.25 < L/d_I < 0.75$ for both turbine and propeller stirrers.

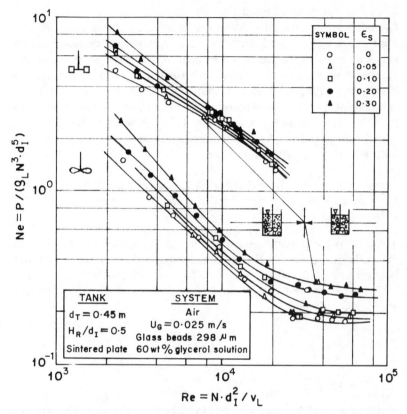

Figure 3   Effect of solid content on power number vs. Reynolds number plot. (After Wiedman et al. [170]).

Their work suggests that beyond a Reynolds number of 20,000, the power number becomes constant. An increase in solid content increases the power consumption, while an increase in gas velocity reduces the power consumption. Kurten and Zehner [88] have studied the effect of gas velocity on the power consumption for suspension of solids. They found that due to simultaneous aeration, a higher power input is required for suspension in the presence of gas.

Although several correlations are proposed for predicting power consumption in agitated gas–liquid systems, limited information is available for three phase slurry systems. Most recently, Albal et al. [3] have evaluated the effect of liquid properties on power consumption for both two and three phase systems.

2.1.3. Suspension of Solid Particles

One of the primary considerations in agitated slurry reactors is to ensure that all the solid particles are in complete suspension. There exists a critical agitation speed to achieve complete suspension. An extensive investigation of complete suspension has been done by Zwietering [178] who stated that suspension can be considered complete if no particle remains at the bottom of the vessel for more than 1 or 2 seconds. Zwietering [178] proposed the following correlation to predict N_m, the minimum rotational speed of agitation required for complete suspension.

$$N_m = \frac{\beta d_p^{0.2} \mu_L^{0.1} g^{0.45} (\rho_p - \rho_L)^{0.45} w'^{0.13}}{\rho_L^{0.55} d_I^{0.85}} \tag{2}$$

where w' is the percentage loading and β is a constant dependent on the ratio (d_T/d_I). All the nomenclature and units for the above equation are described in the list of nomenclature. Nienow [115, 116] suggested that β can be approximated for a disc turbine impeller as:

$$\beta = 2\left(\frac{d_T}{d_I}\right)^{1.33} \tag{3}$$

Albal et al. [3], Baldi et al. [6], Joosten et al. [63], Kolar [83], Kurten and Zehner [88] and Narayan et al. [113] have also studied suspension of solids in stirred vessels. The correlations of Baldi et al. [6] and Zwietering [178] are based on data over a wide range of conditions and are also in good agreement with each other. Therefore, the correlation of Zwietering [178] is recommended for design purposes. Albal et al. [3] have, however, shown that the Zwietering correlation does not work well in an unconventional agitated vessel. More recently, Zlokarnik and Judat [177] studied suspension in a stirred vessel with a self inducing hollow propeller stirrer for air–water–solid (quartz, iron powder and gelena) system. Kurten and Zehner [88] and Wiedmann et al. [170] have also studied suspension of solids in a three-phase slurry system. They found that in the presence of gas bubbles, a higher stirrer speed was required to keep the particles in suspension. This is mainly due to the reduced liquid circulation velocity caused by the reduction in power consumption in the presence of gas bubbles. The minimum stirrer speed for complete suspension also depends on the design and

Table 2 Summary of Studies on Suspension in Agitated Gas–Liquid Reactors

S. No.	Reference	System	Impeller Type	Particle Size, μ
1	Zwietering [178]	Sand–Water, Acetone, Carbon Tetrachloride	Paddle, Six Blade Turbine and Propeller	125–850
2	Nienow [116]		Disc Turbine	—
3	Baldi et al. [6]	Sand–Water, aq $MgSO_4$ soln.	Disc Turbine (8 Blades)	50–545
4	Zlokarnik and Judat [177]	Air–Water–Quartz, Iron Powder	Hollow Propeller Self Induced	—
5	Wiedmann et al. [170]	Air–Water– Glass Beads	Rushton Turbine, Propeller	298, 430
6	Chapman et al. [23]	Air–Water Polystyrene, Glass Powder	Rushton Turbine, Propeller	180–355
7	Albal et al. [3]	Air–Water– Glass Beads	Turbine with six blades conventional and unconventional	100–600

positioning of the stirrer. A brief summary of the important studies on suspension of solids in stirrer vessels is given in Table 2.

Baldi et al [7] observed the concentration profiles of solids suspended in a continuously agitated reactor. The concentration profiles were found to be dependent on stirrer speed and average diameter of the solid particles.

2.1.4. Gas Holdup

Gas holdup is an important hydrodynamic parameter in stirred reactors, because it determines the gas–liquid interfacial area and hence the mass transfer rate. Several studies on gas holdup in agitated gas–liquid systems have been reported [20, 36, 94, 165, 173, 174] and a number of correlations have been proposed. These are summarized in Table 3. For a slurry system, only a few studies have been reported [88, 170]. In general, the gas holdup depends on superficial gas velocity, power consumption, surface tension and viscosity of liquids and the solid concentration. While the dependence of gas holdup on viscosity is mild; for other parameters the following equation has been proposed:

$$\varepsilon_g \sim u_g^{0.36-0.75} P^{0.26-0.47} S_T^{-0.36-0.65} \tag{4}$$

The correlations of Calderbank [20] and Loiseau et al. [94] agree reasonably well with most literature data and are hence recommen-

Table 3 Summary of Correlations for Gas Holdup in Mechanically Agitated Reactors

S. No.	Reference	Correlation for ε_g
1	Calderbank [20]	$\varepsilon_g = \left(\dfrac{u_g}{u_B}\right)^{0.5} + \dfrac{0.0126(P/V_L)^{0.4}L^{0.2}}{S_T^{0.6}}\left(\dfrac{u_g}{u_B}\right)^{0.5}$
2	Loiseau et al. [94]	$\varepsilon_g = 0.011 u_g^{0.36} S_T^{-0.36} L_L^{-0.056}\left(\dfrac{P}{V_L} + \dfrac{P_g}{V_L}\right)^{0.27}$
		For non-foaming systems
		$\varepsilon_g = 0.051 u_g^{0.24}\left(\dfrac{P}{V_L} + \dfrac{P_g}{V_L}\right)^{0.57}$
		For foaming systems
3	Yung et al. [174]	$\varepsilon_g = 6.8 \times 10^{-3}\left(\dfrac{Q_g}{Nd_I^3}\right)^{0.5}\left(\dfrac{\rho_L N^2 d_I^3}{S_T}\right)^{0.65}\left(\dfrac{d_I}{d_T}\right)^{1.4}$

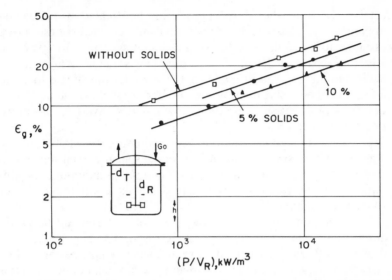

Figure 4   Effect of solid content on gas holdup in stirred reactor-sulfite solution/air/ glass, six blade turbine, $h = 0.2d_T$, $d_R = 0.4d_T$, $d_P = 200$ μ, $u_G = 4$ cm/sec. (After Kurten and Zehner [88]).

ded. These correlations, however, do not account for the effect of solid concentration on ε_g.

Kurten and Zehner [88] have studied the effect of solid content on gas holdup in a stirred vessel with a six blade turbine impeller using air–sulphite solution–glass beads system. They found that ε_g is reduced with an increase in solid concentration (see Fig. 4). In an independent study, Wiedmann et al. [170] reported similar observations on ε_g variation with solid loading. The effect of solid loading was particularly significant at higher gas velocities.

2.2.   Bubble Column Slurry Reactors

Bubble column slurry reactors are extensively used in many industrial processes. The design parameters for bubble columns have also been widely studied and this subject has been reviewed by Mashelkar [99], and Shah et al. [149]. Only the recent advances on this subject will be briefly outlined here.

2.2.1.   Flow Regimes

The flow regimes in a bubble column have been characterized on the basis of the distribution of bubble size and their upward movement [167]. For very small catalyst particles (<50 μm) and low catalyst loading, the slurry phase can be treated as a pseudo homogeneous phase and the hydrodynamic characteristics can be assumed to be similar to that of gas–liquid systems. However, for larger catalyst particles (as encountered in gas–liquid–solid fluidized beds) and higher catalyst loadings, the presence of solids is likely to affect the bubble size distribution and hence the flow regimes [146]. For most practical situations, the flow regimes have been classified into three categories:

1. Quiescent bubble flow regime—in this regime gas bubbles of almost uniform size are distributed equally in the axial and radial directions. The interaction between bubbles is a minimum and this regime is likely to exist only at gas velocities less than 5 cm/s for water-like liquids [34].

2. Heterogeneous bubble flow (or Churn turbulent flow) regime—when the gas flow rate is increased, the bubbles tend to coalesce with each other leading to an unsteady flow pattern. Here large bubbles appear in the system along with small bubbles and this results in a heterogeneous bubble flow regime. The large bubbles travel with higher rise velocities than small bubbles [59].

3. Slug flow regime—in small diameter columns at very high gas flow rates, the diameters of the bubbles formed by coalescence are comparable to the diameter of the column. These large bubbles are stabilized by the column wall leading to the formation of bubble slugs.

The various flow regimes are schematically shown in Fig. 5. For a given gas flow rate and column diameter, the flow regime can be estimated roughly by the flow map shown in Fig. 6. The type of gas distributor, liquid velocity, and physico-chemical properties of the

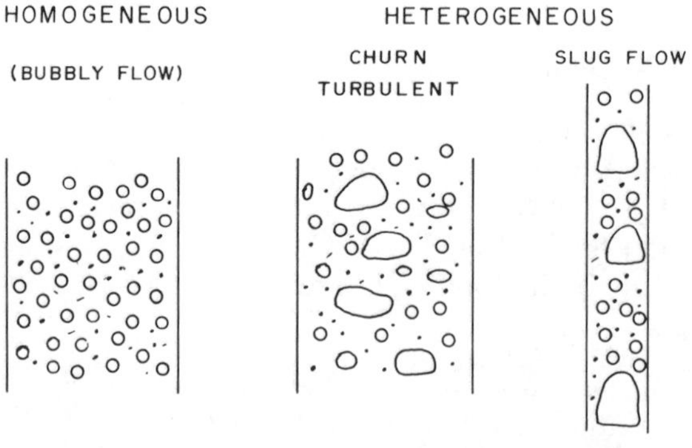

Figure 5   Flow regimes in bubble column slurry reactor. (After Shah et al. [149]).

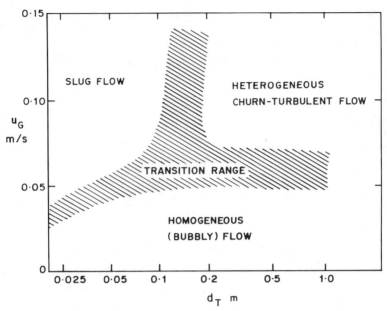

Figure 6   Effects of gas velocity and column diameter on flow regime transition
(after Shah et al. [149]).

gas–slurry system can also influence the flow regime [148]; however, the quantitative knowledge of these effects is presently not available. Kawagoe et al. [78] have classified the flow regimes for foaming systems. Bach and Pilhofer [5] have suggested that the churn-turbulent and slug flow regimes are most commonly encountered in industrial scale bubble columns. The flow regimes affect the design parameters such as gas holdup and gas–liquid interfacial area which in turn affects the performance of the reactor. Therefore, a thorough knowledge of the prevailing flow regime is most essential for the proper design of the reactor.

In a recent report, Kara et al. [72] have studied flow regimes in a three phase, air–water–coal bubble column. They presented the flow regime charts following the method of Darton and Harrison [28] which is based on the drift flux concept. A typical plot of drift flux vs. gas holdup presented by Kara et al. [72] is shown in Fig. 7. Their data fall clearly into two regions: one in which the gas holdup rapidly increases with drift flux and the other, in which the increase is markedly slower. These correspond to the "uniform bubbling" and "churn turbulent" flow regimes respectively. The data in Fig. 8

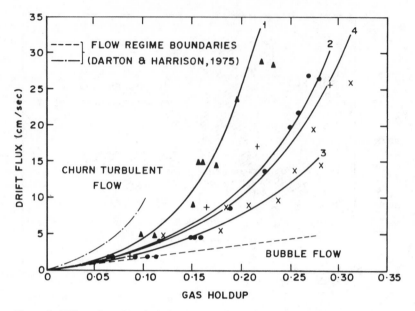

Figure 7 Effect of particle size on flow regime transition: $(U_{SL}) = 0.0 \, \text{m/s}$, 10.5 wt% solids: ($\triangle$) 70 μM; (+) 30 μM (2); ($\times$) 10 μM (3); ($\bigcirc$) air-water (4) (after Kara et al. [72]).

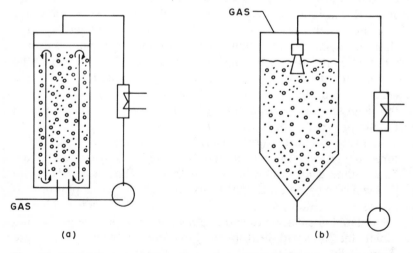

Figure 8 Schematic design of loop slurry reactors: (a) Air-lift loop reactor. (b) Jet loop reactor.

are clearly in the transition regime. In this regime, Kara et al. [72] found that liquid flow rate, solids concentration or particle size had no influence on either the drift flux velocities or gas holdup. They proposed the following form of correlation for drift flux in terms of gas holdup for both bubble and slug flow regimes:

$$v_{cD} = \varepsilon_g u_B (1 - \varepsilon_g)^m \tag{5}$$

with m correlated to slurry and particle Reynolds numbers, and liquid and solid holdup for three phase systems as:

$$m = -\left[1.9 \times 10^{-4} \, \text{Re}_{s1} + 20.6 \, \frac{\varepsilon_s}{\varepsilon_s + \varepsilon_l} + 56.4 \, \text{Re}_p + 2.607 \right] \tag{6}$$

2.2.2. Suspension of Solid Particles

For a bubble column slurry reactor, it is essential to have a knowledge of conditions under which the catalyst particles will be in complete suspension. Two states of suspension can exist for the bubble column slurry reactor: homogeneous suspension where the particles are uniformly distributed throughout the reactor and non-homogeneous suspension where the particles are non-uniformly distributed but in complete suspension. Roy et al. [135] proposed the following correlation to predict minimum gas velocity required for complete suspension in terms of critical solid loading.

$$\frac{w_{max}}{\rho_L} = 6.8 \times 10^{-4} \left(\frac{C_\mu d_T u_g \rho_G}{\mu_G} \right) \left(\frac{s_T \varepsilon_g}{u_g \mu_L} \right)^{-0.23} \left(\frac{\varepsilon_g u_{tp}}{u_g} \right)^{-0.18} \gamma^{-3} \tag{7}$$

Here u_{tp} is the terminal settling velocity of the particles and can be calculated by the correlations given by Levins and Glastonbury [92]. In Eq. [7], C_μ is a viscosity correction factor, defined as:

$$C_\mu = 2.3 \times 10^{-1} - 1.8 \times 10^{-1} \log \mu_L + 1.03 \times 10^{-2} (\log \mu_L)^2 \tag{8}$$

and γ is the wettability factor, which can be taken as unity for most catalysts. The minimum gas velocity required for complete suspension is strongly dependent on catalyst loading and particle size of the solids.

2.2.3. Gas Holdup

Gas holdup in bubble columns has been extensively investigated and several correlations have been proposed. Shah et al. [149] summarized most of the studies on gas holdup in bubble columns. In general, the gas holdup depends on the superficial gas velocity and the prevailing flow regime. The dependence of ε_g on u_g is of the form

$$\varepsilon_g \sim u_g^n \tag{9}$$

The value of n depends on the flow regime. For the bubble flow regime, n in a range of 0.7–1.2 has been observed [5, 15, 33, 47, 107, 136, 144, 175, 176]. In the churn turbulent flow or transition regime, n has been found to be in the range 0.4–0.7, which is consistent with the prediction of Ueyama and Miyauchi [161] based on the application of Nicklin's theory. Shah et al. [149] compared gas holdup data of various investigators for the column diameter, d_T, in a range of 0.075 to 5.5 m and indicated that ε_g is relatively independent of d_T.

Many industrial slurry reactors process organic liquid phases with varying physical properties and are also operated at higher temperatures and pressures. The latter studies on practical systems indicate that the earlier correlations fail to predict ε_g for solutions which contain trace impurities. The dramatic changes in ε_g with an addition of trace impurities cannot be explained based on the changes in the physical properties. A systematic study of gas holdup in mixtures of organic liquids has been carried out by Bhaga et al. [11], Hammer [45], Hammer and Deckert [46] and Hammer and Rahse [47]. Their data could not be correlated with the available literature correlations.

The gas holdup is also dependent on the type of distributor used. Hikita et al. [54] compared data obtained with single and multinozzle spargers with the correlations proposed in literature and found that the discrepancy among the data can be mainly due to differences in the design of gas spargers.

Gas holdup in a bubble column slurry reactor has been studied by Begovich and Watson [8], Deckwer et al. [33], Kara et al. [72], Kato et al. [76], Todt et al. [160] and Ying et al. [172]. In general, it was found that the gas holdup is not significantly affected in the presence of solid particles. Kara et al. [72] found that ε_g decreases with an increase in solid loading and the strong dependence of ε_g on solids concentration and gas velocity was found to diminish with an increase in either of these parameters. They explained that the change in ε_g in the presence of solids is mainly due to the change in the viscosity of the slurry phase. Todt et al. [160] obtained the gas holdup data in a continuous, countercurrent bubble column. They found that ε_g increased about 15% in the presence of solid loading [75] which can be due to the reduction in bubble coalescence in the presence of solids. Kara et al. [72] observed a decrease in gas holdup with an increase in liquid velocity, particularly at high liquid velocity. This result is not in agreement with previous reports except those for three-phase fluidized beds of larger particles [8, 105].

Deckwer et al. [31] examined gas holdup in a Fischer–Tropsch slurry reactor. They found that ε_g decreased with temperature (in the range 180–240°C) in a small diameter column (4.1 cm). In a large diameter column (10 cm), a similar observation was *not* made. They proposed the following correlation

$$\varepsilon_g = 0.053 u_g^{1.1} \pm 0.015 \tag{10}$$

This correlation does not agree with others reported in the literature beyond u_g of 1.5 cm/s and predicts higher values of ε_g.

The above mentioned studies suggest that extensive data on holdup for real systems are necessary to develop a widely applicable correlation. Furthermore, most of the studies discussed above are for average holdup in the column. Hills [58] and Ueyama et al. [162] observed pronounced radial holdup profiles and Kobayashi et al. [81] proposed a correlation for the radial holdup distribution. Similarly, axial variation of gas holdup can be observed at low gas velocities and high conversion of reactant gasses [30, 33]. In future studies, these aspects must be considered in detail.

2.3. Loop Recycle Slurry Reactors

The loop recycle reactor is a novel design which is gaining considerable interest in many three-phase slurry applications. It is a system consisting of a holding vessel for the reactant liquid with catalyst, a circulation pump, an external heat exchanger to remove heat and a self priming nozzle capable of generating a large gas–liquid interface. This novel reactor is characterized by a well circulated flow which can be driven in a fluidized system by a propeller, a jet drive or by "air-lift" drive [12]. Some of the important applications of loop slurry reactors are in hydrogenation of nitroaromatics and vegetable oils [97]. The most popular version of the jet loop reactor is a design developed by the Buss Engineering Company [91]. Schematic diagrams of jet loop and air lift loop reactors are shown in Fig. 8. The key features of the three-phase loop reactors are: (1) high heat and mass transfer and mixing efficiency compared to the stirred reactors; (2) rapid dissipation of heat leading to uniform temperatures; and, (3) increased space-time yield, selectivity and better utilization of catalyst.

In recent years, various aspects of the hydrodynamics of loop reactors have been investigated, though only limited information is available on three-phase loop reactors. Blenke [12] has reviewed the hydrodynamics of loop reactors used in biochemical processes. The industrial loop reactors have been described by Leuteritz [91] and more than 30 three-phase loop reactors have been built for hydrogenation processes in Europe. In this review, important studies on design parameters are outlined.

2.3.1. Flow Regimes and Suspension of Solids

The fluid dynamics and solids suspension in a jet loop reactor were studied by Pfeiffer et al. [127]. The solid distribution was studied by measuring the solid concentration by a newly developed photometric method incorporating a He–Ne gas laser. They found that for the case of suspension flow, the circulation number (defined as the ratio of the volume time flow rate of the slurry in a circulation to that through the nozzle) at equal Reynolds number for the nozzle was lower than that without solids (see Fig. 9) because part of the liquid jet power input is utilized to fluidize the solid particles. They proposed a correlation to predict the minimum Reynolds number $Re_{l\,min}$ required for uniform distribution of solids throughout the reactor as:

$$Re_{l\,min} = 2.3 \times 10^5 \, Fr_s^{0.38} \left(\frac{\rho_s}{\rho_L}\right)^{0.62} \left(\frac{d_P}{d_T}\right)^{0.26} \varepsilon_s^{0.26} \qquad (11)$$

where

$$Fr_s = \frac{u_{tp}^2}{g d_p} \qquad (12)$$

$$Re_l = \frac{u_{l(nozzle)} d_N}{v_L} \qquad (13)$$

Several studies on the liquid recirculation flow pattern and mixing in a jet loop [121], a propeller loop [110, 111, 141] and air-lift loop

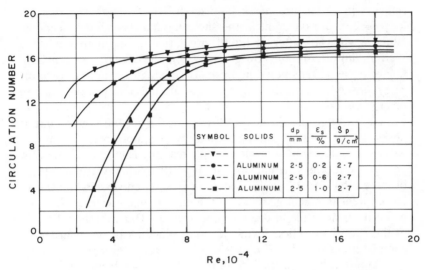

Figure 9   Plots of circulation number vs. Reynolds number. (After Pfeiffer et al.
[127]).

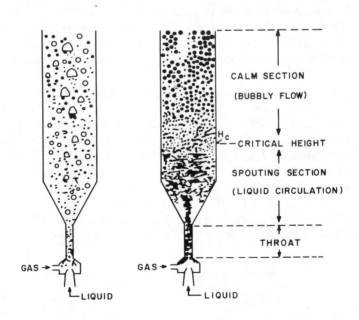

(a) BUBBLE COLUMN REGION (b) JET EJECTOR REGION

Figure 10   Typical flow patterns in jet loop rector. (After Ogawa et al. [121]).

reactors [61, 102] have been published. The flow regimes in a jet loop reactor are different in the spouting and the calming sections. While a vigorous internal liquid circulation exists in the spouting section, uniform bubble flow is achieved in the calming section [124]. Ogawa et al. [121] studied liquid-phase mixing in a jet loop reactor. They observed that the flow pattern in the jet reactor with an ejector varied with Froude number, $Fr = u_l^2/gd_T$, based on liquid velocity through the throat. At low values of u_l, the flow regime is nearly the same as the bubble column region, however, at higher u_l values, a spouting state typical of jet reactors is attained. The flow pattern is shown schematically in Fig. 10. Nishikawa et al. [118] observed that for solid–liquid systems in spouted vessels, the turbulence intensity was large and that the mixing time was about half the mean residence time in the spouting section. Thus, the liquid phase in the spouting section of the jet loop reactor can be regarded as completely mixed. The flow pattern, circulation velocity and power consumption in a propeller loop reactor have been studied by Murakami et al. [110, 111] and Sato et al. [141].

2.3.2. Gas Holdup

The gas holdup in a jet loop reactor has been studied by Hirner and Blenke [60], and Kurten and Zehner [88]. Hirner and Blenke [60] observed that ε_g in a jet loop reactor is independent on Reynolds number based on liquid nozzle and correlated their data as:

$$\varepsilon_g = 2.57 \times 10^{-4} u_g \, Re_l^{1.4} \tag{14}$$

Kurten and Zehner [88] studied the maximum gas holdup function of solid holdup in a loop reactor with downward flow within the draft tube and upwards within the annulus. The maximum gas holdup was found to decrease significantly with an increase in solid loading (see Fig. 11).

Gas holdup in air-lift loop reactors was studied by Blenke [12] Hsu and Dudukovic [61] and Merchuk and Stein [102]. Hsu and Dudukovic [61] proposed the following correlations:

$$\frac{1 - \varepsilon_g}{\varepsilon_g} = 4.6 Z^{0.9} e^{0.47} d_T ; \qquad 2 < d_T < 5 \, \text{cm} \tag{15}$$

and

$$\frac{1 - \varepsilon_g}{\varepsilon_g} = 30 Z^{0.9} e^{0.063} d_T ; \qquad 5 < d_T < 30 \, \text{cm} \tag{16}$$

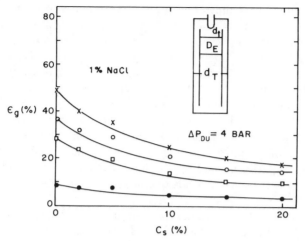

ΔP_{DU} = NOZZLE PRESSURE

D_E = 0·40 d_T d_t ≐ 0·04 d_T , GLASS BEADS 75 TO 150 μ

Figure 11   Maximum gas holdup vs. solid loading in a loop reactor. (After Kurten and Zehner [88]).

where Z is defined as

$$Z = \left(\frac{u_l}{u_g}\right) Fr^{0.36} Re^{0.006} We^{-0.57} \qquad (17)$$

In the above equation, the power on Re is too small to be completely reliable. The reported studies on the hydrodynamics of other types of novel slurry reactors are reviewed by Shah [146] and Shah et al. [151].

3.   MIXING IN SLURRY REACTORS

Slurry reactor performance is dependent on the mixing characteristics of the gas, liquid and solid phases. In this section, recent studies on mixing in various types of reactors are briefly reviewed.

3.1.   Mechanically Agitated Reactors

Mixing in agitated vessels has been extensively studied and the subject is well described by Nagatta [112] and Uhl and Gray [163].

In slurry reactors, it is only important to consider mixing in the presence of gas bubbles and suspended solid particles. In mechanically agitated vessels, the liquid phase is completely backmixed, but the gas and solid phase dispersion may not be complete. Gas phase mixing has been studied by Galor and Resnick [38], Hanhart et al. [48], Van't Reit et al. [164] and Westerterp et al. [169]. It is reported that gas phase mixing depends on the design of the impeller and the nature of bubbles as well as superficial gas velocity. The maximum interaction between the bubbles is found to be near the impeller [164]. Juvekar [67] found that the gas mixing is influenced by the presence of solid particles and the gas phase tends to move in a plug flow. Similar observations have also been made by Komiyama and Smith [85].

Wiedmann et al. [170] have compared the mixing between non-aerated liquids, aerated liquids and slurries in a turbulent flow. They found that the torque required for stirred, aerated liquids is lower than that for nonaerated stirred liquids due to decrease in the density of the gas–liquid mixture. The concentration distribution of the particles in aerated suspension becomes more uniform with increasing impeller speed, whereby the torque is higher than that for aerated liquids but lower than that for nonaerated slurries. For gas–liquid–solid systems, very limited data on dispersion of solids and gas phase are available, and further studies are necessary with different designs and for systems with different physical properties.

3.2. Bubble Column Slurry Reactor

Mixing characteristics of gas–liquid bubble columns have been extensively studied and reviewed by Shah et al. [149, 151]. Backmixing in the liquid phase depends on the column diameter, gas distributor design and gas velocity. The dependence of axial dispersion on liquid velocity and physical properties of the system is, however, not well established. Kara et al. [72] observed that in the range of liquid velocity of industrial interest (<0.03 m/s), the effect is not significant. Correlations for D_L have been proposed by Deckwer [29], Hikita and Kikukawa [57], Joshi and Sharma [66] and Kato and Nishiwaki [77]. The correlations of Deckwer et al. [29] and Hikita and Kikukawa [57] are in close agreement while that of Kato and Nishiwaki [7] predicts slightly higher values. The correlation of Hikita and Kikukawa [57] can be used for practical purposes as it incorporates the influence of liquid properties.

Ohki and Inoue [122] reported a theoretical analysis of backmixing of liquid in bubble columns using a velocity distribution model. Separate correlations for bubble flow and slug flow regimes were proposed. In a recent study, Rice et al. [137] suggested the following dependence of D_L and u_g for different flow regimes.

$$\text{Chain bubbling,} \quad D_L \sim u_g^2$$

$$\text{Bubble flow,} \quad D_L \sim u_g$$

$$\text{Churn-turbulent,} \quad D_L \sim u_g^{1/3}$$

$$\text{Slug flow,} \quad D_L \sim u_g^0$$

The effect of solid particles on the liquid dispersion coefficient has been studied by Kara et al. [72], Kato et al. [75], Kelkar et al. [79] and Ying et al. [172]. Kara et al. [72] measured the axial dispersion coefficient for air–water–coal slurry system and found that D_L was dependent on gas and slurry velocities and particle size but relatively independent of solid concentration. At low slurry velocities, their results agreed reasonably well with the literature correlations (see Fig. 12) but at higher slurry velocities, they observed significant discrepancy with earlier correlations. They proposed an empirical

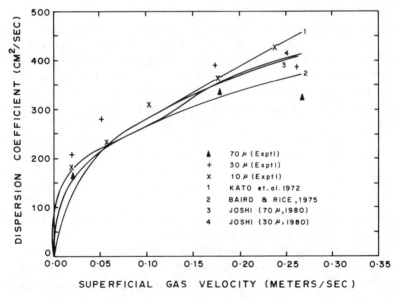

Figure 12 Effect of particle size on air-slurry dispersion and comparison with literature correlations. (After Kara et al. [72]).

correlation for D_{SL} of the form

$$D_{sL} = A u_g^B u_{sl}^C \tag{18}$$

The values of A, B, and C for various particle diameters are reported by Kara et al. [72].

The liquid phase dispersion coefficient in a coal liquefaction reactor under high pressure (2000 psig) and high temperature (450°C) conditions was measured by Panvelkar et al. [125]. They showed that the dispersion coefficient in such a reactor is considerably smaller than the ones reported for air–water systems. This study indicates the need for more dispersion measurements with real systems.

Gas phase dispersion in bubble columns has been studied by Davidson [35], Mangartz and Pilhofer [98], Pilhofer et al. [128], and Seher and Schumacher [145]. It is pointed out by Shah et al. [149] that the data in these reports are not in good agreement. Mangartz and Pilhofer [98] have proposed a correlation for D_G based on data for different types of liquids. Field and Davidson [35] measured D_G in a large diameter bubble column using radioactive tracers.

Dispersion coefficients of solid particles in bubble columns have been measured by Imafuku et al. [62], Kato et al. [75] and Kurten and Zehner [88]. Imafuku et al. [62] measured dispersion coefficients for solid particles using columns with different diameters and heights and air–glycerin solution—solid (glass spheres, ion exchange resin, Cu powder etc.) system. They found that D_S values were the same as that for pure liquids. Kato et al. [75] studied solid distribution in a continuous bubble column reactor using an air–water–glass spheres system. They proposed the following correlation for D_S

$$\text{Pe}_S = \frac{u_g d_T}{D_S} = \left[1 + 0.009 \, \text{Re}_p \left(\frac{u_g}{\sqrt{g d_T}} \right)^{-0.8} \right] \times \frac{13(u_g/\sqrt{g d_T})}{1 + 8(u_g/\sqrt{g d_T})^{0.85}} \tag{19}$$

This correlation is based on the extensive data in three phase bubble columns and is therefore recommended.

Suganuma and Yamanishi [158] proposed a theoretical model to predict the distribution of solids in a bubble column reactor. It was assumed that the particle movement is a random diffusive process and may be characterized by the dispersion coefficient of solids. The particles also move under the gravitational force with a terminal settling velocity of the particles. They derived the following equation for w, the catalyst loading at any point in terms of average catalyst loading w_{avg}.

$$\frac{w}{w_{avg}} = \frac{(u_{tp}L/D_S)\, e^{-u_{tp}x/D_S}}{1 - e^{-u_{tp}L/D_S}} \qquad (20)$$

The axial mixing in gas, liquid and solid phases in novel slurry reactors has been extensively reviewed by Shah et al. [151].

4. GAS–LIQUID MASS TRANSFER

Gas–liquid mass transfer is an important step in three-phase slurry reactions and is generally determined by the overall volumetric mass transfer coefficient, $k_L a_L$. A knowledge of $k_L a_L$ and its dependency on the operating variables is therefore most essential in the design of slurry reactors. Extensive studies for $k_L a_L$ in various types of slurry reactors have been reported in the literature and some of the important results are summarized in this section.

4.1. Mechanically Agitated Reactors

Gas–liquid mass transfer in the absence of solids has been widely studied [16, 20–22, 51, 93, 101, 169, 171]. In these studies, both physical adsorption and chemical methods have been used. Recently Chapman et al. [23] and Niiyama et al. [119], used a dynamic response technique for the measurement of $k_L a_L$ which allows incorporation of the effect of gas residence time distribution. The effect of solid particles on $k_L a_L$ has been studied by Bern et al. [10], Joosten et al. [63], and Slesser et al. [155]. Important studies for $k_L a_L$ in agitated reactors and available correlations are summarized in Table 4. In general, it was observed that $k_L a_L$ is strongly dependent on agitation speed, impeller diameter, linear gas velocity, surface tension and viscosity of the liquid phase. Design of the stirrer, bubbles dynamics and position of the blades also affect $k_L a_L$ values. Bern et al. [10] and Calderbank [20] found that $k_L a_L$ is proportional to $N^{-0.2}$, but Yagi and Yoshida [171] observed that $k_L a_L$ is proportional to $N^{2.2}$. The exponent with respect to impeller diameter is in the range of 1.5–2.0, while that for u_g is in the range of 0.28–0.5. Thus, the correlations reported in these studies differ markedly. This could be due to the variation in system properties

Table 4 Summary of $k_L a_L$ Studies in Mechanically Agitated Reactors

S. No.	Reference	Method	Impeller Type	System	Correlation for $k_L a_L$
1	Calderbank [20]				
2	Mehta and Sharma [101]	Chemical Absorption	Turbine, Propeller	—	
3	Yagi and Yoshida [171]	Physical-desorption of O_2	Six Blade		$0.06 \dfrac{d_T}{d_I^2} \left\{ \dfrac{d_I^2 N \rho_L}{\mu_L} \right\}^{1.5} \left\{ \dfrac{d_I N^2}{g} \right\}^{0.19} \left\{ \dfrac{\mu_L}{\rho_L D} \right\}^{0.5}$
			Turbine		$\left(\dfrac{\mu_L u_g}{S_T} \right)^{0.6} \left(\dfrac{N d_I}{u_g} \right)^{0.32}$
4	Bern et al. [10]	Chemical-Catalytic Hydrogenation in Slurry Reactor	Turbine	Supported Ni Catalyst	$1.099 \times 10^{-2} N^{1.16} d_I^{1.979} u_g^{0.32} v_L^{-0.521}$
5	Komiyama and Smith [85]	Chemical-Catalytic Oxidation in Slurry Reactor	Turbine	Activated Carbon	—
6	Joosten et al. [63]	Physical Desorption of He	Turbine	Glass beads, sugar and polypropylene	—
7	Chandrashekharan and Calderbank [22]	Physical Absorption	Turbine	—	$\dfrac{0.0248}{d_T^4} \left(\dfrac{P}{V_L} \right)^{0.55} (\pi d_T^2 u_g)^{0.551/\sqrt{d_T}}$
8	Niiyama et al. [119]	Dynamic-Absorption			

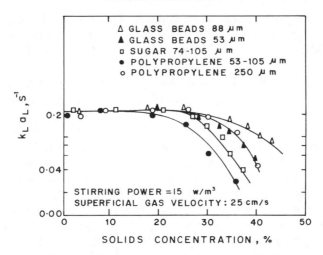

Figure 13   Effect of solid loading on $k_L a_L$ in mechanically agitated reactor. (After Joosten et al. [63]).

and design of the equipment used. Further work for $k_L a_L$ covering systems with widely different physical properties and stirrer designs and in the presence of solids is necessary to develop a generally acceptable correlation.

Joosten et al. [63] studied the effect of solids on $k_L a_L$ and found that up to 10% catalyst loading, values of $k_L a_L$ were not affected by the presence of solid particles; however, at higher concentrations $k_L a_L$ values decrease significantly. Their data for various solid types and loadings are shown in Fig. 13. Bern et al. [10] and Komiyama and Smith [85] measure $k_L a_L$ using a slurry reaction system under gas–liquid mass transfer controlled conditions. They also found that in a low solid concentration range, $k_L a_L$ values were in the same range as that for the gas–liquid systems. Most recently, Albal et al. [3] reported $k_L a_L$ for both conventional and unconventional mechanically agitated vessels. The effects of solid concentration, particle size and nature of solids on $k_L a_L$ were evaluated. The results were in general agreement with those reported by Joosten et al. [63].

4.2.   Bubble Column Reactor

The volumetric gas–liquid mass transfer coefficient, $k_L a_L$, in bubble columns has been extensively studied and this subject has been recently reviewed by Shah et al. [149]. Some of the important studies and correlations are summarized in Table 5. Most of these studies are for gas–liquid systems. Only a few studies on the effect of solids on $k_L a_L$ are reported in literature [68, 76, 114, 154, 155, 177].

The volumetric gas–liquid mass coefficient in a bubble column depends on gas velocity, sparger design, flow regime, physicochemical properties of the systems, solid concentration and catalyst particle size. From the reported studies, it is generally observed that in a low gas velocity range, $k_L a_L$ varies linearly with u_g while at higher flow rates the exponent with respect to u_g is reduced. However, the dependency of $k_L a_L$ on u_g is quite complex and may be due to changing flow regimes with u_g. Henzler [53] found $k_L a_L$ to be proportional to $u_g^{0.11}$, while Mashelkar and Sharma [100] reported $k_L a_L$ to be proportional to $u_g^{0.7}$. Deckwer et al. [29] found $k_L a_L$ to be proportional to $u_g^{0.59}$. Liquid flow rate has shown a negligible influence on $k_L a_L$ values [4, 19, 166].

The $k_L a_L$ values have been found to be almost independent of d_T. Akita and Yoshida [2] found $k_L a_L \sim d_T^{0.17}$ and concluded that $k_L a_L$ is a mild function of d_T, while Gestrich et al. [39] found $k_L a_L \sim d_T^{0.56}$. In contrast, Kastanek [73] and Sharma and Mashelkar [153] found $k_L a_L$ to be independent of d_T. The correlation of Akita and Yoshida [2] is also found to agree with data in large columns reported by Kataoka et al. [74] and hence is recommended for practical purposes.

Sparger design influences $k_L a_L$ mainly in bubble flow regime (i.e. low values of $u_G < 5 \, \text{cm/s}$). It is indicated [33] that with sintered plate distributors, appreciably higher values of $k_L a_L$ are obtained. At higher gas velocities, the effect of sparger design may not be significant. The effect of liquid properties on $k_L a_L$ has also been examined in the literature [149]. The measurements of $k_L a_L$ in non-Newtonian systems has been recently investigated by Godbole et al. [40].

Investigations of the effect of solid particles on $k_L a_L$ [41, 68, 76, 144, 154–156, 177] indicate that the extent of influence of solids on $k_L a_L$ depends on solids concentration, particle size, and liquid–solid density difference. Kato et al. [76] found that at low gas rates ($u_g < 7 \, \text{cm/s}$) $k_L a_L$ decreases with an increase in solid concentration, while at higher gas velocities the effect is negligible. Also, for large size columns ($d_T = 21.4 \, \text{cm}$) the effect of solid particles was insignificant. This could be due to the effect of solids on coalescence of bubbles in the bubble flow regime prevailing in the low gas velocity range, which leads to reduction in interfacial area and

Table 5 Summary of $k_L a_L$ Studies in a Bubble Column Reactor

S. No.	Reference	System	Method	Correlation Proposed
1	Akita and Yoshida [2]	Air–aq-glycerol Air–aq methanol CO_2–water He–water	Physical Desorption	$\dfrac{k_L a_L d_T^2}{D} = 0.6\left(\dfrac{v_L}{D}\right)^{0.5}\left(\dfrac{g d_T^2 \rho_L}{S_T}\right)^{0.65}\left(\dfrac{g d_T^3}{v_L^2}\right)^{0.31}$
2	Kastanek [73]	O_2–water	Physical	$k_L a_L = 2.875 \times 10^{-2} u_g^{0.65}\, \varepsilon_g^{0.35}\,(1 - \varepsilon_g)^{0.65}$
3	Gestrich et al. [39]	—	—	$k_L a_L = 0.0424 u_g^{0.21}\left(\dfrac{L_0}{d_T}\right)^{-0.561}\left(\dfrac{\rho_L S_T^3}{g\mu_L^4}\right)^{0.116} \varepsilon_g$
4	Hikita et al. [55]	O_2–water H_2–water CH_4–water -aq CMC Solution Buffer	Physical	$k_L a_L = \dfrac{14.9 g f}{u_g}\left(\dfrac{u_g \mu_L}{S_T}\right)^{1.76}\left(\dfrac{\mu_L^4 g}{\rho_L S_T^3}\right)^{-0.248}\left(\dfrac{\mu_g}{\mu_L}\right)^{0.243}\left(\dfrac{\mu_L}{\rho_L D}\right)^{-0.604}$
5	Deckwer et al. [29]			$k_L a_L = 0.00315 u_g^{0.59}\, \mu_{\text{eff}}^{-0.84}$
6	Gopal and Sharma [42]	O_2 dithionite Solution Buffer	Chemical	—
7	Kawagoe et al. [78]	Co_2–Na carbonate– bicarbonate Solution	Chemical	$\dfrac{k_L d_B}{D} = 0.975\left(\dfrac{\mu_L}{\rho_L D}\right)^{0.5}\left(\dfrac{g d_B^3 \rho_L^2}{\mu_L^2}\right)^{1/4}$

hence $k_L a_L$, Nguyen-Tien and Deckwer [114] observed that at high liquid velocities and low gas rates, $k_L a_L$ values are slightly higher than those without the presence of solids. With increased gas rates and lower liquid rates, solid distribution becomes non-uniform and $k_L a_L$ values are lower than those without solids. For smaller catalyst particles and lower loadings, the effect of solids on $k_L a_L$ is generally insignificant.

4.3. Loop Recycle Reactor

Studies on the determination of k_L and the interfacial area a_L in jet loop reactors have been reported by Blenke and Hirner [13], and Hirner and Blenke [60]. Blenke and Hirner [13] observed from experiments with the sulfite oxidation system that k_L is only slightly dependent on fluid dynamics in the highly turbulent systems such as jet loop reactor. They observed that over a u_g range of 0.4–6.5 (cm/s), an average value of k_L was 4.6×10^{-2} cm/s. Hirner and Blenke [60] measured interfacial area using sulfite oxidation systems and proposed the following correlations.

Air-Lift Loop Reactor

$$a_L = 4.3 \times 10^3 u_g \qquad (21)$$

Jet Loop Reactor

$$a_L = 5.4 \times 10^3 u_g^{0.4} \left(\frac{P}{V_L} \right)^{0.66} \qquad (22)$$

Here a_L is expressed in $1/m$, u_g in m/s and P/V_L in kW/m^3. The results of Hirner and Blenke [60] for specific interfacial area in a jet loop reactor as a function of specific liquid power input (P/V_L) are shown in Fig. 14. Up to a gas velocity of 1 cm/s and $P > 0$, the gas flow in jet loop reactors is dispersed into small bubbles by the liquid jet leading to a steep rise in a_L values. With an increase in u_g, a_L values increase slowly and reach a maximum which is considered to be a "jet gas loading limit". These authors also observed the overall gas–liquid mass transfer coefficient $k_L a_L$ in loop reactors. For a typical case of highly turbulent flow, and P/V_L of 2 kW/m^3, approximate values of $k_L a_L$ reported for air lift and jet loop reactors are 0.27 and 7.0 s^{-1} respectively. This indicates that very large values of $k_L a_L$ can be achieved in jet loop reactors.

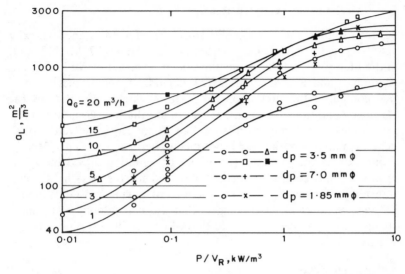

Figure 14 Plots of a_L vs. P/V_R for a jet loop reactor. (After Hirner and Blenke [60]).

Several studies for $k_L a_L$ and a_L in novel slurry reactors such as three-phase fluidized bed reactors and multistage column reactors have been reported. These are extensively reviewed by Shah [146] and Shah et al. [149].

5. LIQUID–SOLID MASS TRANSFER

The liquid–solid mass transfer rate is described in terms of the overal mass transfer coefficient $k_s a_p$ and the concentration difference. In this case, k_s, the liquid film mass transfer coefficient, is an important design parameter. The interfacial area of particles, a_p, for spherical particles is given as

$$a_p = \frac{6w}{\rho_p d_p} \qquad (23)$$

The values of k_s are measured by various methods such as: (a) physical dissolution of sparingly soluble solids with or without reaction; (b) dynamic adsorption from solution and (c) from three phase reactor data. The important studies for each reactor type are summarized in this section.

5.1. Mechanically Agitated Reactor

Several studies on liquid–solid mass transfer in a mechanically agitated reactor have been reported [14, 18, 37, 49, 87, 92, 117, 140] in the literature and this subject has been reviewed by Miller [1] and Chaudhari and Ramachandran [25]. Chaudhari and Ramachandran [25] compared the predictions of k_s from various correlations for identical conditions (see Fig. 15) and found that the variation in the predictions of different correlations was significant. The values of k_s depend mainly on the agitation speed, particle size and physical properties of the system. The effect of agitation speed on k_s is, however, mild compared to that observed for the gas–liquid mass transfer coefficient k_L. The correlations of Levins and Glastonbury [92] and Sano et al. [140] suggest that the dependency of k_s on N changes with particle size, while Boon-Long et al. [14] reported that $k_s \sim N^{0.28}$.

5.2. Bubble Column Reactor

Liquid–solid mass transfer in a three-phase column has been studied by Kamawura and Sasano [71], Kobayashi and Saito [82], Sanger and Deckwer [139] and Sano et al. [140]. A comparison of the

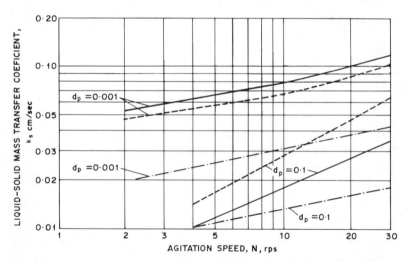

Figure 15 Comparison of correlations for k_s in an agitated reactor; – Levins and Glastonburry [92]; – – – Sano et al. [140]; – · – · – Boonlong et al. [14]. (After Chaudhari and Ramachandran [25]).

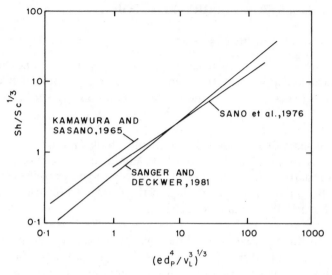

Figure 16   Comparison of correlations for k_s in a bubble column. (After Shah et al. [149]).

predictions of the correlations was made by Shah et al. [149] (see Fig. 16), which shows fairly good agreement between different correlations. The correlation of Sanger and Deckwer [139] is based on data over a wide range of experimental parameters and hence recommended.

6. HEAT TRANSFER

For exothermic reactions in slurry reactors, it is often necessary to design an efficient heat removal system. A knowledge of heat transfer through the reactor walls, to the flowing gas phase and inserted rods, cooling coils etc. is therefore most essential. Heat transfer in mechanically agitated reactors has been studied by Hovas et al. [52], Maerteleire [96], Rao and Murthy [135], Steiff and Weinspach [157] and Zlokarnik [175]. A summary of published data on heat transfer coefficients for both stirred vessels and bubble columns is presented by Steiff and Weinspach [157]. The reported studies of heat transfer coefficients in bubble column slurry reactors have been reviewed by Shah [146], and Shah et al. [149].

7. MODELS FOR SLURRY REACTORS

7.1. Overall Effectiveness Factor

In a three-phase slurry reactor, various mass transfer and surface chemical reaction steps occur in series. For a reaction between a gas-phase reactant A and liquid reactant B on the surface of the catalyst, the following steps are necessary:

1. Transport of A from gas phase to the bulk liquid.
2. Transport of A and B from bulk liquid to the catalyst surface.
3. Intraparticle diffusion of A and B within the pores of the catalyst.
4. Adsorption on the active sites of the catalyst and surface chemical reaction to yield products.

A film theory analysis of three-phase reaction in a slurry reactor has been reported by Sylvester et al. [159] for a pseudo first order reaction and by Chaudhari and Ramachandran [26] and Ramachandran and Chaudhari [129, 131] for nonlinear (power law as well as Langmuir–Hinshelwood type), and zero order kinetics. These authors used the concept of an overall effectiveness factor, which is defined as the ratio of actual rate of reaction to the rate in the absence of mass transfer effects.

$$\eta = \frac{R_A}{w\Omega(A^*)} \tag{24}$$

for a mth order reaction with gas phase reactant A, the solution for η is [131]

$$\eta = \frac{\left(1 - \dfrac{\eta}{\sigma_A}\right)^{(m+1)/2}}{\phi_0} \left[\coth \{3\phi_0(1 - \eta/\sigma_A)^{(m-1)/2}\} \right.$$
$$\left. - \frac{1}{3\phi_0(1 - \eta/\sigma_A)^{(m-1)/2}} \right] \tag{25}$$

where σ_A and ϕ_0 are defined as:

$$\sigma_A = \frac{A^*\left[\dfrac{1}{k_L a_L} + \dfrac{1}{k_s a_p}\right]^{-1}}{w\Omega(A^*)} \tag{26}$$

$$\phi_0 = \frac{R}{3}\left[\frac{(m+1)\rho_p k_m (A^*)^{m-1}}{2D_e}\right]^{1/2} \tag{27}$$

Based on this approach, equations have been developed for L–H kinetics, reaction of two gaseous reactants and also for the case of homogeneous–heterogeneous reaction. These major developments have been reviewed by Chaudhari and Ramachandran [25]. The typical plots for η vs. ϕ_0 are shown in Figs. 17 and 18 for first order and L–H kinetics.

Chaudhari and Ramachandran [26] have pointed out that for a zero order reaction, the rate is unaffected by a change in the concentration gradient as long as the concentration is finite throughout the system. However, there exists a critical gas phase concentration, below which the concentration drops to zero at some point in the catalyst and in this case the mass transfer effects become important. A critical gas phase concentration can be calculated as:

$$(A_g)_{\text{critical}} = \frac{H_A k_0 \rho_p R^2}{6D_e} + w k_0 H_A \left[\frac{1}{k_L a_L} + \frac{1}{k_s a_p}\right] \tag{28}$$

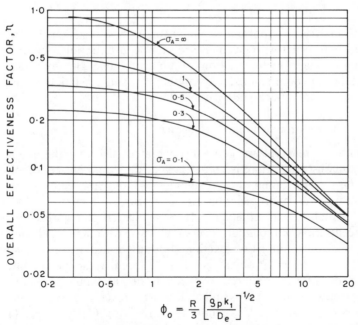

Figure 17 Overall effectiveness factor η vs. ϕ_0 plot for first order reaction. (After Chaudhari and Ramachandran [25]).

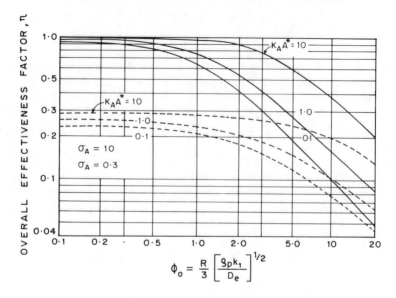

$$\phi_o = \frac{R}{3} \left[\frac{3 p k_1}{D_e} \right]^{1/2}$$

Figure 18 Overall effectiveness factor η vs. ϕ_0 plot for L-H kinetics. (After Chaudhari and Ramachandran [25]).

This analysis is important as many three-phase reactions are found to be zero order with gas phase reactant concentration. Some of the examples are: hydrogenation of dinitrotoluene [1], oxidation of cyclohexane on metal oxide catalysts [103] and ethylation of formaldehyde on a Cu-acetylide catalyst [69].

7.2. Models for Semibatch Reactors

Semibatch reactors are generally operated such that the gas flows continuously and there is no net flow of the liquid phase into or out of the reactor. Some of the examples of industrial semibatch slurry reactors are: hydrogenation of fatty oils, nitrocompounds, and butanediol to cis-butanediol. In a semibatch slurry reactor, the concentration of the liquid reactant changes with time and hence the relative contributions of reaction and mass transfer resistances also vary. The design problem is often to predict the batch time of operation for a required conversion of the liquid reactant. A theoretical analysis of semibatch slurry reactor is presented by Ramachandran and Chaudhari [130, 133] for (1, 0) order, (1, 1) order and also general m, nth order kinetics. The equations for batch time of operations for (1, 0) and (1, 1) order kinetics are:

(1, 0) *Order Reaction*

$$t_B = \frac{B_{li}X_B}{\nu A^*} \left[\frac{1}{k_L a_L} + \frac{1}{k_s a_p} + \frac{1}{\eta_c w k_1} \right] \qquad (29)$$

where

$$\eta_c = \frac{1}{\phi} \left[\coth \left(3\phi - \frac{1}{3\phi} \right) \right] \qquad (30)$$

and

$$\phi = \frac{R}{3} \left[\frac{\rho_p k_1}{D_e} \right]^{1/2} \qquad (31)$$

(1, 1) *Order Reaction*

$$t_B = \frac{B_{li}X_B}{\nu A^*} \left[\frac{1}{k_L a_L} + \frac{1}{k_s a_p} \right] + \frac{B_{li}R^2 \rho_p I}{3 A^* w D_e} \qquad (32)$$

where

$$I = \frac{2}{\phi_i^2} \ln \left[\frac{\phi_i \cosh \phi_i - \sinh \phi_i}{\phi_i \sqrt{1 - X_B} \cosh \phi_i \sqrt{1 - X_B} - \sinh \phi_i \sqrt{1 - X_B}} \right] \qquad (33)$$

and

$$\phi_i = R \left[\frac{\rho_p k_{11} B_{li}}{D_e} \right]^{1/2} \qquad (34)$$

The above equations allow prediction of batch time in the presence of mass transfer resistances. In practice, it is often observed that up to a certain liquid reactant concentration, the rate is zero order and the equations for (1, 0) order can be used, but below a certain concentration, the rate shows first order dependence of the liquid reactant concentration. In this latter region, equations for (1, 1) order should be used.

Chaudhari and Ramachandran [25] have also pointed out that the solubility of the gas-phase reactant may change with change in the composition of the liquid phase. This will affect the batch time for a given conversion. Lemcoff [90] has observed significant solubility changes for hydrogenation of acetone in a slurry reactor, while

Komiyama and Smith [85] reported that the solubility of O_2 is reduced with increasing product concentration in their study on SO_2 oxidation. Chaudhari and Ramachandran [25] derived equations for semibatch reactors incorporating the solubility changes by the following relation:

$$A^* = \frac{A_{gi}}{H_A} = (\gamma_1 + \gamma_2 B_l + \gamma_3 C_l)P_{gA} \tag{35}$$

where γ_1, γ_2, γ_3 are empirical constants.

The following equation for $X_B - t_B$ relationship was developed for the case of negligible pore diffusion effects ($\eta_c \to 1$)

$$\frac{\ln \gamma}{M_A(\gamma_2 - \gamma_3 \nu_p/\nu)} + \frac{\ln \gamma(1 - X_B)}{wk_2\left[\gamma_1 + \gamma_3\left(C_{li} + \frac{\nu_p}{\nu} B_{li}\right)\right]} = \nu P_{gA} t_B \tag{36}$$

where

$$\gamma = \frac{\gamma_1 + \gamma_3[C_{li} + (\nu_p/\nu)B_{li}] + B_{li}[\gamma_2 - \gamma_3(\nu_p/\nu)]}{\gamma_1 + \gamma_3[C_{li} + (\nu_p/\nu)] + B_{li}(1 - X_B)[\gamma_2 - \gamma_3(\nu_p/\nu)]} \tag{37}$$

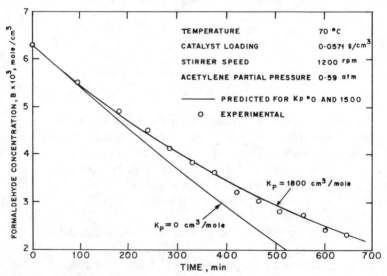

Figure 19   Effect of product absorption on concentration vs. time plot. (After Kale et al. [70]).

Kale et al. [70] have reported a semibatch reactor model for complex kinetics applicable to butanediol synthesis in the absence of mass transfer effects. They demonstrated the use of their model for the analysis of inhibition of the rate due to the product of the reaction and evaluated an adsorption equilibrium constant for the experimental data (see Fig. 19).

7.3. High Pressure Batch Reactors

Batch autoclave reactors under pressure are commonly used for kinetic studies of three-phase slurry reactions. In a batch operation, the pressure of the gas-phase reactant, concentrations of the liquid reactant and products vary simultaneously as a function of time. Thus, in a single batch experiment, it is possible to vary the concentrations of all the species over a wide range. Therefore, the kinetics can be evaluated with a minimum number of experiments. For a reaction between a gas-phase reactant A and a liquid reactant B, the mass balance equations are:

$$-\frac{1}{R_g T}\frac{dp_g}{dt} = \frac{V_L}{V_g}\, w\Omega(P_g, B_l) \qquad (38)$$

$$-\frac{dB}{dt} = \nu w\Omega(P_g, B_l) \qquad (39)$$

From these equations, the following relationship between B_l and P_g can be obtained

$$B_l = B_{li} - \frac{\nu V_g}{V_L R_g T}(P_{gi} - P_g) = B_{li} - \frac{1}{\alpha}(P_{gi} - P_g) \qquad (40)$$

Here $\Omega(P_g, B_l)$ represents the rate of reaction depending on the kinetic model and mass transfer effects. By substituting the relation for B_l in terms of P_g in Eq. (38), we get an expression only in terms of P_g. This equation can be solved for a given set of parameters to predict P_g vs. time data. Similarly, the kinetic parameters can be evaluated by fitting the experimental data with model predictions. In this approach, the direct observations of presure drop vs. time can be used and hence is more accurate for evaluation of kinetic parameters.

Chaudhari et al. [24] have used this approach for the modeling of hydrogenation of butanediol to cis-butanediol. They observed that the intrinsic kinetics were represented by the following expression

$$R_A = w\Omega(P_g, B_l) = \frac{wk\sqrt{P_g/H_A}B_l}{(1 + K_B B_l)^2} \tag{41}$$

in the presence of gas–liquid mass transfer, the relevant mass balance equation is derived as:

$$\frac{dP_g}{dt} = \frac{RTV_L}{V_g}\left[\frac{\{(\bar{\beta}^2/k_L a_L)^2 + \bar{4}\beta^2 P_g/H_A\}^{0.5} - (\bar{\beta}^2/k_L a_L)}{2}\right] \tag{42}$$

where

$$\bar{\beta} = \frac{wk\left[B_{li} - \dfrac{1}{\alpha}(P_{gi} - P_g)\right]}{\left\{1 + K_B\left[B_{li} - \dfrac{1}{\alpha}(P_{gi} - P_g)\right]\right\}^2} \tag{43}$$

Equation (42) can be solved numerically by the Runge–Kutta method

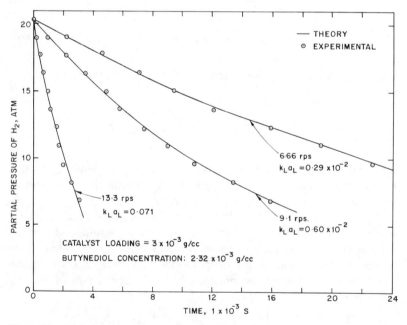

Figure 20   Effect of stirrer speed on hydrogenation rate at 343 K. (After Chaudhari et al. [24]).

to predict P_g vs. time data. The application of this model to evaluate kinetic and mass transfer parameters has been demonstrate by Chaudhari et al. [24] (see Fig. 20).

7.4. Continuous Reactor Models

In continuous slurry reactors, in addition to the mass transfer and kinetic steps, the variation of the reactant concentration and mixing of the various phases need to be considered. Goto and Smith [43] and Goto et al. [44] have analyzed a problem of $(1, 0)$ order reaction in isothermal continuous slurry and trickle bed reactors with uniform distribution of solid particles. The equations for the conversion of liquid reactant B for the case of constant gas phase concentration have been derived for plug flow and backmixed models as

Plug Flow Model

$$X_B = \frac{q\beta_{ls}\alpha_{gl}}{\alpha_{gl} + \beta_{ls}} + \frac{q\beta_{ls}\left[a_{l0} - \dfrac{\alpha_{gl}}{\alpha_{gl} + \beta_{ls}}\right][1 - \exp\,(-[\beta_{ls} + \alpha_{gl}])]}{\beta_{ls} + \alpha_{gl}}$$

$$(44)$$

and

Backmixed Model

$$X_B = q\beta_{ls}\left[\frac{a_{l0} + \alpha_{gl}}{1 + \alpha_{gl} + \beta_{ls}}\right] \qquad (45)$$

In bubble column slurry reactors and mechanically agitated reactors, the backmixed model may be applicable if the following approximate criterion is satisfied.

$$Pe_l L/d_p < 4 \qquad (46)$$

Goto and Smith [43] have developed theoretical models for simultaneous absorption with reaction of two gases (e.g. SO_2 oxidation). Models for plug flow and completely backmixed flow conditions with varying gas phase concentration have been reported. These authors have applied the theory to SO_2 oxidation and some typical results are shown in Fig. 21. An important finding of Goto

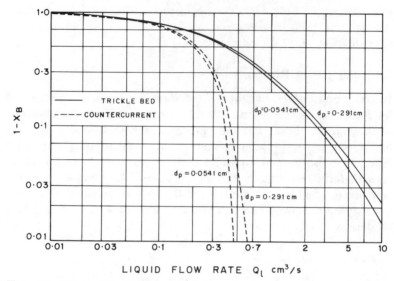

Figure 21   Effect of liquid flow rate on SO_2 removal in a continuous flow slurry reactor. (After Goto and Smith [43]).

and Smith [43] was that the predictions of backmixed and plug flow models were not much different suggesting that the residence time distribution of the gas bubbles has little effect on the reactor performance for sparingly soluble gases (such as O_2 in their case). The models for continuous reactors for other cases considering dispersion in gas and liquid phases are reviewed by Ramachandran and Chaudhari [133].

In many industrially important reactions, nonlinear kinetics is observed and in these cases, the mixing cell model described by Ramachandran and Smith [134] can be used. In this model, a three-phase reactor is viewed as a number of backmixed reactors in series. This method is sufficiently general and can be used for all types of three-phase reactors. Recently Brahme et al. [17] have shown the application of mixing cell model for hydrogenation of glucose in a continuous slurry reactor considering a complex Langmuir–Hinshelwood type of rate expression.

Reactor models for Fischer–Tropsch synthesis have been described by Deckwer et al. [32], Satterfield and Huff [142] and Shah et al. [152] considering the rate limiting step to be first order with respect to H_2 and zero order with respect to CO. The effect of different operating variables on the reactor performance has been discussed based on the theoretical model. A generalized slurry reactor model considering axial dispersion of all three phases has been reviewed by Shah [146].

7.5. Thermal Behavior of Slurry Reactors

Many of the industrial three-phase slurry reactions are highly exo-thermic and therefore it is important to consider the thermal behavior of reactors carrying out such reactions. The intraparticle temperature gradients in slurry reactors may not be significant. Also, in batch or semi-batch reactors, the temperature gradients between the gas phase and catalyst surface may be negligible due to high values of the heat transfer coefficients. However, the bulk liquid temperature is expected to change with residence time, thus affecting the rate of reaction due to changes in the kinetic rate constant and solubility parameters. In a continuous slurry reactor also, a temperature profile may be observed depending on the extent of mixing and exothermicity of the reaction.

For a semibatch reactor with reaction between a gas phase reactant A and liquid phase reactant B, and assuming that intra-particle diffusion effects are negligible, the following equations would describe the material and heat balances [133]:

$$-\frac{dB_l}{dt} = \nu R_A = \nu A^*(T)\left[\frac{1}{k_L a_L} + \frac{1}{k_s a_p} + \frac{1}{wk_2(T)B_l}\right]^{-1} \quad (47)$$

and

$$\left[\varepsilon_g V_R \rho_G C_{pg} + V_R\left(1 - \varepsilon_g - \frac{w}{\rho_p}\right)\rho_L C_{pl} + V_R w C_{ps}\right]\frac{dT}{dt}$$

$$= (-\Delta H)V_R A^*(T)\left[\frac{1}{k_L a_L} + \frac{1}{k_s a_p} + \frac{1}{wk_2(T)B_l}\right]^{-1}$$

$$- U_w A_w(T - T_2) - Q_G \rho_G C_{pg}(T - T_{gi}) \quad (48)$$

with the initial conditions $t = 0$, $B_l = B_{li}$ and $T = T_i$. The above equations can be solved numerically for B_l and T as a function of time. If the temperature change is large, the variation of k_L and k_s with temperature must also be considered.

Shah and Parulekar [150] have reported an axial dispersion model for describing the steady state thermal behavior of an adiabatic three-phase slurry reactor with experimental data. The effect of scaleup variables on the reactor performance and temperature profile were studied. They observed that low liquid and gas velocities and larger reactor length result in higher adiabatic temperature rise. Similarly, lowering the mass transfer resistance results in higher temperature rise. Multiple steady states in slurry reactors have been examined by Parulekar et al. [126]. The thermal behavior of coal liquefaction reactors has also been extensively evaluated by King et al. [80], Ledakowicz et al. [89], Nunez et al. [120] and Shah [146].

As many slurry reactors are operated under conditions such that large temperature gradients exist which can lead to multiple steady states and reactor instability, more work is required in this area.

8. RECOMMENDATIONS FOR FUTURE WORK

The understanding of slurry reactors has advanced considerably in the last few years. However, there are many aspects which need to be further understood. Future work in this area should cover the following topics.

1. Hydrodynamic characteristics such as flow regimes and suspension of solids have been largely studied with air–water–solid systems and at this stage, a precise prediction of flow regimes for many real systems is not possible. The work on systems covering organic liquids, and solids mainly consisting of catalyst supports such as activated carbon, alumina, should be undertaken and the methods of predicting flow regimes need to be developed based on data on systems with widely different physical properties. Flow regimes and suspension for highly viscous slurries and non-Newtonian liquids should be studied.

2. The design parameters such as gas holdup, interfacial area, mass transfer coefficients and dispersion coefficients should be measured for conditions prevailing in industrial reactors. Most studies are reported for ambient temperature and pressure and there is a need for data at higher temperatures and pressures. Also, the effect of various types of solids on these parameters need to be understood. Design data for non-Newtonian fluids and viscous slurries are also lacking. This, of course, involves development of novel experimental techniques for the measurement of design parameters.

3. The understanding of theoretical models for batch and continuous reactors with complex kinetics has also progressed satisfactorily. However, there is now a need to undertake experimental

studies for real systems and understand the additional complexities which are not taken into account in generalized models. For example, the effects of inerts and byproducts on effective diffusivities of reactants adsorption constants, deactivation and solubility changes should be modeled and such complexities need to be understood for practical cases.

4. Many slurry reactors are operated such that isothermal conditions do not prevail. More work is therefore necessary for the understanding of non-isothermal slurry reactors, control strategies and various aspects of stability and multiplicity of such reactors.

5. In complex consecutive reactions, it is often observed that the isolated kinetics determined for each step does not apply to model the entire system indicating that complex adsorption mechanism is involved. Modeling of such systems is most desirable.

NOMENCLATURE

a_L gas–liquid interfacial area per unit volume of reactor, cm^{-1} or m^{-1} (in Eqs. 21 and 22)

a_{l0} dimensionless concentration of A in outlet liquid, mol/cm^3

a_p external area of particles per unit volume of reactor, cm^{-1}

A^* concentration of gas A at the gas–liquid interface, mol/cm^3

A_{gi} concentration of A in inlet gas, mol/cm^3

A_w area of reactor wall available for heat transfer, cm^2

B_l concentration of liquid reactant B, mol/cm^3

B_{li} concentration of reactant B at $t = 0$ in a semibatch reactor, mol/cm^3

C_{pg}, C_{pl} and C_{ps} specific heats of gas, liquid and solid phases respectively, $J/mol/K$

C_μ viscosity correction factor defined by Eq. (8)

d_I diameter of the impeller, cm or m (in Eq. 1)

d_p catalyst particle diameter, cm

d_T diameter of the reactor, cm

D molecules diffusivity in liquid phase, cm^2/s

D_e effective diffusivity of A in liquid filled pore of catalyst, cm^2/s

D_G axial dispersion coefficient of the gas phase, cm^2/s

D_L axial dispersion coefficient of the liquid phase, cm^2/s

D_S axial dispersion coefficient of solid phase, cm^2/s

D_{SL} axial dispersion coefficient of the slurry phase, cm^2/s

e exponential function in Eqs. (15) and (16)

Fr   Froude number

g acceleration due to gravity, cm/s^2

H_A Henry's law constant for solubility of gas A, $cm^3(liquid/cm^3(gas))$

k_L liquid–film mass transfer coefficient, cm/s

k_m pseudo mth order reaction rate coefficient, $(cm^3/g)(cm^3/mol)^{m-1}$

K_B adsorption equilibrium constant for B, cm^3/mol

L height of liquid in mechanically agitated vessel, cm

L_0 aerated liquid height in bubble column, cm

m reaction order with respect to A or constant defined by Eq. (6)

M_A overall gas phase to solid particle mass transfer coefficient, defined as

$$\left[\frac{1}{k_L a_L} + \frac{1}{k_s a_p}\right]^{-1}, 1/s$$

N   speed of agitation, Hz

N_m minimum speed of agitation required for suspension of solid particles, Hz

Ne power number as defined in Fig. 3

n constant in Eq. (9)

P_g pressure in the gas phase, atm

P_{gi} pressure in gas phase at $t = 0$ in a batch reactor, atm

P power consumption for agitation of aerated liquid or slurry, $g\ cm^2/s^3$ or W (in Eq. 1)

P_0 power consumption for agitation of gas free liquid, W

Pe_l liquid phases Peclet number, u_{lL/D_L}

Pe_s Peclet number for solids, $u_g d_T/D_s$

q a parameter defined in Eq. (44) and (45)

$$\left[q = \frac{\nu A^*}{B_{li}} \right]$$

Q_G volumetric flow rate of gas, cm^3/s or m^3/s in
Eq. (1)

Q_l volumetric flow rate of liquid, cm^3/s

$\gamma_1, \gamma_2, \gamma_3$ constants in Eq. (35)

R radius of catalyst particle, cm

R_A rate of reaction of species A per unit volume
of reactor, mol/cm^3/s

R_g gas constant, cm^3 atm/mol/K

Re   Reynolds number

Sc   Schmidt number

Sh   Sherwood number

S_T surface tension of liquid, dyne/cm

t time, s

t_B batch time required to achieve a conversion of
X_B, s

T temperature, K

T_{gi} temperature of the inlet gas, K

T_w temperature of cooling medium, K

u_B terminal bubble rise velocity, cm/s

u_g superficial velocity of gas phase in the reactor,
cm/s or m/s (in Eqs. 21 and 22)

u_{tp} terminal settling velocity of particles, cm/s

u_l superficial velocity of liquid in the reactor,
cm/s

U_{sl} superficial slurry velocity, cm/s

U_w overall wall heat transfer coefficient

V_g volume of gas phase in the reactor, cm^3

V_L volume of liquid phase in the reactor, cm^3

V_R reactor volume, cm^3

w catalyst loading per unit volume of reactor,
g/cm^3

w' catalyst loading, wt%

w_{avg} average catalyst loading in a bubble column,
g/cm^3

w_{max} critical solid loading, g/cm^3

We   Weber number, $u_g^2 \rho_g d_p / S_T$

x axial distance in the reactor, cm

X_B conversion of reactant B

Z constant defined by Eq. (17)

Greek Letters

α_{gl} dimensionless gas–liquid transfer parameter,

$$\frac{k_L a_L L}{u_l}$$

α_{ls} dimensionless liquid–solid mass transfer parameter,

$$\frac{k_s a_p L}{u_l}$$

α_r dimensionless reaction rate parameter, $\dfrac{w k_1 L}{u_l}$

β constant defined by Eq. (3)

β_{ls} a parameter defined as $\alpha_\gamma \eta_c \alpha_{ls} / (\alpha_\gamma \eta_c + \alpha_{ls})$

γ' wettability factor in Eq. (7)

ε_g gas holdup

ε_l liquid holdup

ε_s solid holdup (in percentage in Eq. (11))

η overal effectiveness factor defined by Eq. (25)

η_c catalytic effectiveness factor

ϕ generalized Thiele modulus

ϕ_0 Thiele modulus based on bulk conditions

ϕ_i value of ϕ_0 at $t = 0$ in a semibatch reactor

ρ_G, ρ_L, ρ_p gas, liquid and particle density respectively, g/cm^3

Ω local rate of reaction of A per weight of catalyst, mol/g/s

ν stoichiometric coefficient of reactant B

ν_L kinetic viscosity of liquid, cm^2/s

ν_{CD} drift flux, cm/s

μ_G, μ_L viscosity of gas and liquid phases respectively, $g/cm/s$

σ_A a parameter defined as $M_A A^* / w \Omega(A^*)$

REFERENCES

1. Acres, G. J. K., and Cooper, B. J. (1972). Carbon-supported platinum metal catalysts for hydrogenation reactions—mass transfer effects in liquid phase hydrogenation over Pd/C, *J. Appl. Chem. Biotech.*, **22**, 769.

2. Akita, K., and Yoshida, F. (1973). Gas holdup and volumetric mass transfer coefficient in bubble columns, *Ind. Eng. Chem. Proc. Des. Dev.*, **12**, 76.
3. Albal, R. S., Shah, Y. T., Schumpe, A., and Carr, N. L. (1983). Mass transfer, power consumption and suspension characteristics of surface aerated two and three phase agitated contactor, *Chem. Eng. J.*, **27**, 61.
4. Alvarez-Cuenca, M., Baker, C. G. J., and Bergougnou, M. A. (1980). Oxygen mass transfer in bubble columns, *Chem. Eng. Sci.*, **35**, 1121.
5. Bach, H. F., and Pilhofer, T. (1978). Variation of gas holdup in bubble columns with physical properties of liquids and operating parameters of columns, *Germ. Chem. Eng.*, **1**, 270.
6. Baldi, G., Conti, R., and Alaria, E. (1978). Complete suspension of particles in a mechanically agitated vessel, *Chem. Eng. Sci.*, **33**, 21.
7. Baldi, G., Conti, R., and Gianetto, A. (1981). Concentration profiles for solids suspended in a continuous agitated reactor, *AIChE J.*, **27**, 1017.
8. Begovich, J. M., and Watson, J. S. (1978). Hydrodynamic characteristics of three phase fluidized beds, in *Fluidization, Proc. 2nd Eng. Found. Conf.* (Eds. J. F. Davidson and D. L. Keairns) **190**.
9. Bern, L., Hell, M., and Schoon, N. H. (1975). Kinetics of the hydrogenation of rapeseed oil—II Rate equations of chemical reactions, *J. Am. Oil chem. Soc.*, **52**, 391.
10. Bern, L., Lidefelt, J. O., and Schoon, N. H. (1976). Mass transfer and scale up in fat hydrogenation, *J. Am. Oil Chem. Soc.*, **53**, 463.
11. Bhaga, D., Pruden, B. B., and Weber, M. E. (1971). Gas holdup in a bubble column containing organic liquid mixtures, *Can. J. Chem. Eng.*, **49**, 417.
12. Blenke, H. (1979). Loop reactors, *Adv. Biochem. Eng.*, **13**, 122.
13. Blenke, H., and Hirner, W. (1974). Gas liquid mass transfer in jet loop reactor, *VDI—Berichte*, **218**, 549.
14. Boon-Long, S., Laguerie, C., and Couderc, J. P. (1978). Mass transfer from suspended solids to a liquid in agitated vessels, *Chem. Eng. Sci.*, **33**, 813.
15. Botton, R., Cosserat, D., and Charpentier, J. C. (1978). Influence of column diameter and high gas throughputs on the operation of a bubble column, *Chem. Eng. J.*, **16**, 107.
16. Botton, R., Cosserat, D., and Charpentier, J. C. (1980). Operating zone and scale up of mechanically stirred gas–liquid reactors, *Chem. Eng. Sci.*, **35**, 82.
17. Brahme, P. H., and Doraiswamy, L. K. (1976). Modelling of a slurry reaction. Hydrogenation of glucose on Raney nickel, *Ind. Eng. Chem. Proc. Des. Dev.*, **15**, 130.
18. Brain, P. L. T., Hales, H. B., and Sherwood, T. K. (1969). Transport of heat and mass between liquid and spherical particles in an agitated tank, *AIChE J.*, **15**, 727.
19. Burckhart, R., and Deckwer, W. D. (1976). Der Einflu der Betriebedingungen und die Wirkung Von Electrolytzusatzen auf den Sauerstaffubergang in Blassen-saulen, *Verfahrenstech. (Mainz)*, **10**, 429.
20. Calderbank, P. H. (1958). Physical rate processes in industrial fermentation: Part I—The interfacial area in gas–fluid contacting with mechanical agitation, *Trans. Instn. Chem. Engrs.*, **36**, 443.
21. Calderbank, P. H., and Moo-Young, M. B. (1961). The continuous phase heat and mass transfer properties of dispersions, *Chem. Eng. Sci.*, **16**, 39.
22. Chandrasekharan, K., and Calderbank, P. H. (1981). Further observations on the scale up of aerated mixing vessels, *Chem. Eng. Sci.*, **36**, 819.
23. Chapman, C. M., Gilbilaro, L. G., and Nienow, A. W. (1982). A dynamic response technique for the estimation of gas–liquid mass transfer coefficients in a stirred vessel, *Chem. Eng. Sci.*, **37**, 891.
24. Chaudhari, R. V., Parande, M. G., Ramachandran, P. S., Brahme, P. H., Vadgaonkar, H. G., and Jaganathan, R. (1985). Hydrogenation of butnediol to *cis*-butenediol catalyzed by Pd–Zn–CaCO$_3$: Reaction kinetics and modelling of a batch slurry reactor, *AIChE J.* **31**, 1891.

25. Chaudhari, R. V., and Ramachandran, P. A. (1980). Three phase slurry reactors, *AIChE J.*, **26**, 177.
26. Chaudhari, R.V., and Ramachandran, P.A. (1980). Influence of mass transfer on zero order reaction in a catalytic slurry reactor, *Ind. Eng. Chem. Fundam.*, **19**, 201.
27. Chen, B. H., and McMillan, A. G. (1982). Heat transfer and axial dispersion in batch bubble columns, *Can. J. Chem. Eng.*, **60**, 436.
28. Darton, R. C., and Harrison, D. (1975). Gas and liquid hold-up in three phase fluidization, *Chem. Eng. Sci.*, **30**, 581.
29. Deckwer, W. D., Burckhart, R., and Zoll, G. (1974). Mixing and mass transfer in tall bubble columns, *Chem. Eng. Sci.*, **29**, 2177.
30. Deckwer, W. D., Adler, I., and Zaidi, A. (1978). A comprehensive study on CO_2-interphase mass transfer in vertical concurrent and countercurrent gas flow, *Can. J. Chem. Eng.*, **56**, 43.
31. Deckwer, W. D., Louisi, Y., Zaidi, A., and Ralek, M. (1979). Gas holdup and physical transport properties for the Fischer–Tropsch synthesis in slurry reactor, *AIChE 72nd Annual Meeting, San Francisco, Nov. 25–29.*
32. Deckwer, W. D., Serpemen, Y., Ralek, M., and Schmidt, B. (1982). Modelling the Fischer–Tropsch synthesis in the slurry phase, *Ind. Eng. Chem. Proc. Des. Dev.*, **21**, 231.
33. Deckwer, W. D., Louisi, Y., Zaidi, A., and Ralek, M. (1980). Hydrodynamic properties of the Fischer–Tropsch slurry processes, *Ind. Eng. Chem. Proc. Des. Dev.*, **19**, 699.
34. Fair, J. R., Designing gas-sparged reactors, (1967). *Chem. Eng.*, **74**, 67.
35. Field, R. W., and Davidson, J. (1980). Axial dispersion in bubble columns, *Trans. Inst. Chem. Eng.*, **58**, 228.
36. Foust, H. C., Mack, D. E., and Rushton, J. H. (1944). Gas–liquid contacting by mixers, *Ind. Eng. Chem.*, **36**, 517.
37. Furusawa, T., and Smith, J. M. (1973). Fluid-particle and intraparticle mass transport rates in slurries, *Ind. Eng. Chem. Fundamentals*, **12**, 197.
38. Galor, B., and Resnick, W. (1966). Gas residence time in agitated gas liquid contactor-experimental test of mass transfer model, *Ind. Eng. Chem. Proc. Des. Dev.*, **5**, 15.
39. Gestrich, W., Esenwein, H., and Krauss, W. (1978). Liquid side mass transfer in bubble layers, *Int. Chem. Engng.*, **18**, 38.
40. Godbole, S. P., Schumpe, A., Shah, Y. T., and Carr, N. L. (1984). Hydrodynamics and mass transfer in non-Newtonian solution in a bubble column, *AIChE J.*, **30**, 213.
41. Godbole, S. P., Schumpe, A., and Shah, Y. T. (1983). Hydrodynamics and mass transfer in bubble columns: Effect of solids, *Chem. Eng. Comm.*, **24**, 235.
42. Gopal, J. S., and Sharma, M. M. (1982). Hydrodynamics and mass transfer characteristics of bubble and packed bubble columns with downcomer, *Can. J. Chem. Eng.*, **60**, 353.
43. Goto, S., and Smith, J. M. (1978). Modelling of SO_2 oxidation in three phase slurry and trickle bed reactors, *AIChE J.*, **24**, 286.
44. Goto, S., Watabe, S., and Matsubara, M. (1976). The role of mass transfer in trickle bed reactors, *Can. J. Chem. Eng.*, **54**, 551.
45. Hammer, H. (1979). Blasensaulen-Reaktoren mit Suspendiertem Feststoff: Grundlagen-Auslegung-Anwendung, *Chem. Ing. Tech.*, **51**, 259.
46. Hammer, H., and Deckert, H. (1978). Hydrodynamische Eigenschaften Von Blasensaulen: Blasengro en-Haufigkeitsverteilungen in mit Stickstoff Begasten Gemischen Von *n*-Butane und *n*-Butanol, *Chem. Ing. Tech.*, **50**, 623.
47. Hammer, H., and Rahse, W. (1972). Blasengro en-Haufigkeits-Verteilungen und Phasengrenzflachem in Blasensaulen, *Chem. Ing. Tech.*, **45**, 968.
48. Hanhart, J., Kramers, H., and Westerterp, K. R. (1963). The residence time distribution of the gas in an agitated gas liquid contactor, *Chem. Eng. Sci.*, **18**, 503.

49. Harriott, P. (1962). Mass transfer to particles: Part I—Suspended in an agitated tank, *AIChE J.*, **8**, 93.
50. Hassan, T. M., and Robinson, C. W. (1977). Stirred-tank mechanical power requirement and gas holdup in aerated aqueous phases, *AIChE J.*, **23**, 48.
51. Hassan, R. M., and Robinson, C. W. (1980). Mass transfer coefficients in mechanically agitated gas aqueous electrolyte dispersions, *Can J. Chem. Engng.*, **58**, 198.
52. Havas, G., Deak, A., and Sawinsky, J. (1982). Heat transfer coefficients in an agitated vessel using vertical tube baffles, *Chem. Eng. J.*, **23**, 161.
53. Henzler, H. J., (1980) Begasen Hoherviskoser Flussigkeiten, *Chem. Ing. Tech.*, **52**, 8.
54. Hikita, H., Asai, S., Tanigawa, K., Segawa, K., and Kitao, M. (1980). Gas holdup in bubble columns, *Chem. Engng. J.*, **20**, 59.
55. Hikita, H., Asai, S., Tankgawa, K., Segawa, K., and Kitao, M. (1981). The volumetric liquid–phase mass transfer coefficient in bubble columns, *Chem. Eng. J.*, **22**, 61.
56. Hikita, H., Asai, S., Kikukawa, H., Zaike, T., and Masahiko, Ohue (1981). Heat transfer coefficients in bubble columns, *Ind. Eng. Chem. Proc. Des. Dev.*, **20**, 540.
57. Hikita, H., and Kikukawa, H. (1975). Dimensionless correlation of liquid-phase dispersion coefficient in bubble columns, *J. Chem. Eng. Japan*, **8**, 412.
58. Hills, J. H. (1974). Radial non-uniformity of velocity and voidage in bubble column, *Trans. Instn. Chem. Engrs.*, **52**, 1.
59. Hills, J. H., and Darton, R. C. (1976). The rising velocity of a large bubble in a bubble swarm, *Trans. Inst. Chem. Engrs.*, **54**, 258.
60. Hirner, W., and Blenke, H. (1977). Gas holdup and gas–liquid interfacial area in loop and jet reactors, *Verfahrenstechnik*, **11**, 297.
61. Hsu, Y. C., and Dudukovic, M. P. (1980). Gas holdup and liquid recirculation in gas-lift reactors, *Chem. Eng. Sci.*, **35**, 135.
62. Imafuku, K., Wang, T. Y., Koide, K., and Kubota, H. (1968). Behavior of suspended solid particles in a bubble column, *J. Chem. Eng. Japan*, **1**, 153.
63. Joosten, G. E. H., Schilder, J. G. M., and Janssen, J. J. (1977). Effect of suspended solids on gas–liquid mass transfer coefficients in an agitated reactor, *Chem. Eng. Sci.*, **32**, 563.
64. Jashi, J. B. (1982). Gas phase dispersion in bubble column, *Chem. Eng. J.*, **24**, 213.
65. Joshi, J. B., Pandit, A. B., and Sharma, M. M. (1982). Mechanically agitated gas–liquid reactors, *Chem. Eng. Sci.*, **37**, 813.
66. Joshi, J. B., and Sharma, M. M., (1979). A circulation cell model for bubble columns, *Trans. Inst. Chem. Engrs.*, **57**, 244.
67. Juvekar, V. A. (1976). Studies in gas liquid reactions, Ph.D. (Tech.) Thesis, University of Bombay.
68. Juvekar, V. A., and Sharma, M. M. (1973). Absorption of CO_2 in a suspension of lime, *Chem. Eng. Sci.*, **28**, 825.
69. Kale, S. S., and Chaudhari, R. V. (1978). Ethylation of formaldehyde in an agitated slurry reactor using Cu-acetylide catalyst, Paper presented at 4th National Symposium on Catalysis, at Bombay (Dec. 1978).
70. Kale, S. S., Chaudhari, R. V., and Ramachandran, P. A. (1981). Butynediol synthesis—A kinetic study, *Ind. Eng. Chem. Prod. Res. Dev.*, **20**, 309.
71. Kamuwura, K., and Sasano, T. (1965). Liquid–solid transfer in bubble column, *Kagaku Kogaku*, **29**, 693.
72. Kara, S., Kelkar, G., Shah, Y. T., and Carr, L. W. (1982). Hydrodynamics and axial mixing in a three phase bubble column, *Ind. Eng. Chem. Proc. Dev.*, **21**, 584.
73. Kastanek, F. (1977). Volume mass transfer coefficient in a bubble column, *Coll. Cz. Chem. Comm.*, **42**, 2491.
74. Kataoka, H., Takenchi, H., Nakao, K., Yagi, H., Tadaki, T., Otake, T.,

Miyauchi, T., Washimi, K., Watanabe, K., and Yoshida, F. (1979). Mass transfer in a large bubble column, *J. Chem. Eng. Japan*, **12**, 105.

75. Kato, Y., Nishiwaki, A., Fukuda, T., and Tamaka, S. (1972). The behavior or suspended solid particles and liquid in bubble columns, *J. Chem. Eng. Japan*, **5**, 112.

76. Kato, Y., Nishiwaki, A., Kago, T., Fukuda, T., and Tanaka, S., (1973). Gas holdup and overall volumetric absorption coefficient in bubble columns with suspended solids particles, *Int. Chem. Eng.*, **13**, 562.

77. Kato, Y., and Nishiwaki, A. (1972). Longitudinal dispersion coefficient of a liquid in a bubble column, *Int. Chem. Eng.*, **12**, 182.

78. Kawagoe, K., Inoue, T., Nakao, K., and Otake, T. (1970). Flow pattern and gas holdup conditions in gas-sparged contactors, *Int. Chem. Eng.*, **16**, 176.

79. Kelkar, B. G., Shah, Y. T., and Carr, N. L. (1984). Hydrodynamic and axial mixing in three phase bubble column: Effect of slurry properties, *Ind. Eng. Chem. Proc. Des. Dev.*, **23**, 308.

80. King, W. E. Jr., Moon, W. G., Shah, Y. T., and Carr, N. L. (1984). An experimental study of the ignition behavior of an adiabatic SRC-II coal liquefaction reactor, *Fuel*, **63**, 178.

81. Kobayashi, K., Iida, Y., and Kanegae, N. (1970). Distribution of local void fraction of air–water two phase flow in a vertical channel, *Bulletin of JSME*, **13**, 1005.

82. Kobayashi, T., and Saito, H. (1965). Solid–liquid mass transfer in bubble columns, *Kogaku Kagaku*, **3**, 210.

83. Kolar, V. (1967). Contribution to the theory of suspension and dissolution of granular solids in mechanically mixed liquids, *Coll. Cz. Chem. Commun.*, **32**, 526.

84. Kolbel, H., Hammer, H., and Meis, U. (1965). Hydrogenation and oxidation in gas–liquid reactors with suspended catalysts (in German), *Proceedings on the Third European Symp. on Chem. Ren. Engng.*, p. 115. Pergamon Press, New York.

85. Komiyama, H., and Smith, J. M. (1975). Sulfur dioxide oxidation in slurries of activated carbon, I. Kinetics, *AIChE J.*, **21**, 664.

86. Koritala, S., and Dutton, H. J., (1965). Selective hydrogenation of soybean oil with sodium borohydride-reduced catalysts, *JAOCS*, **43**, 86.

87. Kuboi, R., Komasawa, I., Otake, R., and Iwasa, M. (1974). Fluid and particle motion in turbulent dispersions. III Particle-liquid hydrodynamics and mass transfer in turbulent dispersions, *Chem. Eng. Sci.*, **29**, 659.

88. Kurten, H., and Zehner, P. (1979). Slurry reactors, *Ger. Chem. Eng.*, **2**, 220.

89. Ledakowicz, S., Deckwer, W. D., and Shah, Y. T. (1984). Some aspects of direct coal liquefaction commercial reactor design, *Chem. Eng. Comm.*, **25**, 333.

90. Lemcoff, N. O. (1977). Liquid phase catalytic hydrogenation of acetone, *J. Catalysis*, **46**, 356.

91. Leuteritz, G., Loop reactor gives fast, cool, liquid-phase hydrogenation reactions, *Process Eng.*, **62** (Dec. 1973).

92. Levins, D. M., and Glastonbury, J. R. (1972). Applications of Kolmogoroff's theory to particle-liquid-mass transfer in agitated vessels, *Chem. Eng. Sci.*, **27**, 537.

93. Litmans, B. A., Kukurechenko, L. S., Boilo, I. D., and Tumanov, Yu. v. (1972). Investigation of the liquid phase mass transfer coefficient in baffled bubbling apparatus with mechanical agitation, *Theor. Foundation of Chem. Eng.*, **6**, 690.

94. Loiseau, B., Midoux, N., and Charpentier, J. C. (1977). Some hydrodynamics and power input data in mechanically agitated gas–liquid contactors, *AIChE J.*, **23**, 931.

95. Luong, H. T., and Volesky, B., (1979). Mechanical power requirements of gas–liquid agitated systems, *AIChE J.*, **25**, 893.

96. Maerteleire, E. De (1978). Heat transfer to a helical cooling coil in mechanically agitated gas–liquid dispersions, *Chem. Eng. Sci.*, **33**, 1107.
97. Malone, R. J., (1980). Loop reactor technology improves catalytic hydrogenations, *Chem. Eng. Progr.*, **53** (June 1980).
98. Mangartz, K. H., and Pilhofer, T. H. (1981). Interpretation of mass transfer measurements in bubble columns considering dispersion of both phases, *Chem. Eng. Sci.*, **36**, 1069.
99. Mashelkar, R. A. (1970). Bubble columns, *Brit. Chem. Engng.*, **15**, 1297.
100. Mashelkar, R. A. (1970). Mass transfer in bubble and packed bubble columns, *Trans. Instn. Chem. Engrs.*, **48**, T162.
101. Mehta, V. D., and Sharma, M. M. (1971). Mass transfer in mechanically agitated gas–liquid contactors, *Chem. Eng. Sci.*, **26**, 461.
102. Merchuk, J. C., and Stein, Y. (1981). Local holdup and liquid velocity in air-lift reactors, *AIChE J.*, **27**, 377.
103. Meyer, C., Clements, G., and Balaceanu, J. C. (1965). Initiation of free radical chain reaction by heterogeneous catalysts, *Proceedings 3rd International Congress Catalysis* (Eds. W. H. Schatler et al.), Vol. **1**, 184.
104. Michel, B. J., and Miller, S. A. (1962). Power requirements of gas–liquid agitated systems, *AIChE J.*, **8**, 282.
105. Michelson, M. L., and Ostergaard, K. (1970). Holdup and fluid mixing in gas–liquid fluidized beds, *Chem. Eng. J.*, **1**, 37.
106. Miller, D. N. (1971). Scale up of agitated vessels: Mass transfer from suspended solid particles, *Ind. Eng. Chem. Proc. Des. & Dev.*, **10**, 365.
107. Miyauchi, T., and Shyu, C. N. (1970). Flow of fluids in gas–liquid columns, *Kagaku Kogaku*, **34**, 958.
108. Morooka, S., Uchida, K. and Kato, Y. (1982). Recirculating turbulent flow of liquid in gas–liquid–solid fluidized bed, *J. Chem. Eng. Japan*, **15**, 29.
109. Mounce, W., and Rubin, R. S. (1971). The H-Oil route for hydroprocessing, *Chem. Eng. Progr.*, **67**, 81.
110. Murakami, Y., Hirose, T., Ono, S., Eltoku, H., and Nishijima, T. (1982). Power consumption and pumping characteristics in a loop reactor, *Ind. Eng. Chem. Proc. Des. Dev.*, **21**, 273.
111. Murakami, Y., Hirose, T., Ono, S., and Nishijima, T. (1982). Mixing properties in loop reactor, *J. Chem. Eng. Japan*, **15**, 121.
112. Nagata, S. (1975). *Mixing Principles and Applications*, Kodansha Ltd., Tokyo/Wiley, New York.
113. Narayan, S., Bhatia, V. K., and Guha, D. K. (1969). Suspension of solids by bubble agitation, *Can. J. Chem. Eng.*, **47**, 360.
114. Nguyen-Tien, K., and Deckwer, W. D. (in press). Effect of liquid flow and particle concentration on gas–liquid mass transfer in a bubble column, *Chem. Eng. Sci.*
115. Nienow, A. W. (1968). Suspension of solid particles in turbine agitated baffled vessels, *Chem. Eng. Sci.*, **23**, 1453.
116. Nienow, A. W. (1975). Agitated vessel particle–liquid mass transfer: A comparison between theories and data, *Chem. Eng. J.*, **9**, 153.
117. Nienow, A. W. and Miles, D. (1978). The effect of impeller/tank configurations on fluid-particle mass transfer, *Chem. Eng. J.*, **15**, 13.
118. Nishikawa, M., Kida, F., Kayama, T., and Nagata, S. (1975). Mixing and RTD studies in spouted bed, *Kagaku Kogaku Ronbunshu*, **1**, 363.
119. Niyama, H., Vemura, Y., and Echigoya, E. (1978). Estimation of bubble to liquid mass transfer rate coefficient by transient response technique and by steady state reaction studies, *J. Chem. Eng. Japan*, **11**, 465.
120. Nunez, P., Abichandani, J., Calimli, A., and Shah, Y. T. (1982). Multiple steady states in an adiabatic coal liquefaction reactor—Role of preheater, *Chem. Eng. Commns.*, **13**, 231.
121. Ogawa, S., Kobayashi, M., Tone, S., and Otake, T. (1982). Liquid phase mixing in the gas–liquid jet reactor with jet ejector, *J. Chem. Eng. Japan*, **15**, 469.

122. Ohki, Y., and Inoue, H. (1970). Longitudinal mixing of the liquid phase in bubble columns, *Chem. Eng. Sci.*, **25**, 1.

123. Oldshue, J. Y. (1980). Mixing in hydrogenation processes, *Chem. Eng. Progr.*, **60** (June 1980).

124. Otake, T., Tone, S., and Ogawa, S. (1982). Flow pattern and gas holdup in bubble column with liquid jet ejector, *Kagaku Kogaku Ronbunshu*, **8**, 65.

125. Panvelkar, S. V., Tierney, J. W., and Shah, Y. T. (1982). Backmixing in SRC dissolver, *Chem. Eng. Sci.*, **37**, 1582.

126. Parulekar, S. J., Raguram, S., and Shah, Y. T. (1980). Multiple steady states in adiabatic gas–liquid–solid reactors, *Chem. Eng. Sci.*, **35**, 745.

127. Pfeiffer, W., Blenke, H., and Schelknautz, E. Mu. (1977). Fluid dynamics of suspensions in jet-loop reactors, *Verfahrenstechnik*, **11**, 95.

128. Pilhofer, Th., Bach, H. F., and Mangartz, K. H. (1978). Determination of fluid dynamic parameters in bubble column design, *ACS Symp. Ser.*, **65**, 372.

129. Ramachandran, P. A., and Chaudhari, R. V. (1979). Theoretical analysis of reaction of two gases in a catalytic slurry reactor, *Ind. Eng. Chem. Proc. Des. Dev.*, **18**, 703.

130. Ramachandran, P. A., and Chaudhari, R. V. (1980). Estimation of batch time of a semibatch slurry reactor, *Chem. Engng. J.*, **20**, 75.

131. Ramachandran, P. A., and Chaudhari, R. V. (1980). Overall effectiveness factor of a slurry reactor for non-linear kinetics, *Can. J. Chem. Eng.*, **58**, 412.

132. Ramachandran, P. A., and Chaudhari, R. V. (1980). Predicting performance of three phase catalytic reactors, *Chem. Eng.* (McGraw-Hil) **74** (Dec. 1980).

133. Ramachandran, P. A., and Chaudhari, R. V. (1983). *Three Phase Catalytic Reactors*, Gordon and Breach, New York.

134. Ramachandran, P. A., and Smith, J. M. (1979). Mixing cell model for design of trickle bed reactor, *Chem. Eng. J.*, **17**, 91.

135. Rao, K. B., and Murty, P. S. (1973). Heat transfer in mechanically agitated gas–liquid systems, *Ind. Eng. Chem. Proc. Des. Dev.*, **12**, 190.

136. Reith, R., Renken, S., and Israel, B. A. (1968). Gas holdup and axial mixing in fluid phase of bubble columns, *Chem. Eng. Sci.*, **23**, 619.

137. Rice, P. G., Tuppurajnen, J. H. I., and Hedge, R. M. (1980). Dispersion and holdup in bubble columns, Paper presented at ACS Meeting, Las Vegas.

138. Roy, N. K., Guha, D. K., and Rao, M. N. (1964). Suspension of solids in a bubbling liquid, *Chem. Eng. Sci.*, **19**, 215.

139. Sanger, P., and Deckwer, W. D. (1981). Liquid–solid mass transfer in aerated suspension, *Chem. Eng. J.*, **22**, 179.

140. Sano, Y., Yamaguchi, N., and Adachi, T. (1974). Mass transfer coefficients for suspended particles in agitated vessels and bubble columns, *J. Chem. Eng. Japan*, **7**, 255.

141. Sato, Y., Murakami, Y., Hirose, T., Nashiguchi, Y., Ono, S., and Ichikawa, M. (1979). Flow pattern, circulation velocity and pressure loss in loop reactor, *J. Chem. Eng. Japan*, **12**, 448.

142. Satterfield, C. N., and Huff, G. A. Jr. (1980). Effects of mass transfer on Fischer–Tropsch synthesis in slurry reactors, *Chem. Eng. Sci.*, **35**, 195.

143. Schmidt, H. J. (1970). Hydrog. of triglycerides contg. linoleric acids: II Cont. hydrog. of veg. oils, *J. Am. Oil Chem. Soc.*, **134** (April 1970).

144. Schugerl, K., Lucke, J., and Oels, U. H. (1977). Bubble columns bioreactors, *Adv. Biochem. Eng.*, **7**, 1.

145. Seher, A., and Schumacher, V. (1979). Determination of residence times of liquid and gas phases in large bubble columns with the aid of radioactive tracers, *Ger. Chem. Eng.*, **2**, 117.

146. Shah, Y. T. (1979). *Gas–Liquid–Solid Reactor Design*, McGraw-Hill, New York.

147. Shah, Y. T. (1981). *Reaction Engineering in Direct Coal Liquefaction*, Addison-Wesley, Reading, Mass.

148. Shah, Y. T., and Deckwer, W. D. (in press). *Scaleup in Chem. Process Industries* (Eds. R. Kabel and A. Bisio), John Wiley, New York.
149. Shah, Y. T., Kelkar, B. G., Godbole, S. P. and Deckwer, W. D. (1982). Design parameters estimations for bubble column reactors, *AIChE J.*, **28**, 353.
150. Shah, Y. T., and Parulekar, S. J. (1982). Steady state thermal behavior of an adiabatic three phase slurry reactor—Coal liquefaction under slow hydrogen consumption reaction regime, *Chem. Eng. J.*, **23**, 15.
151. Shah, Y. T., Stiegel, G. J., and Sharma, M. M. (1978). Backmixing in gas–liquid reactors, *AIChE J.*, **24**, 369.
152. Shah, Y. T., Bell, A. T., Stern, D., and Heinemann, H. (1984). Mass transfer imposed constraints on the feed gas composition for F–T synthesis in a bubble column reactor, *Frontiers in Chemical Reaction Engineering* (Eds. L. Doraiswamy and R. Mashelkar), Vol. 1, p. 345, John Wiley, New York.
153. Sharma, M. M., and Mashelkar, R. A. (1968). Absorption and reaction in bubble columns, *Instn. Chem. Engrs. Symp. Ser.*, **28**, 10.
154. Sittig, W. (1977). Unbersuchungen Zur Verbesserung des Stoffaustauschs in Fermentern, *Verfahrestechnik (Mainz)*, **11**, 730.
155. Slesser, C. G. M., Allen, W. T., Cumming, A. R., Pawlowsky, U., and Shields, J. (1968). A study of non-catalytic aspects of catalyst particle influence in slurried bed reactors, *Chem. Reaction Eng.—Proceedings of the Fourth European Symposium Held in Brussels*, **41**.
156. Snyder, J. M., Scholfield, C. R., and Mounts, T. L. (1974). Lab. scale continuous hydrogenation: Copper catalysis, *J. Am. Oil. Chem. Soc.*, **56**, 506.
157. Steiff, A., and Weinspach, P. M. (1978). Heat transfer in stirred and non-stirred gas–liquid reactors, *Ger. Chem. Eng.*, **1**, 150.
158. Suganuma, T., and Yamanishi, T. (1966). Behavior of solid particles in bubble columns, *Kagaku Kogaku*, **30**, 1136.
159. Sylvester, N. D., Kulkarni, A. A., and Carberry, J. J. (1975). Slurry and trickle bed reactor effectiveness factor, *Can. J. Chem. Eng.*, **53**, 313.
160. Todt, J., Lucke, J., Schugerl, K., and Reuken, A. (1977). Gas holdup and longitudinal dispersion in different types of multiphase reactors, *Chem. Eng. Sci.*, **32**, 369.
161. Ueyama, K., and Miyauchi, T. (1979). Properties of recirculating turbulent two phase flow in gas bubble columns, *AIChE J.*, **25**, 258.
162. Ueyama, K., Morooka, S., Koide, K., Kaji, H., and Miyauchi, T. (1980). Behavior of gas bubbles in bubble columns, *Ind. Eng. Chem. Processes. Des. Dev.*, **14**, 492.
163. Uhl, V. W., and Gray, J. B. (1966). *Mixing*, Vol. I, Academic Press, New York.
164. Van't-Reit, K., Boom, J. M., and Smith, J. M. (1976). Power consumption, impeller coalescence and recirculation in aerated vessels, *Trans. Instn. Chem. Engrs.*, **54**, 124.
165. Vlcek, J., Seidl, H., and Kudrna, V. (1969). Holdup of dispersed gas fed continuously into a mechanically agitated liquid, *Coll. Cz. Chem. Comm.*, **34**, 373.
166. Voyer, R. D., and Miller, A. I., (1968). Improved gas–liquid contacting in co-current flow, *Can. J. Chem. Eng.*, **46**, 335.
167. Wallis, G. B. (1967). *One Dimensional Two Phase Flow*, McGraw-Hill, New York.
168. Weissermel, R., and Arpe, H. J. (1976). *Industrialle Organische Chemie*, Verlag Chemie/Weinheim, New York.
169. Westerterp, K. R., Van Diorendonck, L. L., and Dekraa, J. A. (1963). Interfacial areas in agitated gas liquid contactors, *Chem. Eng. Sci.*, **18**, 157.
170. Wiedmann, J. A., Steiff, A., and Weinspach, P. M. (1980). Experimental investigations of suspension, dispersion, power, gas holdup and flooding characteristics in stirred gas–solid–liquid systems, *Chem. Eng. Commun.*, **6**, 245.

171. Yagi, H., and Yoshida, F. (1975). Gas absorption by Newtonian and non-Newtonian fluids in sparged agitation vessels, *Ind. Eng. Chem. Proc. Des. Dev.*, **14**, 488.
172. Ying, D. H., Givens, E. N., and Weimer, R. F. (1980). Gas holdup in gas–liquid and gas–liquid–solid flow reactors, *Ind. Eng. Chem. Proc. Des. Dev.*, **19**, 635.
173. Yoshida, F., and Miura, Y. (1963). Gas absorption in agitated gas–liquid contactors, *Ind. Eng. Chem. Proc. Des. Dev.*, **2**, 263.
174. Yung, C. N., Wong, C. W., and Chang, C. L. (1979). Gas holdup and aerated power consumption in mechanically stirred tanks, *Can. J. Chem. Eng.*, **57**, 672.
175. Zlokarnik, M. (1966). Hollow stirrers for gas–liquid contact-calculation of the attainable mass and heat transfer, *Chem. Ing. Tech.*, **38**, 717.
176. Zlokarnik, M. (1971). Eignung von Einlochboden als Gasverteiler in Blasensaulen, *Chem. Ing. Tech.*, **43**, 329.
177. Zlokarnik, M., and Jukat, H. (1969). Rohe-und Propellerruhrer-eine Wirkungsvolle Ruhrerkombination Zum Gleichzeitigen Begasen und Aufwirbln, *Chem. Ing. Tech.*, **41**, 1270.
178. Zwietering, T. N. (1958). Suspending of solid particles in liquid by agitators, *Chem. Eng. Sci.*, **8**, 244.

Chapter 7

Recent Advances in Trickle Bed Reactors

YATISH T. SHAH

Chemical and Petroleum Engineering Department, University of Pittsburgh, Pittsburgh, Pennsylvania, 15261

Abstract

Trickle bed reactors are widely used in petroleum, chemical and biochemical industries. During the past decade, considerable research on the understanding of various aspects of trickle bed reactors has been published. Shah [112] has presented a critical review of the work published before 1977. The present review outlines important recent advances made on the subject. The review covers the following topics: (a) flow regime boundaries, (b) pressure drop, (c) liquid holdup, (d) radial distribution of liquid, (e) effective wetting of catalyst, (f) interphase mass transfer, (g) interparticle heat transfer, (h) reactor models and (i) scaleup considerations. For each topic, up to date status of knowledge and wherever appropriate, recommendations for further research are outlined.

1. INTRODUCTION

The term "trickle bed reactors" is commonly used for the fixed bed catalytic reactor wherein liquid trickles down over a fixed bed of catalyst with a simultaneous flow of a continuous gas phase. While the trickle flow operation can occur with a countercurrent flow of gas (upwards) and liquid (downwards), in nearly all practical industrial trickle bed operations gas and liquid flow cocurrently downward over a fixed bed of catalyst. Trickle bed reactors are very successful in carrying out numerous hydroprocessing, hydration, amination and oxidation reactions. They are often preferred because of the simple and stable mode of operation, ease of control,

Table 1   Some Industrial Applications of Trickle Bed Reactors

Hydroprocessing Operations
- butadiene synthesis—Rappe process
- acetone to methyl isobutyl ketone
- adiponitrile to hexamethylenediamine
- benzoic acid to cyclohexane–carboxylic acid
- benzene to cyclohexane
- 2-butyne-1,4 diol to 2-butene-1,4 diol and 1,4-butane diol
- γ-butyrolactone to 1,4-butane diol
- "oxo" aldehydes to "oxo" alcohols
- phenol to cyclohexanol
- phenol to cyclohexanone
- purification of C_3 and C_4 olefinic cuts ($C{\equiv}C$) to $C{=}C$
- fatty esters or fatty acids to fatty alcohols
- fulfural to furfuryl alcohol
- glucose to sorbitol
- maleic anhydride to tetrahydro furan
- mesityl oxide to methyl isobutyl ketone or carbinol
- caprolactone to 1,6 hexane diol
- cyclododeca 1,3,5-triene to cyclododecane
- dinitro toluene to diamino toluene
- dimetyl terephthalate to cyclo hexane 1,4-dicarboxylic acid, dimethyl ester
- cyclohexane-1,4-dicarboxylic acid, dimethyl ester to 1,4 dimethylol-cyclohexane
- hydroisometization of butene-1 to butene-2
- reduction of sodium chlorate in (diaphragm cell) aqueous caustic liquor with H_2
- reduction of uranyl (VI) to uranyl (IV) in aqueous solutions with H_2
- deuterium exchange between H_2 and water
- hydrodesulfurization and hydrocracking of petroleum fractions

Hydrations
- hydration of ethylene to ethanol
- propylene to isopropanol
- isobutylene to isobutyl alcohol
- ethylene oxide to glycol

Aminations
- phenols to monoethanol amine and aniline
- hexane diol to hexamethylene diamine
- amination of monoethanol amine to ethylene diamine

Oxidations
- ethylene to ethylene oxide
- oxidations of cyclohexane, ethyl benzene, tetralin, cumene
- oxidation of waste water-biological purification-air oxidation of dilute fumic acid and acetic acid solutions
- air oxidation of ethanol to acetic acid; conversion of primary alcohols to the corresponding sodium salt of the acid
- oxidation of SO_2 to H_2SO_4 in water-absorption of H_2S from stack gases in alkaline solutions; oxidation of aqueous sodium sulphide (black liquor)

Others
- absorption of NO_x (at low ppm) in water
- acrylamide from acrylonitrile
- alkylation of benzene with propylene
- reaction between isobutylene and aqueous formaldehyde (a step in manufacture of synthetic isoprene)

relatively low pressure drop, reasonably good heat transfer characteristics and ease of application to a wide range of feedstocks.
Industrial scale operation is also facilitated by the wide range of
experience with such reactors.

Some practical reaction system which use trickle bed reactors are
listed in Table 1 [64, 112]. As shown in this table, the most common
application is to the variety of hydroprocessing operations. For
these operations, a thin liquid film around the catalyst particle in a
trickle bed reactor allows reactions between a liquid phase reactant
and gaseous hydrogen with small mass transfer resistances at the
gas–liquid and liquid–solid interface. The reactions can, therefore,
be controlled by the intrinsic kinetics. Other usages of trickle bed
reactors are hydration, oxidation and amination reactions.

2. FLOW REGIMES

When gas and liquid flow cocurrently downwards over a fixed bed of
catalyst, a number of flow regimes such as trickle flow, pulsed flow,
spray flow and bubble flow can prevail depending upon the flow
rates of gas and liquid phases, catalyst shape and size and fluid
properties.

In all the works on two-phase downflow over packed beds, before
1975, flow regimes have been considered to be functions of only gas
and liquid flow rates and the flow regime boundaries have been
obtained in terms of superficial mass flow rates of air and water. The
coordinates for the flow maps, showing flow regime boundaries have
been either G_G vs. G_L [1, 3, 8, 102, 105, 137] or G_L/G_G vs. G_G
[133]. Sato et al. [102] noted that the boundary between gas
continuous and pulse flow moves to low liquid flow rates and
returns to high liquid flow rates as the particle size is continuously
increased.

Charpentier and Favier [9] and Morsi et al. [84] introduced the
effects of density, viscosity, and surface tension and the foaming
nature of liquid on the flow regime boundaries and presented flow
maps using Baker's coordinates for gas–liquid flow in an empty
tube. Their results are discussed by Shah [112] and Gianetto et al.
[32]. The latter workers correlated the literature data using a modified Baker's coordinate system which included the bed void fraction. They put the data for both foaming and nonfoaming systems
on the same figure and concluded that the boundaries that predict
the transition from laminar trickle flow (low interaction regime) to
turbulent flow, pulsed, spray or bubble flow (high interaction

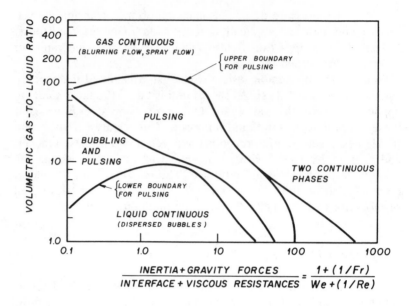

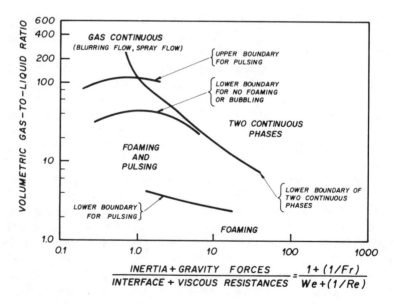

Figure 1   Flow maps: Two-phase downflow through packed beds. (a) Nonfoaming liquids. (b) Foaming liquids.

regime) regime for nonfoaming liquids differ considerably from those for foaming liquids. They concluded that the physical parameters such as density, viscosity and surface tension in Baker's coordinates are not sufficient to define the foaming system. They also noted that packing wettability has a significant effect on the transition from trickling to pulsing flow.

Talmor [130] presented flow maps (Fig. 1) for foaming and nonfoaming liquids using a superficial volumetric gas to liquid ratio versus a ratio of driving force to resistance force. For two-phase downflow, the driving forces are inertia and gravity while the resistance forces are proportional to viscosity and surface tension. The ratio of these two forces can thus be expressed in a dimensionless form as

$$\frac{\text{Driving Force}}{\text{Resistance Force}} = \frac{1 + (1/\text{Fr})}{(1/\text{Re}) + (1/\text{We})} \tag{1}$$

The variables affecting the force ratio are packing size and bed voidage, liquid and gas superficial mass flow rates, densities and viscosities and liquid surface tension. Bed diameter or height does not affect the correlation of flow regimes.

As an extension of the work of Chou et al. [11] on aqueous solutions of alcohol, Clements and Halfacre [14] studied air–isopropanol aqueous solution (0–48 wt%) and compared the flow maps of Charpentier and Favier [9] and that of Talmor [130] against their data. Their study suggests that the flow maps developed by Charpentier and Favier [9] do not handle mixture data well, whereas Talmor's [130] flow maps, with minor modifications, are capable of representing gas-mixed liquid systems very well. These workers have also concluded that interfacial properties exert the dominant influence over flow pattern transitions and foaming in the packed bed. At present, the maps shown in Fig. 1 are recommended for evaluation of the range of trickle flow operation. It should be noted that in many commercial hydroprocessing operations, the trickle bed reactors are operated close to the boundaries for trickle and pulsating flow [105].

3. PRESSURE DROP

Most reported data on pressure drop for trickle bed reactors are for the air–water system. Other gas–liquid systems have also been investigated [15, 16, 62, 121] but more data on hydrocarbon and foaming systems are needed. The data base of Larkin et al. correlation [61] is relatively large and close to the conditions in a commerical reactor and therefore this correlation is often used.

In the correlation of pressure drop data for two-phase flow in packed beds, two approaches have basically been used. In one approach, the variables are lumped into various dimensionless groups, such as Reynolds numbers for the gas and liquid, Weber number, etc. and the pressure drop or friction factor is expressed in terms of these dimensionless groups. Use of Tallmadge or Lock-hardt–Martinelli parameters is included in this approach [133, 121]. The other approach is to correlate the two-phase pressure drop with the pressure drops associated with energy losses for individual phases in single-phase flows under identical conditions.

The correlation of Larkins et al. [61], for cocurrent downflow, is based on the second approach. It correlates the overall two-phase energy loss for the gas and liquid passing through the reactor with the two individual single-phase energy losses as follows:

$$\log_{10}\left(\frac{\delta_{LG}}{\delta_L + \delta_G}\right) = \frac{K_1}{(\log_{10} \bar{x})^2 + K_2} \qquad (2)$$

where δ_G is the gas-phase energy loss, δ_L the liquid energy loss and δ_{LG} the two-phase energy loss and $\bar{x} = (\delta_L/\delta_G)^{1/2}$. The constants K_1 and K_2 have the values of 0.416 and 0.666, respectively.

Larkins et al. [61] defined the two-phase pressure gradient $(\Delta P/\Delta Z)_{LG}$ as follows:

$$\left(\frac{\Delta P}{\Delta Z}\right)_{LG} = \delta_{LG} - \rho_M \qquad (3)$$

where

$$\rho_M = \varepsilon_L \rho_L + (1 - \varepsilon_L)\rho_G$$

Several other workers [1] have used Eq. (3) or modified forms [98, 101] for correlation of their data. Charpentier et al. [8] derived pressure drop correlations by comparing two-phase pressure drop, ΔP_{LG}, with that for single-phase flow. Clements and Schmidt [15, 16, 17] gave a correlation for pressure drop in two-phase cocurrent downflow in terms of Reynolds numbers of gas and liquid phases, gas-phase Weber number and the pressure drop for the gas

Table 2   Pressure Drop Correlations for Two-Phase Concurrent Downward Flow in Packed Beds (After Gianetto et al. [32])

Larkins et al. [61]

$$\log\left(\frac{\delta_{LG}}{\delta_L + \delta_G}\right) = \frac{0.416}{(\log \bar{x})^2 + 0.666}$$

$$0.05 \leqq \bar{x} = (\delta_L/\delta_G)^{0.5} < 30$$

9 mm Raschig rings, 9 mm stoneware spheres and 3 mm glass spheres; nonfoaming systems for all hydrodynamic regimes

Turpin and Huntington [133]

$$\ln f_{LG} = 7.96 - 1.34 \ln Z + 0.0021(\ln Z)^2 + 0.0078(\ln Z)^3$$

$$f_{LG} = \delta_{LG}D_e/2U_G^2\rho_G$$

$$D_e = \frac{2}{3}D_p\frac{\varepsilon}{1 - \varepsilon}$$

$$0.2 < Z = (\mathrm{Re}_G)^{1.167}/(\mathrm{Re}_L)^{0.767} < 500$$

7.5 to 8.1 mm tabular alumina; air–water for all hydrodynamic regimes

Sato et al. [101]

$$\log\left(\frac{\delta_{LG}}{\delta_L + \delta_G}\right) = \frac{0.70}{[\log(\bar{x}/1.2)]^2 + 1} \quad 0.1 < \bar{x} < 20$$

$$\left(\frac{\delta_{LG}}{\delta_G}\right) = 1.30 + 1.85(\bar{x})^{-0.85}$$

2.5 to 24.3 mm glass spheres; air–water for all hydrodynamic regimes

Midoux et al. [74]

$$\left(\frac{\delta_{LG}}{\delta_L}\right)^{0.5} = 1 + \frac{1}{\bar{x}} + \frac{1.14}{(\bar{x})^{0.54}}; 0.1 < \bar{x} < 80$$

3 mm glass and catalyst spheres, 1.8×6 and 1.4×5 mm catalyst cylinders; nonfoaming systems for all hydrodynamic regimes, and foaming systems for the low-interaction regimes

$$\left(\frac{\xi_{LG}}{\xi_L}\right)^{0.5} = 1 + \frac{1}{\bar{x}'} + \frac{6.55}{(\bar{x}')^{0.43}}$$

$$0.05 < \bar{x}' = (\xi_L/\xi_G)^{0.5} < 100$$

$$\xi_{LG} = \frac{1}{\varepsilon}\left[\frac{G_L}{\rho_L} + \frac{G_G}{\rho_G}\right]\left(\frac{\Delta H}{\Delta Z}\right)_{LG} + \frac{G_L + G_G}{\varepsilon\rho_e}$$

$$\xi_L = \frac{G_L}{\varepsilon}\left[\frac{1}{\rho_L}\left(\frac{\Delta H}{\Delta Z}\right)_L + \frac{1}{\rho_e}\right]$$

$$\xi_G = \frac{G_G}{\varepsilon}\left[\frac{1}{\rho_G}\left(\frac{\Delta H}{\Delta Z}\right)_G + \frac{1}{\rho_e}\right]$$

3 mm glass and catalyst spheres and 1.4×5 mm catalyst cylinders; foaming systems for the high-interaction regime

Table 2   (continued)

Specchia and Baldi [121]	$\delta_{LG} = k_1 \dfrac{1 - \varepsilon(1 - \varepsilon_{sL} - \varepsilon_{dL})^2}{\varepsilon^3(1 - \varepsilon_{sL} - \varepsilon_{dL})^3} \rho_G U_G$ $\quad + k_2 \dfrac{1 - \varepsilon(\varepsilon_{sL} - \varepsilon_{dL})}{\varepsilon^3(1 - \varepsilon_{sL} - \varepsilon_{dL})^3} \rho_G U_G^2$

6 mm glass spheres, 5.4 × 5.4 and 2.7 × 2.7 mm glass cylinders, 6.4, 10.3 and 22 mm Raschig rings; foaming and nonfoaming systems for the low-interaction regime

$\ln f_{LG} = 7.82 - 1.30 \ln (Z/\psi^{1.1}) - 0.0573 \ln (Z/\psi^{1.1})^2$

$0.6 < Z/\psi^{1.1} < 500$

$\psi = \dfrac{\sigma_{\text{water}}}{\sigma_L} \left[\dfrac{\mu_L}{\mu_{\text{water}}} \left(\dfrac{\rho_{\text{water}}}{\rho_L} \right)^2 \right]^{1/3}$

6 and 3 mm glass spheres, 5.4 × 5.4 and 2.7 × 2.7 mm glass cylinders, 3.2 × 3.2 mm catalyst cylinders, 6.4, 10.3, and 22 mm Raschig rings; foaming and nonfoaming systems for the high-interaction regime

phase alone. Clements and Halfarcre [13] also gave a theoretical justification for this approach. Theoretical correlations for pressure drop under trickle flow conditions have been proposed by Hutton and Leung [52] and Specchia and Baldi [121].

It is important to note that in most of the reported studies, Reynolds numbers for liquid and gas phases are considered to be the controlling dimensionless groups. However, those studies which use systems other than air–water recognized the role of surface tension on the pressure drop and used either an empirical correction factor [121] or the Weber number [15, 16, 17]. More data over a larger range of Weber number are needed. A brief summary of some of the important correlations for pressure drop reported in the literature is given in Table 2.

4. LIQUID HOLDUP

Total liquid holdup in a fixed bed column is the sum of the operating and static holdup. Static holdup is the amount of liquid in the bed after the liquid inlet is shut off and the column is allowed to drain. The static holdup represents the liquid retained in the pore volume of the catalyst and its packing. Operating holdup is the liquid external to the catalyst particles and it depends on the liquid and gas flow rates, their properties and the nature of catalyst packing. Shah [112] and Gianetto et al. [32] have published reviews of holdup and pressure drop correlations. More recently, Morsi et al. [85] have presented correlations for nonfoaming, foaming, and viscous liquids in the low and high interactions regimes.

The static holdup is often correlated by the Eotvos number Eo $(=\rho_L g D_p^2 / \sigma_L)$, where D_p is the nominal particle diameter and g the gravitational acceleration. The correlation of Van Swaaij et al. [136] indicates that smaller particle diameter and fluid density and larger surface tension give larger static liquid holdup. Generally, for small Eo (Eo < 4) the static holdup remains essentially independent of Eo. For Eo > 4, the static holdup decreases with an increase in Eo. The correlation also indicates that a porous material gives a larger static liquid holdup than a nonporous material.

Basically, two types of correlations for the dynamic or total liquid holdup are reported in the literature. Some investigators have correlated the liquid holdup directly to the liquid velocity and fluid properties by either dimensional or dimensionless relations. In more recent investigations, the liquid holdup is correlated to the Lockhart–Martinelli parameter $\Delta P_L / \Delta P_G$ or an equivalent of it.

For cocurrent downflow, the dimensional correlations [103, 154] are very few. Most of the correlations are in the following dimensionless form [5, 17, 25, 30, 75, 80, 88, 121, 136].

$$\varepsilon_{dL} = \alpha (\mathrm{Re}_L)^\beta (\mathrm{Ga}_L)^\gamma (a_t D_p)^\eta \qquad (4)$$

Values of α, β, γ, and η obtained by different workers are significantly different.

Many correlations using the Lockhart–Martinelli parameter, or an equivalent are reported in the literature [1, 61, 74, 101]. The correlations of Midoux et al. [74] for nonfoaming and foaming liquids are recommended for use in the case of hydrocarbon liquids.

The theoretical models for liquid holdup have been developed by Hutton and Leung [42], Reynier and Charpentier [99], Clements and Schmidt [15], and more recently by Matsuura et al. [68, 69]. Hutton and Leung [52] and Matsuura et al. [68, 69] define the gravity–viscosity regime (at low liquid rates) and the gravity–inertia regime (at high liquid rates) to relate pressure drop and liquid holdup. Similar correlations have also been developed by Specchia and Baldi [121] for both foaming and nonfoaming systems.

Interesting results for pressure drop and liquid holdup in a trickle bed packed with small particles (0.05, 0.1, and 0.18×10^{-2} m) were found by Kan and Greenfield [55, 56]. They observed multiple values of the pressure drop and the total liquid holdup at the same gas and liquid rates in the water–air system. The data were best correlated by the correlations of Specchia and Baldi [121].

More recently, Clements and Schmidt [16, 17] have also studied pressure drop and liquid holdup in a trickle bed packed with various small packings (0.16×10^{-2} and 0.29×10^{-2} m spheres and 0.08×10^{-2} m extrudate). Silicone liquids that covered a wide range of viscosities ($1.5 \times 10^{-3} – 8 \times 10^{-3}$ kg/(m)(s)) and had a surface tension of $18–20 \times 10^{-3}$ N/m were used. Although many of the pressure drop curves were obtained by a series of increasing gas rates followed by a series of decreasing gas rates, no multiple states were observed. While the data, in general, agreed well with those of Kan and Greenfield [55, 56] and Specchia and Baldi [121], they indicated that the effects of physical properties of the liquid for the columns containing small particles should be further investigated. Finally, Mills and Dudukovic [79] correlated dynamic holdup data obtained from dynamic measurements using a tracer [19, 79] and from direct measurements of the flowing liquid in the bed [33] for small particles. The correlation was in terms of various dimensionless numbers and is valid at low gas rates.

The range of variables and physical properties examined in the literature can be roughly stated as $0.3 < Re_L < 3000$, 0.4 mm $<$ $D_p < 50$ mm, $2.6 < a_t D_p < 6.03$, $14 < Ga_L < 320$ for particle shapes consisting of Raschig rings, Berl saddles, spheres, and irregular granules. Both foaming and nonfoaming liquids have been examined. Although there are significant discrepancies in the predictions of various correlations, qualitatively they all indicate that the liquid holdup under trickle flow conditions increases with liquid velocity and is essentially independent of gas flow rate. Although not completely clear, an increase in particle size appears to decrease the liquid holdup. An increase in Galileo number for the liquid also decreases the liquid holdup. The liquid holdup is considered to be independent of surface tension. This conclusion needs to be verified in view of the observed dependence of flow regimes and pressure drop on surface tension.

5.   RADIAL DISTRIBUTION OF LIQUID

For the proper design of a trickle bed reactor, it is essential to have radial distribution of liquid such that all catalyst particles are effectively wetted. This is particularly important for the laboratory reactor where both liquid flow and bed length are small. Three important aspects of the liquid distribution should be considered:

1. The bed depth required for achieving equilibrium liquid distribution (that is, the distribution that does not change with bed depth). This bed depth depends on the gas and liquid rates, physical properties of the liquid, particle size and shape, column diameter and especially the feed distribution at the top of the packed bed.

2. The equilibrium liquid distribution which depends on the gas and liquid rates, physical properties of the liquid and particle size and shape.

3. The higher flow of liquid near the column wall is due to the higher bed porosity at that location. This in turn depends on the liquid physical properties, the particle shape and, particularly, on the ratio of column to packing diameter.

Clearly, a much shorter bed depth is required for reaching equilibrium liquid distribution if a uniform feed distribution rather than a point source is used [92, 93]. If the initial distribution is known, the models presented by Herskowitz and Smith [49] may be employed to calculate this bed depth. The liquid distribution has been usually measured by collecting the liquid in annual collectors located at the bottom of the bed. Some of the important experimental studies reported in the literature are summarized by Herskowitz and Smith [49] and Gianetto et al. [32].

The equilibrium liquid distribution has been found to be essentially uniform at low liquid rates ($U_L < 0.9 \times 10^{-3}$ m/s for water like liquids) and particle sizes commonly used in industrial practice in the low-interaction flow regime [42, 43, 120, 129]. Increasing the liquid superficial velocity beyond 0.01 m/s with no gas flow, Specchia et al. [120] measured a parabolic profile with a minimum at the center of the column. In the same range, Weekman and Myers [137] reported a maximum in the liquid distribution at the center of the column and a minimum in the middle section. Sylvester and Pitayagulsarn [129] found the liquid distribution to be unstable when the liquid superficial velocity was higher than 0.01 m/s and the gas superficial velocity was lower than 0.033 m/s.

The flow of liquid near the wall has been measured as a function of the particle size and shape, the column diameter [18, 42, 43, 93] and the liquid physical properties [94]. Prchlik et al. [92] reported that for $D_c/D_p > 25$ flow of liquid near the wall at equilibrium was less than 10% of the total flow in the bed. Herskowitz and Smith [42, 43] proposed that for $D_c/D_p > 18$ the wall flow is negligible. For

$D_c/D_p < 18$, granular particles cause less wall flow than spherical or cylindrical particles. The influence of particle wettability on the equilibrium wall flow of liquid in a packed column has been recently reported by Patwardhan and Pataskar [91]. A negligible effect of the liquid physical properties on the wall flow measured at equilibrium was found by Prchlik et al. [94]. On the other hand, Herskowitz and Smith measured the negligible wall flow for $D_c/D_p = 12$ when the surface tension was lowered to 38×10^{-3} N/m. For D_c/D_p larger than about 20, the liquid distribution at equilibrium in laboratory and pilot trickle bed reactors, operated in the low-interaction flow regime, is uniform [49]. In the transition regime, a maximum in the center may occur, while in the high-interaction flow regime, the liquid distribution is nearly uniform. No data on the liquid distribution in industrial reactors are available, and more data are needed for organic liquids and, in particular, for foaming liquids.

Radial distribution of liquid has also been modeled by a number of investigators. Herskowitz and Smith [49] reviewed the application of a random-walk mechanism for the trickle flow over a randomly packed bed. The analysis includes an important parameter, D_r, the radial liquid spread coefficient, which in dimensionless form is expressed as $S = 4D_r/D_p$, is the particle diameter. Jameson [53] developed a finite difference model which assumed the bed to consist of discrete spherical particles. In addition to D_r, he introduced a parameter called the wall factor, f_w, which represents the fraction of wall flow that returns to the main part of the bed.

The parameter S mentioned above can be evaluated from liquid distribution [42, 43, 59, 125] or flow tracer distribution data [120]. Herskowitz and Smith [49] showed that the results obtained from various studies are in good agreement. They also indicated that the liquid spreads more quickly in granular packings than with cylinders or spheres. For a particle diameter less than about 0.006 m the radial spreading factor is unity. Changing the surface tension of the liquid had no effect on the radial spreading factor as reported by Herskowitz and Smith [42, 43] and Prchlik et al. [94]. On the other hand, measurements of the liquid distribution performed with a mixture of hydrocarbons yielded a value of D_r, lower by about 25% than the value obtained with water. Thus, the effect of liquid properties on the radial distribution of liquid is as yet not properly understood.

At zero or low gas rates and $U_L < 0.005$ m/s, D_r was found to be constant [49, 120, 127]. At higher values of the liquid superficial velocity, Stanek and Kolar [127] reported that D_r depends on the liquid rate while Specchia et al. [120] reported a constant value of D_r. Specchia et al. [120] also found that, in general, D_r decreases with increasing gas and liquid rates.

Recently Crine et al. [20, 22, 23] and Crine and L'Homme [21] developed a different model based on percolation theory which considers preferential flow paths in the bed. The liquid distribution predicted by this and random walk models was different at low liquid rates but in reasonable agreement at high liquid rates. This model should be simplified and tested experimentally before it can be used for predictions. Additional experimental data, obtained with improved means of measuring the liquid distribution are needed for further testing of the various theoretical models.

6. EFFECTIVE WETTING

As mentioned earlier, in laboratory trickle bed reactors the catalyst particles may not be effectively wetted, causing errors in the evaluation of the intrinsic kinetic rates of the process. The rate of reaction in partially wetted catalyst particles may be higher or lower than that obtained with completely wetted particles, depending on whether the limiting reactant is present only in the liquid or in both flowing phases. A number of interesting experimental systems have been examined in the literature and a few of them are briefly summarized in Table 3.

For packed bed absorbers several correlations have been proposed for the wetting efficiency as a function of liquid and gas superficial flow rates, liquid physical properties, and particle size [107, 108]. As pointed out by Satterfield [105] the use of these results to estimate the wetting efficiency in trickle bed reactors is questionable. Two kinds of wetting must be defined for porous catalysts.

1. Internal wetting, that is, the amount of internal area wetted by the liquid. It is a measure of the active area available for the liquid reactants, and hence of the maximum utilization of the catalyst volume when internal mass transfer limitations are negligible. The active available area is connected with the pore filling, that is the volume of liquid filling the catalyst pores divided by the total pore volume.

2. External effective wetting, that is, the amount of external area that is wetted by the flowing liquid. Although reaction occurs mainly

Table 3   Some Experimental Systems where Effective Wetting is Important

System	Reference	Comments
hydrodesulfurization of gas oil	Bondi [4]	Rate increased with better wetting
hydrotreating of various petroleum fractions desulfurization denitrogenation vanadium and nickel removal etc.	Montagna et al. [82] Henry and Gilbert [41] Mears [72, 73] Paraskos et al. [89] Montagna and Shah [81]	Rate increased with better wetting Better wetting was achieved by increasing bed length and liquid flow rate
denitrogenation of hydrotreated oil	Van Klinken and van Dongen [134]	Same as above
decomposition of H_2O_2 in a methanol/water mixture	Koros [60]	Instrinsic rate was lower for liquid superficial velocity less than 0.5×10^{-2} m/s
hydrogenation of cro-tonaldehyde on a Pd supported catalyst at 31°C	Sedricks and Kenney [109]	Rate increased with poor wetting
hydrogenation of ben-zene at 76°C	Satterfield and Ozel [104]	Rate decreased with increased wetting
hydrogenation of α-methyl styrene in Pd/ Al_2O_2 catalyst	Morita and Smith [83] Herskowitz et al. [44]	Liquid–solid mass transfer resistance may be impor-tant. Lower external wetting enhanced reaction rate. Minimum rate at intermediate liquid velocity
oxidation of sulfur dioxide in liquid water catalyzed by activated carbon	Mata and Smith [67]	Same charcteristics as above system

on the internal area, a partially wetted external area may affect the reaction rate because of limitations to the supply rate of liquid reactants to the solid external surface, and because of variations of the internal diffusion rate of these reactants. In a pellet completely enveloped by moving liquid, the maximum diffusion path of the molecules in the pores is on the order of the pellet radius. When the external area is only partially wetted, the maximum internal path increases and the diffusion rate decreases. As a consequence, the catalyst effectiveness factor would change with the extent of the external wetting. This concept was developed by Colombo et al. [19].

Variations in effective wetting with operating parameters have been examined by a number of investigators using a variety of techniques. Some of the recent literature studies are summarized in Table 4. Two principal methods for measuring effective wetting are: (a) tracer method and (b) the reaction method. According to Herskowitz and Smith [49], the heterogeneous chemcial reaction method has the following advantages:

(1) The external wetting efficiency is clearly defined and may be calculated from a suitable model [49].

(2) The existence of dry regions in the bed may be detected. If the limiting reactant is volatile, such dry regions would decrease the global rate of reaction significantly [48]. The internal wetting efficiency calculated from tracer data may also indicate that dry regions are formed.

Table 4   Summary of Experimental Studies for Effective Wetting

Method/System	Reference	Remarks
Two tracer method, one adsorbable (benzene or naphthalene) one non-adsorbable (heptane)	Schwartz et al. [107, 108] Mills and Dudukovic [79]	Internal wetting efficiency of 0.65 Independent of liquid flow was found
Response curves to a step decrease in inlet concentration of a tracer (KCl and $ZnSO_4$ in aqueous solutions)	Colombo et al. [19]	Both internal and external wetting efficiency were determined Wetting efficiency was defined in terms of effective diffusivity of tracer in particles
Impulse response of a tracer	Mills and Dudukovic [79]	Both internal and external wetting efficiency were determined
Chemical method/dissolution of phthalic anhydride in $KHCO_3$ K_2CO_3 aqueous solution	Specchia et al. [122]	Solubility measurements and physico-chemical constants are needed. The calculation of external wetting strongly depends upon chosen hydro-dynamic model
Chemical method/gaseous reactant rate limiting	Morita and Smith [83] Herskowitz et al. [44] Mata and Smith [67] Herskowitz and Mosseri [48]	Some discrepancy in these data at low liquid flow rates
Tracer method	Sicardi et al. [116, 118, 119]	Effective wetting defined as the square root of the ratio of the effective diffusivities

(3) Since external wetting efficiencies are required for calculations of global rates of reaction, values obtained under reaction conditions may be more reliable.

While the correlation for external wetting efficiency given by Mills and Dudukovic [79] is a reasonable one, it does not correlate the data of Mata and Smith [67] at low liquid flow rates. Normally, the external wetting efficiency ranges from 0.6 to 1.0 so the fractional error in estimating wetting efficiency cannot be large. However, in the case of a fast reaction limited by mass transfer resistances and a volatile reactant, a little change in wetting efficiency changes the global rate of reaction greatly. More data are needed for testing the correlation of Mills and Dudukovic [79], particularly for foaming liquids.

7.   RESIDENCE TIME DISTRIBUTION AND MIXING MODELS

7.1.   Liquid Phase

The liquid-phase residence time distribution (RTD) in a trickle bed reactor has been extensively evaluated by a variety of models in the literature. The mixing models used to correlate RTD contain different numbers of independent parameters. These models are critically reviewed by Shah [113] and are briefly summarized in Fig. 2. It should be noted that the number of parameters listed by Gianetto et al. [32] in Fig. 2 include the liquid holdup which is considered to be an independent parameter. Thus, axial dispersion (or P.D. in Fig. 2) contains two parameters, i.e., axial dispersion coefficient and liquid holdup. Gianetto et al. [32] and Shah [112] show various relationships for multiparameter models. While it is generally believed that the axial dispersion model is unrealistic for correlating RTD for a trickle bed reactor, models with more than three parameters are not easy to use. A three parameter model should be sufficient to accurately describe RTD for the trickle bed reactor. The most up-to-date version of such a model is recently given by Kan and Greenfield [57]. Gianetto et al. [32] and Sicardi et al. [116, 118] also pointed out that for the multiparameter models, the parameters given in different studies are difficult to compare because their values depend on the method of analysis. In practice, the axial dispersion model is still the most widely used for the modeling of a trickle bed reactor. In most commerical trickle bed operations, the liquid phase is believed to move in plug flow, while in the pilot scale

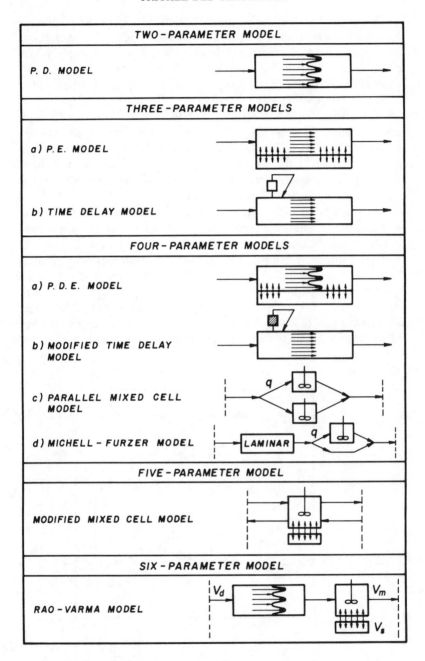

Figure 2   RTD models for trickle bed reactors. (After Gianetto et al. [32]).

operation the liquid phase can be backmixed. As shown in Fig. 3, the axial dispersion in the liquid phase under trickle flow conditions can be an order of magnitude higher than that in a single phase flow.

Correlations of the axial dispersion coefficient as a function of the operating parameters have been published [36, 112, 113]. A comparison between the correlations, carried out by Goto et al. [36], indicates large discrepancies. This is not surprising considering the scatter of the data. Recently Buffham and Rather [6] have investigated the effect of the viscosity on the axial dispersion coefficient. Changing the viscosity from 1.0×10^{-3}(Pa)(s) to 7.5×10^{-3}(Pa)(s) has little effect on the dispersion coefficient.

Kobayashi et al. [58] proposed a correlation, based on RTD data obtained with several aqueous solutions and his correlation predicts that the Peclet number depends strongly on the liquid viscosity, in contrast with the Buffham and Rather [6] results. It also predicts that at a certain value of the Reynolds number, the Peclet number depends strongly on the particle size, a result in disagreement with the data of Hochman and Effron [50] ($D_p = 0.00476$ m) and Schwartz et al. [107] ($D_p = 6.0 \times 10^{-4}$ m). The Buffman and Rather [6] correlation predicts Peclet numbers which agree with the data of

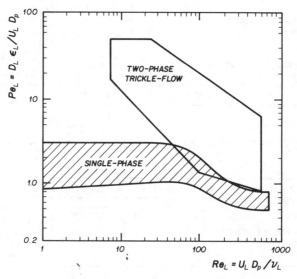

Figure 3 Pe_L and Re_L relations for single-phase and two-phase trickle flow. (After Hofmann [51]).

Hochman and Effron [50] and Schwartz et al. [107]. The Buffham and Rather correlation is recommended by Herskowitz and Smith [49] for estimating the axial dispersion coefficient in the low-interaction regime. In high-interaction regimes, the data of Kobayashi et al. [58] and Lerou et al. [63] indicate that the axial dispersion in the liquid phase is negligible.

7.2. Gas Phase

Just as for the liquid phase, the axial dispersion in gas phase under trickle flow condition is higher than that under single phase flow conditions. The correlation of Hochman and Effron [50].

$$Pe_G = 1.8 \, Re_G^{-0.7} \, 10^{-0.005 \, Re_L} \tag{5}$$

where $Pe_G = U_g D_p / D_G$, $Re_G = G_G D_p / \mu_G (1 - \varepsilon)$, $Re_L = G_L D_p / \mu_L (1 - \varepsilon)$ for the trickle flow condition is recommended.

8. INTERPHASE MASS TRANSFER

Since gas–liquid and liquid–solid mass transfer resistances can be important in trickle bed reactors, these are experimentally evaluated in the literature. It should be noted that these resistances, particularly gas–liquid resistance, are, in general, more important for oxidation reactions than hydroprocessing reactions. Shah [112] has reviewed the correlations reported before 1977. The reviews of Charpentier [10], Goto et al. [36] and Van Landeghem [135] are also available. In trickle bed reactors, when particles are partially wetted, gas–solid mass transfer resistance is generally neglected. In the following discussion, only recent studies are considered.

8.1. Gas–Liquid Mass Transfer

For both hydroprocessing and oxidation operations, gas–film mass transfer resistance is generally considered to be negligible. As pointed out by Shah and Sharma [114], the volumetric mass transfer coefficient $k_L a$ and the interfacial area a are correlated to gas and liquid velocities and fluid and particle properties either by dimensional or dimensionless correlations or by the use of the parameters describing energy losses.

The liquid-side mass transfer coefficient $k_L a$ may be obtained by absorption measurements with or without chemical reaction in the liquid phase. Fukushima and Kusaka [28, 29] measured $k_L a$ and the interfacial area a by oxygen absorption into sodium sulfite solution with cobaltous chloride as a homogeneous catalyst. The data were correlated in terms of the gas and liquid Reynolds numbers, the liquid holdup, the column and particle diameter and a shape factor. These correlations were obtained with relatively large particles (1.28 and 2.54 cm spheres and 1.16 cm Raschig rings). Morsi et al. [86] obtained the data by a chemical method for 0.648 cm Raschig rings and showed that for this particle size the nature of liquid phase has little effect on the interfacial area. Morsi et al. proposed a modification of the Gianetto et al. [31] energy correlation.

The study published by Mahajani and Sharma [65] with 0.4 cm particles (pellets and granular) also indicates that there is little effect of liquid characteristics (organic or aqueous) on the interfacial area. They measured the absorption of CO_2 (from lean gas mixtures) in aqueous solutions of NaOH, MEA in n-butanol, and CHA in p-xylene containing 10% isopropanol. In the gas-continuous flow regime, the interfacial area was a weak function of the liquid and gas rates. Recently, Seirafi and Smith [110] measured $k_L a$ by a dynamic method. The mass transfer coefficients were calculated from the zeroth and the first moments obtained from the analysis of experimental breakthrough curves. The experiments were conducted in a trickle bed reactor in which benzene was absorbed from water on 0.108 cm activated carbon particles. The values of $k_L a$ obtained in this study were in good agreement with those obtained by Mahajani and Sharma [65].

The Fukushima and Kusaka correlation could not predict the interfacial areas measured by Morsi et al. [85] for small packings. The latter measurements were carried out by chemical methods for organic and aqueous solutions with 0.116 cm glass beads and 0.24 cm catalyst ($Co/Mo/Al_2O_3$) spheres. This study indicated that the values obtained for the aqueous solution may not be useful for predicting the interfacial area in organic solutions.

The application of correlations of gas–liquid mass transfer coefficients for organic solutions and small packings was also examined by Turek and Lange [132]. They measured the absorption of hydrogen in α-methyl styrene at low liquid rates $(0 - 4.6 \times 10^{-4} \text{ m/s})$ and for small particles (0.054, 0.09, and 0.3 cm). Their correlation gave much lower values than Goto and Smith correlation [33]. Turek and Lange also calculated much higher $k_L a$ values from rate data obtained from α-methyl styrene hydrogenation on a Pd catalyst. While the recent investigations have been concentrated on the organic systems, the above discussion clearly indicates that further experimental work should be done to obtain reliable correlations for $k_L a$ for organic solutions and small particles which are frequently used in catalytic trickle bed reactors.

Finally, recent studies of Patwardhan [90] have shown that in short trickle bed reactors, a considerable amount of liquid flow near the wall can cause the axial dependence of $k_L a$. This effect should be critically evaluated under a wide range of operating conditions.

8.2. Liquid–Solid Mass Transfer

Liquid–solid mass transfer coefficients in trickle beds have been measured by three methods [49].

(a) Dissolution of slightly soluble solids into the liquid,

(b) Electrochemical reaction; and

(c) Chemical reaction with significant liquid–solid mass transfer resistance. Method (a) has been by far the most widely used. Shah [112, 113], Specchia and Baldi [122] and Satterfield et al. [106] have summarized the work published before 1978. Since then, several important contributions have been reported. These are summarized in Table 5.

For zero gas rate, the data obtained by Goto and Smith [33], Specchia and Baldi [122], Satterfield et al. [106] and Herskowitz et al. [44] are found to be in agreement. However, the study of gas-rate effects on $k_s a_t$ in the gas-continuous flow regime produced marked differences. Specchia and Baldi [122] reported that $k_s a_t$ almost tripled when the gas superficial velocity changed from zero to 0.5 m/s. Over the same range, Satterfield et al. [106] found no effect of the gas rate. These later results are supported by measurements carried out in a trickle bed packed with 0.635 cm Berl saddles coated with a molten mixture of benzoic acid and fluorescent dye [100].

Recently Yoshikawa et al. [140] measured liquid–solid mass transfer coefficients in beds of very small particles (0.46 to 1.3×10^{-3} m). They employed an ion-exchange reaction (Na^+ ions with $R-H^+$ resin particles) followed by an instantaneous irreversible neutralization reaction. The coefficients were only slightly affected by the gas flow rate in the gas-continuous flow regime. Herskowitz and Smith [49] recommend these data for very small particles. For smaller particles, Specchia and Baldi [122] correlation predicts higher values than the other correlations. However, this correlation is based on data measured over a wide range of operating parame-

Table 5   Recent Liquid–Solid Mass Transfer Studies [49]

Reference	Packing		Liquid			Gas	
	Shape	Size, cm	Type	G_L, kg/m^2 s	Type	Type	G_G, kg/m^2 s
Specchia and Baldi [122]	Cylinders of benzoic acid	0.3 and 0.6	Water water with surfactant, water with glycerol	1.6–8.3	Air		0–1.9
Scatterfield et al. [106]	Cylinders of benzoic acid	0.3 and 0.6	Water	0.48–25	Air, helium, argon, nitrogen		0–1.9
Chou et al. [12]	Nickel plated brass spheres	0.78	Water	8.4–23.5	Air		0.12–0.45
Ruether et al. [100]	Berl saddles	0.635	Water	2.1–94.4	Air		0.33–1.77

ters, and it is recommended for estimating $k_s a_t$. More data with small particles and organic liquids should be obtained.

9. HEAT TRANSFER

The first study of radial heat transfer in trickle bed reactors was published by Weekman and Myers [138] for the air–water system. They showed that overall effective thermal conductivity of trickle beds was 200 to 400% larger than for a single-phase liquid flow reactor at the same liquid flow rate. Specchia and Baldi [123] defined the overall thermal conductivity of a bed as

$$\lambda_{er} = \lambda_e^0 + \lambda_{e,g} + \lambda_{e,L} \tag{6}$$

where $\lambda_{e,g}$ and $\lambda_{e,L}$ are the contributions due to flowing gas and liquid. Different correlations for λ_{er} reported in the literature are given in Table 6.

A similar study was recently published by Specchia and Baldi [123] and the data were once again correlated with the use of Eq. (6). However, the first term λ_e^0 was not found to be constant but a function of the bed void fraction and the thermal conductivities of the solid and the gas. Both λ_e^0 and $\lambda_{e,g}$ were derived from a study of gas flow through packed beds [124]. The values of these two terms are much smaller than $\lambda_{e,L}$. The Specchia and Baldi [123] and Hashimoto et al. [40] correlations are in agreement for Peclet numbers for radial heat and liquid transfer. Thus, in principle, the effective radial thermal conductivity may be calculated from the radial liquid spreading factor obtained from liquid distribution data. Matsuura et al. [69] have also studied the effect of the operating parameters on the effective radial thermal conductivity. They proposed a correlation similar to the Hashimoto et al. [40] correlation. These three correlations agree well in the low-interaction regime but not in the high-interaction regime.

The heat transfer coefficient at the wall, h_w, has also been correlated as a function of the Reynolds and Prandtl numbers. Muroyama et al. [87] proposed correlations for the low- and high-interactions regimes. Matsuura et al. [70] developed a mode for h_w which included (a) the apparent wall heat transfer coefficient h_{w0} for a no-flow-situation; h_{w0} is a function of particle size and shape; (b) the apparent wall heat transfer coefficient h_{ws}, which accounts for the heat transfer between the fluid and the wall near the contact

Table 6   Correlations of Radial Heat Transfer Parameters in Trickle Beds [49]

Weekman and Myers [138]

$$\frac{\Lambda_{er}}{\lambda_L} = \frac{2.90 \times 10^{-3}}{\lambda_L} + 0.000285\left(\frac{G_L D_c}{\mu_L}\right) Pr_L$$

Hashimoto et al. [40]

$$\frac{\Lambda_{er}}{\lambda_L} = \frac{1.1 \times 10^{-4}}{\lambda_L} + 0.095\, Re_g\, Pr_g\, \frac{C_{p.g}^* \lambda_g}{C_{p.g} \lambda_L} + (\alpha\beta)_L\, Re_L\, Pr_L$$

Muroyama et al. [87]

$$\frac{h_w D_p}{\lambda_L} = 0.012\, Re_L^{1.7}\, Pr_L^{1/3}$$

low-interaction regime

$$\frac{h_w D_p}{\lambda_L} = 0.092(Re_L'')^{0.8}\, Pr_L$$

high-interaction regime

Matsuura et al. [69],
Matsuura et al. [70]

$$\frac{\lambda_{er}}{\lambda_L} = 1.5 + (\alpha\beta)_g'\, Re_g\, Pr_g\, \frac{C_{p.g}^* \lambda_g}{C_{p.g} \lambda_L} + (\alpha\beta)_L'\, Re_L\, Pr_L$$

Specchia and Baldi [123]

$$\frac{\lambda_{er}}{\lambda_L} = \left\{ \varepsilon + \frac{1-\varepsilon}{0.220\varepsilon^2 + \dfrac{2\lambda_g}{3\lambda_s}} + \frac{Re_g\, Pr_g\, C^*/C_{p.g}}{8.651 + 19.5(D_e/D_c)^2} \right\} \frac{\lambda_g}{\lambda_L}$$

$$+ \frac{\lambda_{eL}}{\lambda_L}$$

$$(\alpha\beta)_L = \left\{ \frac{1}{1.90 + 0.0264(D_e G_L/\varepsilon_{eL}\mu_L)(\mu_L/\mu_0) + 0.197} \right\} \frac{D_e}{D_p}$$

$$h_w = h_{w0} + h_{ws} + \cfrac{1}{\cfrac{1}{h_w^*} + \cfrac{1}{h_{wi,g} + h_{wi,L}}}$$

$$\frac{h_w D_e}{\lambda_L} = 0.057(Re_L'')^{0.89}\, Pr_L^{1/3}$$

low-interaction regime

D_p, cm	$(\alpha\beta)'_g$	$(\alpha\beta)'_L$	$h_{wo}D_p/\lambda_L$	$h_{ws}D_p/\lambda_L$	Range of ε_{dL}
0.12	0.412	$0.201(1 + 2.83 \times 10^{-2}\,\mathrm{Re}_g)$	0.09	$1.10\,\varepsilon_{dL}/\varepsilon$	$0 < \varepsilon_{dL}/\varepsilon < 0.5$
				$21\,\varepsilon_{dL}/\varepsilon - 9.95$	$0 < 2\varepsilon_{dL}/\varepsilon < 1$
0.26	0.334	$0.167(1 + 1.34 \times 10^{-2}\,\mathrm{Re}_g)$	0.21	$1.70\,\varepsilon_{dL}/\varepsilon$	$0 < \varepsilon_{dL}/\varepsilon < 0.38$
				$18.8\,\varepsilon_{dL}/\varepsilon - 6.50$	$0.38 < \varepsilon_{dL}/\varepsilon < 1.0$
0.43	0.190	$0.152(1 + 6.32 \times 10^{-3}\,\mathrm{Re}_g)$	0.43	$3.40\,\varepsilon_{dL}/\varepsilon$	$0 < \varepsilon_{dL}/\varepsilon < 0.25$
				$16.6\,\varepsilon_{dL}/\varepsilon - 3.30$	$0.25 < \varepsilon_{dL}/\varepsilon - 3.30$

$$h_w^* D_p/\lambda_L = \begin{cases} 4.0\,\Pr_g^{1/3}\,\mathrm{Re}_g^{1/2}\,\lambda_g/\lambda_L & \text{low-interaction regime} \\ 2.6\,\mathrm{Re}_L^{1/2}\,\Pr_L^{1/3} & \text{high-interaction regime} \end{cases}$$

$$\frac{h_{wt,g}D_p}{\lambda_L} = 0.5(\alpha\beta)_g\,\mathrm{Re}_g\,\Pr_g\,\lambda_g/\lambda_L\,; \quad \frac{h_{wt,L}D_p}{\lambda_L} = 0.5(\alpha\beta)'_L\,\mathrm{Re}_L\,\Pr_L \quad \text{high-interaction regime}$$

$$\frac{h_{wt,g}D_p}{\lambda_L} = 0.054\,\mathrm{Re}_g\,\Pr_g\,\frac{C^*_{p,g}}{C_{p,g}}\,\frac{\lambda_g}{\lambda_L}\,; \quad h_{wt,L} = 0 \quad \text{low-interaction regime}$$

$$\frac{\lambda_{e,L}}{\lambda_L} = \frac{\varepsilon_{eL}^{0.87}}{0.041}\,\mathrm{Re}_L^{0.13}\,\Pr_L \quad \text{low-interaction regime}$$

$$\frac{\lambda_{e,L}}{\lambda_L} = \frac{\varepsilon_{eL}^{-0.29}}{338}\,\mathrm{Re}_L^{0.325}\,\Pr_L\,(a_t D_p/\varepsilon)^{2.7} \quad \text{high-interaction regime}$$

surface between particles and the wall. This was correlated as a function of the dynamic liquid holdup h_d; and (c) the apparent heat transfer coefficients h_{wg} and h_{wL}, accounting for heat transfer due to radial gas and liquid mixing, respectively.

The true heat transfer coefficient, h_w, correlated as a function of the gas-phase Reynolds and Prandtl numbers for the low-interaction regime, and the liquid-phase dimensionless numbers for the high-interaction regime. These correlations are given in Table 6.

In contrast with these investigations, Specchia and Baldi [123] reported a constant value of $h_w = 2.1 \, kJ/(m^2)(s)(K)$ in the high-interaction regime. According to Herskowitz and Smith [49], the Muroyama et al. correlation is recommended for estimating h_w. This was obtained with small particles normally used in trickle beds. The complex result proposed by Matsuura et al. may not apply for other fluids and for different operating conditions. The Specchia and Baldi proposal is probably not applicable for small packings.

10. REACTOR MODELS

The models for trickle bed reactors can be divided into three categories. First category contains the models for the calculation of catalyst effectiveness factors for partially wetted catalyst. This problem is specific to the catalyst performance under trickle flow conditions. The second category contains the models for laboratory or pilot scale trickle bed reactors. These are generally isothermal models and they consider the roles of various mass transfer resistances, axial dispersion, ineffective catalyst wetting and liquid maldistribution on the reactor performance. The last category contains nonisothermal models, generally needed for the large scale operations and they can include the roles of all of the above phenomena in addition to catalyst aging on the reactor performance.

10.1. Calculation of Catalyst Effectiveness Factor for Partially Wetted Catalyst

The trickle bed reactor offers an unusual situation wherein the catalyst particle can be partially wetted with the liquid. The mass transfer resistances for the wetted particles are, in general, different from those of dry particles. In the last several years, considerable research has been carried out·on the evaluation of overall effectiveness factor of partially wetted catalyst particle. The reported studies are critically reviewed by Herskowitz and Smith [49].

For an isothermal, irreversible first-order reaction, the overall effectiveness factor for slab, cylinder and sphere shape particles can be expressed as

$$\eta_0 = A_1(1 - \exp(-2\phi)) \qquad \text{for slab} \qquad (7)$$

$$\eta_0 = \frac{B_1}{\phi} I_1(2\phi) \qquad \text{for cylinder} \qquad (8)$$

$$\eta_0 = \frac{D_1}{\phi} I_1(3\phi) \qquad \text{for sphere} \qquad (9)$$

where A_1, B_1, and D_1 are functions of the dimensionless parameters α_{gLs}, α_{gs}, f_e, and ϕ. These dimensionless parameters are defined as

$$\frac{1}{\alpha_{gLs}} = \frac{S_x}{V_p} D_{eff}\left(\frac{1}{\varepsilon_p k_s} + \frac{f_e a_t}{k_L a}\right) \qquad (10)$$

$$\alpha_{gs} = \frac{V_p \varepsilon_p k_{gs}}{S_x D_{eff}} \qquad (11)$$

$$\phi = \frac{V_p}{S_x} \sqrt{\frac{\rho_p k_v}{D_{eff}}} \qquad (12)$$

where f_e is the wetting efficiency. All other parameters are defined in the list of the nomenclature. The constants, A_1, B_1, and D_1 are functions of the number of terms in the linear algebraic equations. The details of the solution are given by Mills and Dudukovic [76, 77] and Herskowitz [45–47]. Simpler solutions are also presented by Herskowitz et al. [44], Herskowitz [45–47] and Mills et al. [78]. Capra et al. [7] evaluated the problem of a first-order reaction.

A much simpler approach was developed by Ramchandran and Smith [95, 96] who assumed

$$\eta_0 = f_e \eta_L + (1 - f_e)\eta_g \qquad (13)$$

where η_L and η_G are the effectiveness factors for the liquid- and gas-phase reactions respectively. The solutions obtained from this approximate method for planar, cylindrical and spherical shape particles are summarized in Table 7 [76, 131]. In general, these approximate solutions were found to agree with the rigorous solution within 10% with larger deviations for low values of the Thiele

Table 7 Approximate Expressions for the Overall Effectiveness Factor in Partially Wetted Particles

Geometry	η_o
Slab	$\dfrac{f_e C_L^*}{\dfrac{\phi^2}{\alpha_{gLS}} + \dfrac{\phi}{\tanh \phi}} + \dfrac{1 - f_e}{\dfrac{\phi^2}{\alpha_{gs}} + \dfrac{\phi}{\tanh \phi}}$
Cylinder	$\dfrac{f_e C_L^*}{\dfrac{\phi^2}{\alpha_{gLS}} + \phi \dfrac{I_0(2\phi)}{I_1(2\phi)}} + \dfrac{1 - f_e}{\dfrac{\phi^2}{\alpha_{gs}} + \phi \dfrac{I_0(2\phi)}{I_1(2\phi)}}$
Sphere	$\dfrac{f_e C_L^*}{\dfrac{\phi^2}{\alpha_{gLS}} + \dfrac{3\phi^2}{3\phi \coth(3\phi) - 1}} + \dfrac{1 - f_e}{\dfrac{\phi^2}{\alpha_{gs}} + \dfrac{3\phi^2}{3\phi \coth(3\phi) - 1}}$

modules ($\phi < 1$). Using an approximation similar to that used in Eq. (13), Goto et al. [37] solved the effectiveness factor problem for nonlinear kinetics. A sample calculation illustrating the effects of partial wetting and the external mass transfer resistances on the overall effectiveness factor is given by Herskowitz and Smith [49]. Some criteria for the negligible transport effects in the trickle bed reactors are also given by Lee and Smith [62].

The possibility of partial internal wetting was considered by Dudukovic [27] and Mills and Dudukovic [76] for liquid limiting reactants. Martinez et al. [66] extended these models to include gas-phase reaction in dry zones. Their numerical solution shows that lowering the internal wetting efficiency increases the overall effectiveness factor due to negligible intra-particle mass transfer resistance in dry pores. It should be pointed out that their treatment requires two additional parameters, the fraction of the particle external surface in contact with dry pores and the average depth of the dry pores. The internal wetting efficiency was calculated from these two parameters.

Sicardi et al. [116, 118] proposed to reconcile discrepancies between wetting efficiencies measured by tracer and "chemical" methods by including the effect of stagnant liquid zones. They divided the wetted external surface into a dynamic and a stagnant zone. A nonvolatile reactant in the liquid was assumed to transfer to the solid through dynamic and static areas as well as between the two zones, giving rise to three mass transfer coefficients. These coefficients were evaluated as a function of the liquid rate. Predicted rates of reaction considering only the dynamic areas were lower than those which included the static zones.

The effect of the wetting efficiency on the selectivity of consecutive first-order reactions was examined by Herskowitz [45]. For the case of a volatile reactant and a nonvolatile intermediate product the effect of wetting efficiency on the selectivity was found to be significant.

10.2. Laboratory Reactor Models

Shah [112], Hofmann [51], Goto et al. [36] and Gianetto et al. [32] have reviewed various models for laboratory trickle bed reactors. Goto and Smith [34, 35] developed and solved mass conservation equations for gas and liquid phases and evaluated the roles of various mass transfer processes for a pseudo first-order reaction. While mass transfer resistances are not important for many hydroprocessing operations, they play significant roles in some oxidation reactions. Goto et al. [36] also review a model which includes mass transfer effects for the reactions with both linear and nonlinear kinetics. The resulting nonlinear equations were solved numerically.

The role of axial dispersion on the performance of a trickle bed reactor was first investigated by Mears [72, 73]. He developed the criteria for eliminating axial dispersion effect in hydroprocessing trickle bed reactors. Shah and Paraskos [111] extended his analysis to adiabatic trickle bed reactors for a variety of hydroprocessing reactions. More recently, Ramachandran and Smith [95] proposed a mixing-cell method that assumed that the reaction consists of a number of stirred tanks in series, in each of which the liquid is completely mixed and the gas is in plug flow. The design equations which result from the mass balances in a cell are nonlinear algebraic equations. An analytical solution was presented for a first-order reaction. The degree of backmixing was represented by the number of cells. This model can be modified to include the effects of liquid maldistribution or partial wetting of the particles on the reactor performance.

The models evaluating the role of effective catalyst wetting on the performance of a trickle bed reactor for a variety of hydroprocessing operations have been reviewed by Shah [112]. More recently, Herskowitz [46] used a mixing-cell model to examine the effect of the wetting efficiency on the reactor performance. The model used global rate of reaction for a partially wetted particle [47]. An analytical solution for the outlet concentration of the reactants involved a first-order rate constant, various mass transfer coefficients, the Thiele modulus, the catalyst wetting efficiency, and the bed depth. The model was applied to the oxidation of sulfur dioxide in an aqeuous solution containing activated carbon catalyst. A significant effect of wetting efficiency on the reactor performance was evaluated. Unfortunately, no experimental data are available for further testing of this model.

The effect of liquid distribution on the performance of the reactor has also been investigated by means of a cell model. Hanika et al. [38] employed the Dean and Lapidus [26] approach to examine the effects of the initial liquid distribution, the reaction order, the reaction rate, and the feed rate on the reactor performance. The effects of mass transfer resistances, partial wetting and wall flow were neglected. Herskowitz and Smith [42] proposed a model for the distribution of a tracer (which included effects of the wall flow) that can be extended to include chemical reaction. Recently, Stanek et al. [128] published a more complete model that accounted for liquid maldistribution and heat effects on the reactor performance.

10.3. Non-Isothermal Reactors

The axial temperature profile in an adiabatic trickle bed reactor has been measured by Hanika et al. [39]. They studied the hydrogenation of cyclohexene to cyclohexane, the latter is a relatively volatile solvent. Only a small change in the temperature with bed depth was measured for low concentrations of the reactant in the feed. However, at high concentrations the temperature increased due to the much greater heat release in the reactor. The liquid superficial velocity was very low (9.4×10^{-5} m/s) so that the wetting efficiency was low, enhancing the gas-phase reaction.

Herskowitz and Smith [49] have shown that for most hydroprocessing operations the temperature difference between the catalyst particle and the fluid is small and can be neglected. This temperature difference, however, can be large for reforming reactions.

A temperature rise with bed depth has also been observed in commercial trickle beds that operate adiabatically. The temperature can be controlled by the addition of quenching streams. Shah [112] and Shah et al. [113] summarized several simulation studies of trickle bed reactors with one or more quenches. These studies also include catalyst aging effects. Recently, Yan [139] presented a dynamic model of a trickle bed hydrocracker with a quenching system.

11. SCALEUP CONSIDERATIONS

The proper scaleup of a trickle bed reactor requires a number of careful considerations, some of which are briefly outlined below.

A. Generally there is about a tenfold difference in the liquid space velocity and gas flow rate used in commercial and laboratory trickle bed reactors. One may have different flow regimes in the two cases and if this is the case, the reactor scaleup may be seriously jeopardized. One may encounter situations in which external mass transfer resistance may become important or dominant in the pilot reactor but it may be unimportant in the industrial reactor. There may also be different residence time distributions for gas and liquid phases in the two cases. One may often benefit by replacing expensive catalyst with inexpensive support material of equivalent dimension when solid–liquid or gas–liquid film resistance is important.

B. The performance of a large commercial reactor depends significantly upon the heat balance on the reactor (e.g., hydroprocessing, reforming, etc.). The proper scaleup must be done with the use of a model which adequately considers heat balance on the reactor.

C. Earlier, it was pointed out that proper liquid distribution is important for the effective use of catalyst. The proper radial distribution of liquid for large commercial reactor is an important scaleup consideration.

D. The laboratory reactors should be properly designed so that the effects of ineffective catalyst wetting, radial distribution of liquid, interphase mass and heat transfer resistances, and axial mixing on the reactor performance are either eliminated or properly accounted for. In this manner, one may be able to get the information on intrinsic kinetics from the laboratory reactors which can be used for the design and scaleup of the reactors.

Because of the above considerations, Herskowitz and Smith [49] suggest that the following procedure for the scaleup of a trickle bed reactor:

First, it is important to establish in the laboratory the rate of reaction for single catalyst pellets which will include the effect of wetting efficiency. If there is a soluble gaseous reactant, the rate should account for mass transfer from both the gas- and liquid-covered surfaces of the pellet. These basic rate data then can be used with intrareactor mass and if necessary, energy conservation expressions to design the large scale reactor. This second step should also include the role of liquid distribution. It is clear that the accuracy of scaleup depends significantly upon the reliable estimations of kinetic, hydrodynamic, mixing and transport parameters.

ACKNOWLEDGEMENT

Many helpful discussions and recommendations of Dr. B. Morsi are gratefully acknowledged.

NOMENCLATURE

a Gas–liquid interfacial area for mass transfer (m^{-1})

a_t external area of particle per unit volume of column (m^{-1})

A_1, B_1, D_1 functional constants in Eqs. (7)–(9)

C_{Lb} bulk liquid concentration $(Kmol/m^3)$

C_L solubility in the liquid phase $(Kmol/m^3)$

C_L^* dimensionless concentration in the liquid phase (C_{Lb}/C_L)

Cp specific heat capacity $(Kcal/Kg\,K)$

D_G, D_L axial dispersion for gas and liquid phases $(m^2\,s^{-1})$

D_{eff} effective diffusivity within the catalyst particle $(m^2\,s^{-1})$

D_c column diameter (m)

D_p particle diameter (m)

D_e equivalent particle diameter (m)

D_r radial liquid spread coefficient (m^{-1})

Eo   Eotvos number $(\rho_L g D_p^2/\sigma_L)$

f friction factor as defined in Table 2

f_e wetting efficiency, dimensionless

Fr   Froude number $(a_t G_m^2/\rho_L^2 g)$

g gravitational constant

h_w wall heat transfer coefficient $(kJ/(m^2)(s)(K))$

G_G, G_L gas and liquid superficial mass velocities $(kg\,m^{-2}\,s^{-1})$

Ga   Galileo number $(D_p^3 \rho_L^2 g/\mu_L^2)$

K_1, K_2 constants in Table 2

$k_L a$ volumetric gas–liquid mass transfer coefficient (s^{-1})

$k_s a_t$ liquid–solid mass transfer coefficient (ms^{-1})

k_v intrinsic kinetic constant $(m^3/(kg)(s))$

Pe_G, Pe_L Peclet numbers for gas and liquid phases $(U_G D_p/D_G, \; U_L D_p/D_L)$

Pr_L, Pr_g Prandtl numbers for liquid and gas phases
$(\mu_L C_{PL}/\lambda_L, \mu_g C_{pg}/\lambda_g)$

Re   Reynolds number $(UD_p \rho/\mu)$

S_z surface area of catalyst particle (m^2)

U_G, U_L gas and liquid velocities (m/s)

V_p volume of catalyst particle (m^3)

We   Weber number $(G_L^2/\sigma_L \rho_L a_t)$

Z   parameter defined in Table 2

$\Delta P/\Delta Z$ pressure gradient

$(\Delta H/\Delta Z)_{LG}$,   two-phase, liquid-phase and gas-phase head loss
$(\Delta H/\Delta Z)_G$ per unit length of packing, dimensionless

Greek Letters

δ energy loss

$\bar{x}$ δ_L/δ_G

$\varepsilon_G, \varepsilon_L, \varepsilon_S$ gas, liquid and solid holdups

ε_{eL} total external liquid holdup in Table 6

ε_p void fraction within particle

ε_{sL} static holdup of liquid

ε_{dL} dynamic holdup of liquid

ρ density (Kg/m^3)

ρ_e density of the manometric liquid (generally water) (Kg m^{-3})

σ surface tension

ε bed void fraction

μ viscosity (Kg/(m)(s))

ψ, ζ parameter defined in Table 2

Λ_{er} effective radial thermal conductivity evaluated by Weekman and Myers (kJ/(m)(s)(K))

λ_{er} overall radial thermal conductivity of the bed as defined by Eq. (6) (kJ/(m)(s)(K))

$(\alpha\beta)_L$ parameters defined in Table 6
$(\alpha\beta)'_L$
$(\alpha\beta)'_g$

λ_s thermal conductivity of solid

ν kinematic viscosity (m^2/s)

ϕ Thiele modulus as defined in Eq. (12)

α_{gs} Biot number for mass transfer on the gas-covered surface of the particle (Eq. (11)), dimensionless

α_{gLs} Biot number for mass transfer on the liquid-covered surface of the particle (Eq. (10)), dimensionless

ρ_p particle density (Kg/m^3)

η_o, η_g, η_L catalyst effectiveness factors for overall, gas-phase and liquid-phase reactions respectively

Subscripts

g, G gas phase
L liquid phase
S solid phase
water   related to water
LG two (gas–liquid) phase
M mixture (gas and liquid)

REFERENCES

1. Abbott, M. D., Moore, G. R., and Ralph, J. L. (1967). Paper S-5, 7th ABG Conference.
2. Baldi, G. (1980). Heat transfer in gas–liquid–solid reactors, NATO Institute on Multiphase Chemical Reactors, Vimeiro, Portugal.
3. Beimesch, W. P., and Kesseler, D. P. (1971). Liquid–gas distribution measurements in the the pulsing regime of two-phase concurrent flow in packed beds, *AIChE J.*, 17(5), 1160–1165.
4. Bondi, A. (1971). Handling kinetics from trickle-phase reactors, *Chem. Tech.*, 185 (March 1971).
5. Buchanan, J. E. (1967). Holdup in irrigated ring-packed towers below the loading point, *I&EC Fund.*, 6, 400.
6. Buffham, B. A., and Rather, M. N. (1978). The influence of viscosity on axial mixing in trickle flow in packed beds, *Tr. Inst. Chem. Engrs.*, 56, 266.
7. Capra, V., Sicardi, S., Gianetto, A., and Smith, J. M. (1982). Effect of liquid wetting on catalyst effectiveness in trickle-bed reactors, *Can. J. Chem. Eng.*, 60, 282.
8. Charpentier, J. C., Prost, C., and Legoff, P. (1969), Chute de pression pour des ecoulements a co-courant dans les colonnes a garnissage arrose: Comparision avec le garnissage noye, *Chem. Eng. Sci.*, 24, 1777.
9. Charpentier, J. C., and Favier, M. (1975). Some liquid holdup experimental data in trickle bed reactors for foaming and non-foaming hydrocarbons, *AIChE J.*, 21, 1213–1218.
10. Charpentier, J. C. (1976). Recent progress in two-phase gas–liquid mass transfer in packed beds, *Chem. Eng. J.*, 11, 161.
11. Chou, T. S., Worley, F. L., and Luss, D. (1977). Transition to pulsed flow in mixed-phase cocurrent downflow through a fixed bed, *Ind. Eng. Chem. Pro. Des. Dev.*, 16, 424.
12. Chou, T. S., Worley, F. L., and Luss, D. (1979). Local particle–liquid mass transfer fluctuations in mixed phase cocurrent downflow through a fixed bed in the pulsing regime, *Ind. Eng. Chem. Fund.*, 18, 279.
13. Clements L. D., and Halfacre, G. (1978). Liquid composition and flow regime effects in gas–liquid downflow in packed beds, in *Two Phase Transport and Reactor Safety* (Eds. T. N. Veziroglu and S. Kakac), Vol. 1, pp. 69–78, Hemisphere, Washington, DC.

14. Clements, L. D., and Halfacre, G. (1979). Liquid composition and flow regime effects in gas–liquid downflow in packed beds, AIChE 72nd Annual Meeting, *Eng. J.*, **18**, 151.
15. Clements, L. D., and Schmidt, P. C. (1976). Two-phase pressure drop and dynamic liquid holdup in cocurrent downflow in packed beds, Paper presented at 69th Annual AIChE Meeting, Chicago, IL.
16. Clements, L. D., and Schmidt, P. C. (1980). Two-phase pressure drop in cocurrent downflow in packed beds: Air–silicone oil systems, *AIChE J.*, **26**, 314–317.
17. Clements, L. D., and Schmidt, P. C. (1980). Dynamic liquid holdup in two-phase downflow in packed beds: Air–silicone oil system, *AIChE J.*, **26**, 317–319.
18. Cohen, Y., and Metzner, A. B. (1981). Wall effects in laminar flow of fluids through packed beds, *AIChE J.*, **27**(5), 705–715.
19. Colombo, A. J., Baldi, G., and Sicardi, S. (1976). Liquid–solid contacting efficiency in trickle-bed reactors, *Chem. Eng. Sci.*, **31**, 1101.
20. Crine, M., Marchot, P., and L'Homme, G. A. (1979). Liquid maldistribution in trickle-bed reactors, AIChE 72nd Annual Meeting, San Francisco (Nov. 1979).
21. Crine, M., and L'Homme, G. (1982). *ACS Symp. Ser.*, **196**, 407.
22. Crine, M., Marchot, P., and L'Homme, G. (1979). *Comp. Chem. Eng.*, **3**, 515.
23. Crine, M., Marchot, P., and L'Homme, G. (1980). *Chem. Eng. Comm.*, **7**, 377.
24. Crine, M. (1981). Heat transfer phenomena in trickle-bed reactors, 1981 Annual Meeting of AIChE, New Orleans (Nov. 1981).
25. Davidson, J. F., Cullen, E. J., Hanson, D., and Roberts, D. (1959). Holdup and liquid film coefficient of packed towers, *Trans. Inst. Chem. Eng.*, **37**, 122.
26. Dean, H. A., and Lapidus, L. (1960). A computational model for predicting and correlating the behavior of fixed-bed reactors, *AIChE J.*, **6**, 656.
27. Dudukovic, M. P. (1977). Catalyst effectiveness factor and contacting efficiency in trickle-bed reactors, *AIChE J.*, **23**, 940.
28. Fukushima, S., and Kusaka, K. (1977). Interfacial area and boundary of hydrodynamic flow region in packed column with cocurrent downward flow, *J. Chem. Eng. Japan*, **10**, 461.
29. Fukushima, S., and Kusaka, K. (1977). Liquid-phase volumetric and mass transfer coefficient, and boundary of hydrodynamic flow region in packed column with cocurrent downward flow, *J. Chem. Eng. Japan*, **10**, 468.
30. Gelbe, H. (1968). A new correlation for liquid holdup in packed beds, *Chem. Eng. Sci.*, **23**, 1901.
31. Gianetto, A., Baldi, G., and Specchia, V. (1970). Absorption in packed towers with cocurrent downward high-velocity flows, *Quad. Chem. Ital.*, **6**, 125.
32. Gianetto, A., Baldi, G., Specchia, B., and Sicardi, S. (1978). Hydrodynamics and solid–liquid contacting effectiveness in trickle-bed reactors, *AIChE J.*, **24**(6), 1087.
33. Goto, S., and Smith, J. M. (1975). Trickle-bed reactor performance, Part I. Holdup and mass transfer effects, *AIChE J.*, **21**, 706.
34. Goto, S., and Smith, J. M. (1978). Analysis of three-phase reactors, *AIChE J.*, **24**, 294.
35. Goto, S., and Smith, J. M. (1978). Performance of trickle-bed and slurry reactors, *AIChE J.*, **24**, 286.
36. Goto, S., Levec, J., and Smith, J. M. (1977). Trickle-bed oxidation reactors, *Catal. Rev. Sci. Eng.*, **15**, 187.
37. Goto, S., Lakota, L., and Levec, J. (1981). Effectiveness factors of nth order kinetics in trickle-bed reactors, *Chem. Eng. Sci.*, **36**, 157.
38. Hanika, J., Vychodil, P., and Ruzicka, V. (1978). A cell model of the isothermal trickle-bed reactor, *Coll. Czech. Chem. Commun.*, **43**, 2111.
39. Hanika, J., Sporka, K., Ruzicka, V., and Hrstka, J. (1976). Measurement of axial temperature profiles in an adiabatic trickle bed reactor, *Chem. Eng. J.*, **12**, 193.

334   CONCEPTS AND DESIGN OF CHEMICAL REACTORS

40. Hashimoto, K., Muroyama, K., Fujiyoshi, K., and Nagata, S. (1976). Effective radial thermal conductivity in cocurrent flow of a gas and liquid through a packed bed, *Int. Chem. Eng.*, **16**, 720.
41. Henry, G. H., and Gilbert, J. B. (1973). Scale-up of pilot plant data foɪ catalytic hydroprocessing, *Ind. Eng. Chem. Proc. Des. Dev.*, **12**, 328.
42. Herskowitz, M., and Smith, J. M. (1978). Liquid distribution in trickle-bed reactors, Part I: Flow measurements, *AIChE J.*, **24**, 439.
43. Herskowitz, M., and Smith, J. M. (1978). Liquid distribution in trickle-bed reactors, Part II: Tracer studies, *AIChE J.*, **24**, 450.
44. Herskowtiz, M., Carbonell, R. B., and Smith, J. M. (1978). Effectiveness factors and mass transfer in trickle-bed reactors, *AIChE J.*, **25**, 272.
45. Herskowitz, M. (1981). Wetting efficiency in trickle-bed reactors. The overall effectiveness factor of partially wetted catalyst particles, *Chem. Eng. Sci.*, **36**, 1665.
46. Herskowitz, M. (1981). Effect of wetting efficiency in selectivity in a trickle-bed reactor, *Chem. Eng. Sci.*, **36**, 1099.
47. Herskowitz, M. (1981). Wetting efficiency in trickle-bed reactors: Its effect on the reactor performance, *Chem. Eng. J.*, **22**, 167.
48. Herskowitz, M., and Mosseri, S. (1983). Global rates of reaction in trickle-bed reactors: Effects of gas and liquid flow rates, *Ind. Eng. Chem. Fund.*, **22**, 4.
49. Herskowitz, M., and Smith, J. M. (1983). Trickle bed reactors: A review, *AIChE J.*, **29**, 1.
50. Hochman, J. M., and Effron, E. (1969). Two-phase cocurrent downflow in packed beds, *Ind. Eng. Chem. Fund.*, **8**, 63.
51. Hofmann, H. P. (1978). Multiphase catalytic packed-bed reactions, *Catal. Rev. Sci. Eng.*, **17**, 21.
52. Hutton, B. E. T., and Leung, L. S. (1974). Concurrent gas–liquid flow in packed columns, *Chem. Eng. Sci.*, **29**, 1681.
53. Jameson, J. B. (1966). A model for liquid distribution in packed columns and trickle-bed reactors, *Tr. Inst. Chem. Engrs.*, **44**, T198.
54. Jesser, B. W., and Elgin, J. C. (1943). Liquid holdup in packed towers, *Trans. Inst. Chem. Eng.*, **39**, 277.
55. Kan, K. M., and Greenfield, P. F. (1978). Multiple hydrodynamic states in cocurrent two-phase downflow through packed beds, *Ind. Eng. Chem. Proc. Des. Dev.*, **17**, 482.
56. Kan, K. M., and Greenfield, P. F. (1979). Pressure drop and holdup in two-phase cocurrent trickle flows through beds of small particles, *Ind. Eng. Chem. Proc. Des. Dev.*, **18**, 760.
57. Kan, K. M., and Greenfield, P. F. (1983). A residence-time model for trickle-flow reactors incorporating mixing in stagnant regions, *AIChE J.*, **29**(1), 123–131.
58. Kobayashi, S., Kushiyama, S., Ida, Y., and Wakao, N. (1979). Flow characteristics and axial dispersion in two-phase downflow in packed columns, *Kagaku Kogaku Ronbunshu*, **5**, 256.
59. Kolomaznik, K., Soukup, J., Prchlik, J., Zaletal, V., and Ruzicka, V. (1974). Experimental determination of the spreading factor, *Col. Czech. Chem. Commun.*, **39**, 216.
60. Koros, R. M. (1976). Catalyst utilization in mixed-phase fixed bed reactors, 4th ISCRE, Heidelberg, also published in *Chem. Eng. Sci.*
61. Larkins, R. P., White, R. R., and Jeffrey, D. W. (1961). Two-phase concurrent flow in packed beds, *AIChE J.*, **7**(2), 231–239.
62. Lee, H. M., and Smith, J. M. (1982). Trickle-bed reactors: Criteria of negligible transport effects and of partial wetting, *Chem. Eng. Sci.*, **37**, 223.
63. Lerou, J. J., Glasser, D., and Luss, D. (1980). Packed bed liquid phase dispersion in pulsed gas–liquid downflow, *Ind. Eng. Chem. Fund.*, **19**, 66.
64. L'Homme, G. A. (1979). Chemical engineering of gas–liquid–solid catalyst

reactions, *Proceedings of an International Symposium held at Universite' de Liege, March 1–3, 1978*; also published *CEBEDDC*, Liege, Belgium (1979).

65. Mahajani, V. V., and Sharma, M. M. (1979). Effective interfacial area and liquid side mass transfer coefficient in trickle-bed reactors, *Chem. Eng. Sci.*, **34**, 1425.

66. Martinez, O. M., Barreto, G. F., and Lemcoff, N. O. (1981). Effectiveness factor of a catalyst pellet in a trickle bed reactor, *Chem. Eng. Sci.*, **36**, 901.

67. Mata, A. R., and Smith, J. M. (1981). Oxidation of sulfur dioxide in a trickle-bed reactor, *Chem. Eng. J.*, **22**, 229.

68. Matsuura, A., Akehata, T., and Shirai, T. (1979). Flow pattern of cocurrent gas–liquid downflow in packed beds, *Kagaku Kogaku Ronbunshu*, **5**, 167.

69. Matsuura, A., Akehata, T., and Shirai, T. (1979). Correlation of dynamic holdup in packed beds with cocurrent gas–liquid downflow, *J. Chem. Eng. Japan*, **12**, 263.

70. Matsuura, A., Akehata, T., and Shirai, T. (1979). Effective radial thermal conductivity in packed beds with gas–liquid downflow, *Heat Trans. Jap. Res.*, **8**, 44.

71. Matsuura, A., Akehata, T., and Shirai, T. (1979). Apparent wall heat transfer coefficient in packed beds with downward cocurrent gas–liquid fow, *Heat Trans. Jap. Res.*, **8**, 52.

72. Mears, D. E. (1971). The role of axial dispersion in trickle-flow laboratory reactors, *Chem. Eng. Sci.*, **25**, 1361.

73. Mears, D. E. (1974). The role of liquid holdup and effective wetting in the performance of trickle-bed reactors, *Chem. React. Eng. II, ACS Monograph Ser.*, **133**, 268.

74. Midoux, N., Favier, M., and Charpentier, J. C. (1976). Flow pattern, pressure loss and liquid holdup data in gas–liquid downflow packed beds with foaming and nonfoaming hydrocarbons, *J. Chem. Eng. Japan*, **9**, 350.

75. Michell, R. W., and Furzer, I. A. (1972). Mixing in trickle flow through packed beds, *Chem. Eng. J.*, **4**, 53.

76. Mills, P. L., and Dudukovic, M. P. (1979). A dual-series solution for the effectiveness factor of partially wetted catalysts in trickle-bed reactors, *Ind. Eng. Chem. Fund.*, **18**, 139.

77. Mills, P. L., and Dudukovic, M. P. (1980). Analysis of catalyst effectiveness in trickle-bed reactors processing volatile or nonvolatile reactants, *Chem. Eng. Sci.*, **35**, 2267.

78. Mills, P. L., Erk, E. F., Evans, J., and Dudukovia, M. P. (1981). Some comments on models for evaluation of catalyst effectiveness factors in trickle-bed reactors, *Chem. Eng. Sci.*, **36**, 947.

79. Mills, P. L., and Dudukovic, M. P. (1981). Evaluation of liquid–solid contacting in trickle-bed reactors by tracer methods, *AIChE J.*, **27**, 893.

80. Mohunta, D. M., and Laddha, G. S. (1965). Prediction of liquid phase holdup in random packed beds, *Chem Eng. Sci.*, **20**, 1069.

81. Montagna, A. A., and Shah, Y. T. (1975). The role of liquid holdup, effective catalyst wetting and backmixing on the performance of a trickle-bed reactor for residue hydrodesulfurization, *Ind. Eng. Chem. Proc. Des. Dev.*, **14**, 479.

82. Montagna, A. A., Shah, Y. T., and Paraskos, J. A. (1977). Effect of catalyst particle size on performance of a trickle-bed reactor, *Ind. Eng. Chem. Proc. Des. Dev.*, **16**, 152.

83. Morita, S., and Smith, J. M. (1978). Mass transfer and contacting efficiency in a trickle-bed reactor, *Ind. Eng. Chem. Fund.*, **17**, 113.

84. Morsi, B. I., Midoux, N., and Charpentier, J. C. (1978). Flow pattern and some holdup data in trickle-bed reactors for foaming–nonfoaming and viscous organic liquids, *AIChE J.*, **24**, 357.

85. Morsi, B. I., Midoux, N., Laurent, A., and Charpentier, J. C. (1982). Hydrodynamics and interfacial areas in downward cocurrent gas–liquid flow through fixed beds, influence of the liquid, *Int. Chem. Eng.*, **22**, 142.

336   CONCEPTS AND DESIGN OF CHEMICAL REACTORS

86. Morsi, B. I., Midoux, N., Laurent, A., and Charpentier, J. C. (1980). Interfacial area in trickle-bed reactors: Comparison between ionic and organic liquids and between Raschig rings and small diameter particle, *Chem. Eng. Sci.*, **35**, 1467.
87. Muroyama, K., Hashimoto, K., and Tomita, T. (1977). Heat transfer from wall in gas–liquid cocurrent packed bed, *Kagaku Kogaku Ronbunshu*, **3**, 612.
88. Otake, T., and Okada, K. (1963). Liquid holdup in packed towers, *Kagaku Kogaku*, **17**, 176.
89. Paraskos, J. A., Frayer, J. A., and Shah, Y. T. (1975). The effect of holdup incomplete catalyst wetting and backmixing during hydroprocessing in trickle-bed reactors, *Ind. Eng. Chem. Proc. Des. Dev.*, **14**, 315.
90. Patwardhan, V. S. (1983). Private communications.
91. Patwardhan, V. S., and Pataskar, S. G. (1982). The influence of wettability on the equilibrium flow of liquid in a packed column, *Chem. Eng. J.*, **23**, 145–149.
92. Prchlik, J., Soukup, J., Zapletal, V., and Ruzicka, V. (1975). Liquid distribution in reactors with randomly porous beds, *Coll. Czech. Chem. Commun.*, **40**, 845.
93. Prchlik, J., Soukup, J., Zapletal, V., and Ruzicka, V. (1975). Wall flow in trickle-bed reactors, *Coll. Czech. Chem. Commun.*, **40**, 3145.
94. Prchlik, J., Soukup, J., Zapletal, V., and Ruzicka, V. (1978). Effect of physical properties of the wetting liquid on its distribution in trickel-bed reactors, *Coll. Czech. Chem. Commun.*, **43**, 862.
95. Ramachandran, P. A., and Smith, J. M. (1979). Effectiveness factors in trickle-bed reactors, *AIChE J.*, **25**, 538.
96. Ramachandran, P. A., and Smith, J. M. (1979). Mixing cell method for design of trickle-bed reactor, *Chem. Eng. J.*, **17**, 91.
97. Rao, V. G., and Varma, Y. B. C. (1976). A model for the residence time distribution of liquid phase in trickle-beds, *AIChE J.*, **22**, 612.
98. Reiss, L. P. (1967). Cocurrent gas–liquid contacting in packed columns, *I&EC Process Des. Dev.*, **6**, 486.
99. Reynier, J. P., and Charpentier, J. C. (1971). Holdup prediction in packed columns for cocurrent gas–liquid downflow, *Chem. Eng. Sci.*, **26**, 1781.
100. Ruether, J. A., Yang, C. S., and Hayduk, W. (1980). Particle mass transfer during cocurrent downward gas–liquid flow in packed beds, *Ind. Eng. Chem. Proc. Des. Dev.*, **19**, 103.
101. Sato, Y., Hirose, T., Takahashi, F., and Toda, M. (1973). Pressure loss and liquid holdup in packed bed reactor with cocurrent gas–liquid and downflow in packed beds, *J. Chem. Eng. (Japan)*, **6**(4), 315.
102. Sato, Y., Hirose, T., Takahashi, F., Toda, M., and Hashiguchi, Y. (1973). Pattern and pulsation properties of cocurrent gas–liquid downflow in packed beds, *J. Chem. Eng. (Japan)*, **6**(4), 315.
103. Satterfield, C. N., and Way, P. F. (1972). The role of the liquid phase in the performance of a trickle bed reactors, *AIChE J.*, **18**, 305.
104. Satterfield, C. N., and Ozel, F. (1973). Direct solid catalyzed reaction of a vapor in an apparently completely wetted trickle-bed reactor, *AIChE J.*, **19**, 1259.
105. Satterfield, C. N. (1975). Trickle-bed reactors, *AIChE J.*, **21**, 209.
106. Satterfield, C. N., Ozel, F., van Eek, M. W., and Bliss, G. S. (1978). Liquid–solid mass transfer in packed beds with downward cocurrent gas–liquid flow, *AIChE J.*, **24**, 709.
107. Schwartz, J. G., Weyer, W., and Dudukovic, M. P. (1976). A new tracer method for determinaton of liquid–solid contacting efficiency in trickle-bed reactors, *AIChE J.*, **22**, 894.
108. Schwartz, J. G., Weyer, E., and Dudukovic, M. P. (1976). Liquid holdup and dispersion in trickle-bed reactors, *AIChE J.*, **22**, 953.
109. Sedriks, W., and Kenney, C. N. (1973). Partial wetting in trickle-bed reactors:

The reduction of crotonaldehyde over a palladium catalyst, *Chem. Eng. Sci.*, **28**, 559.

110. Seirafi, H. A., and Smith, J. M. (1980). Mass transfer and adsorption in liquid full and trickle-beds, *AIChE J.*, **36**, 711.

111. Shah, Y. T., and Paraskos, J. A. (1975). Criteria for axial dispersion effects in an adiabatic trickle bed reactor, *Chem. Eng. Sci.*, **30**, 1169.

112. Shah, Y. T. (1979). *Gas–Liquid–Solid Reactor Design*, McGraw-Hill, New York.

113. Shah, Y. T., Singh, C. P. P., and Joshi, J. B. (1981). Gas–liquid–solid reactor design, *Chemical Process Design*, **277**.

114. Shah, Y. T., and Sharma, M. M. (1984). Gas–liquid–solid reactor design, in *Chemical Reaction Engineering Handbook* (Eds. J. Carberry and A. Varma), Marcel Dekker, New York.

115. Sicardi, S., Gerhard, H., and Hofmann, H. (1979). Flow regime transitions in trickle bed reactors, *Chem. Eng. J.*, **18**, 173.

116. Sicardi, S., Baldi, G., Gianetto, A., and Specchia, V. (1980). Catalyst area wetted by flowing and semistagnant liquid in trickle-bed reactors, *Chem. Eng. Sci.*, **35**, 67.

117. Sicardi, S., and Hofmann, H. (1980). Influence of gas velocity and packing geometry on pulsing inception in trickle-bed reactors, *Chem. Eng. J.*, **20**, 251.

118. Sicardi, S., Baldi, G., and Specchia, V. (1980). Hydrodynamic models for the interpretation of the liquid flow in trickle-bed reactors, *Chem. Eng. Sci.*, **35**, 1775.

119. Sicardi, S., Baldi, G., Specchia, V., Mazzarino, I., and Gianetto, A. (1981). Packing wetting in trickle-bed reactors, influence of the gas flow rate, *Chem. Eng. Sci.*, **36**, 226.

120. Specchia, V., Rossini, A., and Baldi, G. (1974). Distribution and radial spread of liquid in two-phase concurrent flows in a packed bed, *Ing. Chim. Ital.*, **10**, 171.

121. Specchia, V., and Baldi, G. (1977). Pressure drop and liquid for two phase concurrent flow in packed beds, *Chem. Eng. Sci.*, **32**, 515.

122. Specchia, V., and Baldi, G. (1978). Solid–liquid mass transfer in cocurrent two-phase flow through packed beds, *Ind. Eng. Chem. Proc. Des. Dev.*, **17**, 362.

123. Specchia, V., and Baldi, G. (1979). Heat transfer in trickle-bed reactors, *Chem. Eng. Commun.*, **3**, 483.

124. Specchia, V., and Baldi, G. (1980). Heat transfer in packed bed reactors with one phase flow, *Chem. Eng. Commun.*, **4**, 361.

125. Stanek, V., and Kolar, V. (1968). Distribution of liquid over a random packing, *Coll. Czech. Chem. Commun.*, **33**, 2636.

126. Stanek, V., and Kolar, V. (1973). Distribution of liquid over a random packing, VII, *Coll. Czech. Chem. Commun.*, **38**, 1012.

127. Stanek, V., and Kolar, V. (1973). Distribution of liquid over a random packing, VIII, *Coll. Czech. Chem. Commun.*, **38**, 2865.

128. Stanek, V., Hanika, J., Hlavacek, V., and Trnka, O. (1981). The effect of liquid flow distribution on the behavior of a trickle-bed reactor, *Chem. Eng. Sci.*, **36**, 1045.

129. Sylvester, N. D., and Pitayagulsarn, P. (1975). Radial liquid distribution in concurrent two-phase downflow in packed beds, *Can. J. Chem. Eng.*, **53**, 599.

130. Talmor, E. (1977). Two phase downflow through catalyst beds: Part I, Flow maps, *AIChE J.*, **23**(6), 868.

131. Tan, C. S., and Smith, J. M. (1980). Catalyst particle effectiveness with unsymmetrical boundary conditions, *Chem. Eng. Sci.*, **35**, 1601.

132. Turek, F., and Lange, R. (1981). Mass transfer in trickle bed reactors, at low Reynolds number, *Chem. Eng. Sci.*, **36**, 569.

133. Turpin, J. L., and Huntington, R. L. (1967). Prediction of pressure drop for

two-phase, two-component concurrent flow in packed beds, *AIChE J.*, **13**(6), 1196.

134. Van Klinken, J., and Van Dongen, R. H. (1980). Catalyst dilution for improved performance of laboratory trickle-bed reactors, *Chem. Eng. Sci.*, **35**, 59.

135. Van Landeghem, H. (1980). Multiphase reactors: Mass transfer and modeling, *Chem. Eng. Sci.*, **35**, 1912.

136. Van Swaaij, W. P. M., Charpentier, J. C., and Villermaux, J. (1969). Residence time distribution in the liquid phase of trickle flow in packed columns, *Chem. Eng. Sci.*, **24**, 1083.

137. Weekman, V. W., and Myers, J. E. (1964). Fluid-flow characteristics of concurrent gas–liquid flow in packed beds, *AIChE J.*, **10**, 951.

138. Weekman, V. W., and Myers, J. W. (1965). Heat transfer characteristics of concurrent gas–liquid flow in packed beds, *AIChE J.*, **11**, 13.

139. Yan, T. Y. (1980). Dynamics of a trickle-bed hydrocracker with a quenching system, *Can. J. Chem. Eng.*, **48**, 259.

140. Yoshikawa, M., Iwai, K., Goto, S., and Teshima, H. (1981). Liquid–solid mass transfer in gas–liquid cocurrent flows through beds of small particles, *J. Chem. Eng. Japan*, **14**, 444.

Chapter 8

Photoreactor Engineering: Analysis and Design

ALBERTO E. CASSANO** ORLANDO M. ALFANO*** and
ROBERTO L. ROMERO***
*INTEC**, Casilla de Correo No. 91, 3000—Santa Fe, Argentina*

PREFACE

This work on photoreactor engineering reports the results obtained with my former students, some of them now full professors, such as: Dr. Horacio A. Irazoqui, Dr. Jaime Cerdá and Dr. Jacinto Marchetti, and others who are still in the process of obtaining their doctor's degrees such as: María A. Clariá, Eliana De Bernardez, Roberto Ré and naturally my two co-authors Orlando M. Alfano and Roberto L. Romero. It is a systematic presentation of the theory and related experimental verification that can be used in the analysis and design of photochemical reactors.

Other people have also made significant contributions to part of the experimental work, namely: Mr. Pedro G. Guarino, Mr. Mario J. J. Didier, Mr. Juan C. García, Mr. Antonio C. Negro, Mr. Omar Brizuela, Mr. José L. Giménez and Mr. J. L. Giombi. Credit must also be given to Miss Yolanda Pereyra for her everlasting patience in typing this manuscript until it reached its final form. Last but not

* Instituto de Desarrollo Tecnológico para la Industria Química. Universidad Nacional del Litoral (U.N.L.) and Consejo Nacional de Investigaciones Científicas y Técnicas (CONICET).
** Scientific and Technological Research Staff Member of CONICET and Professor at U.N.L.
*** Research Assistant from CONICET and U.N.L.

339

least, to Prof. Elsa I. Grimaldi for her unfailing devotion to help us prepare our work in English, which is not our mother tongue.

We have not intended to write a review of all the work done in the field but rather a systematic description of one methodology leading to rigorous, consistent design methods. As such, this monograph condenses several years of work that went from the proposal of appropriate definitions of the problem—extracted from well-known properties of the radiation field—to the present results on optimization of photoreactors for selectivity in consecutive chain reactions.

Since the main objective is to make possible an a priori design of the reactor, no concessions have been made to the use of free or adjustable parameters in the proposed models. Most of the theory has also been experimentally checked and the results—very good in most cases—are also briefly shown.

The task is by no means complete. A long way must still be travelled, especially in the area of heterogeneous photochemical reactors. This monograph also indicates those aspects that need more fundamental research to achieve a better understanding of problems poorly known or as yet unsolved.

A.E.C.

I. THEORETICAL DESCRIPTION OF THE RADIATION FIELD

I.1. Introduction

A rigorous presentation of the problem, from an approach consistently extracted from established definitions in radiation research, was presented by Irazoqui et al. [56, 58]. Some of these ideas will be repeated here as a proper framework for this treatment. The reader interested in a review of the field may resort to several papers produced in the past years [4, 5, 22, 35, 40, 65, 73, 82, 83, 87].

Significance of the Field

The analysis and design of ultraviolet (U.V.) and visible radiation activated reactors have received a sustained interest in the chemical engineering literature for the past two decades. Chlorination and nitrosation reactions, elimination of pollutants and synthesis of pharmaceuticals are some of the main commercial applications which have stimulated this interest. The manufacture of γ-hexachlorocyclohexane was, perhaps, one of the hydrocarbon chlorination reactions of greatest economic significance. Another example is the photosynthesis of vitamin D_2, undoubtedly one of the most relevant photochemical processes of the pharmaceutical industry. The production of antioxidants by addition of hydrogen sulfide to alkenes is another case of a photochemical reaction occurring through a chain mechanism. Among the nitrosation processes of major importance, we can mention that used in the preparation of caprolactam, the Nylon 6 monomer, by photooxidation of cyclohexane.

The main advantages of the photochemical excitation over the thermal route are selectivity and low reaction temperature. This in turn may improve selectivity, allow more favorable equilibrium conditions in exothermic reactions and increase the feasibility of operations in liquid phase, to mention only the most prominent effects. Since they were described in detail by Cassano et al. [22] they will not be discussed here.

One question arises immediately from the previous paragraphs. If there are so many advantages, why is it that photochemical reactors are not widely used? Doede and Walker [40] and Marcus et al. [73] suggested a concise, precise answer: we do not know how to design them. We believe that today this is no longer totally true and for this reason we have decided to publish this monograph.

Problems to be Solved
In 1968 we recognized several areas where lack of information or reliable ways of handling the problem could be observed. They provide key points for further work towards a better and more frequent use of photochemical reactors. They are:

1. Problems related to the U.V. and visible radiation sources, mainly the optimization of the spectral distribution of the radiation energy output in the wavelength range of absorption by one or more reactants. Solution of these problems would improve the utilization of the input energy, thus making the process more profitable. Output of U.V. and visible sources ranges from 2000 to 9000 Å, but certain reactants only absorb in a small wavelength interval (for example from 3000 to 4000 Å). The rest of the energy is lost and could produce undesired secondary reactions. This problem must be tackled with the aid of lamp manufacturers. For a given reaction, the proper selection of sources of photons as well as lamp operating pressures and

transmission characteristics of the lamp walls will certainly improve yield and, what is even more important, help in avoiding undesirable by-products.

2. Problems related to the chemistry of photochemical reactions, particularly the following ones: (1) complex kinetics, generally of the chain type, (2) poisoning and inhibition of these chain reactions and (3) deposits at the wall of the reactor which reduce light transmission and affect yield. These are mostly problems associated with the behavior of atomic and free radical reactions. With the exception of the latter, all of them are common to many complex reactions in conventional reactors.

The last one is a serious disadvantage for the use of photochemical reactors in industrial processes and the main cause of frequent shut-downs for cleaning routines. It is known that some of these difficulties have been overcome by the Toyo Rayon Co. of Japan on their cyclohexane photonitrosation commercial scale process. Some additional work on the subject, proposing novel reactor designs not yet under the stage of providing results of economic significance, was suggested by Lucas [67]. The author proposed a "two-zone segregated reactor" where wall deposits are avoided since the organic material is not in direct contact with radiation. In the meantime, an appropriate mechanical design of the reactor could provide continuous cleaning—by scraping– of the wall where the radiation enters the reactor, thus supplying a practical solution to the problem. Hopefully the time will come when fundamental research on surface chemistry will provide a definite answer to the problem today posed by industrial practice.

On the other hand, in a similar way as it is done with many polymerization reactions, the difficulties caused by the handling of gas-phase chain reactions could be easily solved with careful purification of reactants and effective sealings of reactors (to avoid poisoning due to the presence of radical trapping species). An adequate selectivity for the main reaction—reducing cost in the product stream separation processes—should provide an economic justification for the above-mentioned precautions with reactants and reactors.

3. Problems related to pressurized gas-phase reactors. Since they must have at least one quartz or glass wall, there exist some difficulties in working under pressure. It could be thought of as a materials science problem but mainly a challenge for new reactor designs that could range from multitubular reactors of small

diameter (very often used in industry for heat release reasons) to the introduction of new technologies such as lasers combined with optical fibers.

4. Problems related to the radiation field and its effect on the kinetics and the reactor performance. The radiation field is the only unique characteristic distinction of photochemical reactors which makes them different from thermal or catalytic reactors. The main advantages of photochemical reactors, namely selectivity and the possibility of working at low temperatures, are due to the existence of a radiation field that initiates the reaction in a very particular but precise way. Hence, we thought that progress in our knowledge of photoreactors would come from the study and modelling of the radiation emission and absorption as opposed to the repetition of kinetic studies very similar to the ones that are performed with conventional reactions.

Table I.1   Research Groups—Radiation Models Employed in Homogeneous Systems

Research groups	Radiation models	Period
Hill and coworkers	Incidence models	1965–1975
Dranoff and coworkers	Line source with parallel plane emission Line source with spherical emission	1965–1978
Smith and coworkers	Radial light model Two-dimensional diffuse light model	1965–1974
Williams and coworkers	Radial light model Three-dimensional diffuse light model	1970–1978
Akehata and coworkers	Line source with diffuse emission Two-dimensional diffuse incidence model	1971–1973
Cassano and coworkers	Three-dimensional source with volumetric emission	1973–1984
Foraboschi and coworkers	Line source with spherical emission Three-dimensional source with superficial emission	1959–1983
Yokota and coworkers	Three-dimensional source with diffuse superficial emission	1976–1981

Table I.2 Research Groups—Radiation Models Employed in Heterogeneous Systems

Research groups	Radiation models	Period
Akehata and coworkers	Two-flux models: dilute dispersed-phase model and concentrated dispersed-phase model	1976–1980
Yokota and coworkers	Models based on the light path length distribution Models based on homogeneous media with modified absorption coefficients	1977–1981
Santarelli and coworkers	Emission or incidence models in absorbing-scattering media	1978–1982
Otake and coworkers	Model based on homogeneous media with effective absorption coefficients	1981

Landmarks in Photoreactor Studies

1932: Bhagwat and Dhar. Effects of stirring on the reaction rate. To our knowledge the first engineering contribution to the field.

1955: Doede and Walker. Review paper.

1959: Foraboschi. Dimensional analysis.

1960: Schechter and Wissler. Analytical and elegant solution of a tubular reactor in laminar flow, first-order kinetics and an extremely simple radiation model.

1965–1970: Hill and Felder. Problems of mixing.

1965–1968: Dranoff and coworkers. Scale-up of photoreactors.

1965–1974: Smith and coworkers. Applied kinetics.

Tables I.1 and I.2 are extracted from a previous work of the authors [5] and indicate the main contributions achieved in the field up to 1984. A list of additional bibliography related to photoreactor design is included at the end of this work [97–161]. A critical analysis of most of them can be found in the above-quoted review [5].

I.2. Foundations of the Analysis

Among the properties of the radiation field defined for a given bundle of rays, one has a distinct physical meaning. This is the specific intensity, which must not be confused with the radiation flux density or its vector (Table I.3). The latter is a vector quantity related to the flow rate of energy and the normal area. On the other

Table I.3   Definitions of Radiation Properties

Property	Symbol	Dimensions	
Intensity	I	LSPP (I')	Energy per unit time, unit plane angle and unit length
		LSSE (I'')	Energy per unit time, unit solid angle and unit length
		TDS (I)	Energy per unit time, unit solid angle and unit area
Energy flow rate	E	Energy per unit time	
Energy flux density vector	$\mathbf{q}$	Energy per unit time and unit area	
Energy flux density	$q = \mathbf{q} \cdot \mathbf{n}$	Energy per unit time and unit area	
Energy density	e	Energy per unit volume	
Volumetric rate of energy absorption	e^a	Energy per unit time and unit volume	

hand, the specific intensity does not follow the rules of vectorial algebra and is a point function. In a diactinic medium, and for each pencil of radiation, it must be constant throughout the propagation path of the rays and hence independent of distance. Under these conditions, the specific intensity changes only in absorbing and/or non-homogeneous media (due to absorption, scattering, reflections and refractions) according to the physical characteristics of the substance in which the rays travel. This is not true for the radiation flux density which, in general, may change for those and other reasons, i.e. it may change not only due to the physical properties of the medium of propagation but also because of the geometry of the system, and hence with distance. Only when the cross-sectional area of the bundle is constant (e.g. for collimated beams of radiation) the numerical values of the specific intensity and the flux density are the same.

The only particular requirement to achieve complete consistency in the theory is to accept that the intensity will change its definition according to the dimensions with which the emission of energy is modeled. With this restriction in mind, all other properties such as energy flow rate, energy flux density vector, energy flux density and energy density follow from first principles. Table I.3 summarizes the principal definitions used in this work. An alternative approach to the problem, extracted from the laws of photometry, was proposed by Roger and Villermaux [82, 83]. We consider that our basic properties, extracted directly from well-accepted definitions in the area of radiation engineering and radiation gas dynamics, are more appropriate to photoreactor engineering. This adoption does not claim that the other approach is wrong; in fact it could probably be equally used and should produce, hopefully, the same final results.

The second basic principle that must be applied to the analysis of a photoreactor is related to the absorption of radiation. In general this problem has been solved using a "law" (Lambert's "law"). As it is usually misused, the term "law" has been applied here to what is properly one form of a constitutive equation used to define an attenuation coefficient μ_ν. It is important to investigate the physical and mathematical nature of this "law" of absorption. This will be done here because it must not be confused with a radiation balance, since its use as such is only valid for special cases. The "law" was postulated by Lambert, and it can be derived from molecular theory. It states that, for an absorbing medium, the contribution of absorption to the gradient of the specific intensity is:

$$\frac{dI_\nu}{ds}\bigg|^a = -\mu_\nu I_\nu \tag{I.1}$$

Equation (I.1) is valid for any form of emission with the proper use of the required definition for I. The radiation conservation equation reduces to Eq. (I.1) at steady-state conditions and for non-emitting and homogeneous media. This is so because Lambert's "law" was used in its derivation to formulate the absorption process in the balance of photons. Using index notation, the left-hand side of Eq. (I.1) can be written as:

$$\frac{dI_\nu}{ds} = \frac{\partial I_\nu}{\partial x_i}\frac{dx_i}{ds} = \frac{\partial I_\nu}{\partial x_i}\cos\alpha_i \tag{I.2}$$

where $\cos\alpha_i$ are direction cosines of the ray with respect to a Cartesian orthogonal coordinate system x_i (i.e. x, y, z: Fig. I.1). It is clear that the left-hand side is the directional derivative of I_ν, i.e. the component of the gradient of the scalar function I_ν along the

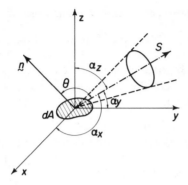

Figure I.1   Geometry for application of Lambert's "law".

direction of propagation of the ray. The right-hand side of Eq. (I.2) shows the three components of this gradient. In a more general way, if ℓ is a unit vector in the direction of each ray, with components $\cos \alpha_i$, Eq. (I.1) can be written:

$$\ell \cdot \nabla I_\nu = -\mu_\nu I_\nu \qquad (I.3)$$

This is the only possible generalization of the absorption constitutive equation which allows it to be used in several coordinate systems. This means that the so-called Lambert's "law" gives the variation of the specific intensity due to absorption along its direction of propagation. In other words, the "law" of absorption is a constitutive equation for absorbing media, which relates the gradient of the specific intensity to the molecular properties of the space through which the ray travels (e.g. the absorption coefficient and molecular density or, equivalently, the attenuation coefficient μ_ν).

A third concept is particularly needed in reaction analysis. When a reaction takes place, an additional property must be considered. Such a property is the local volumetric rate of energy absorption (LVREA) by the reactant which undergoes photoactivation (absorbed energy per unit time and unit volume). It is usually written (although not in this work) as I_a, and has units of einstein (or quantum) per cubic centimeter per second. In a photochemical reaction, the initiation rate will be given by the primary quantum yield times this volumetric rate of energy absorption (e^a). Hence, it is necessary to evaluate e^a locally. The problem will be treated in detail in further sections.

I.3. What Is Photoreactor Analysis and Design?

Out of the four problems described in Section I.1, we believe that with the only possible exception of the wall deposit formations, the central core of a fundamental approach to the subject is the one described in part 4. On these grounds let us state our problem. In general we may have:

1. Several possible radiation sources with:
    1.1. A given energy consumption and output power.
    1.2. The spectral distribution of their output power.
    1.3. Their geometrical dimensions.
    1.4. Their operating conditions.
2. Several possible reactors with:
    2.1. Their proposed shape.
    2.2. In some designs, the transmission characteristics of the wall through which radiation enters the reactor.
    2.3. The spectral distribution of this transmission.
    2.4. Some general and global information about ranges of operating conditions; for instance: pressure, temperature, required production, etc.
3. A reaction with:
    3.1. A kinetic sequence with the following characteristics:
        a. An initiation step which is activated by absorption of radiation.
        b. One or more "dark" reactions.
           From the industrial point of view the most important ones are special types of dark reactions which are recognized with the name of chain reactions. They involve, starting from an atomic species or a free radical formed in the initiation step, the following constituents:
           b.1 One or more propagation steps.
           b.2 One or more termination steps.
    3.2. The radiation absorption characteristics of reactants, inerts and products.
    3.3. The primary quantum yield of the initiation step, i.e. how many atomic or free radical species are effectively produced for each absorbed photon by the desired reactant in the first reaction of the kinetic sequence.
4. Sometimes, a reflecting surface with the following characteristics:
    4.1. Its shape.
    4.2. Its size (at least approximate).
    4.3. Its reflectivity.
    4.4. The spectral distribution of its reflectivity.

Given a tentative irradiating system which could be made up of a radiation source (lamp) and perhaps a reflector, a particular type of reactor (not always imposed) and a reaction to produce the desired product, the design engineer is then faced with the following questions: Which is the best source-reactor system? How much energy will be needed to convert reactants into products? Which are the optimal operating conditions? With which selectivity can the reaction be achieved?

In order to answer these questions, they must be treated in a logical sequence which starts with the proper knowledge of the formulation of the rate of a photochemical reaction. Once one knows how to formulate it, one can progress further with the design problem and its optimal operating conditions. It will be immediately seen that the formulation of the rate of reaction is unavoidably tied to the system used to supply the energy for the initiation step.

As previously stated, when expressing the rate of a photochemical reaction it is necessary to make the distinction between dark and radiation activated steps. To treat the dark reactions one uses the same methodology as for conventional reactors; the main hindrance appears when evaluating the rate of the initiation step. The existence of this very particular initiation step constitutes the main difference between thermal and radiation activated reactions. It is known that the rate of a radiation activated step is related to the LVREA. Consequently, this functional is the important variable in the analysis of photoreactors. The LVREA depends on the radiation field existing in the reaction space; hence, the need to know the radiant energy distribution within the photoreactor. The radiation field is not uniform in space due to several factors. Among them, the attenuation produced by the species absorption is always present. Additional facts, usually also important, are the physical properties and geometrical characteristics of the lamp-reactor system. Consequently, the initiation reaction will be spatially dependent, even in the case of the absence of concentration gradients. This lack of homogeneity in the radiation field is intrinsically irreducible in photochemical reactors.

If one attempts to model a photoreactor, mass, energy and momentum balances should be established. The expression for the conservation of momentum is similar to that of conventional reactors. Due to the energetical features of photochemical reactions (very low "equivalent" activation energy for the global reaction), it is frequent to find that the system can be accurately described with the mass inventory and the energy balance reduced exclusively to radiation; however, they are normally coupled through the concentrations of absorbing species. (If the reactions are very exothermic or endothermic, the complete energy equation of change will be needed.)

The nature of this coupling is well understood. The LVREA is generally a function of the spatial variables, of the concentration of the absorbing species and other physico-chemical parameters. If species (i) which is being considered in the mass balance is an absorbing species, a coupling between the LVREA and the corresponding mass conservation equation will take place. On the other hand, the concentration of species (i) under consideration forms part of an integral expression accounting for the radiation attenuation by absorption (Lambert's constitutive equation). Thus, the general problem of modelling a photoreactor presents a mathematical problem of integro-differential nature.

Many authors have analyzed photosensitized reactions with treatments that considerably simplify the mathematical problem. In this case, since the absorbing species does not suffer changes in concentration, the link "progress of the reaction-attenuation of radiation" is destroyed. The problem is no longer integro-differential and the LVREA becomes a function of the position which can be obtained independently of the information provided by the mass conservation differential equation.

The evaluation of the LVREA (Fig. I.2) is performed stating first a balance of radiant energy at steady state for a homogeneous control volume. For simplicity a non-emitting medium is generally assumed. Afterwards it is necessary to incorporate a radiation source model and Lambert's constitutive equation in the absorption term.

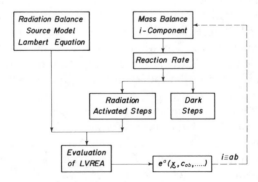

Figure I.2 Evaluation of the local volumetric rate of energy absorption.

The source models have been treated by several research teams which have developed various proposals to predict radiation profiles for different types of photoreactors. Those previous studies may be classified into two well-defined groups: (a) those which assume a given radiation distribution in the vicinity of the reactor (*incidence models*), and (b) those which start from a proposal of a model for the source (*emission models*). There is no way of using incidence models without experimentally adjustable parameters. The reader interested in this approach can find a critical review in our quoted work [5]. However, since we are looking for an a priori design of photoreactors, attention will be paid only to emission models. After presenting emission models we will introduce the proper formulation of the reaction rate for the initiation step which implies the description of the LVREA.

I.4. Radiation Emission Models

Table I.4 summarizes the main characteristics and historical origins of each of the most important emission models. One could add the line model with diffuse emission [1] to complete the available proposals. As it can be immediately seen the work done recognizes first the state of practical availability of U.V. lamps for medium- and large-scale operations: sources are almost invariably cylindrical tubes. In many cases ratio of the radius to length is rather small and this observation has led to two steps in the modelling process:

1. Consideration of the source as a line. Two representations have been proposed:
   1.1. Two-dimensional propagation of the rays; all of them lying

Table I.4   Main Source Emission Models

Source characteristic	Emission characteristic	Model name	Authors
Line source	Parallel planes perpendicular to the lamp axis	LSPP	Harris and Dranoff (1965)
	Spherical and isotropic	LSSE	Jacob and Dranoff (1966)
Three-dimensional source	Volumetric	ESVE	Irazoqui, Cerdá and Cassano (1973)
	Superficial	ESSE	Stramigioli, Santarelli and Foraboschi (1977)

on parallel planes perpendicular to the center line axis of the lamp (LSPP Model).

   1.2. Three-dimensional propagation of the rays, with a more realistic spherical emission (LSSE Model).

2. Consideration of the source as a volume with three dimensions (TDS Models). Two forms of emission have been modelled:

   2.1. A three-dimensional source (extense source) with volumetric emission (ESVE Model).

   2.2. A three-dimensional source (also extense) with superficial emission (ESSE Model).

In the second approach, the lamp is represented as a perfect cylinder. The volumetric emission seems to be the best available representation for U.V. arc sources. One should expect that when fluorescent lamps are used, the ESSE Model should provide better results. The fundamental feature which distinguishes these from prior models is the incorporation of the lamp radius as an additional parameter in the design of a photochemical reactor.

LSPP Model

As shown in Fig. I.3 this case is very ideal. The bundle of rays has the shape of a wedge of height dz and angle of divergence $d\theta$ measured on a plane normal to the z-axis.

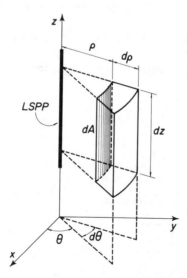

Figure I.3   Geometric representation of the LSPP Model.

The specific intensity must be written as:

$$I'_\nu \, d\nu \overset{\Delta}{=} \frac{dE_\nu}{d\theta \, dz} \tag{I.4}$$

Consequently:

$$|d\mathbf{q}_\nu| \overset{\Delta}{=} \frac{dE_\nu}{dA \cos \theta_n} \tag{I.5}$$

Hence:

$$|d\mathbf{q}_\nu| = \frac{I'_\nu \, d\nu}{\rho} \tag{I.6}$$

and:

$$dE = \int_{\nu=0}^{\nu=\infty} dE_\nu = \left(\int_{\nu=0}^{\nu=\infty} I'_\nu \, d\nu \right) d\theta \, dz = I' \, d\theta \, dz \tag{I.7}$$

It is clear that this model (very attractive because of its simplicity) has very little physical meaning since it requires that every point of the source emits radiation in planes perpendicular to the lamp axis (contradicting the spherical emission of a point source), with zero probability of emission in directions other than those contained in such planes.

Performing a radiation balance in an absorbing medium:

$$d(|d\mathbf{q}_\nu|\rho) = \frac{dI'_\nu}{d\rho}\bigg|^a \, d\rho \, d\nu \tag{I.8}$$

where the superscript "a" indicates absorption. An assumption has been made that the balance is made for steady-state conditions and for a homogeneous non-emitting control volume. We now substitute Lambert's "law" in Eq. (I.8) to obtain

$$\frac{1}{\rho} \frac{d(|d\mathbf{q}_\nu|\rho)}{d\rho} = -\mu_\nu |d\mathbf{q}_\nu| \tag{I.9}$$

Integrating along the radiation path:

$$|d\mathbf{q}_\nu| = |d\mathbf{q}_\nu|_0 \frac{\rho_0}{\rho} \exp\left(-\int_{\rho_0}^{\rho} \mu_\nu \, dp^* \right) \tag{I.10}$$

The flux density must be related to the energy flow rate, in order to be able to introduce the radiation characteristics of the U.V. source in the model. It is straightforward to show that:

$$|d\mathbf{q}_\nu| = \frac{dE_{\nu,0}}{2\pi L_L} \frac{1}{\rho} \exp\left(-\int_{\rho_0}^{\rho} \mu_\nu \, d\rho^*\right) \qquad (I.11)$$

From the radiation balance one can also obtain the LVREA. In general if one identifies the absorbed energy flow rate as dE_ν^a one has:

$$de_\nu^a \triangleq \frac{dE_\nu^a}{dV^a} \qquad (I.12)$$

and

$$e^a = \int_{\nu=0}^{\nu=\infty} de_\nu^a \qquad (I.13)$$

Let us now develop then the value of e^a for the LSPP Model. By definition:

$$dE_\nu\big|_\rho - dE_\nu\big|_{\rho+d\rho} = dE_\nu^a \qquad (I.14)$$

Combining Eqs. (I.4), (I.5) and (I.8) we have:

$$dE_\nu^a = -d(|d\mathbf{q}_\nu|\rho) \, d\theta \, dz = -dI_\nu'\big|^a \, d\nu \, d\theta \, dz \qquad (I.15)$$

Using Eq. (I.1) allows us to write

$$dE_\nu^a = \mu_\nu I_\nu' \, d\nu \, d\rho \, d\theta \, dz = \mu_\nu \frac{I_\nu' \, d\nu}{\rho} \, dV^a \qquad (I.16)$$

since here $dV^a = \rho \, d\rho \, d\theta \, dz$. Finally using Eqs. (I.6) and (I.12) gives

$$de_\nu^a = \mu_\nu |d\mathbf{q}_\nu| \qquad (I.17)$$

and from Eqs. (I.11), (I.13) and (I.17) we obtain

$$e^a = \int_{\nu=0}^{\nu=\infty} \mu_\nu \frac{dE_{\nu,0}}{2\pi L_L} \frac{1}{\rho} \exp\left(-\int_{\rho_0}^{\rho} \mu_\nu \, d\rho^*\right) \qquad (I.18)$$

The attractive feature of this model is that here $\rho = \rho(r)$ only. Even though this model is an ultrasimplified representation of reality it contains many if not all of the more important characteristics of the problem which we list as:

1. In every initiation step of a photochemical reaction, e^a must be present.
2. Even in the simplest case, e^a is a function of position.
3. Even in the most idealized case (Beer's equation), μ_ν is a linear function of concentration and strongly depends upon the wavelength.
4. The energy flow rate $(dE_{\nu,0})$ also has a complicated dependence on the wavelength (spectral distribution of the radiation source output power).

Since e^a is always present in the evaluation of the rate of the initiation step, any conservation equation will have this complex integral in the source or sink term of the mathematical expression representing the behavior of the reactor. This problem will be made much clearer at a later stage of this work.

LSSE Model

Basically, this model assumes that the lamp is a line source in which each one of its points emits radiation in every direction and isotropically; i.e., the source is made up of a finite number of segments ("point" sources), whose contributions are later added up in order to calculate the radiant energy distribution within the reactor [59, 60]. It is then necessary to start from the definitions given for point sources (Fig. I.4):

$$I_\nu^\circ \, d\nu \overset{\Delta}{=} \frac{dE_\nu}{d\Omega} \qquad (I.19)$$

Since Eq. (I.5) is general, regardless of the dimensions of the radiation model, it can be used to obtain:

$$|d\mathbf{q}_\nu| = \frac{I_\nu^\circ \, d\nu}{\rho^2} \qquad (I.20)$$

An important feature of the physics of radiation propagation in homogeneous spaces is illustrated here. The modulus of the radiation flux density vector changes with the inverse of the square of the distance. It follows that:

$$dE = \int_{\nu=0}^{\nu=\infty} dE_\nu = \left(\int_{\nu=0}^{\nu=\infty} I_\nu^\circ \, d\nu \right) d\Omega = I^\circ \, d\Omega \qquad (I.21)$$

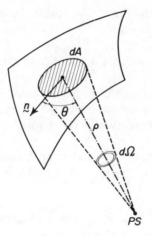

Figure I.4   Geometric representation of the PS Model.

Note that once more, for a given bundle of rays, $I°$ remains constant throughout its path of propagation due to the constancy of dE and $d\Omega$. In a diactinic medium, this is an important property of $I°$ which permits one to define it only by the characteristics of the radiation source.

The LSSE Model assumes that the lamp is a succession of point sources as it is shown in Fig. I.5. The specific intensity can be written as:

$$I''_\nu \, d\nu = \frac{dE_\nu}{dz \, d\Omega} \tag{I.22}$$

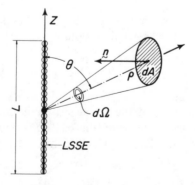

Figure I.5   Geometric representation of the LSSE Model.

and using Eq. (I.5) we get:

$$|d\mathbf{q}_\nu| = \frac{I''_\nu \, d\nu \, dz}{\rho^2} \tag{I.23}$$

With the radiation balance under steady-state conditions and a homogeneous non-emitting control volume (Fig. I.6), we find

$$d(|d\mathbf{q}_\nu| \rho^2) = \frac{dI''_\nu}{d\rho}\bigg|^a \, d\rho \, dz \, d\nu \tag{I.24}$$

Introducing Lambert's equation we can obtain

$$\frac{1}{\rho^2} \frac{d(|\mathbf{q}_\nu| \rho^2)}{d\rho} = -\mu_\nu |d\mathbf{q}_\nu| \tag{I.25}$$

and integrating through the optical path leads to

$$|d\mathbf{q}_\nu| = |d\mathbf{q}_\nu|_0 \frac{\rho_0^2}{\rho^2} \exp\left(-\int_{\rho_0}^\rho \mu_\nu \, d\rho^*\right) \tag{I.26}$$

Finally, we can use Eqs. (I.22) and (I.23) to obtain

$$|d\mathbf{q}_\nu| = \frac{dE_{\nu,0}}{4\pi L_L} \frac{dz}{\rho^2} \exp\left(-\int_{\rho_0}^\rho \mu_\nu \, d\rho^*\right) \tag{I.27}$$

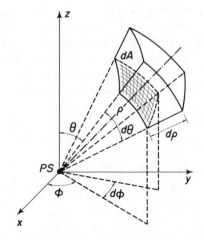

Figure I.6 Radiation balance for point sources.

and integration along the lamp length yields

$$|d\mathbf{q}_\nu| = \frac{dE_{\nu,0}}{4\pi L_L} \int_0^{L_L} \frac{1}{\rho^2} \exp\left(-\int_{\rho_0}^{\rho} \mu_\nu \, d\rho^*\right) dz \qquad (I.28)$$

From the radiation balance we also have:

$$dE_\nu^a = -dI_\nu'' \, d\nu \, dz \, d\Omega \qquad (I.29)$$

and following the same reasoning as before we get:

$$dE_\nu^a = \mu_\nu I_\nu'' \, d\nu \, \frac{dz \, dV^a}{\rho^2} \qquad (I.30)$$

Use of Eqs. (I.23) and (I.12) yields

$$de_\nu^a = \mu_\nu |d\mathbf{q}_\nu| \qquad (I.31)$$

Substituting Eq. (I.27) and integrating along the lamp length and over all wavelengths provides an expression for e^a

$$e^a = \int_{\nu=0}^{\nu=\infty} \mu_\nu \frac{dE_{\nu,0}}{4\pi L_L} \int_0^{L_L} \frac{1}{\rho^2} \exp\left(-\int_{\rho_0}^{\rho} \mu_\nu \, d\rho^*\right) dz \qquad (I.32)$$

The spherical characteristic of the emission makes now $\rho = \rho(r, \theta)$.

ESVE Model

This model considers the U.V. lamp as a rectangular cylinder whose volume is the source of radiation. As an emission model starting from fundamental principles, it was, historically, the first attempt made to predict a priori the properties of the radiation field inside photoreactors without adjustable parameters [57].

From Fig. I.7 and the definition of the specific intensity we have:

$$dI_\nu \, d\nu = \frac{dE_\nu \rho^2}{dA_e \cos\theta_e \, dA_r \cos\theta_r} \qquad (I.33)$$

and the modulus of the flux vector results:

$$|d\mathbf{q}_\nu| = \frac{I_\nu \, d\nu \, dA_e \cos\theta_e}{\rho^2} \qquad (I.34)$$

From this we obtain

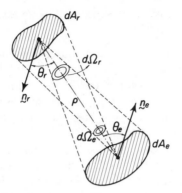

Figure I.7   Geometric representation of the TDS Model.

$$dE = \int_{\nu=0}^{\nu=\infty} dE_\nu = \left(\int_{\nu=0}^{\nu=\infty} I_\nu \, d\nu \right) \frac{dA_e \cos \theta_e \, dA_r \cos \theta_r}{\rho^2} \qquad (\text{I.35})$$

and it can be concluded that:

– For a given bundle of rays in a diactinic medium, I and also $\partial I / \partial \nu = I_\nu$ are constant scalars throughout the direction of propagation and are independent of the distance travelled by the rays; i.e., I and also I_ν are properties only of the source and of the direction of propagation of the electromagnetic wave under consideration.

– An elementary area of emission in a three-dimensional source is similar, but not equal, to a point source (PS). The differential area emits an infinitesimal amount of energy so that the whole source will have a finite emission. On the other hand, the PS requires a finite amount of energy emission from the point.

– The three-dimensional source models do not introduce singularities in the prediction of the radiation flux density at any position in space.

In spite of all these differences, I°, I', I'' and I have common characteristics. All of them are scalar quantities that satisfy the fundamental definition stated before. The specific intensity for a given direction undergoes changes only when an absorption of energy exists (in non-transparent media), or when there are heterogeneities, i.e. interfaces with reflections and refractions, or when scattering occurs.

The ESVE Model is based on the following *assumptions about the emission* (Figs. I.8 and I.9):

(i) The emitters of the radiation source are uniformly distributed inside its volume.

(ii) Each elementary volume inside the source has an isotropic emission, that is, the specific intensity associated with each bundle of radiation coming from the same element of volume is independent of direction.

(iii) Any elementary volume inside the source emits, at any frequency, an amount of energy proportional to its extension.

(iv) The energy emitted from any elementary volume of the lamp is due to spontaneous emission, and each of these differential volumes is transparent to the emission of its surroundings.

The following *assumptions about the source* are considered:

(i) The lamp is a perfect cylinder, bounded by mathematical surfaces with zero thickness. Thus, any bundle of radiation coming from inside does not change its intensity or direction when it crosses this boundary.

(ii) The lamp is long enough and is adequately masked to neglect end effects, that is, the emission characteristics are constant along the z-direction (this assumption does not say anything about the radiation field along the z-direction).

The problem calls for a proposal to obtain the value of dE_ν. It

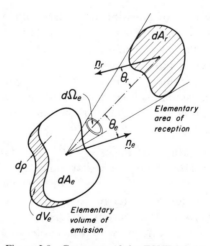

Figure I.8   Geometry of the ESVE Model.

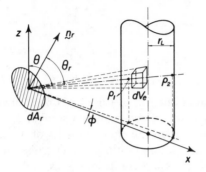

Figure I.9   Variables involved in the theoretical analysis of the ESVE Model.

should relate the characteristics of the radiation field produced by the source with its operating conditions. The emission from the source may be represented by:

$$dE_\nu = N_e P_\nu h\nu \, d\nu \, dV_e \qquad (I.36)$$

This is the rate of energy emission by an elementary volume at frequency ν. In a solid angle $d\Omega_e$ about a given direction we have:

$$dE_\nu = \frac{N_e P_\nu h\nu}{4\pi} \, d\nu \, d\Omega_e \, dV_e \qquad (I.37)$$

Hence:

$$dE_\nu = \frac{N_e P_\nu h\nu \, dA_r \cos\theta_r \, dA_e \cos\theta_e \, d\rho \, d\nu}{4\pi\rho^2} \qquad (I.38)$$

Using Eq. (I.33):

$$dI_\nu = \frac{N_e P_\nu h\nu}{4\pi} \, d\rho \qquad (I.39)$$

and resorting again to Eq. (I.5) yields

$$|d\mathbf{q}_\nu| = \frac{N_e P_\nu h\nu \, d\nu}{4\pi\rho^2} \, (dA_e \cos\theta_e \, d\rho) \qquad (I.40)$$

$$|d\mathbf{q}_\nu| = \frac{N_e P_\nu h\nu \, d\nu}{4\pi\rho^2} \, dV_e \qquad (I.41)$$

Later, it will be convenient to have the definitions

$$\kappa = \int_{\nu=0}^{\nu=\infty} \frac{N_e P_\nu h\nu}{4\pi} d\nu \qquad (I.42)$$

$$|\mathbf{q}| = \kappa \int_\phi \int_\theta \int_\rho \frac{dV_e}{\rho^2} \qquad (I.43)$$

in which the volume integral of Eq. (I.43) must always be extended over the whole volume of the radiation source. The integral of Eq. (I.42) depends only upon the characteristics of the radiation source. Under the previous assumption about the emission (iv), N_e and P_ν are related to the emission coefficient j_ν as follows:

$$j_\nu = \frac{N_e P_\nu h\nu}{4\pi\delta} \qquad (I.44)$$

In Eq. (I.44), δ is the density of the emitting medium, and j_ν is a function of frequency as well as of the state variables of the medium; thus, both N_e and P_ν depend upon the operating conditions of the lamp.

Making use of the radiation balance (Fig. I.10) inside an absorbing control volume and after substituting Lambert's equation, one gets:

$$\frac{1}{\rho^2} \frac{d(|d\mathbf{q}_\nu|\rho^2)}{d\rho} = -\mu_\nu\, d\rho \qquad (I.45)$$

In Eq. (I.45) $d\rho$ is the differential increment of the radiation path length measured along the absorbing medium. On the other hand, in Eqs. (I.38) to (I.40) it was considered inside the emitting control volume.

Integrating Eq. (I.45):

$$|d\mathbf{q}_\nu| = \frac{\rho_0^2}{\rho^2} |d\mathbf{q}_\nu|_0 \exp\left(-\int_{\rho_0}^{\rho} \mu_\nu\, d\rho^*\right) \qquad (I.46)$$

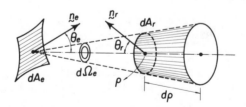

Figure I.10   Radiation balance for extense sources.

with Eq. (I.41) used for $|d\mathbf{q}_\nu|_0$:

$$|d\mathbf{q}_\nu| = \frac{N_e P_\nu h\nu}{4\pi}\ d\nu\ dV_e\left(\frac{1}{\rho^2}\right)\exp\left(-\int_{\rho_0}^{\rho}\mu_\nu\ d\rho^*\right) \qquad (I.47)$$

and following Eq. (I.43):

$$|\mathbf{q}| = \kappa\int_\phi\int_\theta\int_\rho\frac{dV_e}{\rho^2}\exp\left(-\int_{\rho_0}^{\rho}\mu_\nu\ d\rho^*\right) \qquad (I.48)$$

The LVREA can be easily derived now. Recalling Fig. I.10:

$$dE_\nu^a = \mu_\nu I_\nu\ d\nu\ dA_r\cos\theta_r\ d\Omega_e\ d\rho$$

$$dE_\nu^a = \mu_\nu|d\mathbf{q}_\nu|\ dA_r\cos\theta_r\ d\rho = \mu_\nu|d\mathbf{q}_\nu|\ dV^a \qquad (I.49)$$

$$de_\nu^a = \mu_\nu|d\mathbf{q}_\nu| \qquad (I.50)$$

$$e^a = \int_{\nu=0}^{\nu=\infty}\mu_\nu\frac{N_e P_\nu h\nu}{4\pi}\ d\nu\int_\phi\int_\theta\int_\rho\frac{dV_e}{\rho^2}\exp\left(-\int_{\rho_0}^{\rho}\mu_\nu\ d\rho^*\right) \qquad (I.51)$$

From Eq. (I.51) it is clear that the problem is three-dimensional since $\rho = \rho(r, \theta, \phi)$. Equation (I.51) provides the local value (point value) of the volumetric rate of energy absorption at any place in space for the radiation supplied by the source of volume V_e, and the first μ_ν is the attenuation coefficient at the point of absorption. However, the exponential term takes into account the absorption undergone by the radiation bundle in going from ρ_0 to ρ. If μ_ν is a function of position (and in general it is) the value of e^a depends upon the local contribution of μ_ν, but the amount of radiation arriving at such a point for absorption will depend upon all the values of μ_ν for all the points in space through which the ray has passed before (see the exponential term). This is a key point in the analysis of the rate of reaction of a photochemical reactor with non-uniform composition, as it is generally the case.

ESSE Model

The model recognizes the three dimensions of the source whose external surface produces an isotropic emission from each elementary area [88]. The radiation is received by an elementary area of reception dA_r. From the definition of the specific intensity (Fig. I.7), the following equation applies:

$$I_\nu \, d\nu = \frac{dE_\nu \, \rho^2}{dA_e \cos\theta_e \, dA_r \cos\theta_r} \tag{I.52}$$

With Eq. (I.5) and a radiation balance in the absorbing control volume (Fig. I.10) we obtain

$$d(|d\mathbf{q}_\nu|\rho^2) = \left.\frac{dI_\nu}{d\rho}\right|^a d\rho \, d\nu \, dA_e \cos\theta_e$$

Introducing Lambert's equation and integrating leads to

$$|d\mathbf{q}_\nu| = |d\mathbf{q}_\nu|_0 \, \frac{\rho_0^2}{\rho^2} \exp\left(-\int_{\rho_0}^{\rho} \mu_\nu \, d\rho^*\right) \tag{I.53}$$

From this it is clear that the integration of all the radiation power output of the source must be done over its entire surface. Expressing Eq. (I.53) in terms of $I_{\nu,0}$ and the surface integral in terms of the θ and ϕ variables:

$$|d\mathbf{q}_\nu| = I_{\nu,0} \, d\nu \int_\phi \int_\theta \exp\left(-\int_{\rho_0}^{\rho} \mu_\nu \, d\rho^*\right) \sin\theta \, d\theta \, d\phi \tag{I.54}$$

Again, using the radiation balance and following the same steps as before gives

$$de_\nu^a = \mu_\nu I_{\nu,0} \, d\nu \int_\phi \int_\theta \exp\left(-\int_{\rho_0}^{\rho} \mu_\nu \, d\rho^*\right) \sin\theta \, d\theta \, d\phi \tag{I.55}$$

and this can be integrated to obtain .

$$e^a = \int_{\nu=0}^{\nu=\infty} \mu_\nu I_{\nu,0} \, d\nu \int_\phi \int_\theta \exp\left(-\int_{\rho_0}^{\rho} \mu_\nu \, d\rho^*\right) \sin\theta \, d\theta \, d\phi \tag{I.56}$$

Here $I_{\nu,0}$ must be related to the output power of the radiation source to obtain:

$$e^a = \int_{\nu=0}^{\nu=\infty} \mu_\nu \, \frac{dE_{\nu,0}}{8\pi^2 r_L L_L} \int_\phi \int_\theta \exp\left(-\int_{\rho_0}^{\rho} \mu_\nu \, d\rho^*\right) \sin\theta \, d\theta \, d\phi \tag{I.57}$$

In the spherical coordinate system, $\rho = \rho(r, \theta, \phi)$ and the absorption problem is again three-dimensional.

I.5. The Formulation of the Reaction Rate for the Initiation Step

Let us illustrate with the three-dimensional models the way in which the energy absorption can be formulated as a kinetic process. To consider the absorption of energy as a kinetic process, the corresponding state property of the system which allows its quantitative treatment is the previously defined LVREA. As will be shown later, under the conditions of this analysis, the LVREA can be written with the same formalism as that used for the reaction rate equations for conventional chemical reaction steps. Let us look at the TDS Models. For them, the LVREA corresponding to frequency ν at any point in an absorbing medium is:

$$de_\nu^a = \frac{\partial e^a}{\partial \nu}\, d\nu = \mu_\nu \int_\phi \int_\theta |d\mathbf{q}_\nu^{(3)}| = \mu_\nu \int_\phi \int_\theta \frac{\partial}{\partial \nu} |d\mathbf{q}^{(2)}|\, d\nu \qquad (1.58)$$

where $\partial e^a/\partial \nu$ is the frequency distribution function of the LVREA. Since:

$$|d\mathbf{q}_\nu| = I_\nu\, d\nu\, d\Omega_r \qquad (1.59)$$

we can write:

$$\frac{\partial e^a}{\partial \nu} = \mu_\nu \int_\phi \int_\theta I_\nu\, d\Omega_r \qquad (1.60)$$

In Eq. (1.60), the product $I_\nu\, d\Omega_r$ is equal to the frequency distribution function of the energy flux density associated with every bundle of rays coming from each direction (θ, ϕ).

The distribution of the energy density as a function of frequency is:

$$e_\nu = \frac{1}{c} \int_\phi \int_\theta I_\nu\, d\Omega_r = \frac{1}{c} \int_\phi \int_\theta \frac{\partial}{\partial \nu} |d\mathbf{q}^{(2)}| \qquad (1.61)$$

in which c is the velocity of light. From Eqs. (1.64) and (1.65) we have

$$\frac{\partial e^a}{\partial \nu} = c\mu_\nu e_\nu \qquad (1.62)$$

The LVREA for the energy of frequency ν is clearly

$$de_\nu^a = \frac{\partial e^a}{\partial \nu}\, d\nu = c\mu_\nu (e_\nu\, d\nu) \qquad (1.63)$$

$e_\nu \, d\nu$ being the density of energy of frequency ν, which may be considered as moles of photons of frequency ν (einstein) per unit volume, in analogy with mass density. An inspection of Eq. (I.63), taking into account the units of each factor, makes it possible to establish an analogy between the LVREA and a reaction rate equation which is first order with respect to one reactant (in this case the photons). Also, in those cases in which μ_ν is not a function of time, $c\mu_\nu$ plays the role of a pseudo first-order reaction rate constant.

For dilute solutions:

$$\mu_\nu = \sum_i \mu_{i,\nu} = \sum_i \alpha_{i,\nu} C_i \qquad (I.64)$$

where the summation index i denotes each species that absorbs energy of frequency ν. From Eqs. (I.63) and (I.64) we have the relation

$$de_\nu^a = \sum_i (c\alpha_{i,\nu}) C_i (e_\nu \, d\nu) \qquad (I.65)$$

Maintaining the above parallelism with chemical reactions, Eq. (I.65) is a combination of a second-order kinetic constant corresponding to a special reacting system in which a "reactant" (photons of energy of frequency ν) "reacts" simultaneously with other reactants of concentration C_i, $c\alpha_{i,\nu}$ being a second-order reaction rate constant ($\alpha_{i,\nu}$ is a function of P, T, ν and the component i which takes part in the "reaction"). The differential character of the "molar" concentration of energy of frequency ν, $(e_\nu \, d\nu)$, is due to the consideration of ν as a continuous variable, which is implicit in the definition of I_ν and all related distribution functions. If one desires to write the "true" reaction rate equation for a component j which both absorbs energy and undergoes a chemical reaction, a quantum yield must be introduced as follows:

$$R_j = \int_{\nu_{1,j}}^{\nu_{2,j}} \Phi_{j,\nu} \, de_\nu^a \qquad (I.66)$$

where $\Phi_{j,\nu}$ is the quantum yield for the chemical reaction corresponding to component j and frequency ν, and $\Delta\nu = \nu_{2,j} - \nu_{1,j}$ is the frequency range of energy absorption of component j which causes the initiation of the photochemical reaction.

I.6. Applications

To understand the fundamentals of photoreactor analysis it seemed appropriate to follow a progressive study from very simple cases to the most complex ones. The strategy was as follows:

1. To study empty reactors in order to obtain reliable information

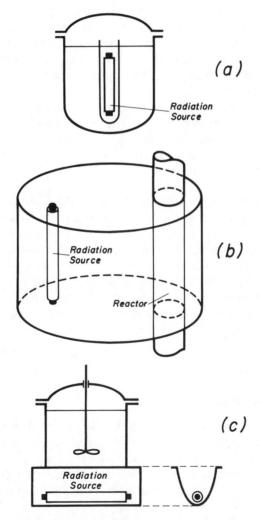

Figure I.11   Reactor geometries. (a) Annular reactor. (b) Tubular reactor inside an elliptical reflector. (c) Cylindrical reactor with parabolic reflector.

regarding the radiation field. We called them "diactinic reactors".

2. To study reactors where almost all the absorption of radiation occurred in a thin layer close to the surface of entrance of the U.V. light. We called them "black body reactors". It means that the layer of reactants through which the radiation enters inside the reacting volume is almost opaque. In this case the attenuation process can be treated in a very simple manner.

3. To study reactors in which the radiation energy can reach all the reacting volume but with very simple reactions.

4. To study reactors of the type indicated in 3 but with complex chain-type kinetics. Extend the analysis to "consecutive" reactions and the identification of parameters for the optimal design of photoreactors.

5. To study photoreactors for heterogeneous systems, i.e., when more than one phase is present. This means to add the complexity of gas–liquid interfaces in the modelling of the absorption-attenuation-reaction phenomena. The work could be later extended to solid–fluid interfaces.

We think that a significant part of most of the problems regarding reactor analysis up to point 4 has been solved. However, some practical problems related to design may need further work. As far as point 5 is concerned the subject is now under development. An increasing emphasis is certainly needed. In our work we have paid attention to the three types of reactors (Fig. I.11): the annular reactor, the tubular reactor inside a cylindrical reflector of elliptical cross section, and the cylindrical reactor irradiated from the bottom by means of a parabolic reflector.

II. DIACTINIC REACTORS

II.1. Annular Photoreactor

Annular photoreactors are an excellent approach to what is, perhaps, the most practical type of photochemical reactor to be used with commercial purposes. The utilization of energy can be the maximum expected and, moreover, they accurately represent the common case of a reaction vessel with a tubular lamp placed at its axis by means of an immersion well.

Figure II.1 shows the main features of the reacting system. The lamp is a cylinder surrounded by an annular photoreactor. The

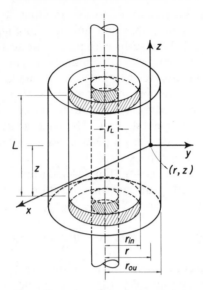

Figure II.1   Annular reactor geometry.

radiation field of interest is limited at the top and bottom positions by opaque zones. The most important parameters and nomenclature of the problem are also shown in the figure.

Model Equations

The model is based upon the assumptions already described for the ESVE Model about the emission and the lamp. To them we must add the following about the reactor [57]:

(i) The reactor is bounded by two concentric cylindrical mathematical surfaces without thickness, that is, no reflection, refraction, or absorption occurs at any of the walls of the reactor.

(ii) The reactor is filled with a transparent fluid, that is, no absorption occurs inside the reaction vessel.

(iii) The opaque zones at the top and bottom parts of the reactor do not reflect or emit radiation.

It is necessary to choose a convenient experimental variable to evaluate the model. This is particularly true in this case, where the important photochemical property (LVREA) is not amenable to local and direct experimental measurement. An appropriate choice of a property directly related to e^a will give, by comparison of the predicted values with the actual ones, a good idea about the model per se and about the validity of its assumptions. For this purpose, the total energy per unit area and unit time impinging on a plane surface was chosen. This surface is placed in such a way that its normal intersects the lamp axis perpendicularly (Fig. I.9). After this choice an experimental verification can be made by means of a properly designed thermopile or photocell so that the mean values of the measured property could represent point values within the accuracy of the experimental error. The solid angle of reception of the sensing unit must also be properly chosen so that every possible emitting direction from the source inside the radiation field boundaries is sought by the sensor. For monochromatic radiation, at a point in space placed in an arbitrary position given by (z, r), this property can be written as:

$$q(z, r) = \mathbf{q}(z, r) \cdot \mathbf{n} = |\mathbf{q}(z, r)| \cos \theta_r \qquad (II.1)$$

θ_r being the projecting angle on the normal to the sensor unit. A second experimental difficulty to be overcome is related to the calibration of the sensing device in absolute units. For this reason, it is convenient to define a relative value of energy (dimensionless). The reference point is chosen where q takes the maximum value and may vary according to the reactor configuration. Thus, the model will be checked by comparing predicted and actual values for the variable:

$$Q = \frac{q}{q_{max}} \qquad (II.2)$$

The total energy per unit area and unit time impinging on the differential area at the point (z, r) from all directions in space and from the whole volume of the lamp can be obtained from:

$$q = \kappa \int_{\phi} \cos \phi \, d\phi \int_{\theta(\phi)} \sin^2 \theta \, d\theta \int_{\rho(\theta,\phi)} d\rho \qquad (II.3)$$

The problem that will always be present in this work is centered on the complexity of the limits for the integrals of Eq. (II.3). To obtain them one starts from the point of reception (z, r) and investigates all the arriving rays with direction (θ, ϕ). One must find those rays which intersect the boundaries of the surface of the cylinder that represents the lamp.

To obtain the *limits for the variable ρ* it is necessary to consider the equation of the boundary surface of the radiation source. In the spherical coordinate system depicted in Fig. I.9, one obtains

$$\rho^2 \sin^2 \theta - 2\rho(\sin \theta \cos \phi)r + (r^2 - r_L^2) = 0 \qquad \text{(II.4)}$$

The two solutions of this quadratic equation are precisely the limits of ρ, i.e., they are the intersections of the ρ coordinate with the front and rear parts of the lamp at any value of θ, ϕ:

$$\rho_{1,2} = \frac{r \cos \phi \mp [r^2 \cos^2 \phi - r^2 + r_L^2]^{1/2}}{\sin \theta} \qquad \text{(II.5)}$$

Integrating for ρ at constant θ and ϕ:

$$q = 2\kappa \int_\phi \int_\theta [r^2(\cos^2 \phi - 1) + r_L^2]^{1/2} \sin \theta \cos \phi \, d\theta \, d\phi \qquad \text{(II.6)}$$

Limits for the variable θ are obtained by considering that the limiting rays coming from the lamp and reaching the generic point (z, r) must satisfy two restrictive conditions. The first one states that if the ray limits the value of θ, its equation (a straight line in space) must have a common solution with the equation of the circumference that defines the opaque zone at the upper and lower parts of the lamp. However, for any plane at constant ϕ, there are two values of θ that satisfy the first restriction for both upper and lower boundaries. Thus, the second restrictive condition arises establishing that this ambiguity is eliminated by choosing the angle corresponding to the intersection of the ray with that portion of the circumference which, limited by the two generatrix lines corresponding to the limiting angles of ϕ, is closer to the generic point (z, r) (Fig. II.2). From Fig. II.1, for the upper and lower boundaries:

$$(L_L - z) = \rho_1 \cos \theta_1 \qquad \text{(II.7)}$$

$$-z = \rho_1 \cos \theta_2 \qquad \text{(II.8)}$$

and using Eq. (II.5), the integration limits as a function of ϕ result:

$$\theta_1(\phi) = \tan^{-1} \left[\frac{r \cos \phi - [r^2(\cos^2 \phi - 1) + r_L^2]^{1/2}}{(L_L - z)} \right] \qquad \text{(II.9)}$$

$$\theta_2(\phi) = \tan^{-1} \left[\frac{r \cos \phi - [r^2(\cos^2 \phi - 1) + r_L^2]^{1/2}}{(-z)} \right] \qquad \text{(II.10)}$$

With these limits, Eq. (II.6) can be integrated at constant ϕ to obtain:

$$q = 2\kappa \int_{\phi} [r^2(\cos^2 \phi - 1) + r_L^2]^{1/2} [\cos \theta_1(\phi) - \cos \theta_2(\phi)] \cos \phi \, d\phi$$

(II.11)

Limits for the variable ϕ can be evaluated employing the following methodology. The limiting rays in the ϕ-direction for any value of angle θ must be tangent to the lamp boundary at points located on two generatrix lines of the cylinder (Fig. II.2). These values can be obtained by imposing a restriction on the values of ρ_1 and ρ_2. The restriction is given by the fact that, at the limiting points, both intersections of the ρ-coordinate with the lamp boundary must coincide, i.e., in a projection of the lamp on the $x-y$ plane, the limiting rays are tangent to its directrix circumference. This means that

$$\rho_1 = \rho_2$$

(II.12)

With Eq. (II.5), the following expression is obtained:

$$r^2 \cos^2 \phi = r^2 - r_L^2$$

(II.13)

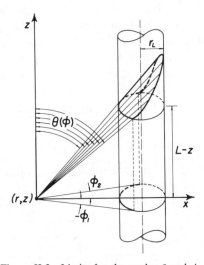

Figure II.2   Limits for the angles θ and ϕ.

and since ϕ can only take values in the closed interval $[-\pi/2, \pi/2]$ we have

$$-\phi_1 = \phi_2 = \cos^{-1}\left[\frac{(r^2 - r_L^2)^{1/2}}{r}\right] \qquad \text{(II.14)}$$

The final expression results:

$$q = 2\kappa \int_{\phi_1}^{\phi_2} [r^2 \cos^2 \phi - r^2 + r_L^2]^{1/2}[\cos \theta_1(\phi) - \cos \theta_2(\phi)] \cos \phi \, d\phi$$

$$\text{(II.15)}$$

We have found, however, an assumption that further simplifies the integration of Eq. (II.15). The approximation is made in order to eliminate the relationship given by Eqs. (II.9) and (II.10) so as to allow a separate integration of variables θ and ϕ. This can be accomplished if it is assumed that the sector of the surface generated by vector ρ, when it intersects the limiting circumference of the opaque zone of the lamp (Fig. II.2), can be substituted by one generated by a vector moving at constant θ (Fig. II.3). This simplification slightly changes the total emitting volume of the lamp, giving a better approach to the exact formulation when the values of θ get closer to $\pi/2$. Also, the larger the lamp, the smaller the error because the differences between the exact and the approximate

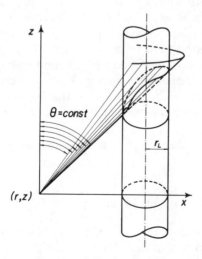

Figure II.3 Simplification of the ESVE Model.

formulation will have a negligible effect on the emission volume of the source. With this change in the limits of θ, the results will be:

$$\cos \theta_1 = \frac{|L_L - z|}{[(L_L - z)^2 + (r - r_L)^2]^{1/2}}$$

$$\cos \theta_2 = \frac{-|z|}{[z^2 + (r - r_L)^2]^{1/2}}$$

(II.16)

and since:

$$\int_{\phi_1}^{\phi_2} [r^2 \cos^2 \phi - r^2 + r_L^2]^{1/2} \cos \phi \, d\phi = \frac{\pi r_L^2}{2r}$$

the final approximate solution is:

$$q = \frac{\kappa \pi r_L^2}{r} \left[\frac{|L_L - z|}{[(L_L - z)^2 + (r - r_L)^2]^{1/2}} + \frac{|z|}{[z^2 + (r - r_L)^2]^{1/2}} \right]$$

(II.17)

The validity of the approximate solution was checked by comparing it with predictions made with Eq. (II.15). The results show that the errors for lamps of practical size are almost negligible.

Results and Conclusions
Radiation profiles were computed for conditions in which similar experimental information was available. The most significant set of data was found in a work related to the scale-up of an annular photoreactor [62] where radiation profiles for different propagating media were reported as a result of direct experimental measurement. The reactor was irradiated with a cylindrical source. The only difference was that the emission from the lamp was actually coming out of a phosphorescent material covering the inner face of the lamp wall. This phosphorescence was activated by the emission from the lamp volume.

In order to plot values of Q, two dimensionless position variables were defined as:

$$\gamma = \frac{r}{r_{in} + 0.14 \, \text{cm}} \qquad \zeta = \frac{z}{(L_L/2)}$$

(II.18)

where r_{in} is the outer radius of the inner wall of the reactor, and 0.14 cm is the thickness of the window of the photocell. When $\gamma = 1$ and $\zeta = 1$, the maximum value of Q is obtained (Q_{max}).

In Fig. II.4, values of Q are plotted as a function of γ, with $\zeta = 1$. It is clear that the best agreement is obtained with the ESVE Model. The air-filled reactor is a fairly good approximation to the conditions under which the calculations for the model were performed since it minimizes all types of refraction effects. The same cannot be said about reflection effects. On the other hand, if the reactor is filled with water, the radiation profiles seem to be modified, indicating that refraction effects may be significant. If this is the case, the model should be improved to include all these forms of distortions.

Figure II.5 shows the variation of the radiation profiles along the z-direction. These results are very significant because they illustrate that the assumption of a z-independent energy distribution inside the reactor must be handled with great care. In fact, for the short distances used in this work, a variation of 43% was found between the top (or bottom) and the middepth portions of the reactor. This effect may be even magnified if the lamp is not long enough to eliminate emission end effects. (The variations reported here with this model are not caused by end effects associated with the emission of the lamp but by the intrinsic geometrical characteristics of the system.)

To analytically investigate the effect of the lamp dimensions we have used the radiation field description given by the simplified

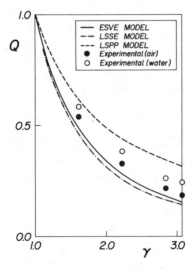

Figure II.4 Dimensionless radiant energy flux profiles. Comparison with experimental data.

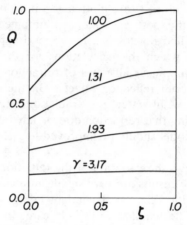

Figure II.5   Radiation energy density profiles along the ζ-coordinate.

version of the ESVE Model (Eq. II.17). Let us use as dimensionless variables [3]:

$$\gamma_L = \frac{r_L}{r_{L,max}} \qquad Q = \frac{q}{q_{max}} \qquad \text{(II.19)}$$

q_{max} being the value of q when $z = L_L/2$, $r = r_{min}$ and $r_L = r_{L,max}$, i.e.

$$q_{max} = \frac{E}{4\pi r_{min}} \left[(L_L/2)^2 + (r_{min} - r_{L,max})^2 \right]^{-1/2} \qquad \text{(II.20)}$$

Here:

$$r_{min} = r_{R,in} + e_w \qquad r_{L,max} = r_{R,in} - e_R$$
$$\Delta e = r_{min} - r_{L,max} = e_w + e_R$$
$$r_{R,in} = \text{outside radius of the reactor inner wall}$$
$$e_w = \text{thickness of the photocell window}$$
$$e_R = \text{thickness of the reactor inner wall}$$

Using the definition

$$k = \left[1 + \left(\frac{2\Delta e}{L} \right)^2 \right]^{1/2} \qquad \text{(II.21)}$$

we obtain:

$$Q = \frac{r_{min}k}{r}\left[1 + \frac{4r_{L,max}^2}{L_L^2}\left(\frac{r}{r_{L,max}} - \gamma_L\right)\right]^{-1/2} \tag{II.22}$$

making $dQ/d\gamma_L = 0$ we find that it can be satisfied with $r = r_L$; since $r_L < r$ always, the radiation field in $z = L_L/2$ is an increasing function of γ_L $(dQ/d\gamma_L > 0)$. The inflection points were obtained from the second derivative with respect to γ_L. It gives:

$$\gamma_L = \frac{r}{r_{L,max}} - \frac{L_L}{8^{1/2}r_{L,max}} \tag{II.23}$$

Since $0 \leqslant \gamma_L \leqslant 1$, we may get an inflection point at

$$8^{-1/2}L_L \leqslant r \leqslant r_{L,max} + 8^{-1/2}L_L \tag{II.24}$$

These predictions are illustrated by the computed results of Q as a function of γ_L for different values of r. In these calculations, the ESVE Model was used (Fig. II.6a). Since r is always greater than r_L, the radiation field does not show extremes and it is monotonously increasing. It can also be observed that whenever the condition given by Eq. (II.24) is fulfilled the inflection point is present (for instance for $r = 4.14$ cm, $r = 5$ cm and $r = 7$ cm). These results are

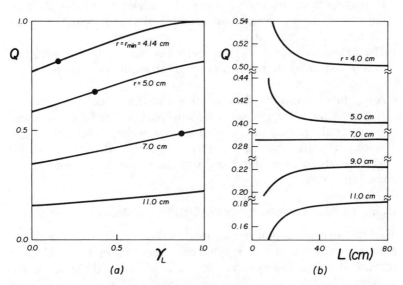

Figure II.6 Radiation energy density profiles. (a) Variation of lamp radius. (b) Variation of lamp length.

for $L_L = 10$ cm; $z = L_L/2$, $r_{R,in} = 4$ cm; $e_w = 0.14$ cm and $e_R = 0$. It is important to remark that the change in lamp radius was made by keeping the output power constant. Given the particular characteristics of the emission model the latter must be stressed.

To analyze the effect of the lamp length a similar procedure was followed. The study was made for $\zeta = 0$ and $\zeta = 1$. Results for the first case are shown in Fig. II.6b. These values were computed for $r_{R,in} = 4.0$ cm and $L_L = 10$ cm with the ESVE Model. As it can be intuitively predicted, the model shows no effects of L when the useful length of the lamp is large enough. For $r = 7$ cm, the insensitivity of Q to all values of L_L is also shown. Comparisons with the results of Eq. (II.17) gave excellent agreement. When the simplified form was used, taking the first derivative of Q with respect to L_L, the expression obtained predicts insensitivity of the radiation field when $r = 2r_{min} - r_L = 7$ cm. For values of $r < 7$ cm, the first derivative will be negative, and vice versa for $r > 7$ cm.

The main conclusions may be summarized in the following way:

– The computed results agree well with published experimental data for a similar lamp-reactor system but, at the same time, show that reflection and refraction effects may not always be negligible and should be included in the formulation of a more general theory. This is particularly true for liquid reacting systems.
– Departing from pre-existing ones, the model introduces all lamp dimensions in its formulation and thus makes it possible to use them as design parameters.
– The extense source model approach also allows the full introduction of reflection, refraction, and wall absorption effects produced by the lamp tube.
– An approximate analytical solution for the model is useful for analyzing the effects of lamp dimensions upon radiation profiles. They should not be disregarded as design parameters; there are values of r_L that produce sensitive changes to the radiation field which is also affected by changes in the source length when L_L is not long enough.

II.2. Tubular Photoreactor with Elliptical Reflector

This system consists of a reactor made with a cylindrical tube placed at one of the focuses of a cylindrical reflector of elliptical cross section. A cylindrical radiation source is placed at the other focus. This particular apparatus is often called an elliptical photoreactor (Fig. II.7).

The possibility of using cylindrical tubular reactors, generally easier to operate than the annular ones as well as the generally accepted concept of the existence of uniform irradiation from outside, has resulted in a rather extensive application of this reactor for laboratory work. The advantage of an almost uniform distribution of radiation inside the cross-sectional area of a photochemical reactor has also been mentioned from the selectivity point of view. This aspect, directly related to a commercial scale unit, may indicate that photoreactors with reflecting devices could also be important for some large-scale operations.

The model was obtained with the following assumptions (Figs. II.7 and II.8):

   (i) The ESVE Model is applicable.
  (ii) The reflector is a perfect elliptical cylinder.
 (iii) The lamp is located so that its center line passes through one of the focuses of the elliptical reflector (F_1) .
  (iv) Specular reflection occurs with an average reflection coefficient that is independent of wavelength and direction (these restrictions could be easily relaxed and are made only because reliable data are not available to be used with the model).
   (v) The medium of propagation inside the elliptical cylinder is transparent to radiation in the frequency range of interest.
  (vi) Reflected radiation comes only from the elliptical reflector, i.e., the top and bottom parts of the cylinder do not reflect radiation.
 (vii) The radiation impinging at any point inside the elliptical cylinder is made up exclusively of two parts: direct and reflected radiation.

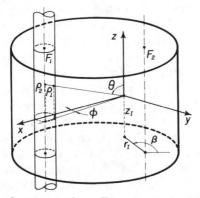

Figure II.7   Geometry and coordinate systems for direct radiation.

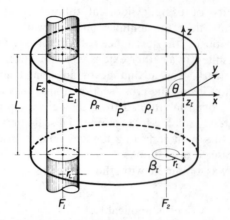

Figure II.8 Geometry and coordinate systems for indirect radiation.

(viii) The reflected radiation impinging at a point is mainly produced by the first reflection in the elliptical mirror, i.e.,
successive reflections are neglected after the first one.

The last assumption needs further analysis. Excluding direct
radiation, a ray could arrive at a point of reception with a given
direction by a single reflection process or as a result of successive
contacts with the elliptical mirror. In the second case, the ray will
describe a broken line which, for points located close to focus F_2,
touches the surface of the reflector at least twice. The question is
whether it is always possible for a ray coming out of the lamp to
undergo so many reflections and still arrive at the point of interest.
The question can be answered by following the ray path in the
reverse way, i.e., starting from the point of reception. A multiply
reflected ray will be effective at a given point in space if and only if,
travelling the reverse trajectory previously described, and in spite of
its progressive vertical displacement, it still intersects the surface of
the lamp boundary that is seen from the point of incidence. Through
the same reasoning, a multiply reflected ray would have no real
existence if this intersection fell outside the length of the lamp. We
may consider that as a first approximation, the second case will
occur, i.e., no energy will be supplied at a point after several
reflections. This will not be true for rays coming out of the lamp
with directions close to a parallel to the normal plane of the
elliptical reflector, since in this case the vertical displacement of the
ray will be small. Further ahead we will see that this assumption

does not always work and that, in some cases at least, the second reflection must be accounted for.

The problem at hand is to find the mathematical relationship between the incident radiant energy flux and the reactor, lamp and reflector dimensions, the optical properties of the reflector and the reactor wall and the physical properties of the radiation coming out of the source [27].

Model Equations

From the formulation of the ESVE Model (see Fig. II.7) at any point I (r_I, β_I, z_I), the *direct radiation* coming from the elementary source volume dV_e with direction (θ, ϕ) is:

$$|\mathbf{q}(r_I, \beta_I, z_I)|_D = \kappa \int_{\phi^D} d\phi \int_{\theta^D(\phi)} \sin \theta \; d\theta \int_{\rho^D(\theta, \phi)} d\rho \qquad (\text{II.25})$$

where point I on the reactor volume is referred to a cylindrical coordinate system located at the center line of the cylindrical reactor, with the origin at the reflector bottom.

Although the discharged arc has an almost uniform intensity along its center line, this is not true in the neighborhood of the lamp anode and cathode. When the reactor is only illuminated by the central middle part of the lamp, an almost constant emission, independent of z, is attained and the end effects can be safely neglected. For this case, the limits of integration for variable ρ may be obtained in a similar way as for the annular reactor. The results, which can be derived from Figs. II.7 and II.9, are:

$$\rho_{1,2}(\theta, \phi) = \frac{\Delta \cos \phi \mp [\Delta^2 \cos^2 \phi - (\Delta^2 - r_L^2)]^{1/2}}{\sin \theta} \qquad (\text{II.26})$$

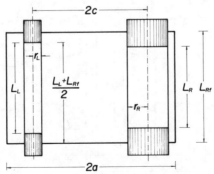

Figure II.9 Variables involved in the determination of the limits for direct radiation.

where:

$$\Delta = [(2c + r_I \cos \beta_I)^2 + r_I^2 \sin^2 \beta_I]^{1/2} \qquad (II.27)$$

On the other hand, when the reactor is irradiated by the whole lamp length, for regions close to the lamp electrodes, the uniform emission with respect to z must be an assumption. In addition, for both ends of the source the upper limit ρ_2 as given by Eq. (II.26) has lost its validity (Fig. II.10). If Eq. (II.26) is used to determine ρ_2, it will always occur that:

$$|\rho_2 \cos \theta| > |1/2(L_{Rf} \pm L_L) - z_I|, \qquad if \ \theta \le \frac{\pi}{2} \qquad (II.28)$$

and the radiation coming from non-existing source volume elements would be considered. To avoid this error, the upper limit must be changed to:

$$\rho_2(\theta) = \frac{1/2(L_{Rf} \pm L_L) - z_I}{\cos \theta}, \qquad if \ \theta \le \frac{\pi}{2} \qquad (II.29)$$

For variable θ and ϕ the limits of integration remain unchanged. The details were described for the annular reactor and the results are:

$$\theta_{1,2}(\phi) = \tan^{-1} \left[\frac{\Delta \cos \phi - (r_L^2 - \Delta^2 \sin^2 \phi)^{1/2}}{1/2(L_{Rf} \pm L_L) - z_I} \right] \qquad (II.30)$$

$$-\phi_1 = \phi_2 = \cos^{-1} \left[\frac{(\Delta^2 - r_L^2)^{1/2}}{\Delta} \right] \qquad (II.31)$$

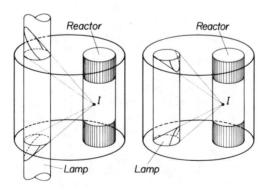

Figure II.10   Schematic representation of upper limit modifications.

For direct radiation the final expression is:

$$|\mathbf{q}(r_I, \beta_I, z_I)|_D = 2\kappa \int_\phi (\Delta^2 \cos^2 \phi - \Delta^2 + r_L^2)^{1/2} [\theta_2(\phi) - \theta_1(\phi)] \, d\phi$$

(II.32)

To simplify this first introduction to the analysis of *reflected radiation*, we will consider the case of $L_L = L_R$ in Fig. II.9. The methodology for reflected radiation is to follow a ray starting from the point of reception in order to identify the point of reflection, and from this (applying the laws of optical physics), to search for the direction of emission. To understand the analysis it is necessary to locate several key points (Fig. II.11). First the point of incidence of the reflected ray (I); secondly, the point of reflection (P); and finally, the two points of intersection of the direction of emission with the lamp cylinder (E_1 and E_2). Any ray coming from the lamp will be indicated by an incidence position vector at the point of reflection (ρ_E), and a reflection position vector at the point of reception (ρ_I). The former has its origin at P and the latter at I. The intersections of ρ_E with the cylinder of the lamp define two vectors ρ_{E_1} and ρ_{E_2}.

At this point we want to compute the value of the modulus of the radiation flux density vector at point I after one reflection. This is given by:

$$|\mathbf{q}(r_I, \beta_I, z_I)|_{In} = \Gamma_{Rf} \kappa \int_{\phi^{In}} d\phi \int_{\theta^{In}(\phi)} \sin \theta \, d\theta \int_{\rho^{In}(\theta, \phi)} d\rho$$

(II.33)

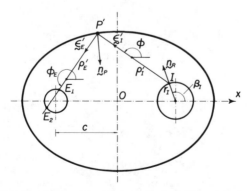

Figure II.11 Variables involved in the theoretical analysis of the indirect radiation.

where the attenuation of the radiant beams on their way to the reactor is taken into account using the reflection coefficient of the elliptical mirror Γ_{Rf}.

The reflection coefficient of the specular surface changes according to the type of reflecting material, the radiation wavelength and the angle of incidence. Some expressions are known to predict Γ_{Rf} as a function of θ and ϕ for different types of reflectors. However, for the reflector commonly used in the elliptical photoreactor (anodic-coated aluminum) we were not able to find a suitable equation to reproduce the experimental information supplied by the manufacturer (normal incidence in the U.V. region). Due to this lack of reliability, it was decided to adopt an average value of the reflection coefficient $\langle \Gamma_{Rf} \rangle$, given by the technical specifications of the Alzac aluminum reflector from ALCOA [2].

Once again the problem is the determination of the *limits of integration*. The unit vector ε_I representing the direction of a ray that impinges at any point I with direction (θ, ϕ) is given by:

$$\varepsilon_I = \sin \theta \cos \phi \mathbf{i} + \sin \theta \sin \phi \mathbf{j} + \cos \theta \, \mathbf{k} \qquad (\text{II.34})$$

This ray has previously rebounded at a point P of the elliptical mirror, whose coordinates referred to point I are:

$$x_P = \rho_I \sin \theta \cos \phi$$

$$y_P = \rho_I \sin \theta \sin \phi \qquad (\text{II.35})$$

$$z_P = \rho_I \cos \theta$$

where

$$\rho_I = \frac{-(hb^2 \cos \phi + ka^2 \sin \phi) + ab[\sigma^2 - (h \sin \phi - k \cos \phi)^2]^{1/2}}{\sigma^2 \sin \theta}$$

$$(\text{II.36})$$

and

$$\sigma^2 = b^2 \cos^2 \phi + a^2 \sin^2 \phi$$

$$h = c + r_I \cos \beta_I \qquad (\text{II.37})$$

$$k = r_I \sin \beta_I$$

The expression for the unit normal vector to the reflector at point P, can be written as follows:

$$\mathbf{n}_P = \frac{\dfrac{x_P + h}{a^2}\mathbf{i} + \dfrac{y_P + k}{b^2}\mathbf{j}}{\left[\dfrac{(x_P + h)^2}{a^4} + \dfrac{(y_P + k)^2}{b^4}\right]^{1/2}} \tag{II.38}$$

The unit vector $\boldsymbol{\varepsilon}_E$ representing the direction of the ray before reflection, i.e. emerging from the lamp, can be obtained from the law of reflection:

$$\mathbf{n}_P \times \boldsymbol{\varepsilon}_E = \mathbf{n}_P \times \boldsymbol{\varepsilon}_I \tag{II.39}$$

Solving this vectorial equation, the following expression for $\boldsymbol{\varepsilon}_E$ is obtained:

$$\boldsymbol{\varepsilon}_E = \sin\theta \cos\phi_E\mathbf{i} + \sin\theta \sin\phi_E\mathbf{j} + \cos\theta\mathbf{k} \tag{II.40}$$

where:

$$\phi_E = \tan^{-1}\left[\frac{(m^2 - 1)\sin\phi + 2m\cos\phi}{2m\sin\phi - (m^2 - 1)\cos\phi}\right]$$

and

$$m = \frac{a^2(y_P + k)}{b^2(x_P + h)} \tag{II.41}$$

The two intersections of direction (θ, ϕ_E) with the bounding surface of the radiation source are precisely the limits of ρ. They are given by

$$\rho_{E_{1,2}} = \frac{D\cos\xi \mp (r_L^2 - D^2\sin^2\xi)^{1/2}}{\sin\theta} \tag{II.42}$$

where:

$$D = [(x_P + h + c)^2 + (y_P + k)^2]^{1/2} \tag{II.43}$$

$$\sin\xi = -\frac{1}{D}[(x_P + h + c)\sin\phi_E - (y_P + k)\cos\phi_E] \tag{II.44}$$

$$\cos \xi = -\frac{1}{D} [(x_P + h + c) \cos \phi_E + (y_P + k) \sin \phi_E] \qquad (\text{II}.45)$$

Integrating for ρ at constant θ and ϕ yields

$$|\mathbf{q}(r_I, \beta_I, z_I|_{\text{In}} = \langle \Gamma_{Rf} \rangle \kappa \int_{\phi^{\text{In}}} d\phi \int_{\theta^{\text{In}(\phi)}} (\rho_{E,2} - \rho_{E,1}) \sin \theta \, d\theta$$

$$(\text{II}.46)$$

The limiting values of the θ-coordinate can be defined by means of:

$$(\rho_I + \rho_{E_1}) \cos \theta_1 = 1/2(L_{Rf} + L_L) - z_I \qquad (\text{II}.47)$$

$$(\rho_I + \rho_{E_1}) \cos \theta_2 = 1/2(L_{Rf} - L_L) - z_I \qquad (\text{II}.48)$$

Using Eqs. (II.36) and (II.42), the integration limits are found to be

$$\theta_{1,2}(\phi) = \tan^{-1} \left[\frac{\rho_I' + D \cos \xi - (r_L^2 - D^2 \sin^2 \xi)^{1/2}}{1/2(L_{Rf} \pm L_L) - z_I} \right] \qquad (\text{II}.49)$$

where:

$$\rho_I'(\phi) = \rho_I(\theta, \phi) \sin \theta \qquad (\text{II}.50)$$

ρ_I' depends only on ϕ. The same can be said for D, ϕ_E and ξ. With these limits, Eq. (II.46) can be integrated at constant ϕ to obtain:

$$|\mathbf{q}|_{\text{In}} = 2\langle \Gamma_{Rf} \rangle \kappa \int_{\phi} (r_L^2 - D^2 \sin^2 \xi)^{1/2} [\theta_2(\phi) - \theta_1(\phi)] \, d\phi \qquad (\text{II}.51)$$

The limiting rays in the ϕ-direction for any value of θ must be tangent to the lamp boundaries, that is:

$$\rho_{E_1} = \rho_{E_2} \qquad (\text{II}.52)$$

With Eq. (II.42) the following expression is obtained:

$$D^2 \sin^2 \xi = r_L^2 \qquad (\text{II}.53)$$

Solving this implicit equation numerically provides the limiting values of ϕ.

At any point I the *total incident energy* is given by:

$$|\mathbf{q}(r_I, \beta_I, z_I)| = |\mathbf{q}|_D + |\mathbf{q}|_{In} \qquad \text{(II.54)}$$

If one desires to obtain point values for any position in space, the LVREA is not amenable to direct experimental verification. (For certain positions, some specially designed actinometric experiments could be used.) Thus, a different quantity must be chosen to check the model. Jacob and Dranoff [61] reported an experimental study of the elliptical reflector system. The radiation impinging on a photocell with its surface always parallel to one of the generatrix lines of the elliptical cylinder was measured. Data were taken at each point, rotating the sensor about its axis and dividing the total plane angle (2π) in four intervals. The results were reported as the average of these four measurements. It was assumed that these results should be related in some way to the radiation energy density. Also, since the paper reports values relative to the point of maximum energy (middepth height at the focal axis where the reactor should be located), it is believed that such values may approximately represent the shape of the radiation profiles.

Cerdá et al. [27] decided that calculations should be made according to:

$$|\mathbf{q}| = \int_\nu \int_\phi \int_\theta \int_\rho |d\mathbf{q}_{\nu,\phi,\theta,\rho}| \qquad \text{(II.55)}$$

The results will be reported relative to the point of maximum energy which is given by

$$Q = \frac{|\mathbf{q}|}{|\mathbf{q}|_{max}} \qquad \text{(II.56)}$$

The integral of Eq. (II.55) indicates that all the energy coming from all directions, impinging per unit time on an area that is always normal to each direction, has been sumed. Thus, the integration is made with the absolute value of the radiation flux density vector. This integral multiplied by the reciprocal of the velocity of light gives the energy density. Hence, Eq. (II.56) gives relative energy densities. This property is important because if the integrand of Eq. (II.55) were multiplied by the attenuation coefficient, the result of the integral would be the LVREA [56]. A correct experimental verification should be made using a thermopile or a photocell with a view angle of 4π and a properly designed size and shape so that the mean value of the measured property would represent point values within the accuracy of the experimental error. Since this type of instrument is not available, the computed results were checked against Jacob and Dranoff's profiles [61].

Results and Conclusions

Figures II.12 and II.13 show computed values of $(|q|/\kappa)$ for both the ESVE Model and the LSSE Model, for $\beta_I = \pi$ and $\beta_I = 0$ (corresponding to the major axis of the ellipse), and for $\beta_I = \pi/2$ and $\beta_I = 3\pi/2$ (corresponding to a line perpendicular to the major axis). These results were calculated at $z_I = L_{Rf}/2$ and compared with experimental data from Jacob and Dranoff [61] where information about the reflector-lamp set up may be found. To make this comparison straightforward, the maximum calculated value of $(|q|/\kappa)$ from the ESVE Model was made to arbitrarily coincide with that corresponding to the experimental data. This had to be done, since only relative values referred to the maximum one were reported in the above-quoted paper.

As it should be expected, the agreement is only fairly good due to the fact that experimental measurements do not correspond exactly to the computed property. The improvement in the prediction of the energy density profiles obtained with the ESVE Model is evident especially if they are compared with those obtained with the LSSE Model. Values predicted with the ESVE Model are closer to the experimental ones. The use of this model also avoids the existence of singularities in the energy density profiles (or related properties) at the foci of the ellipse, as happens with those profiles calculated from line models.

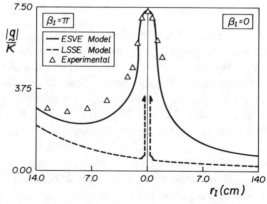

Figure II.12   Radiation energy density profiles along the radial coordinate.

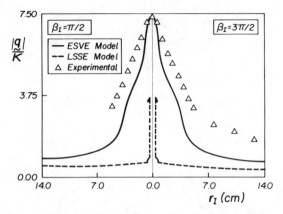

Figure II.13   Radiation energy density profiles along the radial coordinate.

We should also remark that radiation profiles were computed for β_I angles other than those of Figs. II.12 and II.13. The results are shown in Fig. II.14 as energy isodensity curves, and from them we clearly see the validity range and weakness of the assumption of uniform irradiation to the reactor from outside, as well as diffuse irradiation at the reactor zone. The values of energy density corresponding to the contour lines are relative values referred to the maximum, and the distance from the focus is measured in centimeters.

The relative importance of direct $(|\mathbf{q}|_D/\kappa)$ and reflected $(|\mathbf{q}|_{In}/\kappa)$ radiation for $z_I = L_{Rf}/2$ can be evaluated by inspecting Fig. II.15. It can be observed that reflected radiation causes a symmetric ($\beta =$

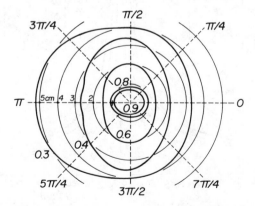

Figure II.14   Radiation energy isodensity curves.

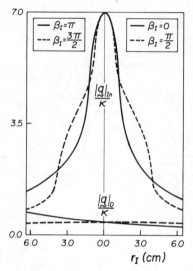

Figure II.15   Direct and indirect radiation energy density profiles along the radial coordinate.

$\pi/2$ and $3\pi/2$), or non-symmetric ($\beta = 0$ and π) bell-shaped contribution to the energy density profile. Thus, it is possible to state that the asymmetry of the energy density profile is not due only to the direct radiation contribution. Direct radiation will progressively lose its significance when the distance between focuses increases regardless of the eccentricity of the ellipse. From Fig. II.16

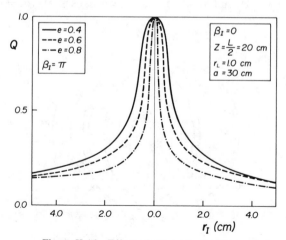

Figure II.16   Effect of ellipse eccentricity.

it can be concluded that for a set of values of characteristic
parameters of a lamp and reflector, the relative energy density
profiles (Q) become narrower as the eccentricity is increased.

Variations of the computed relative energy density values as a
function of z are shown and compared with experimental data [61]
in Fig. II.17. It can be seen that for a short distance close to
middepth, the maximum is almost independent of z. The situation
could be improved if the top and bottom plates of the system also
reflected radiation, but this was not the case for these calculations.

Some conclusions may be obtained from this work:

- With the proposed model, it was possible to obtain good agreement of predicted radiation energy density values with published
  experimental data for a similar lamp-reactor arrangement. Some
  disagreement cannot be avoided because the experimental data do
  not correspond exactly to the radiation energy density, the property computed in this work.
- The formulation of the model introduces all lamp and reflector
  dimensions; thus, the properties of the radiation field can be
  computed without requiring any empirical correction.
- The model avoids the prediction of infinite energy density at the
  center line of the reactor.
- The computed results are significant because they alert those who
  use this reactor to two hypotheses included in previous models.
  They are:
  (i) The assumption of uniform irradiation for all β-directions
      (different approaching directions from outside) is not valid,
      even for a perfect elliptical cylinder. This assumption would

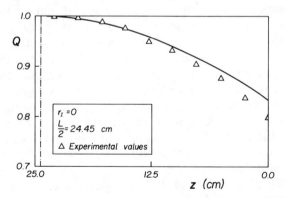

Figure II.17   Radiation energy density profiles. Comparison with experimental data.

be more applicable for a small reactor located in the region very close to the focal line.

(ii) The assumption of constant energy density along the reactor axis is not valid unless a reactor with a small useful length, located at the middepth height position of the system, is used.

– The ideal profiles predicted by line source models give only a poor approximation to the experimental profiles, *even in the case of reporting relative values.*

Limitations of the Linear Models

From previous theoretical developments it is clear that the linear models are much simpler to use than the extense ones. The computing time is considerably reduced and consequently they should always be the starting equations for photoreactor design.

The work of Cerdá et al. [29] and Romero et al. [84] showed that under a wide range of operating variables the LSSE Model provides good results for the annular type of reactors where only direct radiation is involved. In this case, errors are never larger than 15% as can be seen in Section IV.1. On the other hand, the experimental results of Alfano et al. [6, 7] regarding the validity of the linear models when a reflector is used, clearly indicate that simplified models cannot predict the radiation field with an acceptable degree of approximation (see Section II.3). In a recent study [30] the problem was explored from the theoretical viewpoint. The main motivation was the large discrepancies observed (using the LSSE Model and the ESVE Model) when predicted values were compared with experimental results in an elliptical photoreactor for the photo-decomposition of the uranyl oxalate complex (Section IV.2) and the monochlorination of ethane (Section V.2). In short, what was shown in their work is that when a tubular lamp is placed parallel to the generatrix straight line of the cylindrical reflector and is modelled as a line, the source does not "recognize" the curvature of the reflecting surface. Therefore, the concentrating effect of the curved mirror, be it elliptical or parabolic, cannot be predicted. Depending upon the significance of direct radiation on the total radiation flux density, the model can be good (as in the case of the annular reactor), poor (as in the case of the parabolic reflector where errors are of the order of 100%) or inadmissible (as in the case of some elliptical photoreactors where differences are greater than two orders of magnitude).

The work of Clariá et al. [30] was devoted to the case of the elliptical reflector. Without any loss of generality, two assumptions were made: (i) monochromatic radiation and (ii) diactinic medium ($\mu = 0$). The study required the expressions for $|\mathbf{q}|$ given by the LSSE Model and the ESVE Model.

i) LSSE Model:
From Eq. (I.28) with $\mu_\nu = 0$, direct radiation is expressed as

$$|\mathbf{q}|_D = \kappa_L \int_0^{L_L} \frac{dz}{\rho^2} \tag{II.57}$$

with

$$\kappa_L = \frac{E_0}{4\pi L_L} \tag{II.58}$$

Changing coordinates from a cylindrical system located at the radiation source to a spherical one placed at the point of reception, Eq. (II.57) yields the following result

$$|\mathbf{q}|_D = \kappa_L \int_\theta \frac{d\theta}{\Delta} \tag{II.59}$$

where Δ is given by Eq. II.27. From Eq. (II.30), with $\phi = 0$ and $r_L = 0$ one obtains

$$\theta_{1,2} = \tan^{-1}\left[\frac{\Delta}{\dfrac{L_R \pm L_L}{2} - z_I}\right] \tag{II.60}$$

Hence, integrating Eq. (II.59):

$$|\mathbf{q}|_D = \kappa_L \frac{(\theta_2 - \theta_1)}{\Delta} \tag{II.61}$$

where one can notice that the model does not show dependence upon the variable ϕ.

With respect to indirect radiation, one must take the following points into account: all reflected rays must lie on planes that also contain the reactor axis; and a ray which, after reflection, reaches the reactor center line may be projected on a plane that goes through the point of emission and is perpendicular to the lamp axis. The length of this projection is equal to $2a$ (Fig. II.18). Each one of these planes allows us to define a circumference constructed with the locus of all the end points of "rectified" rays (Fig. II.19). This can be done in this way because the reflector is a cylinder and the

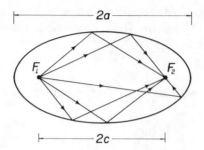

Figure II.18   Trajectory projections of reflected rays.

lamp is assumed to be a line parallel to the generatrix line of the mirror. This circumference may be repeated for each point of emission of the lamp, giving rise to a cylinder of circular cross section of radius equal to $2a$. This cylinder may be thought of as a virtual surface of emission that acts as a substitute for the linear lamp. Taking a planar cross section going through the reactor focal axis of the elliptical cylinder, we finally obtain a situation like the one depicted in Fig. II.20. Points such as those designated by letters A, B, C, D, etc. in Fig. II.19, when transformed into straight lines as was done for one cross section in Fig. II.20, represent the way in which the linear lamp would be seen by an observer located at the reactor center line when the lamp is viewed from different ϕ-directions. The projection of the actual distance travelled by any ray with direction θ on the reactor center line is $2a \cot \theta$.

The key point, as in any emission model, is to know the limits of the values of θ, i.e. out of all those θ-directions arriving at the

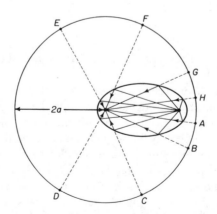

Figure II.19   Locus of end point of rectified rays.

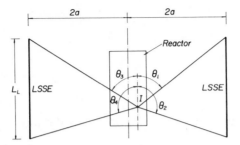

Figure II.20   Planar cross section through the reactor focal axis.

reactor axis, which are really coming out of the "line" source. They can be obtained by comparing the value of $2a \cot \theta$ with the source length. In this way it is straightforward to show that

$$\theta_{1,2} = \tan^{-1}\left[\frac{2a - r_I}{\dfrac{L_R \pm L_L}{2} - z_I}\right] \qquad (II.62)$$

and

$$\theta_{3,4} = \tan^{-1}\left[\frac{2a + r_I}{\dfrac{L_R \pm L_L}{2} - z_I}\right] \qquad (II.63)$$

With these limiting angles, the modulus of the energy flux density vector for indirect radiation, can be obtained as

$$|\mathbf{q}|_{In} = \kappa_L \Gamma_{Rf}\left[\frac{\theta_2 - \theta_1}{2a - r_I} + \frac{\theta_4 - \theta_3}{2a + r_I}\right] \qquad (II.64)$$

The question at this point is: under which conditions can equations (II.60) and (II.61), and (II.62) to (II.64) be obtained as limiting cases of those representing the ESVE Model?

ii) ESVE Model:
   Equations (II.27) and (II.30) to (II.32) give the value for $|\mathbf{q}|_D$. From Eq. (II.32) let us make a power series expansion on the small parameter $\varepsilon = r_L/\Delta$. For $\varepsilon = 0$ (first term of the expansion) we may think of $r_L = 0$ or a radiation source separated from the reactor by an infinite distance ($c \rightarrow \infty$). For both cases the lamp should act as a line. Defining:

$$\omega = \frac{\sin \phi}{\varepsilon} \qquad (II.65)$$

we get:

$$\cos \phi \, d\phi = \varepsilon \, d\omega \qquad (II.66)$$

and the integral given by Eq. (II.32) takes the form:

$$|\mathbf{q}|_D = \frac{2\kappa_L}{\pi \Delta} \int_1^{-1} d\omega \, \Delta\theta(\varepsilon) \frac{(1 - \omega^2)^{1/2}}{(1 - \varepsilon^2 \omega^2)^{1/2}} \qquad (II.67)$$

because $\kappa_L = \kappa \pi r_L^2$.

The Taylor series expansion about $\varepsilon = 0$, after integration gives:

$$|\mathbf{q}|_D = \frac{\kappa_L}{\Delta} \left[\Delta\theta(0) - \frac{8}{3\pi} C_1 \varepsilon + \frac{1}{8} (C_1 + 6C_2 + \Delta\theta(0)) \varepsilon^2 + \cdots \right] \qquad (II.68)$$

where:

$$\Delta\theta(0) = [\theta_2 - \theta_1]_{\varepsilon = 0}$$

$$C_1 = -\cos \theta_2(0) \sin \theta_2(0) + \cos \theta_1(0) \sin \theta_1(0) \qquad (II.69)$$

$$C_2 = \cos \theta_2(0) \sin^3 \theta_2(0) - \cos \theta_1(0) \sin^3 \theta_1(0)$$

As it should be expected:

$$|\mathbf{q}|_D = \frac{\kappa_L}{\Delta} [\theta_2 - \theta_1]_{\varepsilon = 0} \qquad (II.70)$$

is the LSSE Model for direct radiation. If the first three terms of the expansion given by Eq. (II.68) are computed and the results are compared with the ESVE Model we obtain Table II.1. The ratio r_L/c is used as a characteristic parameter. It is clear that in the case of direct radiation, for values of $r_L/c = 0.050$ (and $e = 0.66$), the differences between the LSSE Model and the ESVE Model are never larger than 5%. We should mention that for an annular reactor, the equivalent expression for comparison purposes would be, r_L/r_m with $r_m = 1/2(r_{in} + r_{ou})$. Practical cases would be restricted by

$$0.05 \leqslant r_L/r_m \leqslant 0.15 \qquad (II.71)$$

Table II.1   Direct Radiation: LSSE vs. ESVE Models

r_L	$|\mathbf{q}|_D \times 10^7 (\text{einstein cm}^{-2}\text{s}^{-1})$		
$\overline{c}$	LSSE model	ESVE model	Error (%)
0.018	0.031	0.031	0.00
0.022	0.047	0.047	0.00
0.030	0.083	0.082	1.22
0.033	0.102	0.100	2.00
0.037	0.127	0.123	3.25
0.040	0.143	0.138	3.62
0.050	0.215	0.205	4.88

This means that in some cases the LSSE Model can be used with a relatively small error for an annular reactor.

Considering the modulus of the energy flux density vector for indirect radiation, Clariá et al. [30] proposed to obtain values for which the curvature of the mirror can be neglected. The reason for this is that the line source models applied to the inside of cylindrical reflectors of curved cross section, "see" the mirror as a flat surface. With no dimensions in the direction where the reflector is curved ($\phi = 0$), the lamp model cannot recognize the concentrating effect produced by the elliptical or parabolic shape of the specular surface.

Let us compare, after reflection, the convergent angle to the incident point when we have: a) a curved surface ($\Delta\phi_c$), and (b) a plane surface ($\Delta\phi_P$); and let us do it for the points of maximum curvature (B and D) and minimum curvature (A and C) of the elliptical reflector (Fig. II.21). Since $\Delta\phi_P = \Delta\phi_c - \Delta\theta$, if we define:

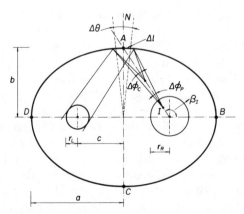

Figure II.21   Convergent angles considering plane and curved surfaces.

$$R_{max} = \frac{a^2}{b} \quad \text{(for } A \text{ and } C\text{)} \quad \text{(II.72)}$$

$$R_{min} = \frac{b^2}{a} \quad \text{(for } B \text{ and } D\text{)} \quad \text{(II.73)}$$

it is clear that for points such as A and C:

$$\Delta \ell = R_{max} \, \Delta \theta = \frac{a}{b} \, \Delta \phi_c (a - r_I)$$

$$\Delta \theta = \Delta \phi_c \frac{(a - r_I)}{a}$$

and

$$\Delta \phi_P = \Delta \phi_c \left[1 - \frac{a - r_I}{a} \right] \quad \text{(II.74)}$$

For points where the curvature is a maximum (B and D):

$$\Delta \phi_P = \Delta \phi_c \left[1 - \frac{a}{b^2} (a - c - r_I) \right] \quad \text{(II.75)}$$

It is now clear that for points such as A and C:

$$\Delta \phi_c \rightarrow \Delta \phi_P \quad \text{when } r_I \rightarrow a \quad \text{(II.76)}$$

and for points such as B and D:

$$\Delta \phi_c \rightarrow \Delta \phi_P \quad \text{when } r_I \rightarrow a - c \quad \text{(II.77)}$$

Conditions (II.76) and (II.77) indicate that the curvature could be neglected when the point under analysis is very close to the surface of reflection. This is an impossible situation because: (i) no reactor with $r_R \rightarrow a$ can be placed inside the elliptical cavity, and (ii) if the condition is fulfilled with point I very close to point A for example, it simultaneously contradicts the criteria for point C that also imposes the same requirement (opposite side of the mirror).

However, more precise conclusions can be reached if the analysis is pursued still further. The authors assumed that it would be possible to work with a reactor located very close to the mirror and that the reflector consists of only one half of the curved surface, i.e. points such as C would not be reflecting radiation. Although this is a very impractical scenario, it is very useful for reaching more conclusive quantitative results.

Let us choose a point where $\beta_I = 0$ and delete the left-hand side of the surface from A to B (Fig. II.21). Let us further set $r_I = a - c$ in the expression for the ESVE Model. Then, Eq. (II.51) becomes exactly equal to Eq. (II.32) with the inclusion of Γ_{Rf} as a coefficient, and the limits of integration given by (II.49) and (II.53) become equal to (II.30) and (II.31). This means that under these extreme conditions reflected radiation behaves like direct radiation but affected by the attenuation introduced by the reflection coefficient. Consequently, it is now possible to introduce the expansion already derived for direct radiation:

$$|\mathbf{q}|_{\text{In}} = \frac{\kappa_L \Gamma_{Rf}}{\Delta} \left[\Delta\theta(0) - \frac{8}{3\pi} C_1 \varepsilon + \frac{1}{8} (C_1 + 6C_2 + \Delta\theta(0)) \varepsilon^2 + \cdots \right]$$

$$(\text{II.78})$$

For points very close to the mirror, after reflection, we have

$$\Delta = 2c + r_I + 2(a - c - r_I) = 2a - r_I \qquad (\text{II.79})$$

Introducing Eq. (II.79) into Eq. (II.78) and taking the first term of the expansion yields

$$|\mathbf{q}|_{\text{In}} = \frac{\kappa_L \Gamma_{Rf}}{2a - r_I} \Delta\theta(0) \qquad (\text{II.80})$$

For the particular case of having only half of the mirror reflecting radiation, Eq. (II.80) is the version of LSSE Model (Eq. II.64) for indirect radiation. If $r_I/(a - c)$ is adopted as a characteristic parameter, Table II.2 may be obtained. It can be seen that when $r_I/(a - c)$ tends to zero both values coincide. For values of about 0.96, the error is close to 6%. However, practical values are between 0 and 0.25. In these cases, even with the half-mirror idealization, differences are larger than two orders of magnitude.

The authors concluded that for most practical purposes the linear models can be used when only direct radiation is involved. When reflecting curved surfaces are present, only the extense source type of models can be used. The finite size of the radius of the tubular lamp must be incorporated into the model so as to permit the volume (or the surface) of emission to recognize the curvature of the mirror and its concentrating effects.

Table II.2 Indirect Radiation: LSSE vs. ESVE Model (half of the elliptical mirror)

$\dfrac{r_I}{a-c}$	$|\mathbf{q}|_{\text{In}} \times 10^8$(einstein cm^{-2} s^{-1})(*)		Error (%)
	LSSE model	ESVE model	
0.996	1.673	1.680	0.4
0.994	1.669	1.681	0.7
0.992	1.664	1.682	1.1
0.987	1.653	1.685	1.8
0.981	1.640	1.688	2.9
0.975	1.626	1.692	3.9
0.969	1.613	1.694	4.8
0.962	1.600	1.700	5.9

(*) $r_L = 1$ cm, $a = 50$ cm, $\beta_I = 0°$ and $c = 10$ cm.

II.3. Cylindrical Photoreactor Irradiated from the Bottom

A certain number of important photochemical reactions share the common characteristic of having products and reactants with highly corrosive or dissolving properties. A typical example is the chlorination of hydrocarbons where the presence of chlorine, hydrochloric acid and chlorinated solvents creates difficulties in the selection of materials for equipment and seals. In the case of gas–liquid systems, a vigorous stirring is indispensable and consequently it is important to emphasize this aspect of the reactor. In these cases it would be desirable to avoid the introduction of the radiation source inside the reactor. In order to isolate the reaction system from the radiation source (this would simplify the well-known problems of wall deposits which are generally more severe at the source walls) a perfectly stirred tank reactor was designed for continuous, semi-

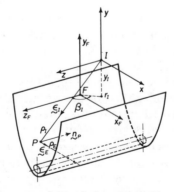

Figure II.22 Geometry of the tubular source with parabolic reflector.

batch or batch operation, irradiated at the bottom surface outside
the reactor. To do so, a tubular source of U.V. radiation was located
at the focal axis of a cylindrical reflector of parabolic cross section
(Fig. II.22).

The radiation field produced by the tubular source placed at the
focal axis of the parabolic reflector was obtained with three different
emission models: the LSPP Model, the LSSE Model and the ESVE
Model. For each one of them, equations were developed to calculate the modulus of the radiant energy flux density vector and for
the component of this vector along the normal to a surface on the
x–z plane [6, 7].

Model Equations

A fixed coordinate system (x_F, y_F, z_F) was defined with the origin
at point F located at the center of the bottom of the empty
cylindrical reactor with y_F being its center line (Fig. II.22). An
auxiliary coordinate system, parallel to the former, was also needed.
Its center was located at the point of incidence designated by I in
Fig. II.22 and it was moved all over the space under analysis.

Table II.3 includes the final equations provided by the LSPP and
the LSSE Models for the direct and indirect energy flux density
considering only one reflection. A distinction was made between the
modulus of the vector and the component along the normal to a
surface parallel to the x–z plane. The corresponding expressions
derived from the ESVE Model are included in Tables II.4 and II.5.

Due to the shape of the experimental device which was being
modelled (Fig. II.23), there were totally and partially irradiated
zones within the reaction space. Consequently, for each incidence

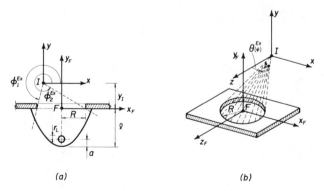

(a)   (b)

Figure II.23   Extreme angles for ϕ and $\theta(\phi)$.

Table II.3 Direct and Indirect Radiation Flux Density Expressions—Line Models

LSPP model

$$|\mathbf{q}|_D(\mathbf{x}_I) = \frac{E}{2\pi L_L \rho'(\mathbf{x}_I)}$$

$$q_D(\mathbf{x}_I) = \frac{E(-\sin\phi)}{2\pi L_L \rho'(\mathbf{x}_I)}$$

$$|\mathbf{q}|_{In}(\mathbf{x}_I) = \frac{E}{2\pi L_L}\frac{\langle\Gamma_{Rf}\rangle}{\rho'_I(\mathbf{x}_I) + \rho'_E(\mathbf{x}_I)}$$

$$q_{In}(\mathbf{x}_I) = \frac{E}{2\pi L_L}\frac{\langle\Gamma_{Rf}\rangle}{\rho'_I(\mathbf{x}_I) + \rho'_E(\mathbf{x}_I)}$$

LSSE model

$$|\mathbf{q}|_D(\mathbf{x}_I) = \frac{E}{4\pi L_L \rho'(\mathbf{x}_I)}[\theta_2^D(\mathbf{x}_I) - \theta_1^D(\mathbf{x}_I)]$$

$$|\mathbf{q}|_{In}(\mathbf{x}_I) = \frac{E}{4\pi L_L}\frac{\langle\Gamma_{Rf}\rangle}{\rho'_I(\mathbf{x}_I) + \rho'_E(\mathbf{x}_I)}[\theta_2^{In}(\mathbf{x}_I) - \theta_1^{In}(\mathbf{x}_I)]$$

$$q_D(\mathbf{x}_I) = \frac{E\sin\phi}{4\pi L_L \rho'(\mathbf{x}_I)}[\cos\theta_2^D(\mathbf{x}_I) - \cos\theta_1^D(\mathbf{x}_I)]$$

$$q_{In}(\mathbf{x}_I) = \frac{E}{4\pi L_L}\frac{\langle\Gamma_{Rf}\rangle}{\rho'_I(\mathbf{x}_I) + \rho'_E(\mathbf{x}_I)}[\cos\theta_2^{In}(\mathbf{x}_I) - \cos\theta_1^{In}(\mathbf{x}_I)]$$

$$\theta_{1,2}^D = \tan^{-1}\left[\frac{\rho'}{\pm(L_L/2) - r_I\cos\beta_I}\right]$$

$$\theta_{1,2}^{In} = \tan^{-1}\left[\frac{\rho'_I + \rho'_E}{\pm(L_L/2) - r_I\cos\beta_I}\right]$$

Auxiliary expressions

$$\rho' = [(\ell - a + y_I)^2 + (r_I\sin\beta_I)^2]^{1/2}$$

$$\rho'_I = \frac{4a(y_I + \ell) - (r_I\sin\beta_I)^2}{4a}$$

$$\rho'_E = \{[(y_I + \ell - a) - \rho'_I]^2 + (r_I\sin\beta_I)^2\}^{1/2}$$

Table II.4   Direct Radiation Flux Density Expressions—ESVE Model

ESVE model	$$|\mathbf{q}|_D(\mathbf{x}_I) = \kappa \int_{\phi_1^D(\mathbf{x}_I)}^{\phi_2^D(\mathbf{x}_I)} \Delta\rho'(\phi, \mathbf{x}_I)[\theta_2^D(\phi, \mathbf{x}_I) - \theta_1^D(\phi, \mathbf{x}_I)]\, d\phi$$ $$q_D(\mathbf{x}_I) = \kappa \int_{\phi_1^D(\mathbf{x}_I)}^{\phi_2^D(\mathbf{x}_I)} \Delta\rho'(\phi, \mathbf{x}_I)[\cos\theta_2^D(\phi, \mathbf{x}_I) - \cos\theta_1^D(\phi, \mathbf{x}_I)]\sin\phi\, d\phi$$
Source limit angles	$$\left.\begin{array}{c}\phi_1^D(\mathbf{x}_I)\\[2pt]\phi_2^D(\mathbf{x}_I)\end{array}\right\} \text{roots from: } F^D(\phi) = r_L^2 - [r_I \sin\beta_I \sin\phi$$ $$- (y_I + \ell - a)\cos\phi]^2 = 0$$ $$\theta_{1,2}^D(\mathbf{x}_I) = \tan^{-1}\frac{[F^D(\phi)]^{1/2} + (y_I + \ell - a)\sin\phi + r_I \sin\beta_I \cos\phi}{r_I \cos\beta_I \mp (L_L/2)}$$
Auxiliary expression	$$\Delta\rho'(\phi, \mathbf{x}_I) = 2\{r_L^2 - [r_I \sin\beta_I \sin\phi - (y_I + \ell - a)\cos\phi]^2\}^{1/2}$$

point I, we had to determine which portion of the parabolic reflector or source was capable of illuminating the point. From the mathematical point of view, the integration interval $(\theta_i^{\text{Int}}, \phi_i^{\text{Int}})$ for each point I had to be adjusted. In order to do this, the angles $(\theta_i^{\text{Ex}}, \phi_i^{\text{Ex}})$ determined by the circular hole of the lamp-reflector cover system had to be calculated and afterwards compared with the limiting angles (θ_i, ϕ_i) defined for the integration of the source volume. Table II.6 provides the extreme angles for variables ϕ and θ, as well as the selection criteria to choose the appropriate integration limits. In order to consider the effectively irradiated zones in the presence of the cover of the emitting system, it was necessary to employ as coordinates (r_i, β_i, y_i) the system represented by (r_I, β_I, y_I) and use as radius (R_i) the radius of the circular hole of the cover (R_R), according to the notation employed in Table II.6.

Computing separately $|\mathbf{q}|_D$ and $|\mathbf{q}|_{\text{In}}$ for each model, the total energy arriving at a given reception point I was obtained; i.e.:

$$|\mathbf{q}|_T = |\mathbf{q}|_D + |\mathbf{q}|_{\text{In}} \qquad (\text{II.81})$$

The resolution of the equations corresponding to the energy flux density for the three models employed was performed numerically.

Results and Conclusions
Figure II.24 represents the dimensionless modulus of the energy flux density vector (Q) for the three models as a function of the azimuthal coordinate β_I. Different heights in parametric form were considered and in every case the radial coordinate was maintained constant. The value corresponding to the indirect or reflected

Table II.5　Indirect Radiation Flux Density—ESVE Model

ESVE model	$\|\mathbf{q}\|_{\mathrm{In}}(\mathbf{x}_I) = \kappa \langle \Gamma_{Rf} \rangle \int_{\phi_1^{\mathrm{In}}(\mathbf{x}_I)}^{\phi_2^{\mathrm{In}}(\mathbf{x}_I)} \Delta \rho_E'(\phi, \mathbf{x}_I)[\theta_2^{\mathrm{In}}(\phi, \mathbf{x}_I) - \theta_1^{\mathrm{In}}(\phi, \mathbf{x}_I)]\, d\phi$
	$q_{\mathrm{In}}(\mathbf{x}_I) = \kappa \langle \Gamma_{Rf} \rangle \int_{\phi_1^{\mathrm{In}}(\mathbf{x}_I)}^{\phi_2^{\mathrm{In}}(\mathbf{x}_I)} \Delta \rho_E'(\phi, \mathbf{x}_I)[\cos \theta_2^{\mathrm{In}}(\phi, \mathbf{x}_I) - \cos \theta_1^{\mathrm{In}}(\phi, \mathbf{x}_I)] \sin \phi \, d\phi$
Source limit angles	$\left. \begin{array}{c} \phi_1^{\mathrm{In}}(\mathbf{x}_I) \\ \phi_2^{\mathrm{In}}(\mathbf{x}_I) \end{array} \right\}$ roots from: $F^I(\phi) = r_L^2 - \{[r_I \sin \beta_I + x_p(\phi)] \sin \phi_E(\phi) - [y_I + \ell - a + y_p(\phi)] \cos \phi_E(\phi)\}^2 = 0$
	$\theta_{1,2}^{\mathrm{In}}(\phi, \mathbf{x}_I) = \tan^{-1} \dfrac{[F^I(\phi)]^{1/2} + [y_I + \ell - a + y_p(\phi)] \sin \phi_E(\phi) + [r_I \sin \beta_I + x_p(\phi)] \cos \phi_E(\phi) - [x_p^2(\phi) + y_p^2(\phi)]^{1/2}}{r_I \cos \beta_I \mp (L_L/2)}$
Auxiliary expression	$\Delta \rho_E'(\phi, \mathbf{x}_I) = 2\{r_L^2 - [(r_I \sin \beta_I + x_p(\phi)) \sin \phi_E(\phi) - (y_I + \ell - a + y_p(\phi)) \cos \phi_E(\phi)]^2\}^{1/2}$

Table II.6 Determination of Effectively Irradiated Zones

Extreme angles	Integration limit angles
$\phi_{1,2}^{Ex} = \tan^{-1}\left[\dfrac{y_i}{r_i \sin \beta_i \pm R_i}\right]$	$\phi_1^{Int} = \max\{\phi_1^{Ex}, \phi_1\}$ $\phi_2^{Int} = \min\{\phi_2^{Ex}, \phi_2\}$
$[F_i^2(\theta, \phi) + R_i^2 - r_i^2]^{1/2} - F_i(\theta, \phi)$ $- y_i \dfrac{(1 - \sin^2\theta \sin^2\phi)}{\sin\theta \sin\phi} = 0$	$\theta_1^{Int}(\phi) = \max\{\theta_1^{Ex}(\phi), \theta_1(\phi)\}$
$F_i(\theta, \phi) = r_i \sin\beta_i \sin\alpha(\theta, \phi)$ $\quad\quad\quad + r_i \cos\beta_i \cos\alpha(\theta, \phi)$	
$\sin\alpha(\theta, \phi) = \dfrac{\sin\theta \sin\phi}{(1 - \sin^2\theta \sin^2\phi)^{1/2}}$	$\theta_2^{Int}(\phi) = \min\{\theta_2^{Ex}(\phi), \theta_2(\phi)\}$
$\cos\alpha(\theta, \phi) = \dfrac{\cos\theta}{(1 - \sin^2\theta \sin^2\phi)^{1/2}}$	

radiation was also indicated. The direct radiation may be obtained by difference.

The first conclusion that may be drawn is that the system is symmetric with respect to axes x_F and z_F. For example, equal values are obtained for angles $\pi/3$, $2\pi/3$, $4\pi/3$ and $5\pi/3$. For that reason the problem will be analyzed only for angles β_I located in the first quadrant. A second important conclusion is that the radiant energy distribution is strongly dependent on the azimuthal coordinate for small heights ($\zeta_I = 0.31$), while for large values of ζ_I variations are negligible. It is useful to take this into account because in some

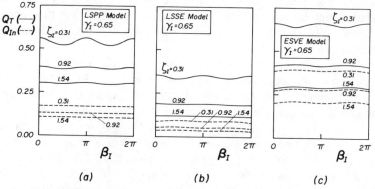

Figure II.24 Effect of the azimuthal coordinate on the radiation field (a) LSPP Model. (b) LSSE Model. (c) ESVE Model.

cases the system could be modelled without considering the azimuthal variable. This reduces a three-dimensional problem to a two-dimensional one with the subsequent simplification in mathematical and computational complexity.

It should also be noted that with the ESVE Model direct radiation constitutes about 30% to 40% of the total, this percentage being larger when the coordinate ζ_I is smaller. This is a very important result with respect to the behavior of photoreactors with cylindrical reflectors of elliptical cross section. On the other hand, the simple models (LSSE and LSPP) predict much larger percentages: 70% to 75% for the LSSE model and 65% to 70% for the LSPP model. This may be explained if one takes into account that in both cases indirect radiation may be represented by an "imaginary" line source (the specular image of the lamp) parallel to the real source, and located on a horizontal plane under the reflector ($\phi = 3\pi/2$). Since the distance from the point of incidence to the line "source" is generally larger for indirect radiation than for direct radiation, the result is a greater flux density of the latter. This problem will be dealt with in detail later in this section.

It should also be mentioned that the ESVE Model predicts a maximum for indirect radiation and a minimum for direct radiation at $\beta_I = \pi/2$. According to the relative contribution of each one of them resulting from the axial position of the point of reception (ζ_I), total flux density curves with minima ($\zeta_I = 0.31$) or maxima ($\zeta_I = 0.92$ and $\zeta_I = 1.54$) for changes in the azimuthal coordinate may be obtained. The LSSE Model predicts the same trends as the ESVE Model, but as a consequence of a larger percentage of direct radiation (70–75%), a total radiation field with minima at $\beta_I = \pi/2$ results. The LSPP Model predicts the same variations as the LSSE Model, but in this case the determining factor is that the former shows a reflected energy distribution constant with the azimuthal variable.

Figure II.25 represents the dimensionless radiant energy field as a function of the axial coordinate ζ_I for three different angles and for a fixed value of the radial coordinate. As expected, at increasing heights a considerable decrease of the flux density may be observed; the previously explained dependence of Q on β_I for low ζ_I is again observed. The percentages corresponding to direct radiation are in this case: 30% to 50% for the ESVE Model, 70% to 80% for the LSSE Model, and 65% to 70% for the LSPP Model, the trend already pointed out for changes in heights being maintained.

Figure II.26 shows variations of Q as a function of the radial

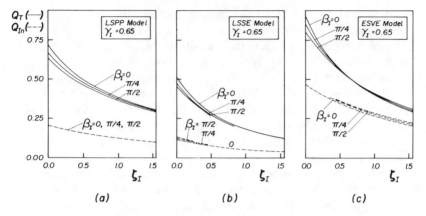

Figure II.25   Effect of the axial coordinate on the radiation field. (a) LSPP Model.
(b) LSSE Model. (c) ESVE Model.

coordinate (γ_I) for a fixed value of the axial coordinate and for
three different angles. The general trend indicates that for larger γ_I,
smaller values of the energy flux density are obtained. In the central
zone of the reaction space (small γ_I) the variation is very small as a
consequence of: (a) greater emitting system-incidence point distance
for $\beta_I = \pi/2$, (b) more separation of the reception point from the
maximum radiation zone (central) for $\beta_I = 0$, and (c) both effects for
$\beta_I = \pi/4$. Close to the maximum reactor radius, the decrease of the
radiation field is abrupt. This effect may be explained by the passage
of zones illuminated by the total emitting system to zones partially
irradiated by the system.

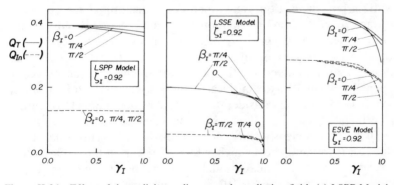

Figure II.26   Effect of the radial coordinate on the radiation field. (a) LSPP Model.
(b) LSSE Model. (c) ESVE Model.

It should be stressed that simple models are not capable of predicting certain tendencies shown by the ESVE Model. The LSPP Model recognizes neither the radial variations of the reflected radiant energy for any angle under consideration nor the direct radiation field for $\beta_I = 0$. The LSSE Model, on the other hand, predicts a constant radial distribution of the reflected flux density for $\beta_I = \pi/2$ except in the vicinity of the reactor wall where the existence of totally or partially illuminated zones cannot be attributed to the source model itself. Here the distortion is produced by the cover system.

To perform an *experimental verification* of the ability of the three models to predict the radiation field, a simple test was devised. It provided a precise procedure to check the relative significance of direct and total radiation at different positions of the reacting space. Several measurements were performed to obtain values of the radiation flux density under two different conditions: (1) the radiation probe receiving the sum of direct and indirect radiation and (2) the probe receiving only direct irradiation from the source.

An Eppley Thermopile (Circular shape, 3/8" diameter, Bi–Ag type, water-jacketed, quartz-windowed) connected to a potentiometric recorder was used. Although the measurements cannot be considered as strictly point values, the orientation and finite size of the thermopile receiving area was accounted for in all theoretical calculations.

The experimental procedure started with the source stabilization time and with an adequate positioning of the thermopile within the spatial position under investigation. Output voltage recordings were then performed. Afterwards, the same measurements were carried out isolating the reflector with a copper sheet completely covered with carbon deposition. This operation transformed the reflecting surface to a close approximation to a black body. Results of the ratio between both measurements (with and without cover) are listed in the last column of Table II.7. Tables II.3, II.4 and II.5 list the equations obtained for the projection of the flux density vector upon the normal to the surface of the sensor device for the three source models under consideration. The thermopile was horizontally placed (plane $x_F - z_F$), thus, its normal coincides with the $-y_F$ axis.

These results clearly show that the error that occurs when using the ESVE Model is never higher than 15% for all the values under consideration. The linear models, on the other hand, introduce important discrepancies ($>100\%$), confirming their incapacity to predict certain characteristics of the radiation field (see Section

Table II.7 Direct Radiation Percentages

Dimensionless cylindrical coordinate	Contribution of direct radiation to the total radiation			
	Prediction of LSPP model (%)	Prediction of LSSE model (%)	Prediction of ESVE model (%)	Experimental value (%)
$\zeta_l = 1.85$ $\gamma_l = 0.43$ $\beta_l = \pi/4$	65.1	69.0	27.7	29.2
$\zeta_l = 1.85$ $\gamma_l = 0.26$ $\beta_l = 0$	65.4	69.3	28.1	30.8
$\zeta_l = 1.85$ $\gamma_l = 0.46$ $\beta_l = 0$	65.4	69.2	28.0	31.2
$\zeta_l = 1.85$ $\gamma_l = 0.67$ $\beta_l = 0$	65.4	69.1	27.7	31.6

II.2). Some experimental difficulties in the ability of the device to exclude indirect radiation led us to conclude that the experimental values should have always been slightly higher than the theoretical predictions. This has also been consistently verified as shown in Table II.7.

From the studies reported in this work, it can be concluded that:

- The predictions of the models indicate a decrease of the energy flux density when the axial coordinate increases. The radial coordinate shows a similar effect but presents unimportant variations in the reactor central zone and significant changes in the vicinity of its walls. Azimuthal variations indicate a remarkable sensitivity of the radiation field for low heights and negligible changes above certain values of the axial coordinate.
- It should be observed that the inevitably poor physical representation of the emission process by the line source models results in their inability to predict certain variations of the radiant field which are physically clear. In this case, the above-mentioned difficulty is surely magnified by the existence of a reflecting curved surface incorporated in the emitting device.
- The symmetry found in the radiation field with respect to axes x_F and z_F should also be noted. This allows the reduction of the system analysis to only one quadrant.
- The relative contributions to the total radiation field supplied by the energy that arrives directly at the point of incidence, or after reflection on the parabolic mirror, were studied. This analysis shows the significance of direct radiation for this system. Consequently, it cannot be neglected as it could in the case of photoreflectors of elliptical cross section.
- Through the use of a thermopile, a simple experimental test was carried out which proved the inability of the linear models to predict the relative contribution of direct radiation to the total and the good agreement with the predictions of the ESVE Model.

III. "BLACK BODY" REACTORS

We call "black body" reactors those systems where the reaction takes place in a very thin layer attached to the surface through which the radiation enters the reactor. It means that the absorption coefficient and/or the concentration of the absorbing species are so high that after a very short distance the energy is reduced below 1% of the incident value.

A careful experimental work fulfilling these operating conditions was done by Williams and coworkers. No fundamental theoretical analysis was reported, but their result seemed very useful for a comparison with theory. Experiments were done with a chemical actinometer of very strong absorption in tubular reactors with an elliptical reflector. An actinometer is a well-known and simple reaction whose quantum yield is precisely known (No. of molecules reacted/No. of quanta of radiation absorbed). Measuring the product conversion in a "black body" reactor is equivalent to measuring the amount of energy that reaches the surface of radiation incidence. Measurements were done in two different ways:

1. By radiating the whole reactor, as a way to measure the yield of the system to concentrate the energy at the reactor surface.
2. With longitudinal "windows" to study the angular distribution of energy.

We applied our model to both situations. The main body of theory was already presented in the previous section for the case of tubular diactinic reactors inside the elliptical reflector.

Radiation Efficiencies
The energy yield of a photoreactor can be defined as the ratio between the amount of desired product obtained in the reactor and the amount of energy required to operate the radiation source. In other words, it is the probability that a given amount of energy, supplied to the lamp, reaches the reactor and becomes absorbed by the reactant to yield the desired product.

For a better analysis of the factors influencing the photoreactor energy yield, the process can be imagined by extending the idea proposed by Hancil et al. [48], as a sequence of steps, each one with a probability of occurrence amenable to measurement. They are defined as:

– The probability that the lamp power consumption results in an energy emission in the desired region of the radiant energy spectrum. It is related to the emission spectrum and energy yield of the lamp:

$$\eta_1 = \text{(Energy emitted in } \Delta\lambda)/\text{(Energy consumption by the source)}$$

– The probability that this portion of the radiation energy spectrum be available for absorption in the reaction volume when the latter is assumed to be a black body:

$\eta_2 = $ (Energy available for absorption in $\Delta\lambda$)/(Energy emitted in $\Delta\lambda$)

– The probability that this amount of available energy for absorption in the reaction volume be absorbed in the actual reactor by the desired reactant. It is related to the radiation absorption of reactants, products and inerts:

$\eta_3 = $ (Energy absorbed by proper reactant in $\Delta\lambda$)/(Energy available for absorption in $\Delta\lambda$)

– The probability that the desired reactant, excited by the absorbed energy, undergo the expected transformation into the desired product. It is usually known as overall product quantum yield:

$\eta_4 = $ (Amount of the desired product)/(Energy absorbed by the proper reactant in $\Delta\lambda$)

The product of these four probabilities gives the value of the energy efficiency of the photoreactor. It is clear that these concepts will be applicable to any photochemical reactor. While a proper design of a process will take into account the optimization of all these probabilities, the quality of the device as a concentrator of radiant energy on the reaction surface will depend exclusively upon the value of η_2.

To achieve a better understanding of the physical meaning of η_2 it should be thought that the reaction volume used for its direct experimental determination must operate as a black body. Thus, the numerator of η_2 gives, for each lamp–reflector–reactor arrangement, the value of the maximum energy available for absorption at the reactor. As such, it excludes nearly all effects directly ascribable to the reactant or to the reaction. Its value depends explicitly upon: (1) the geometrical characteristics of the lamp, the reflector and the reactor; and (2) the physical properties of the reactor and reflector materials. For these reasons, it is more useful to compare different sizes and shapes of reactor configurations.

Since the question under analysis is the eventual usefulness of the elliptical reflector, the work was primarily devoted to study η_2, which is called incidence efficiency [28]. Its value was predicted with the aid of Irazoqui et al.'s model [57] and the results compared with the available measurements.

The experimental determination of the incidence efficiency requires some special considerations. Let us write:

$$\eta = \eta_1 \eta_2 \eta_3 \eta_4 \qquad (III.1)$$

The reaction surface can be considered as a black body if, and only if, the actinometric reaction has a value of η_3 as close to one as possible. For this reason, an actinometer with practically 100% absorption for all the spectral range of interest must be used. The value of η_1 can be obtained from the manufacturer's specifications; η_4 is a well-known value for an actinometric reaction; hence, the incidence efficiency can be directly computed from experimental measurements of the total efficiency η. Thus, the value of η_2 will only be distorted by the possible inaccuracies of the kinetic information and measurements, as well as in equipment specifications and construction, and those of the model used for the analysis of the data.

Model Equations

The problem consists in finding the functional relationship between the incidence efficiency of an elliptical photoreactor and its optical and geometrical characteristics, i.e. to relate the incident radiant energy flux upon the limiting surface of the reaction volume (reaction surface) with the reactor, lamp and reflector dimensions, as well as with the optical parameters of the reflector and the reactor wall.

According to the ESVE Model [57], the total direct energy per unit area and unit time impinging (from all directions in space and from the whole volume of the lamp) on the differential area at point I, can be obtained by:

$$q_D(r_R, \beta_I, z_I) = \kappa \int_\phi \int_\theta \int_\rho \sin\theta \cos\theta_n \, d\rho \, d\theta \, d\phi \qquad (III.2)$$

where:

$$\cos\theta_n = \sin\theta \cos(\psi_I - \phi) \qquad (III.3)$$

$$\psi_I = \tan^{-1}\left[\frac{2c \sin\beta_I}{r_R + 2c \cos\beta_I}\right] \qquad (III.4)$$

and ψ_I is the angle between the x-axis and the normal vector to the cylindrical surface at point I. The integration limits for variables ρ, θ and ϕ were obtained in Section II.2 (see Eqs. II.26 to II.31).

The total energy per unit area and unit time with only one reflection is given by:

$$q_{\text{In},1}(r_R, \beta_I, z_I) = \kappa \langle \Gamma_{Rf} \rangle \langle Y_R \rangle \int_\phi \int_\theta \int_\rho \sin \theta \cos \theta_n \, d\rho \, d\theta \, d\phi$$

$$\text{(III.5)}$$

Again, Section II.2 showed the expressions of the integration limits for $q_{\text{In},1}$ (Eqs. II.35 to II.53).

Looking at Fig. III.1 it is clear that the integration interval of the variable ϕ so computed may include radiant beams that have already reached the reaction surface as direct radiation. When this is the case, since they have been absorbed by the reaction volume, they cannot undergo any subsequent reflection process. Their contributions must not be considered and it is necessary to delete them from the integration interval. As shown by Fig. III.1, the lamp and the reactor boundaries define a zone on the reflector where the first reflection process cannot occur. As long as the reaction volume behaves as a black body, this region of the reflector operates as an umbra zone. All those directions that are inside the sub-interval defined by angles ϕ_{U_1} and ϕ_{U_2} must be eliminated. It is rather straightforward to show that

$$\phi_{U_{1,2}} = \tan^{-1} \left[\frac{\rho_U \sin \alpha_{1,2} + y_{T_{1,2}} - k}{\rho_U \cos \alpha_{1,2} + x_T - h} \right]$$

$$\text{(III.6)}$$

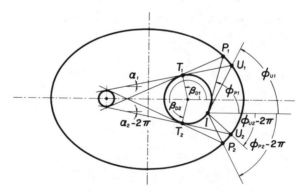

Figure III.1   Umbra and penumbra zones on the elliptical reflector.

where:

$$x_T = \frac{r_R(r_L - r_R)}{2c} + c$$

$$y_{T_{1,2}} = \pm [r_R^2 - (x_T - c)^2]^{1/2}$$

$$\alpha_{1,2} = \tan^{-1}\left\{ \pm \frac{r_R - r_L}{[4c^2 - (r_R - r_L)^2]^{1/2}} \right\} \qquad \text{(III.7)}$$

In its turn, ρ_U must be computed using Eq. (II.36) with h, k and ϕ replaced by x_T, y_{T_1} and α_1.

At the same time, lamp and reactor boundaries also define two partially shadowed zones on the reflector, limited by angles ϕ_{U_1}, ϕ_{P_1}, ϕ_{U_2}, ϕ_{P_2}. They are equivalent to penumbra zones. When these zones fall inside the integration interval corresponding to variable ϕ, one would be adding contributions of rays that have already reached the reaction volume as direct radiation. An incident ray will be able to actually reach the penumbra zone provided it has not previously intersected the reaction volume as direct radiation. The intersection will occur if:

$$r_R^2 > [(x_P + r_R \cos \beta_I) \sin \phi_E - (y_P + r_R \sin \beta_I) \cos \phi_E]^2 \qquad \text{(III.8)}$$

and the ray will have to be excluded.

The energy emerging from the source can also arrive at the reaction surface after two or more reflections (Fig. III.2). Cerdá et al. [27] qualitatively postulated that the energy contribution due to the radiant bundles reaching the reaction volume after more than

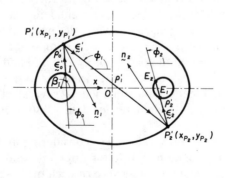

Figure III.2   Indirect radiation with two or more reflections.

one rebounding on the elliptical mirror should not be very important. However, a quantitative treatment computing this type of contribution was accomplished [28] employing a methodology similar to the one used for one reflection. It is only an algebraic complication with no conceptual additions to the problem and we will not include the details here.

The total energy per unit time arriving at the reaction surface is:

$$E = r_R \int_z \left[\int_{\beta_{D_1}}^{\beta_{D_2}} q_D \, d\beta + \sum_n \int_0^{2\pi} q_{\mathrm{In},n} \, d\beta \right] dz \qquad \text{(III.9)}$$

where the integration limits are given by:

$$\beta_{D_1} = \frac{\pi}{2} + \sin^{-1}\left(\frac{r_R - r_L}{2c}\right)$$

$$\beta_{D_2} = \frac{3\pi}{2} - \sin^{-1}\left(\frac{r_R - r_L}{2c}\right) \qquad \text{(III.10)}$$

$$z_{1,2} = \frac{L_{Rf} \mp L_R}{2}$$

Finally, the photoreactor incidence efficiency is:

$$\eta_2 = \frac{E}{E_L} \qquad \text{(III.11)}$$

where E_L is the radiation power irradiated by the source in $\Delta\lambda$ and can be obtained from the manufacturing specifications.

Results and Conclusions

Theoretical predictions made according to Eqs. (III.9) and (III.11) were compared with available experimental information. For this purpose, useful data congruent with all model descriptions and limitations discussed in the previous sections were published by Williams [93], based on the papers of Zolner [95] and Ragonese and Williams [80]. Their estimations of efficiencies were based upon the measurement of kinetic data obtained in several elliptical photoreactors, using the potassium ferrioxalate solution as actinometric reacting media.

Predicted values were calculated considering both direct and indirect incident radiation; the latter after one and two reflecting mechanisms. The relative significance of each of these contributions is shown in Table III.1, where the first column represents different

Table III.1 Contribution of Direct and Indirect Radiation to the Total Incidence Power

Data	Direct radiation		Indirect radiation (n = 1)		Indirect radiation (n = 2)		Total incidence
	Power × 10^{-6} (einstein/s)	%	Power × 10^{-6} (einstein/s)	%	Power × 10^{-6} (einstein/s)	%	Power × 10^{-6} (einstein/s)
1	0.096	5.09	1.680	89.12	0.109	5.78	1.885
2	0.312	5.20	5.313	88.98	0.347	5.82	5.972
3	0.250	5.11	4.347	89.07	0.284	5.82	4.881
4	0.390	6.43	5.258	86.95	0.400	6.62	6.048
5	0.289	5.85	4.363	88.50	0.278	5.65	4.930
6	0.487	7.89	5.274	85.64	0.398	6.47	6.159
7	0.680	7.93	7.314	85.46	0.565	6.61	8.559

values of lamp and reactor lengths, reactor diameter, and lamp power emission. It is clear that the first reflection makes the most significant contribution to the total energy reaching the reactor. It also seems clear that successive reflections after the second one can be generally neglected. Finally, it can be seen that direct radiation, though important because it distorts the uniform irradiation as far as the x-coordinate is concerned, does not represent more than 5–8% of the total radiation reaching the reactor. A similar conclusion can be reached for the indirect radiation impinging after two reflections.

Table III.2 shows a comparison of predicted and experimental values of η_2. Good, and in some cases remarkable, agreement can be observed. If one takes into account the errors involved in the measurement of the kinetic data, the possible imperfections in the constructions of both the reactor and the reflector, as well as the difficulties to attain the precise positioning of the reactor and source at the reflector focal center lines, one must remark that the model accurately represents the behavior of these experimental devices.

Conclusions, taking into account the geometry considered, may be summarized as follows:

– The ESVE Model was used to predict the elliptical reflector-tubular photoreactor incidence efficiency. As far as it can be judged from the available experimental information, this model predicts well all published data.
– The available energy for absorption in the reaction volume is principally provided by reflected radiation after only one reflection. In fact, this contribution represents from 85 to 90% of the total energy reaching the reactor. The lower value corresponds to the case when the L_L/L_R ratio approaches one.
– The energy contributions due to direct radiation and indirect radiation with two reflections are less important. The magnitude

Table III.2 Comparison of Predicted and Experimental Incidence Efficiency Values

Data	Experimental value (%)	Prediction according to ESVE model (%)	% error
1	4.63	5.70	23.11
2	16.13	18.06	11.96
3	12.65	14.76	16.67
4	15.02	18.29	21.77
5	19.15	22.19	15.87
6	27.84	27.72	−0.43
7	38.05	38.52	1.23

of each of these contributions ranges from 5 to 8% of the total
energy reaching the reactor.
- The energy reaching the reactor volume, by radiation beams that
undergo more than two reflections, can be safely neglected.

Azimuthal Asymmetries
The assumption of a uniform angular distribution of the radiation
flux density has been widely used in laboratory research with
elliptical photoreactors. Zolner III and Williams [96] studied the
problem and found that the uniformity hypothesis is fulfilled only
for small values of the reactor diameter–source diameter ratio. In
order to evaluate the radiation flux density at different angular
positions, the authors masked the reactor with an opaque plastic
cover in order to prevent the introduction of radiation on the entire
reactor external area. The radiation, entering through a thin longitudinal slot on the mask, was totally absorbed by an actinometric
solution of high absorptivity (potassium ferrioxalate), circulating in
the reactor. Rotating the mask so as to put the slot at different
angular positions and measuring in each case the reaction produced
by the absorbed energy, the angular distribution of energy flux
density was determined.

To correlate the experimental results, the authors [96] proposed
the following simple empirical equation:

$$\frac{q(\bar{\beta})}{q_M} = 1 - \delta \cos \bar{\beta} \qquad \text{(III.12)}$$

where $\bar{\beta}$ is the angle corresponding to the medium point of the slot,
δ is an experimental parameter and q_M is the angularly averaged
radiation flux density. In order to study the influence of the
geometrical parameters of the elliptical reflector–photoreactor system upon the angular distribution of the radiation flux density, De
Bernardez and Cassano [37] used an extension of the emission
model developed by Cerdá et al. [27]. The results obtained with the
ESVE Model were verified against the experimental values of
Zolner III and Williams [96].

Model Equations
The radiation flux density reaching the whole slot, assigning its
value to $\bar{\beta}$, is given by:

$$q(\bar{\beta}) = \frac{1}{\Delta \beta \, L_R} \int_{\beta_1}^{\beta_2} \int_{z_1}^{z_2} q(r_R, \beta, z) \, dz \, d\beta \qquad \text{(III.13)}$$

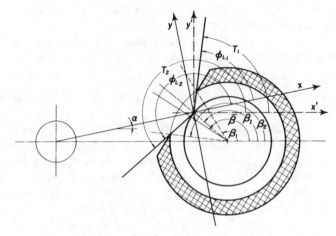

Figure III.3   Limiting angles of the lighted slot.

where $q(r_R, \beta, z)$ is obtained by adding the contributions of direct and indirect radiation given by Eqs. (III.2) and (III.5), respectively.

Due to the presence of the mask, a point I placed on the surface of the reactor, and in the lighted slot, can receive radiation only between the limits given by ϕ_{L_1} and ϕ_{L_2} in Fig. III.3. By comparing the values of (ϕ_1, ϕ_2) which were established by Cerdá et al. [27] with (ϕ_{L_1}, ϕ_{L_2}), the real integration limits of variable ϕ for direct radiation were defined. The detailed calculations of these limits can be found in the quoted paper [37].

Again, the presence of an opaque reactor in the indirect radiation path gives rise to the appearance of umbra and penumbra zones in the reflector. Their characteristics and limits were also analyzed by Cerdá et al. [28] and their results were incorporated to the equations.

Finally, the expression of q_M to obtain the dimensionless magnitude q/q_M, is:

$$q_M = \frac{1}{2\pi} \int_0^{2\pi} q(\bar{\beta}) \, d\bar{\beta} \qquad \text{(III.14)}$$

Results and Conclusions

Figures III.4a, b and c show the distribution of radiation flux density when the reactor radius is decreased. In polar coordinates, they are a plot of q/q_M vs. $\bar{\beta}$. They also show the experimental results of Zolner III and Williams [96]. The radiation flux density is greater for points located close to the lamp, and smaller for those situated

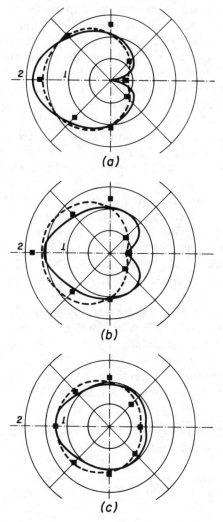

Figure III.4 Angular distribution of the radiation flux density. Predictions of the model (—). Least square fitting of experimental data (----). (a) $r_R = 1.25$ cm. (b) $r_R = 0.75$ cm. (c) $r_R = 0.15$ cm.

far from it. This is due to the overlapping of two effects: (a) the arrival of direct radiation (only upon the points situated in front of the lamp), and (b) the existence of umbra and penumbra zones, giving rise to non-uniformities in indirect radiation, especially for those points most distant from the lamp.

A good agreement between the experimental values and the predictions performed with the ESVE Model may be noticed. This is quite remarkable if one considers the errors already listed in the previous section. As the reactor radius decreases, the value of δ also decreases, and hence, the angular distribution of radiation flux density tends to be uniform. The same conclusion may be drawn if one observes Figs. III.4a, b and c. This occurs primarily so because as the reactor radius increases, the effect of the existence of umbra and penumbra zones on the indirect radiation path is greater, thus producing steeper asymmetries in the angular distribution of indirect radiation flux density, and bringing about the same effect upon the total radiation.

Figure III.5 shows the values of parameter δ for a given value of the dimensionless distance between focuses ($c^* = 50$) varying the reflector eccentricity and the dimensionless reactor radius (γ_R). This figure also shows that for small values of γ_R (of about 0.1) the δ parameter tends to zero and is practically independent of the elliptical reflector eccentricity. Increasing γ_R the eccentricity effect is more noticeable and it may be observed that, for high eccentricities, the asymmetries in the angular distribution of the radiation flux density are considerably increased. In these instances, in spite of the fact that in $\bar{\beta} = 0.0$ the radiation flux density must be positive, it may occur that δ takes values higher than 1. In these cases, in general, the radiation flux density is practically negligible.

The main conclusions related to azimuthal asymmetries are the following:

– A good agreement between published data and the ESVE Model predictions was observed.

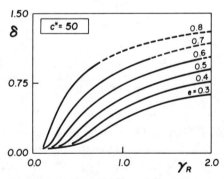

Figure III.5   Angular distribution of the radiation flux density. Reactor radius and reflector eccentricity effects.

– Uniform angular distribution is obtained only if one works with a
reactor diameter/lamp diameter ratio lower than 0.5 and with an
eccentricity as low as possible (of about 0.4). If one wishes to
work with higher eccentricities, the ratio should be lower than 0.2.

– The modelling of this device could present difficulties for commercial scale operations, since in these cases one generally works with
reactors whose diameters are larger than those of the lamp, and
this implies the existence of strong azimuthal asymmetries. However, this experimental device is very appropriate for laboratory or
bench scale research work, where one operates reactors which can
easily fulfill the conditions for symmetry.

IV. HOMOGENEOUS REACTORS WITH SIMPLE KINETICS

In this Section and in the following ones, we will consider those
reactors with intermediate optical thickness. Photoreactors with
negligible radiant energy attenuation (Section II) or with very high
optical thickness (Section III) were already analyzed.

This section will be concerned with different types of reactors but
with rather simple kinetics, such as polymerizations, isomerizations
or photosensitized reactions. It is possible to distinguish two cases:

1. Photosensitized reactions. The absorbing species acts as an energy transfer agent, reacts and after a cycle is restored. Its original
   concentration does not change. If the other reacting species do
   not absorb light, the radiation field is independent of the extent
   of the reaction. We can say that "there is no coupling between
   the attenuation of radiation (caused by absorption) and the
   extent of the reaction".
2. Non-photosensitized reactions. The absorbing species reacts and
   its concentration changes with the extent of the reaction. Then,
   the attenuation of radiation is related to the distribution of
   concentration in space. Two different situations may occur:
   2.1. A very well stirred reactor, where the attenuation process
        is, at most, a function of time.
   2.2. A reactor with space dependent concentrations, where the
        problem of the coupling "attenuation of radiation-extent of
        the reaction" arises.

This coupling is very complicated as shown in Fig. IV.1. The
LVREA is not only a point function of position ($e^a = e^a$ [$\mathbf{x}_I$,

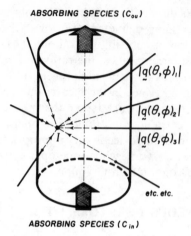

ABSORBING SPECIES (C_{ou})

$|q(\theta,\phi)_i|$

$|q(\theta,\phi)_2|$

$|q(\theta,\phi)_3|$

etc. etc.

ABSORBING SPECIES (C_{in})

Figure IV.1   Schematic representation of the coupling "attenuation of radiation—extent of the reaction".

parameters]) but also a functional of all the optical properties of the space through which every ray of radiation reaching point I has travelled. The consequence of this is that there exists a very close association between the extent of the reaction in the whole reactor (which controls the reactant absorbing properties) and the absorption rate at each point. Thus, to know the reaction rate at each point I we must know the value of $|q_\nu|$ at the point; but the value of $|q_\nu|$ at each point depends on the concentration of the absorbing species along the trajectory of every ray reaching point I from the whole space of reaction. Finally, to know the concentration of the absorbing species at each point in the space of reaction we must have the reaction extent at every point (or what is the same, the reaction rate at every point). Therefore, to solve this link "extent of the reaction-attenuation of radiation" an iterative procedure must be superimposed on the numerical technique used to solve the system of mass conservation equations. Starting from an initial value of concentration in the whole reactor one must solve the problem until a "steady-state" concentration of reactants is achieved. This is done by repeating the finite difference calculations as many times as needed until a prescribed minimum error is reached.

This is only part of the problem because the attenuation phenomenon (as well as the radiation ray contributions) is a three-dimensional process; however, with the hypothesis of angular symmetry, the mass balances are two-dimensional. Without losing rigor, the problem can be solved by using a cylindrical projection of all concentration values on the rectangular mesh used in the numerical solution [84].

IV.1. Annular Photoreactors

Polymerization Reaction
Cerdá et al. [29] studied the liquid-phase vinyl acetate polymerization, photosensitized by di-ter-butyl peroxide. They used the following sequence to represent the kinetics of the reaction:

$$\text{In} + h\nu \xrightarrow{\Phi e^a} 2R\cdot$$

$$R\cdot \xrightarrow{k_i} M_1\cdot$$

$$\cdots\cdots\cdots\cdots\cdots$$

$$M_n\cdot + M \xrightarrow{k_p} M_{n+1}$$

$$M_n\cdot + M_m\cdot \xrightarrow{k_{tc}} P_{m+n}$$

$$M_n\cdot + M_m\cdot \xrightarrow{k_{td}} P_m + P_n$$

The authors employed the following hypotheses: (i) kinetic steady-state assumption for intermediate radicals, (ii) long chain kinetic approximation, and (iii) k_p and k_t independent of the polymer chain length. The rate of polymerization was given by:

$$R = k_p C_M \left[\frac{R_i}{2k_t} \right]^{1/2} \tag{IV.1}$$

where:

$$k_t = k_{tc} + k_{td}$$

$$R_i = 2f\Phi e^a$$

$$f = \left[\frac{\text{Free radicals } (R\cdot) \text{ successfully used in initiating the polymerization reaction}}{\text{Free radicals } (R\cdot) \text{ formed in the initiation reaction}} \right] \tag{IV.2}$$

Model Equations
The photopolymerization reaction was carried out in an annular well-stirred reactor (Fig. II.1). If the mass conservation equation and an averaged value of the volumetric rate of energy absorption

(VREA) over the reactor volume are used, the exit monomer conversion is given by

$$X_M = \frac{k_p \Theta_R \left[\frac{f\Phi}{k_t}\right]^{1/2} \langle (e^a)^{1/2} \rangle}{1 + k_p \Theta_R \left[\frac{f\Phi}{k_t}\right]^{1/2} \langle (e^a)^{1/2} \rangle} \qquad \text{(IV.3)}$$

From Eq. (IV.3) one can see that the average value of the square root of the VREA must be evaluated. This was done using three different radiation models: the LSPP Model, the LSSE Model, and the ESVE Model. Using Eqs. (I.18), (I.32) and (I.51), the expressions for $\langle (e^a)^{1/2} \rangle$ are given by:

$$\langle (e^a)^{1/2} \rangle_{LSPP} = \left[\frac{2\pi E\mu L_R^2}{L_L V_R^2}\right]^{1/2} \int_{r_{in}}^{r_{ou}} (r)^{1/2} \exp\left[-\frac{\mu}{2}(r - r_{in})\right] dr$$

$$\text{(IV.4)}$$

$$\langle (e^a)^{1/2} \rangle_{LSSE} = \frac{2\pi}{V_R} \int_0^L \int_{r_{in}}^{r_{ou}} \left\{\frac{E\mu}{4L_L \pi r}\right.$$

$$\left. \times \int_{\theta_1}^{\theta_2} \exp\left[-\mu \frac{(r - r_{in})}{\cos \theta}\right] d\theta \right\}^{1/2} r \, dr \, dz$$

$$\text{(IV.5)}$$

$$\langle (e^a)^{1/2} \rangle_{ESVE} = \frac{2\pi (2\kappa\mu)^{1/2}}{V_R} \int_0^{L_R} \int_{r_{in}}^{r_{ou}} r[F(r, z)]^{1/2} \, dr \, dz \qquad \text{(IV.6)}$$

where:

$$F(r, z) = \int_\phi \int_{\theta(\phi)} [r^2 \cos^2 \phi - (r^2 - r_L^2)]^{1/2} \exp[-\mu\sigma(\theta, \phi)] \, d\theta \, d\phi$$

$$\text{(IV.7)}$$

$$\sigma(\theta, \phi) = \frac{r \cos \phi - [r^2 \cos^2 \phi - (r^2 - r_{in}^2)]^{1/2}}{\sin \theta} \qquad \text{(IV.8)}$$

and $\sigma(\theta, \phi)$ is the length inside the reactor travelled by the incident ray with direction (θ, ϕ). The integration limits for Eq. (IV.7) were already presented in Section II.1.

Results and Conclusions

For the three models, Eq. (IV.3) was applied to five different reactors with the only difference being the reactor radius. Other reactor parameters were constant; i.e., the reactor length, the reactor volume and the average residence time.

Table IV.1 shows the computed results using the three emission models. It is clear that for high values of r_{in} the LSPP Model shows significant differences with the other two models. On the other hand, no clear distinction can be seen between the LSSE Model and the ESVE Model. This very good approximation is only true when direct radiation, exclusively, is involved. When a reflecting surface is part of the reaction system as in the case of the elliptical reflector, this may not be true. The problem has been treated in detail in Section II.2.

Isomerization Reaction

Romero et al. [84] studied the behavior of a continuous annular reactor irradiated with a tubular ultraviolet source placed at its axis, and analyzed the case of a chemical reaction occurring in homogeneous phase with a three-step kinetics. Considering three different emission models of radiation source, they investigated and evaluated the influence of all significant parameters of the system on the reactant, intermediate species and product concentration profiles, as well as their effect on the final conversion.

The chosen reaction sequence represents radiant energy absorption by a reactant A and a subsequent reaction of the activated molecule to produce a product B. The activated species A^* may also be destroyed by a termination step involving, for example, collision with an inert M present in constant concentration:

$$A + h\nu \xrightarrow{\Phi e^a} A^*$$

$$A^* \xrightarrow{k_2} B$$

$$A^* + M \xrightarrow{k_3} A + M$$

Table IV.1 Monomer Exit Conversion According to the Different Radiation Models

Radiation models	% Conversion				
	Reactor internal radius (cm)				
	2.5	5.0	10.0	20.0	50.0
LSPP model	28.33	28.14	26.43	22.90	17.09
LSSE model	27.24	27.13	25.11	20.31	11.94
ESVE model	27.43	27.35	25.33	20.49	12.03

Model Equations

The system was studied under the following assumptions: (i) steady state; (ii) negligible thermal effects; (iii) unidirectional, incompressible, laminar flow regime; (iv) Newtonian fluid; (v) azimuthal symmetry; (vi) negligible axial diffusion when compared to the convective flux along that direction; (vii) constant and identical diffusion coefficients (isomerization); (viii) negligible reflection and refraction phenomena on the walls; (ix) monochromatic radiation; and (x) non-permeable reactor walls. The dimensionless material balance for any species may be written:

$$U(\gamma)\,\frac{\partial \Psi_i}{\partial \zeta} = \frac{1}{\text{Pe Ge}}\,\frac{1}{\gamma}\,\frac{\partial}{\partial \gamma}\left(\gamma\,\frac{\partial \Psi_i}{\partial \gamma}\right) + \Omega_i \qquad \text{(IV.9)}$$

with the boundary conditions:

$$\Psi_i(0, \gamma) = \begin{cases} 0; & i \neq A \\ 1; & i = A \end{cases}$$

$$\frac{\partial \Psi_i}{\partial \gamma}\,(\zeta, 1) = \frac{\partial \Psi_i}{\partial \gamma}\,(\zeta, 1/K) = 0 \qquad \text{(IV.10)}$$

In order to determine the expression for the local reaction rates for each species, the following assumptions were made: (i) no dark initiation reaction is present; (ii) the walls of the reactor are inert; (iii) the absorption coefficient α and the primary quantum yield Φ are constant. According to these assumptions, the dimensionless rates are:

$$\Omega_A = K_3 \Psi_{A^*} - \frac{J}{\text{Ge}}\,\Psi_A \Theta_\ell$$

$$\Omega_B = K_2 \Psi_{A^*} \qquad \text{(IV.11)}$$

$$\Omega_{A^*} = \frac{J}{\text{Ge}}\,\Psi_A \Theta_\ell - (K_2 + K_3)\Psi_{A^*}$$

where Θ_ℓ has a different expression for each of the radiant energy distribution models used in the analysis.

It should be noted that inside the reactor there are regions at which the radiant energy bundles do not arrive. These are regions between the internal radius and the straight lines SLI and SLII respectively (Fig. IV.2a):

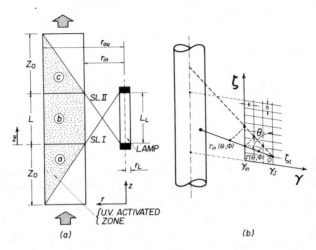

Figure IV.2   (a) Geometry of the annular reactor. Irradiated zones. (b) Mechanism of attenuation.

$$\zeta_{\text{SLI}} = \frac{\gamma_{\text{ou}} - \gamma}{\gamma_{\text{in}} - \gamma_L} \qquad \zeta_{\text{SLII}} = \frac{\gamma + \gamma_{\text{ou}} - (\gamma_{\text{in}} + \gamma_L)}{\gamma_{\text{in}} - \gamma_L} \qquad \text{(IV.12)}$$

Even though dark secondary chemical reactions, and consequently diffusion, occur in these zones, there is no production of radiation activated species; thus in these zones Θ_ℓ is considered null in the solution of the corresponding mass balance equations.

The annular photoreactor involves only direct radiation. Hence, the authors decided to employ linear models together with the ESVE Model. Performing a radiant energy balance and using the Lambert–Beer equation for each model the following dimensionless expressions for the generic term Θ_ℓ, were obtained:

$$\Theta_{\text{LSPP}} = \frac{1}{\gamma} \exp\left(-\int_{\gamma_{\text{in}}}^{\gamma} \varepsilon \Psi_A \, d\bar{\gamma}\right) \qquad \text{(IV.13)}$$

$$\Theta_{\text{LSSE}} = \frac{1}{2\gamma} \int_{\theta_1}^{\theta_2} \exp\left(-\int_{\Gamma_{\text{in}}(\theta)}^{\Gamma(\theta)} \varepsilon \Psi_A \, d\bar{\Gamma}\right) d\theta \qquad \text{(IV.14)}$$

$$\Theta_{\text{ESVE}} = \frac{1}{2\pi\gamma_L^2} \int_{\phi_1}^{\phi_2} \Delta(\phi) \int_{\theta_1(\phi)}^{\theta_2(\phi)} \exp\left(-\int_{\Gamma_{\text{in}}(\theta,\,\phi)}^{\Gamma(\theta,\,\phi)} \varepsilon \Psi_A \, d\bar{\Gamma}\right) d\theta \, d\phi$$

$$\text{(IV.15)}$$

where:

$$\Delta(\phi) = 2(\gamma_L^2 - \gamma^2 \sin^2 \phi)^{1/2}$$

$$\Gamma_{in} = 1 \qquad\qquad \gamma = \frac{r}{r_{in}}$$

$$\Gamma_{in}(\theta) = \frac{1}{\sin \theta} \qquad\qquad \Gamma(\theta) = \frac{\gamma}{\sin \theta} \qquad\qquad \text{(IV.16)}$$

$$\Gamma_{in}(\theta, \phi) = \frac{1}{\sin \theta(\phi)} \qquad\qquad \Gamma(\theta, \phi) = \frac{\gamma}{\sin \theta(\phi)}$$

In order to evaluate the limiting angles $\theta_i(\phi)$ in Eqs. (IV.14) and (IV.15), one must consider that each zone will have a different pair of integration values. The expression for the limits is:

$$\theta_i(\phi) = \tan^{-1}\left[\frac{\gamma \cos \phi - (P - \gamma^2 \sin^2 \phi)^{1/2}}{(\zeta_0 + Q)} Ge \right] \qquad i = 1, 2$$

$$\text{(IV.17)}$$

where P and Q are given in Table IV.2. The ϕ-limits for the ESVE Model are given by the following expressions:

$$-\phi_1 = \phi_2 = \cos^{-1}\left[\frac{(\gamma^2 - \gamma_L^2)^{1/2}}{\gamma} \right] \qquad\qquad \text{(IV.18)}$$

A numerical complexity that must be treated carefully is the introduction of the three-dimensional attenuation into the two-dimensional model of the reactor. The problem can be solved by analyzing the projection of the different attenuating planes on the two-dimensional reacting plane (rectangular mesh).

In order to analyze the LSPP Model, we must consider that the absorption occurs through a unidimensional attenuation along a straight line. We can easily make the attenuation line coincide with the horizontal line of the computational mesh. For each point, the LSSE Model takes into account all the rays contained in the mesh

Table IV.2   Definitions of P and Q

Zone	Defined by	$\theta_1(\phi)$		$\theta_2(\phi)$	
		P	Q	P	Q
a	$\zeta_{SLI} \leq \zeta < \zeta_0$	γ_L^2	$1 - \zeta$	γ_{in}^2	$-\zeta$
b	$\zeta_0 \leq \zeta \leq \zeta_0 + 1$	γ_L^2	$1 - \zeta$	γ_L^2	$-\zeta$
c	$\zeta_0 + 1 < \zeta \leq \zeta_{SLII}$	γ_{in}^2	$1 - \zeta$	γ_L^2	$-\zeta$

plane that go through the reactor center line. In this case the
radiation convergence to a point in the reaction space is a two-
dimensional phenomenon. The ESVE Model introduces a three-
dimensional convergence phenomenon; consequently, rays not be-
longing to the r–z plane should also be considered. This makes it
necessary to evaluate the real distance travelled by each energy
bundle and the sequence of concentrations found along its path. To
know the exact path they travel within the reaction space, one may
resort to trigonometric relationships. Considering the azimuthal
symmetry, it is possible to find the required information performing
a cylindrical projection of the rays on the plane of the mesh (Fig.
IV.2b).

The numerical solution computes through the mesh in the flow
direction. When solving the equations at a point, bundles of radiation going through regions already calculated (upstream section)
must be computed with those bundles crossing regions of the reactor
whose correct concentration has not yet been calculated (down-
stream section). The problem was described at the beginning of this
section where the iterative procedure was explained. This complexity is avoided only if the LSPP model is used.

Results and Conclusions

In order to analyze the results for the three species involved in the
process, a representative set of parameters for a gaseous reacting
mixture was taken. Figure IV.3a shows the radial profiles of A^* and
B for three axial positions ($\bar{\zeta}$). Radial variations for B show a
minimum in the medium zone of the reacting space. This situation is
the result of the magnitude acquired by the radiant field and the

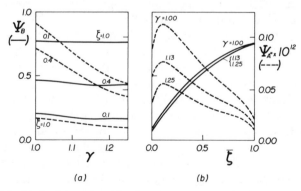

Figure IV.3   (a) Radial profiles of A^* and B for three axial positions. (b) Axial
profiles of A^* and B for three radial positions.

residence time at each point. Due to the attenuation effect, the radiation flux density decreases with the radius; hence, for regions near the inner radius the effect of a strong radiation field and that of a long residence time are added, which explains the high conversions attained. The same situation is repeated near the outer radius, even though with a weaker radiation field due to attenuation. In this case, conversions are large but smaller than those reached in the inner radius. In the central zone there appear low conversions as a consequence of shorter residence times and of a partially attenuated radiation field.

Figure IV.3b shows the axial profiles of A^* and B for three different radii. Axial variations of Ψ_B do not require further explanation, since they show larger conversions for increasing axial lengths. Besides, as it has already been said, greater concentrations of the product are reached in γ_{in} than in γ_{ou} which in turn are greater than those for the central region. Regarding A^*, one may conclude that this species behaves in the same way as the LVREA.

After the initial calculations, the authors carried out an extensive parametric study for a wide range of values. The major difficulty lay in the fact that there is no way of completely isolating the different dimensionless parameters, as Foraboschi [43] anticipated in 1959. For instance, the reactant absorptivity appears in the initiation rate and in the attenuation of radiation, the concentration of the absorbing species enters into the reaction kinetics and the attenuation of radiation, and the reactor radius affects all conventional reactor properties and the attenuation of radiation. However, some conclusions could be drawn and stress was put on the regions of higher sensitivity to the variations of operating conditions. The whole analysis can be found in the quoted paper, but here we will look into some of the "non-conventional" results.

In order to take into account the combined effect of the radial mixing and the optical density, Fig. IV.4a represents $\langle \Psi_B \rangle$ at the reactor outlet vs. Pe for different values of the absorption number ε (due to variations of the diffusion coefficient $D_{i,m}$ and the absorption coefficient α, respectively). It can be noted that for low and intermediate values of ε, the radial mixing has little influence. Nevertheless, as ε increases, this no longer occurs. It can be observed that the effect of a high absorption coefficient clearly depends on the value of the operating Peclet number. When the diffusion coefficient is high enough, the system physically "averages" the activation caused by radiation, although it may be produced in a narrow zone of the reactor near the region of incidence

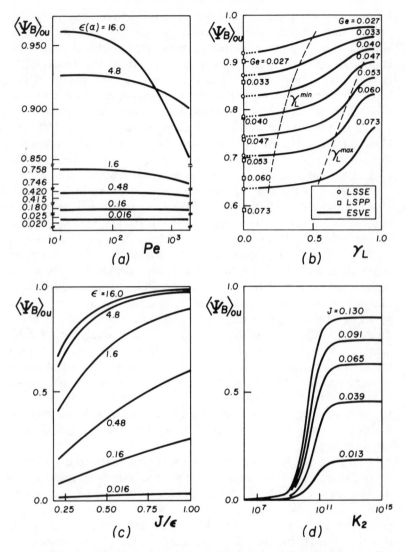

Figure IV.4   (a) Combined effect of the radial mixing and the absorption number. (b) Effect of the radiation source radius. (c) Effect of the radiation source power output and the absorption number. (d) Effect of the kinetic constant K_2 and the kinetic initiation number J.

of radiation. For certain values of the Peclet number, up to certain limits, a strong absorption will produce an increase in the final conversion.

On the other hand, if the radial mixing is poor, a strong absorption may lead to a drastic decrease of conversion because a significant portion of the reacting feed may go through the reactor without "realizing" the presence of radiation, thus it is clear that the conversion goes through a maximum. Here it is very important to note that an optimal value of ε can be found, i.e., a value of ε which produces a maximum outlet conversion. Several authors overlooked this effect since their determination of concentration profiles involved a previous cross-sectional area average of the LVREA. Consequently, this average disguised the possibility of recognizing that, in some cases, the reactor was almost "opaque". Hence, they concluded that as ε increases, the conversion always increases.

The effect of the radiation source radius is shown in Fig. IV.4b, which represents $\langle \Psi_B \rangle$ vs. γ_L using Ge as a parameter with constant equivalent diameter. The results of the LSPP and LSSE Models are also included. Each curve implies an increase in γ_L to the maximum value of $\gamma_L = 1$. However, such a distance cannot always be achieved and, in fact, the two broken lines depicted in the figure indicate the practical lower and upper limits. Three well-defined regions can be detected: (1) when γ_L is small the system has low sensitivity to changes in the source radius. Physically the reacting mixture is relatively far from the source and does not "sense" changes in lamp radius; (2) when γ_L is large, the system is again scarcely sensitive. The source is close enough to the reacting mixture and therefore, further increments in the lamp radius result in slight changes in conversion; (3) for intermediate values of γ_L, the system is very sensitive and slight variations in its value produce significant changes in conversion. These effects are magnified when Ge increases.

Interesting conclusions may be drawn regarding the source dimensions if one restricts the analysis to the practical region enclosed by the broken lines. Although for high values of Ge the curves present significant variations in the third region, in practice the remote possibility of using a source with such a large radius makes the situation almost indifferent to the lamp radius. For intermediate values of Ge the broken lines enclose a great part of the third region of the curve. This means that the radiation source radius may acquire considerable importance in such a situation and lead to significant errors if it is not taken into account. For low values of Ge the useful area mainly covers the second region of the curve, and again γ_L loses importance.

The existence of the regions indicated in (1) and (2), particularly for low values of Ge, do not imply that either the LSPP or the LSSE

Model will predict the proper conversion. This is illustrated, for example, by the case of Ge = 0.027. However, one can say that the LSSE Model is reliable when Ge is high and γ_L is low.

The effect of the radiation source output power should be analyzed by simultaneously considering the optical thickness of the system. Figure IV.4c shows this effect using as abscissa the ratio J/ε which is proportional to the power of the source and independent of α. It should be noted that when ε is low, the system becomes almost insensitive to changes in the input power. As a result of the low absorption of the reacting mixture, the value of the source power is not important. When ε is large, at first the conversion increases with the input power, but later the system shows low sensitivity due to a "saturation" of the absorption phenomenon. Consequently, when ε is either low or high enough, an increase in the source power is not the best route to increase conversion.

Figure IV.4d represents the influence of K_2 on $\langle \Psi_B \rangle$ for different values of J (using Φ as a parameter). Two regions may be distinguished: (i) for lower and higher values of K_2, the reactor has low sensitivity to changes in K_2; (ii) for intermediate values of K_2, the system is very sensitive to its variations. Within this intermediate range, any simulation of the reactor will require the knowledge of a precise value of K_2, particularly if Φ is very high. This is especially important for chain reactions where correct values of the kinetic constants are difficult to obtain. On the other hand, if the situation corresponds to case (i), a precise determination of K_2 is not important. Summarizing, one may conclude that: (1) for low values of K_2, variations in either Φ or K_2 will produce small changes in the prediction of the output conversion, (2) for large values of K_2, only changes in Φ will influence the prediction of the output conversion since those made on K_2 would be of no significance, (3) for intermediate values of K_2 changes in both Φ and K_2 are important. However, in this range K_2 has a stronger effect.

The main conclusions, which the authors pointed out about the parametric study, were the following:

– Conversion may present a maximum according to the values of some parameters of the system, such as the Peclet number and the absorption number. This is an interesting situation since it indicates the possibility of acting on some of the design variables to achieve the optimal conversion.
– It was possible to delineate regions in which the linear models are reliable. However, the use of linear models outside these regions

may lead to significant errors (low and intermediate values of Ge, and intermediate and high values of γ_L).
- The output conversion may be significantly increased with increasing lamp output power, but only within certain ranges of ε.
- It is possible to find zones where the system is very sensitive to variations in K_2 and Φ.
- There exist ranges of high sensitivity of the design parameters of an annular photoreactor where their variation considerably affects the final product conversion. In general, these ranges are associated with pairs of parameters due to the high, unavoidable coupling existing between the dimensionless numbers governing this type of reactors.

Actinometric Reaction

De Bernardez and Cassano [38] investigated the possibility of using the uranyl oxalate actinometer for continuous photoreactors under extreme conditions of radiation gradients. With this approach they simultaneously checked the quality of the ESVE Model and the ability of a chemical actinometer to evaluate the radiation field inside a practical photoreactor in an absolute manner.

Details about the actinometer can be found in several papers [18, 44, 51, 66, 91]. Most results have been obtained in generally well stirred, batch reactors. The reactant is a complex formed with uranyl sulfate (or uranyl nitrate) and oxalic acid (or sodium oxalate) in aqueous dilute solutions. For its use as an actinometric system the recommended concentrations are: $0.01\,M$ UO_2SO_4 and $0.05\,M$ $C_2O_4H_2$, although concentrations ten-fold smaller can also be used. The oxalic acid decomposition overall quantum yield, for oxalic acid conversions below 70–80%, is well known. However, it is strongly recommended that the extent of the reaction should never exceed 20% because beyond that conversion it cannot be assumed that the system behaves as a photosensitized reaction. On the other hand, in order to be able to carry out the analysis of the reaction products with the required accuracy, conversion must be higher than 5%.

The main overall photochemical reactions are the following:

$$UO_2^{2+} + H_2C_2O_4 + h\nu \rightarrow UO_2^{2+} + CO_2 + CO + H_2O$$

$$UO_2^{2+} + H_2C_2O_4 + h\nu \rightarrow UO_2^{2+} + CO_2 + HCOOH$$

We know that under controlled conditions, the reaction has the following characteristics: (1) zero order with respect to the oxalic acid concentration, (2) temperature coefficient close to unity, (3) small effects of added electrolytes, (4) constant attenuation coefficient, and (5) constant quantum yield in the pH range from 1 to 6. In spite of all these advantages it must be stressed that the true nature of the chemical complexes formed with the oxalic acid and the uranyl ions is poorly known. In addition, it is known that the reaction mechanism has a strong dependence on the pH of the reacting solution. Heidt et al. [51] pointed out that at low pH (close to 0) the CO produced/oxalate decomposed molar ratio is approximately 1, clearly establishing that under these conditions the first reaction is the only significant one. This ratio decreases when the pH is in the range from 1 to 3 showing an increasing significance of the second reaction. For pH values higher than 6 there is evidence that the first reaction no longer occurs.

Model Equations
The authors employed the same assumptions listed in Section IV.1.2. Therefore, the mass balance equation for oxalic acid is given by Eqs. (IV.9) and (IV.10). Considering the kinetic characteristics of the reacting system described before, R_{ox} may be written [19, 75]:

$$R_{ox} = -\int_{\nu=0}^{\nu=\infty} H(C_{ox}) \Phi_{ox,\nu} e_\nu^a \, d\nu \qquad \text{(IV.19)}$$

where $\Phi_{ox,\nu}$ is an overall quantum yield and the Heaviside function H indicates that the reaction is of order zero with respect to the oxalic acid concentration.

The LVREA expression provided by the ESVE Model, for monochromatic radiation in a homogeneous medium, is given by Eq. (IV.15). It is clear that the reactor can be safely modelled as operating with monochromatic radiation since 85% of the radiation output power of the lamp falls into a single wavelength at $\lambda = 2537$ Å. Besides, the absorption characteristics of the uranyl oxalate complex show the following significant variations:

λ (Å)	μ (cm^{-1})
2250	16.71
2950	1.96
3450	0.17

Based on the criterion drawn for heterogeneous systems [94] and the characteristics of the lamp used, the authors modelled the absorption process with the equations valid for homogeneous media, because the value of the attenuation coefficient of the pure liquid is higher than $0.4 \, \text{cm}^{-1}$.

With respect to the evaluation of the limiting angles $\theta_i(\phi)$ and ϕ_i, and considering the existence of the reactor wedges, the authors used the same methodology employed in Section IV.1.2. Besides, the existence of lamp wedges was considered using a function f defined by:

$$f(\theta, \phi) = \frac{(z_j - z) \tan \theta - r \cos \phi + (r_L^2 - \sin^2 \phi)^{1/2}}{2(r_L^2 - \sin^2 \phi)^{1/2}} \qquad \text{(IV.20)}$$

where:

$$z_j = z_{L2} \text{ for the upper lamp wedge}$$

$$z_j = z_{L1} \text{ for the lower lamp wedge}$$

Hence, considering the f-function the LVREA expression becomes:

$$e_\nu^a = \frac{E_\nu \mu_\nu}{4 \pi V_L} \int_{\phi_1}^{\phi_2} \Delta(\phi) \int_{\theta_1}^{\theta_2} f(\theta, \phi) \exp\left(-\int_{\Gamma_{\text{in}}(\theta,\phi)}^{\Gamma(\theta,\phi)} \mu_\nu \, d\bar{\Gamma}\right) \qquad \text{(IV.21)}$$

The numerical procedure followed in the solution, particularly to compute the three-dimensional attenuation process, was described in detail in the previous section and will not be repeated here.

Once the oxalic acid concentration field inside the reactor was known, the authors calculated the average exit conversion, to compare it with the experimental results. The average exit conversion was computed as follows:

$$\langle X_{\text{ox}} \rangle = 1 - \frac{\langle C_{\text{ox}} \rangle}{C_{\text{ox}}^{\circ}} \qquad \text{(IV.22)}$$

where:

$$\langle C_{\text{ox}} \rangle = \frac{1}{F_v} \int_{r_{\text{in}}}^{r_{\text{ou}}} 2 \pi r v_z(r) C_{\text{ox}}(r, L_R) \, dr \qquad \text{(IV.23)}$$

Experimental Study

The reaction was carried out in two continuous annular reactors of different sizes. The irradiated length of the reactors was limited by the presence of two masks which made it possible to vary the effective reactor length. The dimensions of the reactors used were:

	L_R (cm)	r_{in} (cm)	r_{ou} (cm)
Reactor No. 1	30	2.3	2.7
Reactor No. 2	30	2.3	3.6

The radiation was provided by a General Electric G15T8 [47], low-pressure germicidal lamp, designed to emit almost all the radiant energy at 2537 Å. Due to the shorter wavelength of the emissions, the reactor inner wall was made of quartz, Suprasil quality. The initial concentration of the solution was 0.001 M uranyl sulfate and 0.005 M oxalic acid.

Results and Conclusions
Figures IV.5a and b show the comparison between the experimental results and the theoretical predictions estimated using the ESVE Model. This comparison is made for both reactors. The solid line indicates the theoretical values obtained using the maximum U.V. output power (after 100 hours of operation). The dashed line corresponds to the calculations using the average U.V. output power through life. Both values are nominal specifications provided by the

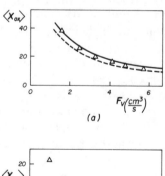

(a)

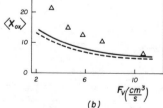

(b)

Figure IV.5 Comparison between theoretical predictions and experimental data for constant attenuation coefficient. (a) Reactor No. 1. (b) Reactor No. 2.

lamp manufacturer. It should be expected that experimental values fall below the solid line, at least for conversion values lower than 20%, which is the maximum recommended value for the actinometric reaction. Figure IV.5a (reactor with short radiation path length) seems to reproduce at least approximately the expected results. It is clear that the predictions are no longer valid for Fig. IV.5b (reactor with longer radiation path length).

In a continuous reactor, the flow characteristics generate the presence of unavoidable concentration gradients, mainly in the radial direction where the LVREA variations are more noticeable. Hence, although the average exit conversion may be lower than 20%, it is possible that in certain regions inside the reacting volume the local conversion exceeds this value, mainly in those regions close to the entrance of light. It is clear that it is almost impossible to avoid the operation of the reactor with very high conversions in regions closer to the inner wall (the exception would be the impractical case of using a very low optical thickness).

Surely when most of the oxalic acid has been consumed, we may have at least three sources of discrepancies: (1) the attenuation coefficient may be a function of composition, (2) a change in the reaction mechanism may occur, and (3) the value of the quantum yield could be affected when the conversion is too large. It seems plausible that the main reason for the disagreement between experimental and theoretical results shown in Figs. IV.5a and b could be the invalidation of the hypothesis of constant attenuation coefficient when conversion values are higher than 20%. Since the uranyl ion concentration remains constant during a large extent of the reaction, variations in the attenuation coefficient should result from changes in the oxalic acid concentration.

Absorbance of solutions of various concentrations of oxalic acid but constant concentration of uranyl ion were spectrophotometrically measured at different wavelengths and Fig. IV.6 shows the results. It should be noticed that the attenuation coefficient remains

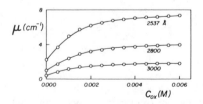

Figure IV.6.   Absorbance of different oxalic acid concentrations.

almost constant when the oxalic acid concentration is greater than 0.004 M. Hence, the hypothesis of constant attenuation coefficient is satisfied only if the reaction proceeds without reaching conversions greater than 20% (this situation is really significant at short wavelengths).

Considering the new absorption data, theoretical predictions were repeated. Since the attenuation coefficient is now a function of the oxalic acid concentration, there is a coupling between the mass balance equation and the evaluation of the radiant energy field. Figures IV.7a and b show results calculated with a variable attenuation coefficient and provide surprisingly good corroboration of the expected results. The point that lies above the solid line corresponds to the operating conditions of a reactor where oxalic acid conversions close to one are produced in significant portions of the reactor volume. If this is the case, departures from the initial assumptions as indicated in (2) and (3) will gain significance. Thus for very high values of the mean residence time the predictions of the model should show considerable differences with the experimental values.

Finally, from the previous analysis, the authors arrived at the following conclusions:

– Special care should be taken when the uranyl oxalate actinometer is used to verify the radiation field inside a continuous, segregated

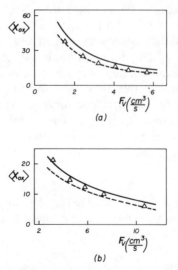

Figure IV.7   Comparison between theoretical predictions and experimental data for variable attenuation coefficient. (a) Reactor No. 1. (b) Reactor No. 2.

flow reactor where it is impossible to avoid operation at very high conversions in certain regions of the reacting volume. If one desires accurate predictions, and the evaluation of the LVREA is performed under the hypothesis of constant attenuation coefficient, it must be assured that the oxalic acid conversion does not exceed the maximum permissible value of 20% at any point inside the reactor volume. The situation is particularly critical at low wavelengths.

- The reactor modelling for this system may be carried out allowing the attenuation coefficient to be a function of the oxalic acid concentration. In this case, local conversion values must be lower than 70% due to the lack of reliable quantum yield information beyond this value.

- When the reaction is used to evaluate the radiation field, i.e. no model is used to compute the radiation properties, the system requires very good stirring to avoid high local values of conversion.

IV.2. Tubular Photoreactor with Elliptical Reflector

Clariá et al. [31] modelled a tubular photoreactor with an elliptical reflector in order to predict the average exit conversion under different operating conditions. For the experimental verification they chose a simple reaction to avoid the introduction of kinetic complexities into the scope of the study. This included: (i) the analysis of the quality of radiation field predictions, (ii) the errors introduced by azimuthal asymmetries when a two-dimensional modelling is adopted, and (iii) the applicability of the elliptical photoreactor to obtain reliable kinetic data in laboratory and bench-scale reactors. To accomplish this they used the uranyl oxalate photodecomposition and an elliptical photoreactor where angular asymmetries may be present. Careful experimental results were compared with theoretical predictions in order to reach conclusions about the above-mentioned questions.

Model Equations
Using the standard assumptions listed in the isomerization reaction and including azimuthal symmetry, the dimensionless mass balance for the oxalic acid can be written as:

$$U(\gamma) \frac{\partial \Psi_{ox}}{\partial \zeta} = \frac{1}{\text{Pe Ge}} \frac{1}{\gamma} \frac{\partial}{\partial \gamma} \left(\gamma \frac{\partial \Psi_{ox}}{\partial \gamma} \right) + \Omega_{ox} \qquad \text{(IV.24)}$$

$$\Psi_{ox}(0, \gamma) = 1 \tag{IV.25}$$

$$\frac{\partial \Psi_{ox}}{\partial \gamma} (\zeta, 0) = \frac{\partial \Psi_{ox}}{\partial \gamma} (\zeta, 1) = 0 \tag{IV.26}$$

where Ω_{ox} is given by Eq. (IV.19) with the corresponding characteristic scale factors.

In order to obtain the LVREA expression, it is necessary to consider that the radiation reaches the reactor by direct and reflected radiation. It has been shown [28] that for typical reactors indirect radiation with only one reflection accounts for a very large fraction of all radiation contributions to the reactor (from 80 to 90%). Thus, the authors modelled the reactor with single reflected indirect radiation only. This approach yielded:

$$e^a = \frac{C_{ab}}{2\pi V_L} \int_{\nu=0}^{\nu=\infty} \Upsilon_{R,\nu} \Gamma_{Rf,\nu} \alpha_\nu E_\nu \int_\phi d\phi \int_\theta (r_L^2 - D^2 \sin^2 \xi)^{1/2}$$

$$\times \exp\left[-\alpha_\nu C_{ab}(\rho_I - \rho_0)\right] d\theta\, d\nu \tag{IV.27}$$

where all the variables involved in Eq. (IV.27) were discussed in Section II.2.

In Eq. (IV.27) four design parameters are wavelength dependent. The upper and lower bounds of the integral for ν must be obtained for each particular case according to the overlapping influence of all of them. The lower limit will be given by the lower wavelength of lamp emission, reactant (and/or product) absorption, reflector reflection coefficient and wall reactor transmission (whichever the largest). The upper limit is given by the upper wavelength limits of the same properties (whichever is the smallest). According to the experimental operating conditions we can define two limits ν_1 and ν_2 in place of 0 and ∞ respectively. From Eq. (IV.27), the frequency integral can be written:

$$I = \int_{\nu_1}^{\nu_2} \Upsilon_{R,\nu} \Gamma_{Rf,\nu} \alpha_\nu E_\nu \exp\left[-\alpha_\nu C_{ab}(\rho_I - \rho_0)\right] d\nu \tag{IV.28}$$

In practice, the reflection coefficient ($\Gamma_{Rf,\nu}$) for specular type aluminum sheets, the transmission coefficient of the reactor wall ($\Upsilon_{R,\nu}$) and the absorption coefficient for the absorbing species (α_ν) are continuous functions of ν. On the other hand, most U.V. lamps provide an emission spectrum which is discontinuous (see, for example, emission lines of mercury arc lamps by Hanovia, General

Electric, etc.). One can write:

$$E_\nu = E_\sigma \delta(\nu - \nu_\sigma) \qquad \text{(IV.29)}$$

with δ being the Dirac function. Then, the authors substituted the polychromatic lamp by a finite number of monochromatic ones, each one of them emitting the output power corresponding to each line of emission σ. Then, Eq. (IV.28) can be written as:

$$I = \sum_{\sigma=1}^{\sigma=n} \int_{\nu_1}^{\nu_2} \Upsilon_{R,\nu} \Gamma_{Rf,\nu} \alpha_\nu E_\nu \delta(\nu - \nu_\sigma) \exp\left[-\alpha_\nu C_{ab}(\rho_I - \rho_0)\right] d\nu$$

$$\text{(IV.30)}$$

$$I = \sum_{\sigma=1}^{\sigma=n} \Upsilon_{R,\sigma} \Gamma_{Rf,\sigma} \alpha_\sigma E_\sigma \exp\left[-\alpha_\sigma C_{ab}(\rho_I - \rho_0)\right] \qquad \text{(IV.31)}$$

Finally, the LVREA expression can be obtained by Eqs. (IV.27) and (IV.31):

$$e^a = \frac{C_{ab}}{2\pi V_L} \int_\phi d\phi \int_{\theta(\phi)} (r_L^2 - D^2 \sin^2 \xi)^{1/2}$$

$$\times \sum_{\sigma=1}^{\sigma=n} \Upsilon_{R,\sigma} \Gamma_{Rf,\sigma} \alpha_\sigma E_\sigma \exp\left[-\alpha_\sigma C_{ab}(\rho_I - \rho_0)\right] d\theta \qquad \text{(IV.32)}$$

Experimental Study
The experimental set-up is shown in Fig. IV.8. Tanks A, B, C and D were $0.2\,\text{m}^3$ each in order to assure enough reactant solution for a

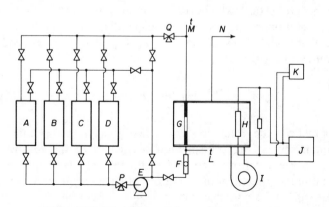

Figure IV.8   Flow-sheet of the experimental device.

one-day run. All parts in contact with the reacting solution were made of Pyrex glass (or Suprasil quartz for the reactor wall) and Teflon. With the exception of the irradiated zone of the reactor (L_R) all other parts of the system were totally isolated from light.

The lamp designated by H was a GE-UA-3 Uviarc Lamp from General Electric Co. [46] which operated with an input power of 360 watts. It had an arc length of 15.2 cm and its nominal diameter was 1.9 cm. The spectral distribution of its output power (mainly in the U.V. and visible range) was provided by the manufacturer. The first 200 hours of operation were used to put the lamp in the most stable state of its average lifetime. The reflector was made of aluminum sheet, specularly finished with Alzac treatment, provided by ALCOA [2]. The dimensions were: $a = 30$ cm, $c = 20$ cm and $L_{Rf} = 43$ cm. It was mounted on two aluminum plates, 1.2 cm thick, precisely machined to the desired elliptical shape. Special care was taken to have the lamp and the reactor tubes very accurately located at the reflector focal axis.

Three reactors were used:

	No. 1	No. 2	No. 3
L_R (cm)	12.0	10.0	10.0
r_R (cm)	0.2	0.5	1.0

where L_R and r_R are nominal length and nominal inside radius respectively. Inlet concentrations were sampled at P and outlet concentrations at Q. Once the steady-state exit conversion measured in Q was achieved, every run was repeated three times. Conversions were expressed as:

$$\langle X_{exp} \rangle = \frac{\langle C_{ox} \rangle_{in} - \langle C_{ox} \rangle_{ou}}{\langle C_{ox} \rangle_{in}} = 1 - \langle \Psi_{ox} \rangle_{ou} \qquad (IV.33)$$

For each point the three experimental values were averaged and compared with predicted values according to:

$$\langle X_t \rangle = 1 - 4 \int_0^1 (1 - \gamma^2)\Psi_{ox}(\gamma, 1)\gamma \, d\gamma \qquad (IV.34)$$

Results and Conclusions
The prescribed conditions for conversion gave rise to some limitations in the possible operating flow rate. Also, when the standard actinometric solution was used, the reactor operated almost as an

opaque body because all the absorption of radiation was produced in a thin layer close to the wall. This resulted in a very small averaged exit conversion and a large amount of bubble production at the wall. Hence a ten-fold smaller concentration was used since the reported quantum yields are still valid [44].

Figure IV.9 shows a compendium of all the experimental data: $\% \langle X_{exp} \rangle$ vs. $\% \langle X_t \rangle$. It is clear that for the small reactor diameter predictions are good indeed and the error is never larger than 15%. However, for medium and large diameter reactors the situation is different. Deviations are between 15 and 35% for the reactor with $r_R = 0.5$ cm and from 45 to 50% for the reactor with $r_R = 1.0$ cm. An explanation is needed for all these agreements and deviations. In order to find it let us first look at Figs. IV.10a, b, c and d. Here the solid lines are theoretical predictions and the experimental points are represented by triangles and circles. The reactor with $r_R = 0.2$ cm shows remarkable good results regardless of the magnitude of the oxalic acid local concentration. Looking at the ratio r_R/r_L and the reflector eccentricity, one can find that $r_R/r_L = 0.2$ and $e = 0.67$. In this case the system operates under conditions where azimuthal asymmetries can be safely neglected [37].

That is not the case of the other two reactors, but in the one of intermediate size we find some surprising results. The concentrated solution gave experimental data consistently above the theoretical predictions but presented only very small deviations from them (about 15% in the worst cases). The low concentration solution produced larger deviations (30%). Going into a more detailed investigation of the intermediate diameter reactor one can

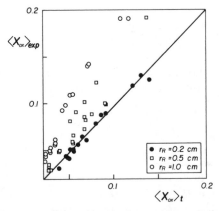

Figure IV.9   Compendium of the experimental results.

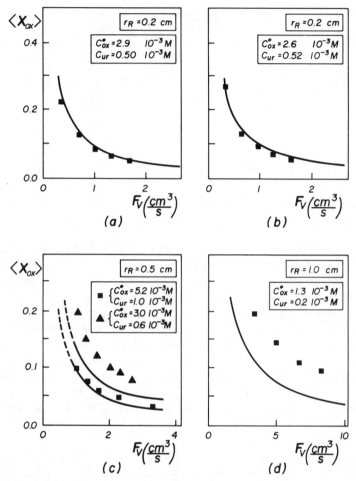

Figure IV.10 Comparison between theoretical predictions and experimental data. Effect of the volumetric flow rate on the average exit conversions.

obtain a plot of theoretical radial concentration profiles using the reacting solution concentration as a parameter. The results are shown in Fig. IV.11. When the solution is concentrated $(C_{ur} = 2.0 \times 10^{-3}\,\text{M}$ and $C_{ur} = 1.0 \times 10^{-3}\,\text{M})$ conversions at any point never exceed 25%. In all other cases the combination of: (1) the radiant energy flux density radial gradients, (2) the parabolic velocity profiles and [3] the required mean residence time for achieving the prescribed range of averaged exit conversions, produced very large local values of X close to the

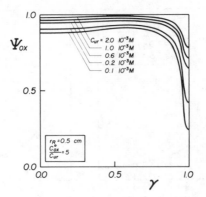

Figure IV.11   Theoretical radial concentration profiles as a function of uranyl ion concentration.

reactor wall. If the conversion exceeds 20% we know that some of the reaction parameters reported by the literature are not valid any more. In addition we have previously seen that when the attenuation coefficient is a function of the oxalic acid concentration and the latter changes significantly close to the reactor wall, theoretical predictions with $\mu_\nu = $ constant must fall below the experimental data (Section IV.1). This effect is systematically shown in all these comparisons. At the same time, when the uranyl oxalate complex concentration is rather high, radiation "penetrates" only a small fraction of the total reactor volume. Under these conditions azimuthal asymmetries have a weaker effect. Not many differences would be observed with respect to a modelling using a three-dimensional approach.

Consequently we may think that azimuthal asymmetries, changes in the value of μ_ν and possible invalidation of the values of $\Phi_{ox,\nu}$ (and in this order) may provide an explanation for the intermediate diameter reactor behavior. On the other hand, the good agreement found in the small diameter reactor is due to the fact that it operates with angular symmetry and that changes in the attenuation coefficient for this system are less significant than those observed in Section IV.1. The spectral distribution of the radiation output power for both lamps is very different (power emission at short wavelengths is not dominant here) and the oxalic acid concentration effect results less influential.

To obtain reasonable conversions (>5%) within the practical possibilities of the system, the large diameter reactor had to be operated at low actinometric solution concentrations. Therefore, in these runs angular asymmetries and high conversions at the reactor wall certainly introduce large errors in the model predictions. Moreover, when the reacting solution has low optical thickness and the reactor diameter is large, very likely second reflections will have significant contributions. To simplify computations, Clariá et al. [31] performed all their calcuations with the hypothesis of angular symmetry at $\beta = 0$. This point provides minimum values of the radiation field. Actual values must always be greater than predictions due to the contributions of other angular positions. The same reasoning can be extended to the expected results if the second reflection is included.

Further research work (using oxalic acid concentration-dependent attenuation coefficients, three-dimensional modelling and inclusion of the second reflection) is needed to confirm these explanations [81]. However, good evidence in support of our hypothesis can be found in the results plotted in Fig. IV.12. It is clear that the inclusion of angular asymmetries (with the addition of direct radiation and the inclusion of other positions in the reactor with higher values of the first reflection contributions) and second reflections, will increase the value of the radiation field. Since the decrease in the oxalic acid concentration produces a decrease in the value of μ, the radiation field inside the reactor will necessarily be stronger. The three factors will contribute to move the solid line of predicted values towards the experimental results. The figure also shows that equivalent mean residence times for the same radiation source and

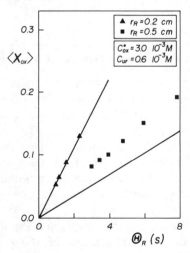

Figure IV.12 Effect of the mean residence time on the averaged exit conversion.

the same reactant concentration produce very different yields, depending upon the "optical density" of the reactors. This is a characteristic behavior of photoreactors that must be stressed once more.

From the results obtained in the three experimental reactors one may conclude that:

– When azimuthal asymmetries can be neglected, a two-dimensional modelling of the tubular reactor can be safely performed. This will increase the applicability of the elliptical photoreactor for laboratory work particularly for much more complex reactions like chain-type kinetics.
– When conditions of negligible azimuthal asymmetries cannot be fulfilled, the modelling of the elliptical photoreactor cannot be performed with a two-dimensional mass balance equation. A three-dimensional radiation and concentration balances should be employed. For commercial scale applications where r_R/r_L will always be larger than 0.5 and $e \geqslant 0.6$, the reactor modelling cannot be simplified.
– When a non-negligible fraction of the reactor volume is operating with local values of the conversion higher than 20% (even if the averaged exit conversion is lower) the uranyl oxalate reaction will not behave as a true photosensitized reaction. This may preclude its use for segregated flow reactors and restricts its safe application to perfectly stirred photoreactors. Undoubtedly this is an additional reason to encourage the use of more recommended potassium ferrioxalate actinometer.

IV.3. Cylindrical Photoreactor Irradiated from the Bottom

For the system previously discussed in Section II.3 on diactinic media, we decided to carry out an experimental verification of the ESVE Model predictions [8, 9]. Absolute and "almost point" values of the radiation field were obtained. To do so, a microreactor was used with the possibility of being precisely located at different positions within the cylindrical space under investigation (Fig. IV.13). This microreactor was operated differentially in a batch recycling system, which allowed us to follow the conversion of the reaction as a function of time in the tank (Fig. IV.14). From the comparison of the experimental results and the theoretical predictions for each location of the microreactor, the quality of the model may be rigorously tested. To carry out this test a well-known,

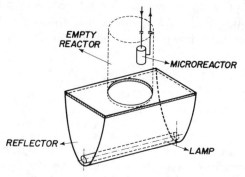

Figure IV.13 Geometry of the microreactor inside the empty cylindrical reactor.

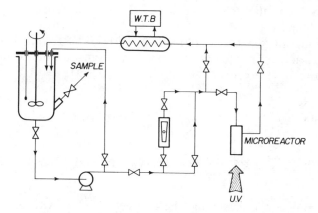

Figure IV.14 Flow-sheet of the experimental device.

kinetically simple photochemical reaction was used: the photo-decomposition of uranyl oxalate.

Model Equations
The material balance for the recycling reactor (differential operation in the reaction volume and high velocity of recirculation to assure perfect mixing in the total system) is given by the following expression [24–26]:

$$\frac{dC_i}{dt} = \frac{V_{MR}}{V_T} \langle R_i(\mathbf{x}, C_{ab}) \rangle = -\frac{V_{MR}}{V_T} \sum_{\Delta\lambda} \Phi_{ox,\lambda} \langle e_T^a(\mathbf{x}, C_{ab}) \rangle_{\Delta\lambda}$$

(IV.35)

where the expression of the reaction rate given by Eq. (IV.19) is used. Integrating Eq. (IV.35) with respect to time and making the resulting expression dimensionless, one obtains:

$$\Psi_i^{(j+1)} = \Psi_i^{(j)} - \frac{t_T \left[\sum_{\Delta\lambda} \Phi_{ox,\lambda} \langle e_T^a(\mathbf{x}, C_{ab}) \rangle_{\Delta\lambda} \right]}{C_i^{(0)}} \frac{V_{MR}}{V_T^{(j)}} [\tau^{(j+i)} - \tau^{(j)}]$$

(IV.36)

in which the variation of the total volume of the system due to successive samplings has been considered by means of the superscript (j). In order to complete the analysis, Eq. (IV.36) requires the evaluation of the LVREA as a function of the spatial variables of the system. To do so, it is necessary to consider the radiant energy arriving at each point of the microreactor, be it directly from the source or indirectly after a reflection on the parabolic mirror.

Considering the ESVE Model, the direct LVREA evaluated at point $\mathbf{x}$, and computing all contributions from the total volume of the source one obtains

$$e_D^a(\mathbf{x}, C_{ab}) = \alpha C_{ab} \kappa \langle Y_{MR} \rangle \int_{\phi_1^D(\mathbf{x})}^{\phi_2^D(\mathbf{x})} \Delta\rho'(\mathbf{x}, \phi)$$

$$\times \int_{\theta_1^D(\mathbf{x},\phi)}^{\theta_2^D(\mathbf{x},\phi)} \exp[-\alpha C_{ab} \sigma(\mathbf{x}, \theta, \phi)] \, d\theta \, d\phi \qquad (IV.37)$$

The expression representing the indirect LVREA with one reflection, after computing the total volume of the source and including an average reflection coefficient is:

$$e_{In}^a(\mathbf{x}, C_{ab}) = \alpha C_{ab} \kappa \langle \Gamma_{R_f} \rangle \langle Y_{MR} \rangle \int_{\phi_1^{In}(\mathbf{x})}^{\phi_2^{In}(\mathbf{x})} \Delta\rho_E'(\mathbf{x}, \phi)$$

$$\times \int_{\theta_1^{In}(\mathbf{x}, \phi)}^{\theta_2^{In}(\mathbf{x}, \phi)} \exp[-\alpha C_{ab} \sigma(\mathbf{x}, \theta, \phi)] \, d\theta \, d\phi \qquad (IV.38)$$

where a linear dependence of the attenuation coefficient with concentration is used.

The limiting angles for direct and indirect radiation and the expressions of $\Delta\rho'$ and $\Delta\rho_E'$ (Eqs. IV.37 and IV.38) are those already indicated in Tables II.4 and II.5 respectively. According to

Fig. IV.15, the attenuation path for any position of point I and any ray with direction (θ, ϕ) is given by:

$$\sigma(\mathbf{x}, \theta, \phi) = \rho(\mathbf{x}, \theta, \phi) - \rho_0(\mathbf{x}, \theta, \phi) = \frac{-y_{IR}}{\sin \theta \sin \phi} \quad \text{(IV.39)}$$

Again, it is necessary to consider the existence of totally or partially irradiated zones inside the microreactor. In this case, however, the extreme angles given by the cover of the emitting system should be incorporated together with those determined by the opening of the quartz circular window. The angles corresponding to the cover are those already indicated by Fig. II.23, while those of the microreactor are defined by employing similar geometrical criteria. Hence, the evaluation (see Table II.6) is done using as coordinates (r_i, β_i, y_i), the system given by $(r_{IR}, \beta_{IR}, y_{IR})$ and for the radius (R_i) the one corresponding to the quartz window (R_{MR}).

It is possible to obtain the total LVREA adding the contributions of direct and indirect radiation, i.e.:

$$e_T^a(\mathbf{x}, C_{ab}) = e_D^a(\mathbf{x}, C_{ab}) + e_{In}^a(\mathbf{x}, C_{ab}) \quad \text{(IV.40)}$$

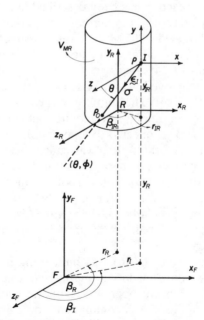

Figure IV.15 Microreactor coordinate systems.

Finally, the local values of e_T^a must now be averaged over the whole microreactor volume:

$$\langle e_T^a(\mathbf{x}, C_{ab}) \rangle = \frac{1}{V_{MR}} \int_{V_{MR}} e_T^a(\mathbf{x}, C_{ab}) r_{IR} \, dr_{IR} \, d\beta_{IR} \, dy_{IR} \qquad (IV.41)$$

Replacing the average values of the total LVREA in Eq. (IV.36), one obtains the final theoretical expression for the temporal evolution of the oxalic acid concentration.

In a previous section we saw that the values of α, Φ, and E present significant variations over the wavelength range of interest; hence, it is necessary to incorporate the polychromatic character of the radiation used. With this in mind, the average rate of reaction was evaluated for different values of $\Delta\lambda$. As a result it was found that significant contributions fall between 2200 and 4200 Å only. This range was included in Eq. (IV.36) considering five characteristic intervals of the emission process of the source. Finally, the experimental verification of the model was carried out by means of concentration-time curves (Eq. IV.36).

Experimental Study
The uranyl oxalate reaction has been described in previous sections. For the experimental runs the authors employed uranyl sulfate 0.01 M and oxalid acid 0.05 M. It was possible to follow the concentration of oxalic acid as a function of time with conventional titration techniques [9, 74].

The radiation employed was a GE-UA-3 Uviarc Lamp from General Electric Co. [46]. The stability in the emission of the lamp was determined by an Eppley thermopile (similar to the one described in Section II.3) connected to a potentiometric recorder, and the operating voltage of the source was continuously monitored. In order to exhaust the ozone generated by the U.V. radiation and to cool the lamp, a stream of controlled room-temperature circulating air was used. A cylindrical reflector of parabolic cross section made of aluminum sheet specularly finished with Alzac treatment was used. The total length of the reflector was 15.8 cm and its focal distance was 2.1 cm.

The microreactor (Fig. IV.16) was built using four Teflon parts with a circular window made of quartz and transparent to the radiation. The reaction chamber was cylindrical and had connections corresponding to the incoming and outgoing actinometric streams. An accurate positioning device allowed one to fix the

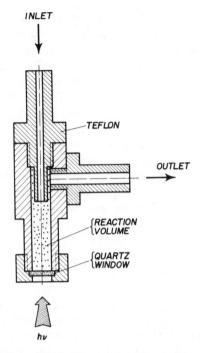

Figure IV.16   View of the microreactor.

microreactor in space and place it precisely with respect to the coordinate system used (x_F, y_F, z_F). The same mechanism allowed the simultaneous use of two microreactors in order to take advantage of the existing symmetry conditions and reduce the corresponding running time.

The experimental device was completed with a stirred collecting tank, a cooler, a recirculating pump and a rotameter (Fig. IV.14). The collecting tank was provided with a Teflon cover with the corresponding connections for sampling, actinometric solution circulation, placement of the thermometer and stirring with a variable speed controller. The heat exchanger kept the solution temperature constant. The fluid recirculation was obtained by a special Flex-I-Liner Vanton pump and the stability of the flow rate was controlled with a Matheson rotameter. All the elements of the apparatus that were in contact with the solution were built of Teflon, PVC, Kel-F, or Pyrex glass, and the complete system was protected from the action of light by means of an isolating black tape (except the microreactor quartz window).

The experimental procedure began with the stabilization of the whole system which required a minimum of four hours. During this period the microreactor window was protected from the action of the radiant energy, while the degree of stabilization of the lamp was followed by means of the thermopile readings and the continuous indication of the source voltimeter. During the last hour, the actinometric solution was charged and recirculated to attain the temperature and flow rate required in the fluid. Once this was achieved, the initial sample was taken and the photodecomposition of the oxalic acid was initiated by uncovering the quartz window of the microreactor. To obtain the experimental concentration-time curves, samples were taken at prespecified time intervals. Using the two symmetrically positioned microreactors simultaneously, the total time of reaction employed was nine hours, always remaining within the allowed conversion ranges.

Results and Conclusions
In order to experimentally verify the symmetry found in the theoretical study of the radiation field, the concentration-time results were compared for different runs by successively placing only one microreactor at symmetrical positions. Differences were never larger than 1%. Because of these results the analysis of the problem was performed only for angles within the first quadrant.

To carry out the comparison between the theoretical values of the radiation field and the results obtained from the experimental device, the slope of the Eq. (IV.36) representing the dimensionless material balance was used (m). Its variation was analyzed as a function of the space variables defining the position of the microreactor in the cylindrical space under study (β_R, ζ_R and γ_R).

Figures IV.17a, b and c illustrate the radiation field as a function of the azimuthal coordinate for a fixed height and for three different radial positions. This representation requires one to rotate the position of the microreactor in the first quadrant in a fixed axial position ($\zeta_R = 0.31$) and for three different radial positions ($\gamma_R = 0.41$, $\gamma_R = 0.65$ and $\gamma_R = 0.89$, respectively). The solid line indicates the theoretical predictions for the ESVE Model, and it can be observed that there is good correspondence with the experimental results. The maximum deviation is never greater than 20%. The theoretical predictions show smooth variations of the radiation field with the angular coordinate, and in all cases the largest value is reached for $\beta_R = 0$. We know that for higher values of ζ_R this difference between the maximum and minimum value of (m) is

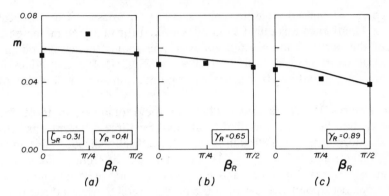

Figure IV.17   Effect of the azimuthal coordinate on the radiation field.

negligible, which would allow one to model the system by only employing the coordinates ζ_R and γ_R.

Figures IV.18a, b and c illustrate (m) as a function of ζ_R for a fixed angular position and for three different values of the radius. Physically, this study requires one to move the reactor vertically with $\beta_R = \pi/2$ and with γ_R equal to 0.41, 0.65 and 0.89, respectively. Again, good correspondence is observed between the theoretical predictions and the experimental results, the maximum deviation being about 20%. If one observes, in the same figure, how the radiation field is modified with the spatial coordinates, we can first observe a decrease of (m) as the axial variable increases, as a result of the increase of the microreactor-emitting system distance. On the other hand, if one simultaneously observes the three figures mentionend above, one may notice in the theoretical results that they

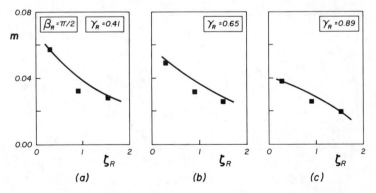

Figure IV.18   Effect of the axial coordinate on the radiation field.

change from concave to convex as γ_R increases. The physical explanation of this effect should be looked for in a gradual increase of the partially illuminated zones as the radius increases and the axial coordinate decreases. Hence, in Fig. IV.18c the effect is maximized and yields the distortion pointed out for small values of ζ_R.

Figures IV.19a, b and c illustrate the radial variations of the radiation field for a specified height and for three different azimuthal coordinates. This requires one to radically displace the microreactor located at $\zeta_R = 0.31$ and at three different angular locations: $\beta_R = 0$, $\beta_R = \pi/4$ and $\beta_R = \pi/2$. Again we should note the good correspondence between the model predictions and the experimental results. As explained in Section II.3, the radiation field decreases as the radial position increases. On the other hand, in Fig. IV.19a we should notice a systematic trend in which the experimental results are always somewhat lower than the theoretical predictions. For $\beta_R = 0$ the radial displacement of the microreactor is parallel to the lamp axis, hence a large fraction of the reflected radiation should again get across the source walls to reach the microreactor. Since the lamp is not totally transparent to the reflected rays as assumed by the theoretical model, one should obtain experimental conversions slightly lower than the predicted ones.

Several important conclusions related to the experimental verification of the ESVE Model should be stressed:

– It should be pointed out that the verification has been carried out with almost "point" values different from macroscopic actino-

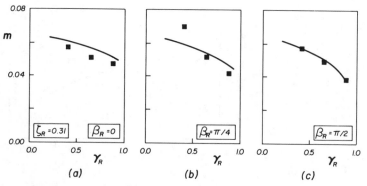

Figure IV.19   Effect of the radial coordinate on the radiation field.

metric measurements frequently found in the literature. The reason for this lies in the fact that for photoreactor design a local value of the radiation field is what is needed. One should notice that the frequent use of macroscopically averaged energy values inside the system has led to partial or erroneous conclusions in several previous studies.

– Another aspect that should be noted about the experimental verification is related to the absolute mesurement of the radiation field. It could happen that models of different reliability indicate similar spatial variations, and this would produce similar relative results while, actually, the predictions on an absolute basis could be totally different.

V. HOMOGENEOUS REACTORS WITH COMPLEX KINETICS

In the last section we have shown that for simple reactions it is possible to carry out an a priori design of photochemical reactors with no adjustable parameters. However, very often a photochemical process will be much more attractive if it involves a chain reaction. In this section we present the work done on a priori design of homogeneous photoreactors with chain kinetics.

A crucial test of the modelling of a photoreactor should include the study of two main difficulties: (1) the prediction of the reactor behavior in non-photosensitized chain reactions, and (2) the modelling of a reactor with a rather complex radiation field. To achieve this, we have studied the gas-phase photochemical chlorination of methane in a continuous annular reactor, and then the gas-phase photochemical chlorination of ethane in a tubular cylindrical reactor with a reflector of elliptical cross section.

V.I. Annular Photoreactors

The gas-phase photochemical chlorination of methane was modelled in a continuous annular reactor by using the complete chain reaction scheme [39]. The annular reactor was chosen because this type of reactor provides a very practical system for medium- to large-scale operations. Some simplifications were introduced to reduce the number of chemical species with negligible effect on the final results. It was felt that in order to account for the several variables affecting the reactor performance, an optimization problem could

Table V.1   Reaction Mechanism

Initiation step	$Cl_2 \xrightarrow{\Phi e^a} 2Cl\cdot$
Propagation steps	$CH_4 + Cl\cdot \rightleftarrows CH_3\cdot + HCl$ $CH_3\cdot + Cl_2 \rightleftarrows CH_3Cl + Cl\cdot$ $CH_3Cl + Cl\cdot \rightleftarrows CH_2Cl\cdot + HCl$ $CH_2Cl\cdot + Cl_2 \rightleftarrows CH_2Cl_2 + Cl\cdot$ $CH_2Cl_2 + Cl\cdot \rightleftarrows CHCl_2\cdot + HCl$ $CHCl_2\cdot + Cl_2 \rightleftarrows CHCl_3 + Cl\cdot$ $CHCl_3 + Cl\cdot \rightleftarrows CCl_3\cdot + HCl$ $CCl_3\cdot + Cl_2 \rightleftarrows CCl_4 + Cl\cdot$
Homogeneous termination steps	$Cl\cdot + CH_3\cdot \rightarrow CH_3Cl$ $Cl\cdot + CH_2Cl\cdot \rightarrow CH_2Cl_2$ $Cl\cdot + CHCl_2\cdot \rightarrow CHCl_3$ $Cl\cdot + CCl_3\cdot \rightarrow CCl_4$ $Cl\cdot + Cl\cdot + M \rightarrow Cl_2 + M$ $R_i\cdot + R_i\cdot \rightarrow R_i - R_i$ $R_i\cdot + R_j\cdot \rightarrow R_i - R_j$
Heterogeneous termination steps	$Cl\cdot + Wall \rightarrow Products$ $R_i\cdot + Wall \rightarrow Products$

be set, the goal being to find the optimal operating conditions for a prescribed selectivity in the reactor.

For the photochemical chlorination of methane, the general mechanism is shown in Table V.1. A great deal of the work done in the field has yielded the values of the kinetic parameters for each elementary reaction. Chiltz et al. [34] compiled activation energy and frequency factor data measured in methane photochlorination experiments. Kinetic data for the homogeneous termination reactions involving different organic free radicals are not available in the literature. However, Kurtz [63] estimated the respective kinetic constants for these reactions using a collision frequency average method. The values obtained are of the expected order of magnitude. For the wall termination reactions, the rate constants can be calculated using kinetic theory, with recombination coefficients taken from the literature [34, 77].

Model Equations

The authors [39] employed the following assumptions: (i) steady state, (ii) negligible thermal effects, (iii) Newtonian fluid, (iv) incompressible flow, (v) axial laminar flow, (vi) azimuthal symmetry, (vii) negligible axial diffusion when compared with the convective flux, and (viii) diffusive fluxes calculated by Stefan–Maxwell relations. The dimensionless mass balance equations are expressed as:

$$U(\gamma)\frac{\partial \Psi_i}{\partial \zeta} = \frac{1}{\text{Ge}}\frac{1}{\gamma}\frac{\partial}{\partial \gamma}\left[\frac{\gamma}{\text{Pe}_i}\frac{\partial \Psi_i}{\partial \gamma}\right] + \Omega_i \qquad (V.1)$$

with

$$\Psi_i(0, \gamma) = \begin{cases} \Psi_{i,\text{in}} & i = \text{reactant} \\ 0 & i \neq \text{reactant} \end{cases} \qquad (V.2)$$

$$\frac{\partial \Psi_i}{\partial \gamma}\left(\zeta, \frac{K}{1-K}\right) = \frac{\partial \Psi_i}{\partial \gamma}\left(\zeta, \frac{1}{1-K}\right) = \begin{cases} 0 & i \neq \text{free radical} \\ \Gamma_i \Psi_i & i = \text{free radical} \end{cases}$$

$$(V.3)$$

To solve this coupled system of integro-differential equations a numerical method is required. The integral characteristic is given by the evaluation of the LVREA which is calculated at each point in the reactor by integrating over the three spatial dimensions because the bundles of energy arrive from every direction; thus, the concentration field inside the whole reactor is needed to evaluate the attenuation of radiation at each single point. The calculating procedure has been described in detail in Section IV.

The numerical complexity of the problem is mainly due to the following effects: (i) the great number of chemical species, considering the non-chloromethane compounds which are formed during some homogeneous and heterogeneous termination reactions, and (ii) the iterative scheme required for calculating the radiation and concentration fields inside the reactor.

In order to reduce this complexity some simplifications of the problem may be introduced. Heterogeneous reactions may have little or no effect upon the reactor performance due to the geometrical characteristics of the reactor and the flow conditions used in the design. To analyze this effect, a representative range of values of the design parameters was taken, and the mass balance equations were solved within this range with and without considering heterogeneous reactions. Results indicate that these termination reactions do not affect the concentration profiles for the stable species, not even for points close to the reactor wall. Differences in radial and axial concentration profiles are less than 0.1%. Hence, the first simplification of the problem will be to disregard the heterogeneous termination steps in the mechanism. This means that the reactor wall is non-permeable for all chemical species.

A second simplification of the problem consists of the analysis of the incidence of the homogeneous termination reactions which lead to the formation of chloroethanes. By solving the mass balance equations within the prescribed range with and without considering these reactions, it was found that the amount of chloroethanes produced is negligible when compared to the amount of the main products, the difference being three or more orders of magnitude.

In a photoreactor of practical optical thickness, radiant energy flux density profiles are unavoidable. Hence concentration profiles are always present. Unless the work is done in a well-stirred reactor (in which case one could probably assume a uniform concentration field at least for stable species), one cannot apply the steady-state approximation for the whole reactor but only for point equations. This procedure has been called "micro steady-state" approximation (MSSA). This third simplification can be applied to short-lived intermediates such as Cl· and organic free radicals. By setting $\Omega_i = 0$ for those species, only algebraic equations must then be solved. They remain coupled with the remaining partial integro-differential equation (PIDE) through the reaction rate terms where concentrations of atomic chlorine or organic radicals are involved. This, a third simplification of the problem may be to apply the local or microscopic steady-state approximation. A procedure similar to that described before showed that the local molar distribution of products is the same for both conditions.

With these simplifications, the complex chain mechanism with more than 30 species and elementary steps can be reduced to the following simplified kinetic scheme:

$$CH_4 + Cl_2 \rightleftarrows CH_3Cl + HCl$$

$$CH_3Cl + Cl_2 \rightleftarrows CH_2Cl_2 + HCl$$

$$CH_2Cl_2 + Cl_2 \rightleftarrows CHCl_3 + HCl \tag{V.4}$$

$$CHCl_3 + Cl_2 \rightleftarrows CCl_4 + HCl$$

The reaction rate expressions, which can be derived following Temkin's technique [89], become a non-linear function of: (1) concentration of the stable species, (2) individual kinetic constants of each step of the true mechanism and (3) the LVREA. The final expressions are shown in Table V.2.

To evaluate the LVREA which characterizes the photochemical process, the ESVE and ESSE Models were used depending upon the type of lamp under study. The numerical complexity introduced

Table V.2 Final Reaction Rate Expressions

Reaction No.	Reaction rate expressions
R_1	$(k_2 C_{CH_4} - k_2' K_1 C_{HCl}) C_{Cl}.$
R_2	$(k_4 C_{CH_3Cl} - k_4' K_2 C_{HCl}) C_{Cl}.$
R_3	$(k_6 C_{CH_2Cl_2} - k_4' K_3 C_{HCl}) C_{Cl}.$
R_4	$(k_8 C_{CHCl_3} - k_8' K_4 C_{HCl}) C_{Cl}.$

*Where K_i values are:

$$K_1 = \frac{k_2 C_{CH_4} + k_3' C_{CH_3Cl}}{k_2' C_{HCl} + k_3 C_{Cl_2} + k_{10} C_{Cl}.}$$

$$K_2 = \frac{k_4 C_{CH_3Cl} + k_5' C_{CH_2Cl_2}}{k_4' C_{HCl} + k_5 C_{Cl_2} + k_{11} C_{Cl}.}$$

$$K_3 = \frac{k_6 C_{CH_2Cl_2} + k_7' C_{CHCl_3}}{k_6' C_{HCl} + k_5 C_{Cl_2} + k_{12} C_{Cl}.}$$

$$K_4 = \frac{k_8 C_{CHCl_3} + k_9' C_{CCl_4}}{k_8' C_{HCl} + k_9 C_{Cl_2} + k_{13} C_{Cl}.}$$

and $C_{Cl}.$ is obtained from the implicit equation:

$$\Phi e^a - (k_{10} K_1 + k_{11} K_2 + k_{12} K_3 + k_{13} K_4 + k_{14}) C_{Cl}^2. = 0$$

by the iterative scheme required for computing the models cannot be reduced, but the machine time can be lessened by using the simplified extense source model [84]. For preliminary studies, the work was also done with frequency-averaged properties instead of performing the integration over the entire frequency range. Those properties are calculated as follows:

$$E = \int_\nu E_\nu \, d\nu \qquad (V.5)$$

$$\bar{\alpha} = \frac{1}{E} \int_\nu \alpha_\nu E_\nu \, d\nu \qquad (V.6)$$

This is only an approximation and the rigorous treatment was employed in the final steps of the problem solution.

Selectivity Studies

Before undertaking the optimization problem it was necessary to perform some selectivity studies in order to: (a) obtain some information about the lamp for the production of the required chloroderivative, and (b) find a starting set of variables for the optimization problem. Preliminary studies will be illustrated using the ESVE Model only.

Selectivity was defined using the flow rate averaged exit concentration as follows:

$$S_i = \frac{\langle \Psi_i \rangle}{\sum_{j=1}^{4} \langle \Psi_j \rangle} \quad i = 1, 2, 3, 4 \tag{V.7}$$

where:

$$\langle \Psi_i \rangle = \int_{K/(1-K)}^{1/(1-K)} \gamma U(\gamma)\Psi_i(\gamma, 1)\, d\gamma \tag{V.8}$$

and the subscript i indicates the number of chlorine atoms in the chloromethane derivative. Exit reactant conversions were calculated by:

$$\langle X_{Cl_2} \rangle = 1 - \langle \Psi_{Cl_2} \rangle \tag{V.9}$$

$$\langle X_{CH_4} \rangle = 1 - \langle \Psi_{CH_4} \rangle F \tag{V.10}$$

where F is the chlorine to methane feed molar ratio.

Figure V.1 shows a plot of the selectivity for the four chloro-derivatives and conversion of the two reactants as a function of F keeping the chlorine feed fraction constant at 50%. The authors [39] pointed out that it was possible to obtain a high methyl chloride selectivity (close to one) by working at very low values of F (less than one). To reach these small values of F, the chlorine fraction in

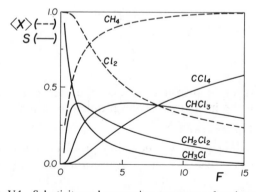

Figure V.1   Selectivity and conversion curves as a function of F.

the feed must be reduced below 50% with which, in spite of the high chlorine conversion (close to one), the amount of methyl chloride produced is very low. On the other hand, the methane conversion is low and a great amount of the reactant passes through the reactor without reacting. A recommended procedure for methyl chloride production would be to use a methane recycle to improve both global methane conversion and methyl chloride production. In addition, high carbon tetrachloride selectivities can be obtained by working at very high values of F (higher than 10). Methane conversion is high but chlorine conversion is very low and so is carbon tetrachloride production. To improve both chlorine conversion and carbon tetrachloride production it would be useful to work with a recycle of chlorine. It was not possible to achieve very high selectivities for the intermediate chloroderivatives (methylene chloride and chloroform) but, at least for the range of parameters chosen, their selectivity values reached a maximum for a unique value of F.

These preliminary results showed the necessity of performing further studies for the intermediate products in order to find optimal operating conditions for their production. On the other hand, the production of CH_3Cl and CCl_4 was almost totally controlled by the composition of the feed as it occurs in any conventional thermal reactor.

In order to choose an appropriate emission system for each chloroderivative production, it is useful to analyze the incidence of the reaction source characteristics upon the selectivity and conversion of the reaction. To obtain a first approximation to the problem the relative incidence of E and α was studied by plotting selectivities and conversions as a function of J/ε, at different values of ε. Variations in J/ε were achieved by changing the total energy output. The starting composition of feed F for each intermediate chloroderivative production was chosen from Fig. V.1.

Figure V.2 shows the results for the CH_2Cl_2 production, when the feed composition value (F) is 1.25. Figure V.2a shows the variations of CH_2Cl_2 selectivity as a function of J/ε, at different values of ε. Figures V.2b and c show the resulting chlorine and methane conversions for the same conditions. Since the LVREA is a linear function of the lamp energy output, it should be expected that reactant conversions be favored at large values of J/ε. However, those figures show that this conversion increment is rather low for values of J/ε greater than 0.001. Similarly, the CH_2Cl_2 selectivity shows low sensitivity to changes in J/ε (Fig. V.2a). In spite of the fact that

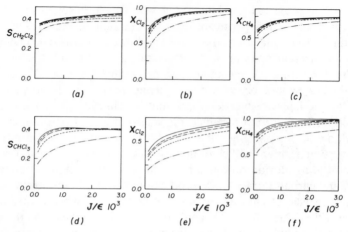

Figure V.2   Effect of the J/ε on the selectivity and conversion. Key: $\varepsilon = 0.1$ (—), $\varepsilon = 0.5$ (– – –), $\varepsilon = 1.0$ (–·–·–), $\varepsilon = 2.0$ (- - - -), and $\varepsilon = 5.0$ (–··–··–).

one might be tempted to use high-power lamps, these results show that the methylene chloride production can be carried out using photochemical lamps of low energy output, with the resulting reduction in operating costs. Similarly, Figs. V.2a, b and c show that ε must be as low as possible to increase CH_2Cl_2 selectivity and reactant conversions. To achieve low values of ε, the average chlorine absorption coefficient must be low. Considering the chlorine absorption spectrum, this condition is obtained by working at wavelengths smaller than 2800 Å or greater than 3500 Å. Low output power with emission in the far ultraviolet range of the spectrum is fulfilled by germicidal lamps which have an isotropic volumetric emission. The second alternative may be furnished by "black-light" lamps which need the use of a superficial diffuse emission model. In the second case, one can avoid the necessity of making the inner reactor wall of quartz thus making the process even more economical.

To choose a suitable emission system for chloroform production a new analysis was carried out at $F = 5$. The results, reported in terms of $CHCl_3$ selectivity, Cl_2 conversion and CH_4 conversion vs. J/ε at constant values of ε are plotted in Figs. V.2d, e and f respectively. Reactant conversions are favored when J/ε is large. Due to the large value of F the effect is very small for CH_4 but significant for Cl_2. However, this rather high value of F causes a different behavior regarding the $CHCl_3$ selectivity. As it is shown in Fig. V.2d, when

J/ε is greater than 0.001 the large Cl_2 conversion obtained produces a decrease in the $CHCl_3$ selectivity for the benefit of the CCl_4 production. From this figure it is clear that an optimal value of J/ε exists and that recirculation of unreacted Cl_2 will be advisable. From Figs. V.2d, e and f, it is possible to conclude that a small value of ε will also be needed. All these requirements can be accomplished with lamps of the types previously described when analyzing CH_2Cl_2 selectivity.

Figures V.3a and b show the effect of the radiation path length on methylene chloride and chloroform, respectively. Increasing the value of $(r_{ou} - r_{in})$, at constant values of r_{in} and Θ_R, causes a larger reacting volume at constant energy input into the system. This reduces exit conversion of both reactants (chlorine and methane). The final result is the utilization of chlorine to produce compounds with the minimum possible chlorine substitution, i.e. a decrease in the CH_2Cl_2 selectivity for the benefit of the CH_3Cl production in the first case (Fig. V.3a) and a decrease of the $CHCl_3$ selectivity for the benefit of the CH_2Cl_2 and CH_3Cl production in the second case (Fig. V.3b). For the chosen operating conditions a maximum value of the $CHCl_3$ selectivity is observed for the second case. This is a result of the combination of a strong decrease in the CCl_4 selectivity with an increase in the CH_3Cl and CH_2Cl_2 selectivities.

Figures V.3c and d show plots of the molar distribution of product and reactant conversions as a function of Θ_R, for $F = 1.25$ and $F = 5$

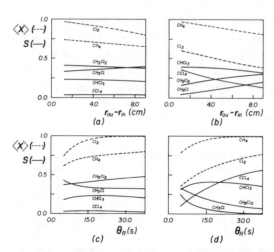

Figure V.3   Selectivity and conversion curves. Effect of the radiation path length: (a) $F = 1.25$, (b) $F = 5.0$. Effect of the mean residence time: (c) $F = 1.25$, (d) $F = 5.0$.

respectively. Since the value of $(r_{ou} - r_{in})$ does not change, we can achieve variations in the mean residence time (at constant reactor length) only by changing the volumetric flow rate. As expected, reactant conversions are favored at large values of Θ_R. When $F = 1.25$ the methylene selectivity increases when Θ_R increases; the CH_3Cl selectivity decreases and the $CHCl_3$ and CCl_4 fractions go through a maximum due to the small excess of chlorine in the feed composition. On the other hand, when $F = 5$ the increase in the reactant exit conversions produces a beneficial effect on the CCl_4 production while the chloroform selectivity goes through a maximum and CH_3Cl and CH_2Cl_2 fractions always decrease. Qualitatively, these results would have also been expected in a thermal reactor.

Optimization Problem and Conclusions
The optimization problem consists of the determination of the design and operating conditions which result in maximum selectivity for each of the intermediate chloromethane derivatives (CH_2Cl_2 and $CHCl_3$). As a result of the previous preliminary and qualitative analysis it is clear that we know: (i) in both cases a low energy input would be sufficient, (ii) for both reactions, the feed molar concentration ratio, the mean residence time inside the reactor and the characteristic radiation path length are the most important design variables to affect selectivity, and (iii) the spectral distribution of the radiation power output could provide additional improvements in the intermediate stable species selectivity.

To reduce the magnitude of the optimization problem we may resort to some realistic constraints which will be provided by the availability of lamps. For a continuous tubular reactor we will certainly need tubular lamps of an adequate length to provide sufficiently long mean residence times. In addition, they should be of low energy consumption and have an output in the wavelength range between 2500 and 5000 Å. In the market we can find the following:

Lamp	Power input (watts)	Power output (watts)	Main Range of λ (Å)	Length (m)
Germicidal[1] 1	15	3.6	2537	0.46
Black light 1	15	2.0	3200–4200	0.46
Germicidal 2	30	8.3	2537	0.915
Black light 2	30	5.4	3200–4200	0.915

Thus, the problem can be stated as follows: given this choice of lamps, obtain the best design to produce methylene chloride or chloroform with the maximum selectivity. The problem will be illustrated here with chloroform.

The independent variables are the feed molar ratio $Cl_2 : CH_4$, the characteristic radiation path length $(r_{ou} - r_{in})$ and the mean residence time (Θ_R). Some restrictions are imposed on the permissible values that these variables may take, according to plausible operating ranges. Additional constraints concerned with the minimum acceptable exit reactant conversions are also established. Thus, the complete constrained optimization problem can be written as:

$$\text{Maximize } S = \frac{\mathbf{W}^T\mathbf{G}}{\mathbf{E}^T\mathbf{G}} \tag{V.11}$$

subject to:

$$a_i \leqslant v_i \leqslant b_i \qquad i = 1, 2, 3 \tag{V.12}$$

$$\mathbf{D}^T\mathbf{x} \geqslant \mathbf{D}^T\mathbf{p} \tag{V.13}$$

and:

$$h_k(\mathbf{v}, \mathbf{n}, \mathbf{t}) = 0 \qquad k = 1, \dots, 4 \tag{V.14}$$

where S is the selectivity function, and $\mathbf{G}$ is given by:

$$\mathbf{G} = [\langle \Psi_i \rangle]^T \qquad i = 1, \dots, 4 \tag{V.15}$$

with $\langle \Psi_i \rangle$ defined by Eq. (V.8). Let "m" be the compound whose selectivity we wish to maximize, then:

$$\mathbf{W} = [W_k]^T \qquad W_k = \begin{cases} 1 \text{ for } k = m \\ 0 \text{ for } k \neq m \end{cases} \qquad k = 1, \dots, 4 \tag{V.16}$$

$$\mathbf{E} = [1\ 1\ 1\ 1]^T \tag{V.17}$$

Constraints (V.12) define the permissible values that the independent variables may take and:

$$\mathbf{x} = [F(r_{ou} - r_{in})\Theta_R]^T \tag{V.18}$$

In order to decrease the cost of the product streams purification and the reactant consumption, a minimum acceptable value of one of the reactant exit conversions is imposed. Let "n" be this reactant, and p_n its minimum acceptable value of conversion, then in constraint (V.13) we have:

$$\mathbf{D} = [d_j]^T \qquad d_j = \begin{cases} 0, & j = n \\ 0, & j \neq n \end{cases} \qquad j = 1, 2 \qquad \text{(V.19)}$$

$$\mathbf{p} = [p_j]^T \qquad p_j = \begin{cases} p_n, & j = n \\ 0, & j \neq n \end{cases} \qquad j = 1, 2 \qquad \text{(V.20)}$$

Finally, equality constraints (V.14) represent the solution of the species mass balance equations which provide a relationship between the independent variables $\mathbf{v}$ and the dependent variables $\mathbf{u}$ necessary to evaluate the objective function S. Here $\mathbf{u}$ is defined by:

$$\mathbf{u} = [\Psi_k]^T \qquad k = 1, \dots, k \qquad \text{(V.21)}$$

and the remaining design and operating conditions, such as lamp parameters, temperature, pressure, inert concentration, etc., are included in $\mathbf{t}$.

The characteristics of the optimization problem stated by (V.11) to (V.14) allow us to use a non-linear unconstrained optimization technique, modified to account for constraints (V.12) and (V.13). Since the derivatives of the objective function cannot be evaluated directly, we used Powell's method which generates conjugate directions of search with no need for further information. A penalty is imposed on the value of the objective function if constraint (V.13) is violated. Considering the penalty, the objective function becomes:

$$B = S + \delta(\mathbf{D}^T\mathbf{x} - \mathbf{D}^T\mathbf{p}) \qquad \text{(V.22)}$$

where

$$\delta > 0 \qquad \text{if } \mathbf{D}^T\mathbf{x} - \mathbf{D}^T\mathbf{p} > 0 \qquad \text{(V.23)}$$

$$\delta < 0 \qquad \text{if } \mathbf{D}^T\mathbf{x} - \mathbf{D}^T\mathbf{p} < 0 \qquad \text{(V.24)}$$

During the search, constraints (V.12) are ignored until one or more of them is violated; then the current point returns to the feasible region and the constraints are again ignored. Since the behavior of the objective function is unknown, the search should have been performed starting from several initial points. However, the preliminary selectivity studies provide a close initial estimate of the optimal point avoiding the necessity of exploring several solutions.

Table V.3 summarizes the results of the optimal design problem; including the starting solution (provided by Figs. V.1 and V.3) and lower and upper bounds required by restrictions (V.12).

Comparing the results obtained with lamps of equal input energy and size, but different output energy and spectral distribution, it follows that:

- Germicidal lamps emit at a wavelength range where chlorine light absorption is poor. Hence, greater equivalent radiuses can be used; but to obtain high methane conversions the system is forced to use high values of the chlorine to methane feed molar ratio. This considerably reduces chloroform production.
- In the wavelength range of emission of black-light lamps, chlorine light absorption is stronger, hence the required equivalent radiuses are smaller than those obtained with germicidal lamps. Chloroform selectivity and production and methane conversion are greater than the ones achieved with germicidal lamps, thus indicating that in spite of the differences in power output (up to more than 50%) what really counts is the quality of the supplied radiation instead of the quantity. It must be pointed out that all these advantages of black-light lamps are obtained with a smaller equivalent radius.

Comparing the results obtained with lamps of similar spectral distribution it follows that:
- Selectivity and conversion increase when the reactor and lamp lengths, the energy emitted per unit length, and the mean residence time are increased.
- Moving from 15 to 30 W of energy input in the germicidal lamps we can get only a 57% increase in chloroform production combined with a 33% increase in selectivity. Unless separation costs are very important, the duplication of the energy demand does not seem justified.
- Duplicating the input energy of the black-light lamps it is possible to achieve 15% increase in selectivity with 120% increase in chloroform production.
- The minimum permissible value of methane conversion cannot be achieved; however, it should be expected that working with radiation sources with greater energy input (which are also longer) the desired methane conversion may be obtained.

Table V.3 Values of Parameters and Solution of the Optimization Problem

Parameters	Germicidal 1	Black light 1	Germicidal 2	Black light 2
Starting solution:				
F	5.0	5.0	5.0	5.0
$r_{ou} - r_{in}$ (cm)	3.0	3.0	3.0	3.0
θ_R (s)	7.5	7.5	7.5	7.5
Range:				
F	1–12	1–12	1–12	1–12
$r_{ou} - r_{in}$ (cm)	1–8	1–8	1–8	1–8
θ_R (s)	5–17	5–17	5–35	5–35
Solution:				
S_{CH_3Cl}	25.5	8.5	11.3	3.9
$S_{CH_2Cl_2}$	34.0	21.0	25.0	18.9
S_{CHCl_3}	30.3	41.1	40.4	47.0
S_{CCl_4}	10.2	29.4	23.3	30.2
X_{Cl_2}	14.5	33.0	21.8	60.3
X_{CH_4}	75.9	91.8	89.8	95.8
CHCl$_3$ production $\left(\dfrac{gr}{s} \cdot 10^3\right)$	4.7	5.9	7.4	12.8
F	11.8	8.5	11.4	5.3
$r_{ou} - r_{in}$ (cm)	2.0	1.2	2.0	1.4
θ_R (s)	17.0	17.0	35.0	35.0

The results shown in Table V.3 leave little room for doubt regarding the choice of the system: for chloroform production the best solution is provided by a 30 W black-light lamp. Finally, it can be pointed out that in order to increase the total amount of chloroform produced, a battery of several lamp-reactor arrangements (each one of them operated with low energy input) should be used.

The authors concluded that:

– In the particular case of the chlorination of methane the following simplifications provide results similarly valid to those of the rigorous modelling: (a) Disregard of several homogeneous and all heterogeneous termination steps. (b) Application of the MSSA to atomic and free radical species.
– As in thermal reactions the reactant feed molar ratio is a key parameter in selectivity.
– When the exit conversion of one of the reactants is added to the definition of the optimization problem, the results about the mean residence time indicate the trivial answer (use the maximum possible value).
– The characteristic radiation path length provides an additional design parameter for optimization of the selectivity.
– The lamp output power spectral distribution proved to have a stronger effect on selectivity and conversion than the effect produced by the energy input to the reactor.
– The selective photochlorination of methane does not require high energy radiation sources but a reactor made of several tubes irradiated with lamps of low energy output.
– The authors also solved the problem of methylene chloride selectivity. This system showed a much higher sensitivity to process variable changes. In fact for the case of CH_2Cl_2, the final results of selectivity and output production provide values significantly different from those obtained for chloroform.

V.2. Tubular Photoreactor with Elliptical Reflector

Clariá et al. [32, 33] studied the gas-phase photochemical monochlorination of ethane in a tubular cylindrical reactor. They chose a cylindrical reflector of elliptical cross section with a tubular lamp located at one of the focal axes. Special motivations for this choice can be found in the quoted paper. A general reaction sequence was postulated (Table V.4) which includes all possible reactions within

Table V.4   Reaction Mechanism

Initiation step	$Cl_2 \xrightarrow{\Phi e^a} 2Cl\cdot$
Propagation steps	$Cl\cdot + EtH \underset{k_{p-2}}{\overset{k_{p2}}{\rightleftharpoons}} ClH + Et\cdot$
	$Cl_2 + Et\cdot \underset{k_{p-3}}{\overset{k_{p3}}{\rightleftharpoons}} ClEt + Cl\cdot$
Homogeneous termination steps	$Cl\cdot + Cl\cdot + M \xrightarrow{k_{t4}} Cl_2 + M$
	$Cl\cdot + Et\cdot \xrightarrow{k_{t5}} ClEt$
	$Et\cdot + Et\cdot \xrightarrow{k_{t6}} C_4H_{10}$
Heterogeneous termination steps	$Cl\cdot + W \xrightarrow{k_{w7}} Products$
	$Et\cdot + W \xrightarrow{k_{w8}} Products$

the experimentally explored conditions ($T < 600$ K and low pressures).

Noyes and Fowler [76] indicated that when the atomic chlorine concentration is relatively large (high Cl_2 concentration and high LVREA), the dominating termination step is the atomic chlorine recombination. However, when the concentration of Cl_2 decreases, a first-order termination step is observed. Two explanations were proposed for this experimental observation: (1) the first-order termination is the result of atomic chlorine deactivation with some impurity (atomic or radical scavenger) which is present at almost constant equilibrium concentration, (2) a second mechanism would include deactivation of atomic species by collisions with the wall. Under controlled experimental conditions this second explanation would be much more plausible.

Regarding this point, the authors proposed to work at atmospheric pressure. Computational experiments were performed including all possible wall reactions as boundary conditions for the mass conservation equations, and the exit product and reactant concentration distribution were compared with the same system without wall reactions. The results were used to introduce further simplifications and this will be shown in the next sections. The true nature of reaction (6) (see Table V.4) is of no significance for the purposes of this work. Be it a disproportionation reaction or a recombination reaction, the final products for all practical purposes will be almost indistinguishable. Chlorinations are long chain reactions and products from termination reactions are almost impossible to detect. Therefore, for this work the authors adopted a recombination reaction with the appropriate kinetic parameter. Pritchard et al. [79] suggested that reactions (-2) and (-3) be excluded because their activation energies were much higher than the direct reactions. Reverse reactions were not taken into account in the numerical procedure.

Model Equations
De Bernardez and Cassano [37] developed design criteria under which angular asymmetries can be neglected. The key parameter was the ratio r_R/r_L. The established conditions (small ellipse eccentricity, large value of c and $r_R/r_L \ll 1$) can be easily achieved in bench scale equipment and they were taken into account for the design of the experimental reactor employed by Clariá et al. [32, 33]. Under these conditions the mass balances can also be simplified because they may be written in two dimensions only. The following assumptions were employed: (i) steady state, (ii) streamline, incompressible flow, (iii) negligible thermal effects, (iv) Newtonian fluid with constant physical properties ($D_{i,m}$, ρ, μ, α, etc.), (v) angular symmetry, and (vi) negligible axial diffusion. Using these assumptions the dimensionless mass conservation equations may be written:

$$U(\gamma) \frac{\partial \Psi_i}{\partial \zeta} = \frac{1}{Ge} \frac{1}{\gamma} \frac{\partial}{\partial \gamma} \left[\frac{\gamma}{Pe_i} \frac{\partial \Psi_i}{\partial \gamma} \right] + \Omega_i \qquad (V.25)$$

where for simplicity the following notation is used: $Cl_2(\Psi_1)$, $Cl\cdot(\Psi_2)$, $EtH(\Psi_3)$, $ClEt(\Psi_4)$, $Et\cdot(\Psi_5)$, $HCl(\Psi_6)$ and $C_4H_{10}(\Psi_7)$. The boundary conditions are:

$$\Psi_1(0, \gamma) = \Psi_{1,in} = 1$$

$$\Psi_3(0, \gamma) = \Psi_{3,in} \qquad (V.26)$$

$$\Psi_j(0, \gamma) = \Psi_{j,in} = 0 \qquad (j = 2, 4-7)$$

$$\frac{\partial \Psi_i}{\partial \gamma} (\zeta, 0) = 0 \qquad (V.27)$$

$$\frac{\partial \Psi_j}{\partial \gamma}(\zeta, 1) = 0 \qquad (j = 1, 3, 4, 6 \text{ and } 7) \qquad \text{(V.28a)}$$

$$\frac{\partial \Psi_\ell}{\partial \gamma}(\zeta, 1) = \text{Pe}_\ell \Omega_\ell \qquad (\ell = 2, 5) \qquad \text{(V.28b)}$$

where Ω_ℓ is the rate of heterogeneous termination. The dimensionless reaction rate expressions are shown in Table V.5. In this table, the following definitions have been used:

$$\Omega_{\text{init}} = \frac{\Phi e^a L_R}{\langle v_z \rangle C_{1,\text{in}}} \qquad \text{(V.29)}$$

$$K_m = \frac{k_m L_R C_{1,\text{in}}}{\langle v_z \rangle} \qquad (m \neq 4) \qquad \text{(V.30)}$$

$$K_{t_4} = \frac{k_{t_4} L_R C_{1,\text{in}}^2}{\langle v_z \rangle} \qquad \text{(V.31)}$$

$$K_{w,r} = \frac{k_{w,r}}{\langle v_z \rangle} \qquad (r = 7, 8) \qquad \text{(V.32)}$$

To complete the reactor modelling, an expression to evaluate the LVREA in Eq. (V.29) is needed. Since we have a reflecting surface, the modulus of the radiation energy flux density should be obtained by adding the contributions of both directly incident radiation and indirectly incident radiation with one or more reflections at the mirror wall. At this point the authors introduced another simplific-

Table V.5   Dimensionless Reaction Rates

Ω_1	$-\Omega_{\text{init}} - K_{p3}\Psi_1\Psi_5 + K_{t4}\Psi_2^2\Psi_8$
Ω_2	$2\Omega_{\text{init}} - K_{p2}\Psi_2\Psi_3 + K_{p-2}\Psi_5\Psi_6 + K_{p3}\Psi_1\Psi_5 - 2K_{t4}\Psi_2^2\Psi_8 - K_{t5}\Psi_2\Psi_5$
Ω_3	$-K_{p2}\Psi_2\Psi_3 + K_{p-2}\Psi_5\Psi_6$
Ω_4	$K_{p3}\Psi_1\Psi_5 + K_{t5}\Psi_2\Psi_5$
Ω_5	$K_{p2}\Psi_2\Psi_3 - K_{p-2}\Psi_5\Psi_6 - K_{p3}\Psi_1\Psi_5 - K_{t5}\Psi_2\Psi_5 - 2K_{t6}\Psi_5^2$
Ω_6	$K_{p2}\Psi_2\Psi_3 - K_{p-2}\Psi_5\Psi_6$
Ω_7	$K_{t6}\Psi_5^2$
$\Omega_2(w)$	$-K_{w7}\Psi_2(w)$
$\Omega_5(w)$	$-K_{w8}\Psi_5(w)$

ation to the problem. It is known that for many elliptical photo-reactors, indirect radiation with only one reflection accounts for 80 to 95% of the total radiation [28]. Therefore, the value of the LVREA at any point I inside the reactor, considering the indirect radiation with only one reflection, is:

$$e^a = \frac{2\langle \Gamma_{Rf} \rangle \langle Y_R \rangle}{4\pi V_L} C_1 \int_{\nu=0}^{\nu=\infty} \alpha_\nu E_\nu \, d\nu \int_\phi d\phi \int_{\theta(\phi)} [r_L^2 - D^2 \sin^2 \xi]^{1/2}$$

$$\times \exp\left[-\int_{\rho_0(\theta,\phi)}^{\rho_I(\theta,\phi)} \alpha_\nu C_1 \, d\rho\right] d\theta \tag{V.33}$$

The integration limits were carefully derived [28] and will not be included here (see Section II.2). Similarly, the methodology used to consider the polychromatic radiation was described in detail in Section IV.2.

For all the preliminary studies and most of the computed results, the authors employed the information supplied by the specific bibliography on photochemistry and the operating characteristics and dimensions of the source-reflector system. The primary quantum yield needs a more careful analysis. Generally, absorption of light by diatomic molecules results in the dissociation of the absorbing molecule into two atomic species. Under these conditions, the primary quantum yield is defined as [75]:

$$\Phi_\nu = \left[\frac{\begin{array}{c}\text{number of molecules dissociated}\\ \text{by the primary process}\end{array}}{\begin{array}{c}\text{number of quanta of frequency}\\ \nu \text{ absorbed by the molecule}\end{array}}\right] \tag{V.34}$$

The primary process has the possibilities indicated in Table V.6. By inspection of the potential curves for the molecule of Cl_2 [52] it is easy to recognize that Cl_2 is very unstable and immediately dissociates into two atomic species. Thus, reaction (2) will be the dominant process with the only possible exception of (4). Both produce the same product. Hence, it can be safely assumed that $\Phi_\nu = 1$. This behavior occurs over the total range of continuous absorption by chlorine. Therefore, the result can be immediately extended to $\Phi = 1$, independent of wavelength.

Two aspects can be further analyzed in order to achieve a simpler mathematical model: the problem of the wall reactions and the possibility of applying the MSSA. To do so the system of PIDE was

Table V.6   Primary Process

1. Absorption	$Cl_2 + h\nu \rightarrow Cl_2^*$	
2. Spontaneous dissociation	$Cl_2^* \rightarrow 2Cl\cdot$	
3. Fluorescence	$Cl_2^* \rightarrow Cl_2 + h\nu$	
4. Induced dissociation by collision	$Cl_2^* + M \rightarrow 2Cl\cdot + M$	
5. Deactivation by collisions	$Cl_2^* + M \rightarrow Cl_2 + M$	
6. Deactivation with wall collisions	$Cl_2^* + W \rightarrow Cl_2 + W$	
7. Reaction with other molecules	$Cl_2^* + A \rightarrow Products$	

solved under the following conditions:

1. With wall reactions.
2. Without wall reactions.
3. With the full set of PIDE.
4. With the MSSA.

To solve the equations, one is faced again with the link "extent of the reaction-attenuation of radiation". The numerical procedure was explained in detail in Section IV for photoreactors with simple kinetics.

Comparing the results obtained with and without wall reactions, Fig. V.4 shows radial profiles for atomic chlorine concentrations obtained with the boundary conditions given by Eqs. (V.28a and b) and those obtained using B.C. (V.28a) for all species ($j = 1$ to 7), respectively. It can be seen that close to the reactor wall there is a small effect. It may be worthwhile to point out that in their computations [32, 33] they were using values for k_{w7} and k_{w8} closer to the maximum provided by the kinetic theory (molecular collisions with a wall), which is an upper bound for such a process. However, no appreciable changes in the concentration profiles for stable molecules were observed. Table V.7 shows the averaged exit conversion for stable species with and without wall reactions. It is clear that wall reactions have no influence on the reactor modelling, at least for this particular kinetic system. Since the reactor surface/reactor volume ratio is rather large for this case ($2/r_R = 1000 \text{ m}^{-1}$)

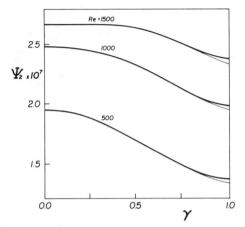

Figure V.4 Atomic chlorine radial concentration profiles. With (—) and without (—) wall reactions.

the authors concluded that for all practical purposes (at atmospheric or higher pressures) wall reactions can surely be neglected, particularly because a commercial scale reactor will have a ratio of A_R/V_R which is always smaller than the one used here (perhaps up to two orders of magnitude). Consequently, wall reactions were excluded (see Table V.4). This meant that all reactor walls were considered impermeable and B.C. (V.28a) can be used for $i = 1$ to 7.

Figure V.5 shows atomic chlorine axial concentration profiles, with and without the MSSA, for $\gamma = 0$ using Re as a parameter. Recalling the order of magnitude of these concentrations (10^{-12} to 10^{-13} mole/cm^3) it is clear that the MSSA holds. The most unfavorable conditions were found for higher Re (Re = 1500). The plot shows significant differences when $\zeta \to 0$ which is what one might have expected since reactants must spend some time inside the

Table V.7   Averaged Exit Conversions

Re	$\langle \Psi_4 \rangle_{ou}$			Error %
	Exact	Without wall reactions	MSSA	
500	43.74	43.79	44.71	2.2
1000	25.35	25.40	26.65	5.1
1200	22.25	22.29	23.44	5.4
1500	14.94	14.97	16.01	7.2

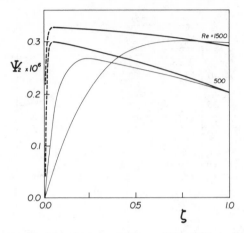

Figure V.5   Atomic chlorine axial concentration profiles. With (—) and without (—) MSSA.

reactor to achieve the "almost" constant concentration of inter-mediates. It is also seen that the MSSA is only a very good "approximation" and that in fact, intermediate species follow a pattern behavior close to the one corresponding to the stable species from which they were originated (Cl· with Cl_2 and Et· with EtH).

To compare the quality of the MSSA the authors computed:

$$t_{SS} = \frac{\zeta_{SS} L_R}{\langle v_z \rangle} \qquad (V.35)$$

where t_{SS} is the mean residence time needed to reach a "quasi" steady-state concentration for an intermediate species, and ζ_{SS} is the value of the axial coordinate where it is achieved. The larger the value of t_{SS}, the poorer the approximation. When Re is large due to a large value of $\langle v_z \rangle$, ζ_{SS} must also be large since L_R is fixed and t_{SS} is constant for a given reaction. Hence for high values of $\langle v_z \rangle$ the MSSA will deviate more from the exact result. Surely a large fraction of the reactor length will be calculated under an assumption that is not valid; not even as a good first approximation. For these cases, under the worst conditions, ζ_{SS} can be more than 50% of the reactor length. All these comments about Cl· may be repeated for Et·.

For a value of $\zeta = 0.5$, Fig. V.6 shows atomic chlorine radial concentration profiles. The influence of the parabolic velocity profile is quite clear and for $\gamma \to 0$ the Cl· concentration never reaches its

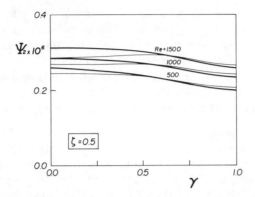

Figure V.6   Atomic chlorine radial concentration profiles. With (——) and without (—) MSSA.

proper SS value. When $\gamma \rightarrow 0$, since $v_z = 0$, the value of t_{SS} tends to infinity. For practical applications, what is important is to know how much of the concentration of a key compound (some stable species) differs when it is calculated with the exact solution and when the MSSA is used. This quantity is given by:

$$\left[\frac{\Psi_1^{Exact} - \Psi_1^{SS}}{\Psi_1^{Exact}} \right]_{\substack{\gamma=1 \\ \zeta=0.5}} \qquad (V.36)$$

and results for different values of Re (by changing $\langle v_z \rangle$ alone) are shown on Table V.8. What is even more important to know is how much these local discrepancies affect the averaged product exit conversion. Values for $\zeta = 1$ are also shown in Table V.7, where:

$$\text{Error } \% = \left[\frac{[\langle \Psi_4 \rangle_{ou}]^{Exact} - [\langle \Psi_4 \rangle_{ou}]^{SS}}{[\langle \Psi_4 \rangle_{ou}]^{Exact}} \right] \times 100 \qquad (V.37)$$

These results indicate that even under the most unfavorable conditions (higher values of Re) the error is never larger than 8%.

Table V.8   Exact and MSSA Solutions

Re	$\Psi_1^{Exact}(\gamma = 1, \zeta = 0.5)$	$\Psi_1^{SS}(\gamma = 1, \zeta = 0.5)$	Error %
500	0.6428	0.6425	0.03
1000	0.7702	0.7669	0.43
1200	0.7999	0.7933	0.83
1500	0.8559	0.8427	1.54

Consequently the MSSA can be incorporated into the reactor modelling, considerably reducing the complexity of the problem. This will be even more significant if the reacting system operates at high conversions of chlorine. Then, two- and three-chlorinated compounds will appear, and the reaction mechanism will involve a larger number of species and steps. The MSSA may still be used to obtain a numerical solution in a reasonable computing time.

Experimental Study

Figure V.7 shows a schematic flow-sheet of the experimental apparatus. Nitrogen (99.998%) and ethane (99.0%) were carefully purified. Chlorine, Matheson HP grade (99.5%) was dried with concentrated H_2SO_4 and then condensed in a monel gas cylinder at 70 KPa and $-60°C$. Then it was vented to eliminate any non-condensable gases. After reaching ambient temperature, the small cylinder was connected to the chlorine feed tube in the system. Nitrogen, ethane and chlorine flow rates were measured with mass flowmeters previously calibrated.

All parts in contact with chlorine or the reacting mixture were made of glass, quartz, Teflon or monel metal. The reactor was made of quartz Suprasil quality. Since the reactor tube must be cleaned very often and positioning at the focal axis of the ellipse is very critical, a special adjusting device was included. The lamp was a General Electric G30T8 (Germicide) with 30 watts of nominal

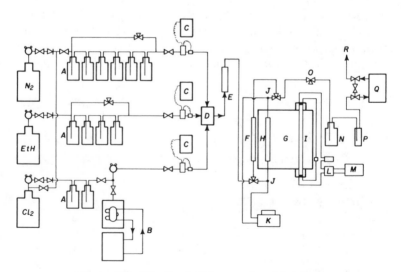

Figure V.7   Flow-sheet of the experimental device.

input power. Its operation was continuously monitored (intensity, voltage and input power) and its input voltage was permanently controlled to assure that 30 watts were always fed to the lamp. According to the manufacturer's specifications, for stabilizing purposes, the lamp was used after a minimum of 100 hours of operation. The reflector was made of aluminum sheet, specularly finished with Alzak treatment. With the exception of the irradiated length of the reactor, all other parts of the system were painted black to avoid any possible effect produced by laboratory light. An additional tube (dark reactor) with the same hydrodynamic characteristics as the one made of quartz was placed in a by-pass position to improve the control of the steady-state operation and check for dark reactions (they were found negligible and observed values were always smaller than the experimental error).

The experimental procedure allowed provisions for very careful control and measurement of the reactant and product streams as well as full assurance that the stable operation of the lamp and reactor systems was achieved. The temperature was carefully controlled at the inlet and outlet reactor ends. When chlorine conversion was lower than 50%, temperature differences were never larger than 2 C. Beyond 50% conversion, temperature differences as large as 5 C were observed. Every experimental point required about eight hours of continuous operation. The details can be obtained from the quoted paper.

Results and Conclusions

Figures V.8a and b give a compendium of the experimental results for two reactors. The chlorine molar fraction was changed between 0.5% and 7%, ethane between 0.5% and 14% and nitrogen always provided the make up gas to have atmospheric pressure. The Reynolds number was maintained approximately at 650 and the nominal temperature was always kept at 25 C.

For the small reactor, the chlorine conversion was below 20%. For the large one it reached values up to 60%. Under no conditions, were differences between predicted and experimental values greater than 15%. Dichlorinated compounds (1,1-ethane dichloride and 1,2-ethane dichloride) were observed in the G.C. analysis only at very high conversions, but the amounts were almost negligible. It should be recalled that the authors always worked with an excess of ethane and with a low chlorine molar fraction in the feed.

Figures V.9a and b show predicted (solid line) and experimental values illustrating the effect of chlorine concentration in the feed on

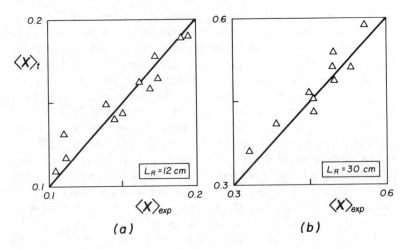

Figure V.8 Compendium of the experimental results.

the averaged exit chlorine conversion. Similarly, Figs. V.10a and b depict the effect of the molar fraction of ethane on the product yield.

All reported results show extremely good agreement between predictions and experimental results. It is interesting to remark that the total mass balance was also satisfied with an error never larger than 3%. All these agreements indicate two important results: (i) the microkinetic constants reported in the literature are very good

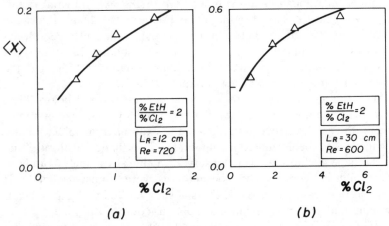

Figure V.9 Effect of the inlet chlorine percentage on the averaged exit chlorine conversion.

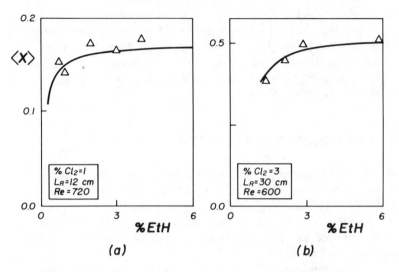

Figure V.10 Effect of ethane mole fraction on the product yield.

for reactor design purposes, and (ii) the proposed radiation and reactor models provide a reliable design methodology.

From the reported work it can be concluded that:

– A homogeneous photochemical reactor for chain-type reactions can be rigorously modelled from first principles. All one needs is: (i) microkinetic information about the reaction, (ii) physical properties of reactant and products, which should include optical parameters, (iii) operating characteristics and dimensions of the radiation source, which should include spectral distribution of the lamp power output, (iv) optical properties of the surface of entrance of radiation, and (v) optical properties and dimensions of reflecting devices if they were used.

– To design a photochemical reactor, an iterative procedure will be generally needed in order to obtain the best combination of reactor parameters-radiation emission system characteristics (with or without reflectors). The main reason lies on the unavoidable interactions between the properties of the energy supplier (lamp) and energy receptor and user (reactor and reacting species).

– Wall reactions can be safely neglected in most practical situations.

– Under controlled conditions, the micro steady-state approximation can be safely used, thus considerably simplifying the design of the photoreactor.

– To obtain good agreement between model predictions and experimental results, careful purification of reactants and systematic cleaning of the reactor wall through which the radiation enters, is indispensable. This experience provides some areas of further research work for industrial scale applications, particularly for more stable wall conditions.

VI.   HETEROGENEOUS REACTORS

Alfano and Cassano [10, 11] studied a semi-continuous stirred tank reactor irradiated from the bottom, using the emitting system shown in Sections II.3 and IV.3. This is a typical gas–liquid heterogeneous system (Fig. VI.1). The photochlorination of trichloroethylene and pentachloroethane was analyzed, in carbon tetrachloride solution at atmospheric pressure.

The reaction sequence including all the possible steps is shown in Table VI.1. Globally, the overall reaction scheme can be represented by the following consecutive reactions:

$$C_2HCl_3 \xrightarrow{+Cl_2} C_2HCl_5 \xrightarrow[-HCl]{+Cl_2} C_2Cl_6 \qquad (VI.1)$$

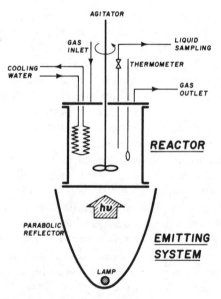

Figure VI.1   Schematic diagram of the heterogeneous reactor.

Table VI.1   Reaction Mechanism

Initiation
step

$$Cl_2 \xrightarrow{\Phi e^a} 2Cl\cdot$$

Propagation
steps

$$C_2HCl_3 + Cl\cdot \underset{k_2'}{\overset{k_2}{\rightleftarrows}} C_2HCl_4\cdot$$

$$C_2HCl_4\cdot + Cl_2 \underset{k_3'}{\overset{k_3}{\rightleftarrows}} C_2HCl_5 + Cl\cdot$$

$$C_2HCl_5 + Cl\cdot \underset{k_4'}{\overset{k_4}{\rightleftarrows}} C_2Cl_5\cdot + HCl$$

$$C_2Cl_5\cdot + Cl_2 \underset{k_5'}{\overset{k_5}{\rightleftarrows}} C_2Cl_6 + Cl\cdot$$

Termination
steps

$$2Cl\cdot \xrightarrow{k_6} Cl_2$$

$$2C_2HCl_4\cdot \xrightarrow{k_7} \text{Products}$$

$$2C_2Cl_5\cdot \xrightarrow{k_8} \text{Products}$$

$$Cl\cdot + C_2HCl_4\cdot \xrightarrow{k_9} \text{Products}$$

$$Cl\cdot + C_2Cl_5\cdot \xrightarrow{k_{10}} \text{Products}$$

$$C_2Cl_5\cdot + C_2HCl_4\cdot \xrightarrow{k_{11}} \text{Products}$$

where the first step is an addition reaction and the second is a substitution reaction. Information about the reacting system may be obtained from the data provided by the specific literature on gas- and liquid-phase chlorinations [36, 64, 78, 92].

Model Equations
The theoretical analysis of the problem must take the following aspects into account: the radiant energy distribution inside the photoreactor (Section II.3), the mixing states of the reacting species, the gas absorption rate, and the reaction rate.

The existence of a non-uniform radiation field produces a non-uniform reaction rate. Hence, even if one assumes perfectly mixed stable reactants and products, concentration gradients of free radicals may be present. To consider these mixing effects, a comparative analysis between the mean lifetime of all reacting species and a characteristic mixing time is performed. Two mixing states can be mainly considered [41, 42]:

i) Mixing state 1: Perfect mixing for stable species and free radicals.

ii) Mixing state 2: Perfect mixing for stable species and no mixing for free radicals.

A third possible mixing state considers no mixing for both types of species, but will not be taken into account because it would just mean a poorly designed reactor. A fourth case based on perfect mixing for free radicals and no mixing for stable species has obviously no physical sense. The mean lifetime of the stable species may be less (third state) or greater (second state) than the characteristic mixing time. In the first state, the mixing time is less than the mean lifetime of the stable species and that of the free radicals. So, both species are considered in a perfect mixing state (Fig. VI.2). In the second and third states the free radicals are immobile. One can say that they are born, live and die in the same place.

Tables VI.2 and VI.3 indicate the array definitions and the dimensionless mass balance equations in the liquid bulk for the models generated by considering mixing states 1 and 2. The number of chlorine atoms in each molecule was employed as subscript, i.e.: $HCl(C_1)$, $Cl_2(C_2)$, $C_2HCl_3(C_3)$, $C_2HCl_5(C_5)$, $C_2Cl_6(C_6)$, $Cl\cdot(\hat{C}_1)$, $C_2HCl_4\cdot(\hat{C}_4)$ and $C_2Cl_5\cdot(\hat{C}_5)$. A system of integro-differential equations with specified initial values was obtained. Due to the complexity of the problem, some approximations had to be used in order to find the solution of the system. In addition to that, the

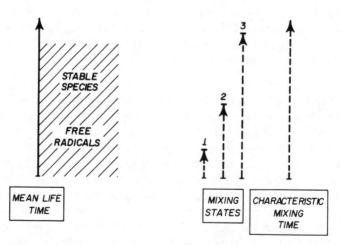

Figure VI.2 Comparative analysis between the mean lifetime and the characteristic mixing time.

Table VI.2 Definitions for the Mass Balances

Name	Symbol		Definition
Stable species concentration array	Dimensional	$\mathbf{C}(t)$	$(\langle C_1 \rangle, \langle C_2 \rangle, \langle C_3 \rangle, \langle C_5 \rangle, \langle C_6 \rangle)^T$
	Dimensionless	$\boldsymbol{\Psi}(\tau)$	$(\Psi_1, \Psi_2, \Psi_3, \Psi_5, \Psi_6)^T$
Free radical concentration array	Dimensional	$\hat{\mathbf{C}}(t)$ or $\hat{\mathbf{C}}(\mathbf{x}, t)$	$(\hat{C}_1, \hat{C}_4, \hat{C}_5)^T$
	Dimensionless	$\hat{\boldsymbol{\Psi}}(\tau)$ or $\hat{\boldsymbol{\Psi}}(\mathbf{x}, \tau)$	$(\hat{\Psi}_1, \hat{\Psi}_4, \hat{\Psi}_5)^T$
Stable species reaction rate	Dimensional	$\mathbf{R}(\mathbf{x}, t)$	$(R_1, R_2, R_3, R_5, R_6)^T$
	Dimensionless	$\boldsymbol{\Omega}(\mathbf{x}, \tau)$	$(\Omega_1, \Omega_2, \Omega_3, \Omega_5, \Omega_6)^T$
Free radical reaction rate	Dimensional	$\hat{\mathbf{R}}(\mathbf{x}, t)$	$(\hat{R}_1, \hat{R}_4, \hat{R}_5)^T$
	Dimensionless	$\hat{\boldsymbol{\Omega}}(\mathbf{x}, \tau)$	$(\hat{\Omega}_1, \hat{\Omega}_4, \hat{\Omega}_5)^T$
Absorption rate	Dimensional	$\mathbf{N}_{GL}(t)$	$(0, k_L A_v(\langle C_2 \rangle^i - \langle C_2 \rangle), 0, 0, 0)^T$
	Dimensionless	$\mathbf{N}(\tau)$	$(0, K_L(\Psi_2^i - \Psi_2), 0, 0, 0)^T$

expressions for the absorption rate $\mathbf{N}(\tau)$ and the reaction rate $\boldsymbol{\Omega}(\mathbf{x}, \tau)$, must be found to complete the mathematical description of the process.

The vector $\mathbf{N}(\tau)$ represents the absorption rates into the bulk liquid of reactants. If one considers that in the liquid phase only gaseous chlorine is absorbed, the dimensionless expression for each component of $\mathbf{N}(\tau)$ is given by:

$$N_i(\tau) = 0 \qquad (i \neq 2)$$
$$N_2(\tau) = k_L[\Psi_2^i(\tau) - \Psi_2(\tau)] \qquad \text{(VI.2)}$$

Table VI.3 Dimensionless Mass Balance Equations

Model	Mixing states		Integro-differential equations	Initial conditions
Model 1	Stable species	Perfect mixing	$\dfrac{d\boldsymbol{\Psi}(\tau)}{d\tau} = -\langle \boldsymbol{\Omega}(\mathbf{x}, \tau) \rangle + \mathbf{N}(\tau)$	$\boldsymbol{\Psi}(0) = \boldsymbol{\Psi}_0$
	Free radicals	Perfect mixing	$\dfrac{d\hat{\boldsymbol{\Psi}}(\tau)}{d\tau} = -\langle \hat{\boldsymbol{\Omega}}(\mathbf{x}, \tau) \rangle$	$\hat{\boldsymbol{\Psi}}(0) = \mathbf{0}$
Model 2	Stable species	Perfect mixing	$\dfrac{d\boldsymbol{\Psi}(\tau)}{d\tau} = -\langle \boldsymbol{\Omega}(\mathbf{x}, \tau) \rangle + \mathbf{N}(\tau)$	$\boldsymbol{\Psi}(0) = \boldsymbol{\Psi}_0$
	Free radicals	No mixing	$\dfrac{\partial \hat{\boldsymbol{\Psi}}(\mathbf{x}, \tau)}{\partial \tau} = -\hat{\boldsymbol{\Omega}}(\mathbf{x}, \tau)$	$\hat{\boldsymbol{\Psi}}(\mathbf{x}, 0) = \mathbf{0}$

where

$$K_L = \frac{t_T}{(1/k_L A_v)} = \left[\frac{\text{Total time of reaction}}{\text{Absorption time}} \right] \qquad \text{(VI.3)}$$

$$\Psi_2^i = \frac{C_2^i}{C_{\text{Ref}}^\circ} = \left[\begin{array}{c} \text{Dimensionless concentration of chlorine} \\ \text{at the gas–liquid interface} \end{array} \right] \qquad \text{(VI.4)}$$

Toor and Marcello's film-penetration theory [90] was used to determine the reaction regime. For this case, it can be shown that the reaction in the liquid phase is slow enough to be negligible during the lifetime of the liquid surface elements ($t_{\text{Dif}} \ll t_{rc}$). This condition was defined as a "slow reaction régime" [13–15]; if this is the case, the chemical absorption coefficient is equal to the physical absorption coefficient. Hence, the enhancement factor is:

$$I \triangleq \frac{k_L}{k_L^\circ} \cong 1 \qquad \text{(VI.5)}$$

In order to evaluate Ψ_2^i, a well-mixed gas phase was assumed. Therefore, the outlet mole fraction of chlorine was calculated using the mass balance for the gas phase under steady-state conditions. The expression for Ψ_2^i is given by:

$$\Psi_2^i(\tau) = \frac{1}{2} \left[\frac{a}{K_L} + b + \Psi_2(\tau) \right]$$
$$- \left\{ \left[\frac{a}{K_L} + b + \Psi_2(\tau) \right]^2 - 4b \left[y_{2,\text{in}} \frac{a}{K_L} + \Psi_2(\tau) \right] \right\}^{1/2}$$

$$\text{(VI.6)}$$

where:

$$a \triangleq \frac{F_{\text{in}} t_T}{V_L C_{\text{Ref}}^\circ} = \left[\frac{\text{Total moles of gas at the inlet}}{\text{Initial moles of liquid reactant}} \right] \qquad \text{(VI.7)}$$

$$b \triangleq \frac{P}{H C_{\text{Ref}}^\circ} = \left[\frac{\text{Maximum concentration of chlorine in the liquid}}{\text{Initial concentration of liquid reactant}} \right]$$

$$\text{(VI.8)}$$

Finally, from Eqs. (VI.3) and (VI.5) to (VI.8) one can evaluate $N(\tau)$ using Eq. (VI.2).

i) Model 1:

If the steady-state approximation is applied to the free radicals ("macroscopic" SSA) the expression of $\langle \hat{\Omega}(\mathbf{x}, \tau) \rangle$ is null (see Table VI.3). Introducing the kinetic information and separating the contributions of the initiation, termination and propagation steps, the following equation is obtained:

$$\mathbf{A}(\tau) \cdot \hat{\mathbf{\Psi}}(\tau) \quad + \quad \mathbf{R}(\tau) \cdot \mathbf{T} \cdot \hat{\mathbf{\Psi}}(\tau) \quad - 2\langle \mathbf{g}(\mathbf{x}, \tau) \rangle = 0$$

(VI.9)

(Propagation rates)   (Termination rates)   (Initiation rates)

where the definitions given by Table VI.4 were employed. If the initiation and termination rates are neglected [45], the system of equations (VI.9) may be written as:

$$\mathbf{A}(\tau) \cdot \hat{\mathbf{\Psi}}(\tau) = 0 \qquad\qquad \text{(VI.10)}$$

From Eqs. (VI.10) it is possible to get $\hat{\Psi}_4(\tau)$ and $\hat{\Psi}_5(\tau)$, and then introducing these concentrations into Eq. (VI.9) we can obtain the expression for $\hat{\Psi}_1(\tau)$. The final results are:

$$\hat{\Psi}_4(\tau) = \frac{\hat{\Psi}_1(\tau)}{\alpha} \qquad\qquad \text{(VI.11)}$$

$$\hat{\Psi}_5(\tau) = \frac{\hat{\Psi}_1(\tau)}{\beta} \qquad\qquad \text{(VI.12)}$$

$$\hat{\Psi}_1(\tau) = \left[\frac{\Psi_2}{K_6 + \dfrac{K_7}{\alpha^2} + \dfrac{K_8}{\beta^2} + \dfrac{K_9}{\alpha} + \dfrac{K_{10}}{\beta} + \dfrac{K_{11}}{\alpha\beta}} \right]^{1/2}$$

$$\times \left[\left\langle \sum_\lambda J_\lambda \Theta_{\text{ESVE},\lambda}(\mathbf{x}, \tau) \right\rangle^{1/2} \right] \qquad\qquad \text{(VI.13)}$$

where the following definitions were employed:

$$\alpha = \alpha(\tau) = \frac{K_3 \Psi_2 + K_2'}{K_2 \Psi_3 + K_3' \Psi_5} \qquad\qquad \text{(VI.14)}$$

$$\beta = \beta(\tau) = \frac{K_5 \Psi_2 + K_4' \Psi_1}{K_4 \Psi_5 + K_5' \Psi_6} \qquad\qquad \text{(VI.15)}$$

Using the same matrix representation for the stable reaction rates

Table VI.4   Arrays for the Reaction Rate Expressions

Name	Symbol	Definition
Free radical concentration array	$\mathbf{R}(\tau)$ or $\mathbf{R}(\mathbf{x}, \tau)$	$\begin{bmatrix} \dot{\Psi}_1 & 0 & 0 \\ 0 & \dot{\Psi}_4 & 0 \\ 0 & 0 & \dot{\Psi}_5 \end{bmatrix}$
Propagation reaction array	$\mathbf{A}(\tau)$	$\begin{bmatrix} K_2\Psi_3 + K_3'\Psi_5 + K_4\Psi_5 + K_5'\Psi_6 & -K_2' + K_3\Psi_2 & -K_4'\Psi_1 + K_5\Psi_2 \\ -K_2\Psi_3 + K_3'\Psi_5 & K_2' + K_3\Psi_2 & 0 \\ -K_4\Psi_5 + K_5'\Psi_6 & 0 & K_4'\Psi_1 + K_5\Psi_2 \end{bmatrix}$
Termination constant array	$\mathbf{T}$	$\begin{bmatrix} 2K_6 & K_9 & K_{10} \\ K_9 & 2K_7 & K_{11} \\ K_{10} & K_{11} & 2K_8 \end{bmatrix}$
Initiation reaction array	$\mathbf{g}(\mathbf{x}, \tau)$	$\left[\Psi_2 \sum_\Lambda J_\Lambda \Theta_{\mathrm{ESVE},\Lambda}(\mathbf{x}, \tau), 0 \ldots 0 \right]^T$
Termination reaction array	$\mathbf{t}(\tau)$ or $\mathbf{t}(\mathbf{x}, \tau)$	It varies according to the product formed in the termination steps.
Propagation reaction array	$\mathbf{B}(\tau)$	$\begin{bmatrix} -K_4\Psi_5 & 0 & K_4'\Psi_1 \\ -K_3'\Psi_5 + K_5'\Psi_6 & K_3\Psi_2 & K_5\Psi_2 \\ K_2\Psi_3 & -K_2' & 0 \\ K_3'\Psi_5 + K_4\Psi_5 & -K_3\Psi_2 & -K_4'\Psi_5 \\ K_5'\Psi_6 & 0 & -K_5\Psi_2 \end{bmatrix}$

(see Table VI.4) yields:

$$\langle \mathbf{\Omega}(\mathbf{x}, \tau) \rangle = \mathbf{B}(\tau) \cdot \hat{\mathbf{\Psi}}(\tau) + \langle \mathbf{g}(\mathbf{x}, \tau) \rangle - \mathbf{t}(\tau) \qquad \text{(VI.16)}$$

where $\hat{\mathbf{\Psi}}(\tau)$ is given by Eqs. (VI.11) to (VI.15), and the other variables may be calculated from the available kinetic information and the predictions for the radiation field.

ii) Model 2:

Here we can only apply the steady-state methodology for point concentration expressions. By carrying out the steady-state approximation locally for the free radicals (or MSSA) and using the same procedure as above, we obtain:

$$\underset{\text{(Propagation rates)}}{\mathbf{A}(\tau) \cdot \hat{\mathbf{\Psi}}(\mathbf{x}, \tau)} + \underset{\text{(Termination rates)}}{\mathbf{R}(\mathbf{x}, \tau) \cdot \mathbf{T} \cdot \hat{\mathbf{\Psi}}(\mathbf{x}, \tau)} - \underset{\text{(Initiation rates)}}{2\mathbf{g}(\mathbf{x}, \tau)} = \mathbf{0}$$

$$\text{(VI.17)}$$

where each term is a function of position and time. The long chain approximation can be applied again:

$$\mathbf{A}(\tau) \cdot \hat{\mathbf{\Psi}}(\mathbf{x}, \tau) = \mathbf{0} \qquad \text{(VI.18)}$$

Employing the same procedure and using Eqs. (VI.18) and (VI.17), the free radical concentrations as a function of position and time are:

$$\hat{\Psi}_4(\mathbf{x}, \tau) = \frac{\hat{\Psi}_1(\mathbf{x}, \tau)}{\alpha} \qquad \text{(VI.19)}$$

$$\hat{\Psi}_5(\mathbf{x}, \tau) = \frac{\hat{\Psi}_1(\mathbf{x}, \tau)}{\beta} \qquad \text{(VI.20)}$$

$$\hat{\Psi}_1(\mathbf{x}, \tau) = \left[\frac{\Psi_2}{K_6 + \dfrac{K_7}{\alpha^2} + \dfrac{K_8}{\beta^2} + \dfrac{K_9}{\alpha} + \dfrac{K_{10}}{\beta} + \dfrac{K_{11}}{\alpha\beta}} \right]^{1/2}$$

$$\times \left[\sum_\lambda J_\lambda \Theta_{\text{ESVE},\lambda}(\mathbf{x}, \tau) \right]^{1/2} \qquad \text{(VI.21)}$$

Finally, the volumetric-averaged values required by the stable species mass balance are:

$$\langle \hat{\Psi}_4 \rangle (\tau) = \frac{\langle \hat{\Psi}_1 \rangle (\tau)}{\alpha} \tag{VI.22}$$

$$\langle \hat{\Psi}_5 \rangle (\tau) = \frac{\langle \hat{\Psi}_1 \rangle (\tau)}{\beta} \tag{VI.23}$$

$$\langle \hat{\Psi}_1 \rangle (\tau) = \left[\frac{\Psi_2}{K_6 + \dfrac{K_7}{\alpha^2} + \dfrac{K_8}{\beta^2} + \dfrac{K_9}{\alpha} + \dfrac{K_{10}}{\beta} + \dfrac{K_{11}}{\alpha\beta}} \right]^{1/2}$$

$$\times \left\langle \left[\sum_\lambda J_\lambda \Theta_{ESVE,\lambda}(\mathbf{x}, \tau) \right]^{1/2} \right\rangle \tag{VI.24}$$

Considering the reaction rate of the stable species, one can write:

$$\langle \mathbf{\Omega}(\mathbf{x}, \tau) \rangle = \mathbf{B}(\tau) \cdot \langle \hat{\mathbf{\Psi}}(\mathbf{x}, \tau) \rangle + \langle \mathbf{g}(\mathbf{x}, \tau) \rangle - \langle \mathbf{t}(\mathbf{x}, \tau) \rangle \tag{VI.25}$$

where $\langle \hat{\mathbf{\Psi}}(\mathbf{x}, \tau) \rangle$ is given by Eqs. (VI.22) to (VI.24), and all other variables are known. To complete the theoretical development of the reaction rate it is necessary to introduce the initiation rate expression (Table VI.4). The dimensionless initiation rate is:

$$\sum_\lambda J_\lambda \Psi_2(\tau) \Theta_{ESVE,\lambda}(\mathbf{x}, \tau) \tag{VI.26}$$

The ESVE Model will be used to compute the direct radiation impinging at point $\mathbf{x}$ at time τ. Considering the whole volume of the lamp the resulting expression is:

$$\Theta^D_{ESVE,\lambda}(\mathbf{x}, \tau) = \frac{\langle Y_R \rangle}{4\pi r_L} \int_{\phi_1^D(\mathbf{x})}^{\phi_2^D(\mathbf{x})} \Delta\rho'(\mathbf{x}, \phi)$$

$$\times \int_{\theta_1^D(\mathbf{x},\phi)}^{\theta_2^D(\mathbf{x},\phi)} \exp\left[-\frac{\mu_\lambda^*(\tau) y_I}{\sin\theta \sin\phi} \right] d\theta \, d\phi \tag{VI.27}$$

where the limiting angles and other functions can be found in Section II.3. Taking into account only one reflection, the indirect radiation at incident point $\mathbf{x}$ and time τ, according to the ESVE Model, can be written as:

$$\Theta_{ESVE,\lambda}^{In}(\mathbf{x}, \tau) = \frac{\langle \Gamma_{Rf} \rangle \langle Y_R \rangle}{4\pi r_L} \int_{\phi_1^{In}(\mathbf{x})}^{\phi_2^{In}(\mathbf{x})} \Delta \rho_E'(\mathbf{x}, \phi)$$

$$\times \int_{\theta_1^{In}(\mathbf{x}, \phi)}^{\theta_2^{In}(\mathbf{x}, \phi)} \exp \left[-\frac{\mu_\lambda^*(\tau) y_I}{\sin \theta \sin \phi} \right] d\theta \, d\phi \qquad (VI.28)$$

Again the additional required information can be found in Section II.3. Finally, the total radiant energy is obtained by adding both contributions:

$$\Theta_{ESVE,\lambda}(\mathbf{x}, \tau) = \Theta_{ESVE,\lambda}^{D}(\mathbf{x}, \tau) + \Theta_{ESVE,\lambda}^{In}(\mathbf{x}, \tau) \qquad (VI.29)$$

In Eqs. (VI.27) and (VI.28) a modified attenuation coefficient (μ_λ^*) was employed. It takes into account the radiation field distortions due to the existence of gas bubbles. This coefficient has been correlated as a function of the optical properties of the liquid phase, the gas hold-up, and the specific interfacial area. Yokota et al. [94] proposed the following correlation for a modified value of μ:

$$\mu^* = \begin{cases} \mu \left[1 + \left(\dfrac{3.6}{d_b} \right)^{0.66} \varepsilon_G \right]; & \mu < 0.4 \, \text{cm}^{-1} \\ \mu; & \mu \geq 0.4 \, \text{cm}^{-1} \end{cases} \qquad (VI.30)$$

Finally, it should be mentioned that in Eqs. (VI.27) and (VI.28) two integrals (θ and ϕ) must be evaluated for each position and each time. In addition, the integration over the reactor volume (see Eqs. VI.13 and VI.24) and the consideration of polychromatic radiation through the summation on the wavelengths must also be incorporated. Thus, for every time in a computational run, the initiation rate term in the system of integro-differential equations required the solution of six nested integrals.

Experimental Study
The experimental device is schematically represented in Fig. VI.3. The gas inlet was a mixture of N_2–Cl_2 which can be fixed in the desired ratio. Nitrogen gas (99.99%) was bubbled continuously through a solution of sodium hydrosulfite (7) and afterwards passed through concentrated sulfuric acid (8) to remove oxygen and water, respectively. Chlorine gas (99.5%) was scrubbed in concentrated sulfuric acid (8) to eliminate the water content. Flow rates were measured with mass flowmeters (12), and then both feed gases were passed through a mixer (15).

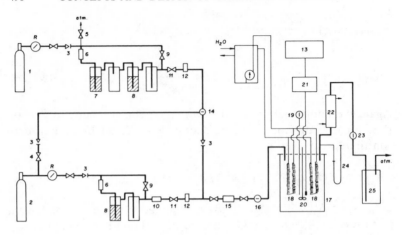

Figure VI.3   Flow-sheet of the experimental device.

The reactor is a cylindrical vessel constructed of Pyrex glass (17). At the bottom, a circular optically clear quartz plate (20) was located to provide a transparent wall for the radiation coming from the emitting system. The reactor cover had a central hole where we allocated the stuffing box for the stirrer shaft (especially constructed to avoid gas leaks). The tank was also equipped with a sample valve (19), a thermometer (21) and a gas inlet and condenser (22). The reaction temperature was controlled by circulating constant temperature water through two Pyrex glass coils (18), while the reactor pressure was measured with a sulfuric acid manometer (24). Gases left the reactor and passed through a sodium hydroxide scrubber (25) which removed the unreacted chlorine.

The experimental procedure began with the stabilization of the radiant energy source (four hours). About two hours before each chlorination run, nitrogen was turned on to remove oxygen and moisture from the device. A check for leaks was then performed. After the reactor was filled with the liquid, nitrogen was bubbled through it to remove any oxygen and water that could still be present. Then chlorine and nitrogen streams were established at a predetermined concentration ratio. The reaction time started from the moment at which the emitting system cover was removed. Liquid samples were taken from the photoreactor at regular intervals, transferred to a flask containing potassium iodide solution, and titrated with sodium thiosulfate. The organic phase and the aqueous phase were separated, and the composition of the hydrocarbon mixture was determined by gas chromatography using an internal standard method. In general, once the cover was removed, each run required not more than two hours. Afterwards, the lamp and the gas flows were turned off, and the whole system was washed by a nitrogen stream. Finally, the reactor was emptied through the sample port.

Results and Conclusions

At first, we shall present the experimental results corresponding to the addition and substitution steps. Afterwards, we shall include a comparison between experimental values and the theoretical predictions for both mixing states [10, 11].

Figures VI.4a and b show the results corresponding to the chlorination reaction of trichloroethylene under different conditions of mechanical agitation and chlorine mole fraction in the feed. In Fig. VI.4a we present the pentachloroethane and chlorine concentrations vs. time at increasing stirrer speeds. Larger conversions were observed when agitation was increased; the chlorine bulk concentration was practically negligible ($\cong 10^{-3}$ M). In addition it must be pointed out that both curves are nearly straight lines suggesting a constant reaction rate. To discard any possible gas side film resistance, the same operating conditions were employed but with pure chlorine in the feed. The results are shown in Fig. VI.4b. An increase in the stirring speed produced less noticeable effects. All

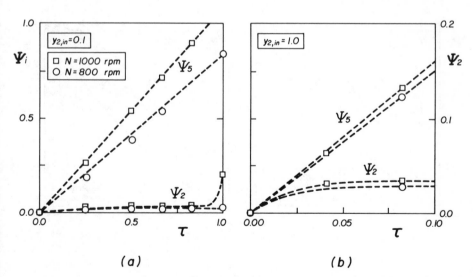

Figure VI.4   Experimental results of trichloroethylene chlorination. (a) $y_{2,\text{in}} = 0.1$. (b) $y_{2,\text{in}} = 1.0$.

these observations confirmed the existence of a diffusional sub-régime for the addition reaction.

Figures VI.5a and b portray the experimental results of pentachloroethane chlorination at increasing stirrer speeds and employing two different types of reactor bottom plates. The dimensionless concentrations of chlorine and pentachloroethane vs. time with pure chlorine in the feed are plotted. Differing from previous behavior, the experimental Ψ_i vs. τ curves were not on a straight line, i.e. clearly the reaction rate was time dependent. Conversion was not modified at increasing mechanical agitation, as can be observed in Fig. VI.5a. Moreover, chlorine bulk concentrations greater than those presented in Fig. VI.4 were found. It must be noted that the Ψ_2-scale in Fig. VI.4 is significantly stretched with respect to the one in Fig. VI.5. In Fig. VI.5b the effect of two reactor bottom plates with different transmission characteristics was studied. It can be observed that the reactor conversion with a Pyrex bottom was slightly lower than that obtained with the quartz one. It means that the reactor behavior depends on the reaction rate, and consequently on the radiation field.

It should be noticed that all these results differ considerably from those obtained with the addition reaction (diffusional subrégime). It can be concluded that the substitution reaction is under a kinetic

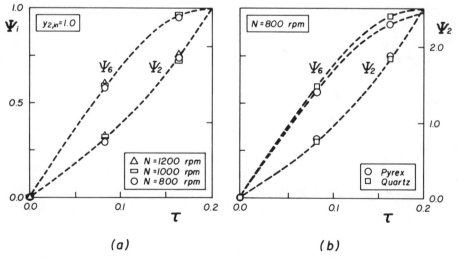

Figure VI.5 Experimental results of pentachloroethane chlorination. (a) Effect of mechanical agitation. (b) Effect of two reactor bottom plates.

subrégime, and hence it is suitable to verify the ESVE Model predictions.

In the following analysis a comparison between theoretical (Models 1 and 2) and experimental results is presented. The consecutive chlorination reactions of trichloroethylene and pentachloroethane, respectively, under different operating conditions are studied. Figures VI.6a and b show the dimensionless concentration-time curves for low and intermediate chlorine mole fractions in the feed. The solid and broken lines indicate the predictions of the ESVE Model; they show good agreement with the experimental results, mainly when Model 2 is employed. It must be pointed out that the agreement observed for the addition reaction is only ascribable to the knowledge of the mass transfer rates. This is not the case when the substitution reaction is considered, where the kinetic information is crucial for a priori predictions.

The general trend of the concentration-time curves corresponds to the typical behavior of consecutive reactions ($A \rightarrow B \rightarrow C$). In Fig. VI.6a, very close predictions between Model 1 and Model 2 can be observed. It means that when the absorption of radiation is not very large, the mixing states are unimportant in the description of photoreactor behavior. However, Fig. VI.6b shows that both mixing models produce different results when (as in the case of the

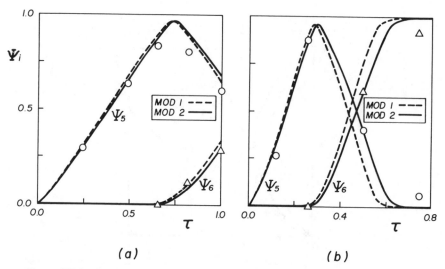

(a)   (b)

Figure VI.6 Comparison between model predictions and experimental data of trichloroethylene chlorination. (a) $y_{2,in} = 0.14$. (b) $y_{2,in} = 0.40$.

substitution reaction) the absorption by the reacting medium is considerably increased.

A very interesting aspect related to the pentachloroethane select-ivity is to be noted. Since the first step is diffusion controlled and the second one is kinetically dominated with a relatively small reaction rate, the addition occurs at a constant absorption rate and a negligible amount of hexachloroethane is formed. Hence the penta-chloroethane concentration curve falls on a straight line until nearly all the trichloroethylene is transformed into pentachloroethane. Thus, if one freezes the reaction at $\tau \cong 0.75$ (Fig. VI.6a) or $\tau \cong 0.30$ (Fig. VI.6b), it is possible to obtain almost a 100% selectivity. This particular time can be modified according to the inlet chlorine molar fraction, as can be seen by comparison between Figs. VI.6a and VI.6b.

Figure VI.7 portrays the theoretical predictions and the experi-mental results starting from pentachloroethane. This allows us to study the substitution reaction alone. Again, one can observe good agreement between the Model 2 predictions and the experimental values. In fact, this conclusion about the mixing state is similar to that obtained for the photochlorination of benzyl chloride [12].

One can see that the mixing model predictions are very different. Due to the high bulk chlorine concentrations, large absorption of

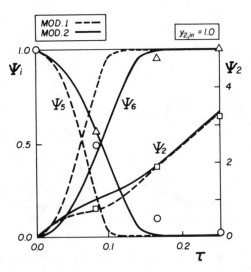

Figure VI.7 Comparison between model predictions and experimental data of pentachloroethane chlorination.

radiation occurs and the differences between Model 1 and Model 2 are significantly larger than those previously observed. In this case, the consideration of the appropriate mixing state is absolutely necessary. Obviously, both models present similar predictions after a complete conversion has been reached ($\tau > 0.16$).

The main conclusions may be summarized as follows:

- Two limiting mixing states may be defined. As a consequence, the reaction rate is a function of "the square root of the initiation rate volumetric average" or "the volumetric average of the initiation rate square root" (Models 1 and 2 respectively).
- The differences in predicted conversions between both models are a function of the magnitude of radiant energy absorption. Both models coincided only when the optical density of the reacting medium is negligible. In addition, the conversions predicted by Model 1 were always larger than those provided by Model 2.
- The gas–liquid reaction occurs in the slow reaction régime. As a consequence, the chemical absorption coefficient is equal to the physical absorption coefficient. Therefore, for this reaction system, the chlorine absorption rate is not spatially dependent. This is a great simplification for the reactor modelling.
- The existence of a diffusional subrégime for the addition reaction was experimentally verified. Thus, the photoreactor behavior was independent of the kinetics of the chemical reaction and consequently of the radiation distribution inside the reactor.
- The chlorination of pentachloroethane (substitution reaction) was also experimentally studied, and it was found that it proceeds under the kinetic subrégime. Therefore, the photoreactor behavior was kinetically dependent; consequently, the accurate knowledge of the radiation field was indispensable.
- The consecutive reactions: trichloroethylene → pentchloroethane → hexachloroethane, were studied. The diffusion to kinetic subrégime change, and the relatively small value of the substitution rate, allowed us to achieve nearly a 100% selectivity for the intermediate product. In addition, the dimensionless time to reach such selectivity may vary according to the inlet chlorine mole fraction.
- A comparison between the predictions of Models 1 and 2 and the experimental values was performed. Good agreement with both reactions was only observed when considering free radicals in a non-mixing state. It means that Model 2 is the one better for representing the actual behavior of the reactor system.

ACKNOWLEDGEMENTS

The authors are grateful to CONICET and U.N.L. (Santa Fe—Argentina) for their support in producing this work.

NOMENCLATURE

a = ellipse semimajor axis, cm

A = area, cm^2

Av = interfacial area per unit volume, cm^{-1}

b = ellipse semiminor axis, cm

c = half distance between foci, cm or velocity of light, $cm\,s^{-1}$

C = concentration, $mole\,cm^{-3}$

d_b = bubble diameter, cm

D = diffusion coefficient, $cm^2\,s^{-1}$

e = ellipse eccentricity, dimensionless

e^a = local volumetric rate of energy absorption, einstein $s^{-1}\,cm^{-3}$

E = energy flow rate, einstein s^{-1}

F = chlorine to methane feed molar ratio, dimensionless

F_v = volumetric flow rate, $cm^3\,s^{-1}$

Ge = geometric number, dimensionless

h = Planck's constant, joule s

H = Henry's constant, $atm\,cm^3\,mole^{-1}$

I = intensity, einstein $s^{-1}\,cm^{-2}\,sr^{-1}$

I° = intensity for a point source with spherical emission, einstein $s^{-1}\,sr^{-1}$

I' = intensity for a line source with emission in parallel planes, einstein $s^{-1}\,cm^{-1}\,rad^{-1}$

I'' = intensity for the line source with spherical emission, einstein $s^{-1}\,cm^{-1}\,sr^{-1}$

J = kinetic initiation number, dimensionless

k_i = reaction rate constant for i-step, s^{-1} (first order) or $cm^3\,mole^{-1}\,s^{-1}$ (second order)

k_L° = physical absorption coefficient, $cm\,s^{-1}$

k_L = chemical absorption coefficient, $cm\,s^{-1}$

K = ratio of the inner radius to outer radius, dimensionless

K_i = reaction rate constant for i-step, dimensionless

K_L = ratio of reaction time to absorption time, dimensionless

L = length, cm

m = slope of the mass balance defined by Eq. (IV.36), dimensionless

$\mathbf{n}$ = unit normal vector

N = absorption rate, dimensionless

N_e = number of emitters per unit volume of source, cm^{-3}

N_{GL} = absorption rate, mole $cm^{-2} s^{-1}$

P = total pressure, atm

Pe = Peclet number, dimensionless

P_ν = probability density distribution function for emission per unit time and per unit frequency, dimensionless

$\mathbf{q}$ = radiation flux density vector, einstein $cm^{-2} s^{-1}$

$q = \mathbf{q} \cdot \mathbf{n}$, radiation flux density, einstein $cm^{-2} s^{-1}$

Q = radiation flux density, dimensionless

r = radial coordinate, cm

R = reaction rate, mole $cm^{-3} s^{-1}$

Re = Reynolds number, dimensionless

s = linear coordinate, cm

S = selectivity, dimensionless

t = time, s

U = velocity, dimensionless

v = velocity, cm s^{-1}

V = volume, cm^3

x, y, z = rectangular Cartesian coordinates, cm

X = fractional conversion, dimensionless

Greek Letters

α = absorption coefficient, $cm^2 mole^{-1}$

β = cylinder coordinate, rad

γ = radial coordinate, dimensionless

Γ = reflection coefficient, dimensionless

δ = experimental parameter, dimensionless

$\boldsymbol{\varepsilon}$ = unit vector representing the direction of a ray

ε = absorption number, dimensionless

ε_G = volume fraction of bubbles in reactor, dimensionless

ζ = axial coordinate, dimensionless

η = photoreactor energy yield, dimensionless

θ = angular coordinate, rad

Θ_R = residence time, s

κ = characteristic property of the lamp emission, einstein $cm^{-3} s^{-1} sr^{-1}$

λ = wavelength, cm

μ = attenuation coefficient, cm^{-1}

ν = frequency, s^{-1}

ρ = spherical coordinate, cm

τ = time, dimensionless
Υ = reactor wall transmittance, dimensionless
ϕ = spherical coordinate, rad
Φ = primary quantum yield, mole einstein^{-1}
Φ_i = overall quantum yield for i-species, mole einstein^{-1}
Ψ = concentration, dimensionless
Ω = reaction rate, dimensionless; or solid angle, sr

Subscripts

ab = absorbing species
D = direct radiation
Dif = diffusion value
e = emission value
exp = experimental value
E = property of an emerging ray from the source
ESVE = relative to the Extense Source Model with Volumetric Emission
i = i-species
in = relative to the inner wall of the reactor or inlet condition
init = initiation value
I = incident point or an incident ray property
In = indirect radiation
IR = relative to the bottom of the microreactor
L = lamp property or relative to liquid phase
LSPP = relative to the Line Source Model with emission in Parallel Planes
LSSE = relative to the Line Source Model with Spherical Emission
m = relative to a multicomponent mixture
max = maximum value
min = minimum value
MR = microreactor property
n = relative to the normal
$^{\circ}$ = wall property
ou = relative to the outer wall of the reactor or outlet condition
ox = relative to oxalic acid
P = denotes point of reflection or penumbra zone
r = reception point
rc = characteristic reaction time
R = reactor property
Ref = reference value
Rf = reflector property

SS = relative to the steady-state approximation
t = theoretical value
T = denotes total value or point of tangency of the direction
tangent to the lamp and reactor boundaries with the
reactor boundary
ur = relative to the uranyl ion
U = denotes umbra zone.
w = denotes a wall reaction
z = relative to the z-axis
1 = lower limit of integration
2 = upper limit of integration
λ = wavelength dependence
ν = frequency dependence

Superscripts

a = absorbed value
D = direct radiation
Ex = extreme value
i = interface value
In = indirect radiation
Int = integration value
$\circ$ = initial value
$'$ = values projected on the $x–y$ plane
$*$ = modified property

Special Symbols

$\langle\ \rangle$ = average value
$\hat{}$ = denotes a free radical

REFERENCES

1. Akehata, T., and Shirai, T. (1972). Effect of light-source characteristics on the performance of circular annular photochemical reactor, *J. Chem. Eng. Japan*, **5**, 385.
2. ALCOA (1964). Aluminium Company of America, Technical Bulletin.
3. Alfano, O. M., Romero, R. L., and Cassano, A. E. (1980). Internal report, INTEC.
4. Alfano, O. M., Romero, R. L., and Cassano, A. E. (1981). Modelado del transporte y absorción de energía en reactores activados por radiación, *Cuadernos del CAMAT*, **11**, 63.
5. Alfano, O. M., Romero, R. L., and Cassano, A. E. (1984). Modelling of radiation transport and energy absorption in photoreactors, *Adv. in Transp. Processes*, **IV**, 201.

6. Alfano, O. M., Romero, R. L., and Cassano, A. E. (1984). Radiation field inside a cylindrical photoreactor irradiated from the bottom: Theory and experiments, in *Frontiers in Chemical Reaction Engineering*, Vol. 1, pp. 506–522, Wiley Eastern Limited, New Delhi.
7. Alfano, O. M., Romero, R. L., and Cassano, A. E. (1985). A cylindrical photoreactor irradiated from the bottom. I. Radiation flux density generated by a tubular source and a parabolic reflector, *Chem. Engng. Sci.*, **40**, 2119.
8. Alfano, O. M., Romero, R. L., and Cassano, A. E. (1984). A cylindrical photoreactor irradiated from the bottom. II. Local volumetric rate of energy absorption with polychromatic radiation. Theory, submitted for publication.
9. Alfano, O. M., Romero, R. L., Negro, C. A., and Cassano, A. E. (1984). A cylindrical photoreactor irradiated from the bottom. III. Measurements of absolute values of the local volumetric rate of energy absorption. Experiments with polychromatic radiation, submitted for publication.
10. Alfano, O. M., and Cassano, A. E. (1984). Modelling of a two-phase photoreactor irradiated from the bottom. I. Theory, to be published.
11. Alfano, O. M., and Cassano, A. E. (1984). Modelling of a two-phase photoreactor irradiated from the bottom. II. Experiments, to be published.
12. Andre, J. C., Niclause, M., Tournier, A., and Deglise, X. (1982). Industrial photochemistry II: Influence of the stirring of unstable radical species on the kinetics of long chain photochemical reaction, *J. Photochem.*, **18**, 57.
13. Astarita, G. (1966). Regimes of mass transfer with chemical reaction, *Ind. Eng. Chem.*, **58**, 18.
14. Astarita, G. (1967). *Mass Transfer with Chemical Reaction*, Elsevier, New York.
15. Astarita, G., Savage, D. W., and Bisio, A. (1983). *Gas Treating with Chemical Solvents*, John Wiley, New York.
16. Bhagwat, W. V., and Dhar, N. R. (1932). Influence of stirring on the velocity and temperature coefficient of photochemical reactions, *J. Indian Chem. Soc.*, **335**.
17. Boval, B., and Smith, J. M. (1970). Heterogeneous reactions in the photochlorination of propane, *AIChE J.*, **16**, 553.
18. Brackett, F. P. Jr., and Forbes, G. S. (1933). Actinometry with uranyl oxalate at $\lambda\lambda$ 278, 253 and 208 mμ. Including a comparison of periodically intermittent and continuous radiation, *J. Amer. Chem. Soc.*, **55**, 4459.
19. Calvert, J. G., and Pitts, J. N. Jr. (1966). *Photochemistry*, John Wiley, New York.
20. Cassano, A. E., and Smith, J. M. (1966). Photochlorination in a tubular reactor, *AIChE J.*, **12**, 1124.
21. Cassano, A. E., and Smith, J. M. (1967). Photochlorination of propane, *AIChE J.*, **13**, 915.
22. Cassano, A. E., Silveston, P. L., and Smith J. M. (1967). Photochemical reaction engineering, *Ind. Eng. Chem.*, **59**, 18.
23. Cassano, A. E., Matsuura, T., and Smith, J. M. (1968). Batch recycle reactor for slow photochemical reactions, *Ind. Eng. Chem. Fundamentals*, **7**, 655.
24. Cassano, A. E. (1968). Ph.D. Thesis, University of California, Davis, California.
25. Cassano, A. E. (1968). Uso de actinómetros en reactores tubulares continuos, *Rev. Fac. Ing. Qca.*, **XXXVII**, 469.
26. Cassano, A. E. (1980). The rate of reaction: A definition or the result of a conservation equation?, *Chem. Eng. Educ. Winter*, **14**.
27. Cerdá, J., Irazoqui, H. A., and Cassano, A. E. (1973). Radiation fields inside an elliptical photoreflector with a source of finite spatial dimensions, *AIChE J.*, **19**, 963.
28. Cerdá, J., Marchetti, J. L., and Cassano, A. E. (1977). Radiation efficiencies in elliptical photoreactors, *Lat. Am. J. Heat Mass. Transf.*, **1**, 33.

29. Cerdá, J., Marchetti, J. L., and Cassano, A. E. (1978). The use of simple radiation models for the case of direct irradiation of photochemical reactors, *Lat. Am. J. Chem. Eng. Appl. Chem.*, **8**, 15.
30. Clariá, M. A., Irazoqui, H. A., and Cassano, A. E. (1984). The use of linear and extense source models in photoreactor design, to be published.
31. Clariá, M. A., Irazoqui, H. A., and Cassano, A. E. (1984). Modelling and experimental validation of the radiation field inside an elliptical photoreactor, submitted for publication.
32. Clariá, M. A., Irazoqui, H. A., and Cassano, A. E. (1984). A priori design of the photochemical monochlorination of ethane. I. Theory, submitted for publication.
33. Clariá, M. A., Irazoqui, H. A., and Cassano, A. E. (1984). A priori design of the photochemical monochlorination of ethane. II. Computed and experimental results, submitted for publication.
34. Chiltz, G., Goldfinger, P., Hyubrechts, G., Martens, G., and Verbeke, G. (1963). Atomic chlorination of simple hydrocarbon derivatives in the gas phase, *Chem. Rev.*, **63**, 355.
35. Costa, J., Esplugas Vidal, S., and Parejo, C. (1975). Reactores fotoquímicos, *Ing. Química (Madrid)*, Febrero, **37**.
36. Daiton, F. S., Lomax, D. A., and Weston, M. (1957). The kinetics of the gas-phase photochlorination of trichloroethylene, *Trans. Faraday Soc.*, **53**, 460.
37. De Bernardez, E., and Cassano, A. E. (1982). Azimuthal asymmetries in tubular photoreactors, *Lat. Am. J. Heat Mass Transf.*, **6**, 333.
38. De Bernardez, E., and Cassano, A. E. (1985). A priori design of a continuous annular photochemical reactor. Experimental validation for simple reactions, *J. Photochem.*, **30**, 285.
39. De Bernardez, E., and Cassano, A. E. (1984). Optimal selection of the radiation source for consecutive-chain type-reactions in a continuous photoreactor, submitted for publication.
40. Doede, C. M., and Walker, C. A. (1955). Photochemical Engineering, *Chem. Engng.*, February, **159**.
41. Felder, R. M., and Hill, F. B. (1969). Mixing effects in chemical reactors—I. Reactant mixing in batch and flow systems, *Chem. Engng. Sci.*, **24**, 385.
42. Felder, R. M., and Hill, F. B. (1970). Mixing effects in chemical reactors. Chain reactions in batch and flow systems, *Ind. Eng. Chem. Fundam.*, **9**, 360.
43. Foraboschi, F. P. (1959). La conversione in un reattore fotochimico continuo, *Chem. Ind. (Milan)*, **41**, 731.
44. Forbes, G. S., and Heidt, L. J. (1934). Optimum composition of uranyl oxalate solutions for actinometry, *J. Am. Chem. Soc.*, **56**, 2363.
45. Gavalas, G. R. (1966). The long chain approximation in free radical reaction systems, *Chem. Eng. Sci.*, **21**, 133.
46. General Electric Co. (1959). Tech. Bull. No. L-S-104.
47. General Electric Co. (1966, 1967). Tech. Bull. No. T-P-122.
48. Hancil, V., Schorr, V., and Smith, J. M. (1972). Radiation efficiency of photoreactors, *AIChE J.*, **18**, 43.
49. Harano, Y., and Smith, J. M. (1968). Design of tubular flow reactor for photochemical systems, *AIChE J.*, **14**, 584.
50. Harris, P. R., and Dranoff, J. S. (1965). A study of perfectly mixed photochemical reactors, *AIChE J.*, **11**, 497.
51. Heidt, L. J., Tregay, G. M., and Middleton, F. A. Jr. (1970). Influence of pH upon the photolysis of the uranyl oxalate actinometer system, *J. Phys. Chem.*, **74**, 1876.
52. Herzberg, G. (1950). *Molecular Spectra and Molecular Structure. I. Spectra of Diatomic Molecules*, 2nd ed., Van Nostrand–Reinhold Company.
53. Hill, F. B., and Felder, R. M. (1965). Effects of mixing on chain reactions in isothermal photoreactors, *AIChE J.*, **11**, 873.

54. Hill, F. B., and Reiss, N. (1968). Nonuniform initiation of photoreactions. II. Diffusion of reactive intermediates, *Can. J. Chem. Eng.*, **46**, 124.
55. Hill, F. B., Reiss, N., and Shendalman, L. H. (1968). Nonuniform initiation of photoreactions. III. Reactant diffusion in single-step reactions, *AIChE J.*, **14**, 798.
56. Irazoqui, H. A., Cerdá, J., and Cassano, A. E. (1972). The radiation field for the point and line source approximation and the three-dimensional source models: Applications to photoreactions, paper presented at the 1972 Dallas *AIChE* meeting.
57. Irazoqui, H. A., Cerdá, J., and Cassano, A. E. (1973). Radiation profiles in an empty annular photoreactor with a source of finite spatial dimensions, *AIChE J.*, **19**, 460.
58. Irazoqui, H. A., Cerdá, J., and Cassano, A. E. (1976). The radiation field for the point and line source approximation and the three-dimensional source models: Applications to photoreactions, *Chem. Eng. J.*, **11**, 27.
59. Jacob, S. M., and Dranoff, J. S. (1966). Radial scale-up of perfectly mixed photochemical reactors, *Chem. Eng. Prog. Symp. Ser.*, **62**, 47.
60. Jacob, S. M., and Dranoff, J. S. (1968). Design and analysis of perfectly mixed photochemical reactors, *Chem. Eng. Prog. Symp. Ser.*, **64**, 54.
61. Jacob, S. M., and Dranoff, J. S. (1969). Light intensity profiles in an elliptical photoreactor, *AIChE J.*, **15**, 141.
62. Jacob, S. M., and Dranoff, J. S. (1970). Light intensity profiles in a perfectly mixed photoreactor, *AIChE J.*, **16**, 359.
63. Kurtz, B. E. (1972). Homogeneous kinetics of methyl chloride chlorination, *Ind. Eng. Chem. Proc. Des. Dev.*, **11**, 332.
64. Leermakers, J. A., and Dickinson, R. G. (1932). The photochlorination of tetrachloroethylene in carbon tetrachloride solution, *J. Am. Chem. Soc.*, **54**, 4648.
65. Legan, R. W. (1982). Ultraviolet light takes on CPI role, *Chem. Engng.*, **95**.
66. Leighton, W. G., and Forbes, G. S. (1930). Precision actinometry with uranyl oxalate, *J. Am. Chem. Soc.*, **52**, 3139.
67. Lucas, G. (1971). A new concept in oximation by the nitrosyl chloride molecule, *Information Chimie*, **16**, 33.
68. Matsuura, T., Cassano, A. E., and Smith, J. M. (1969). Acetone photolysis: Kinetic studies in a flow reactor, *AIChE J.*, **15**, 495.
69. Matsuura, T., and Smith, J. M. (1970). Light distribution in cylindrical photoreactors, *AIChE J.*, **16**, 321.
70. Matsuura, T., and Smith, J. M. (1970). Photodecomposition kinetics of formic acid in aqueous solutions, *AIChE J.*, **16**, 1064.
71. Matsuura, T., and Smith, J. M. (1970). Kinetics of photocomposition of dodecyl benzene sulfonate, *Ind. Eng. Chem. Fundam.*, **9**, 252.
72. Matsuura, T., and Smith, J. M. (1971). Photodecomposition of detergent solutions at high conversions, *Ind. Eng. Chem. Fundam.*, **10**, 316.
73. Marcus, R. J., Kent, J. A., and Schenck, G. O. (1962). Industrial photochemistry, *Ind. Eng. Chem.*, **54**, 20.
74. Murov, S. L. (1973). *Handbook of Photochemistry*, Marcel Dekker, New York.
75. Noyes, W. A. Jr., and Leighton, P. A. (1941). *The Photochemistry of Gases*, Reinhold Publications, New York.
76. Noyes, R. M., and Fowler, L. (1951). Mechanisms of chain termination in chlorine atom reactions, *J. Am. Soc.*, **73**, 3043.
77. Noyes, R. M. (1951). Wall effects in photochemically induced chain reactions, *J. Am. Chem. Soc.*, **73**, 3039.
78. Poutsma, M. L., and Hinman, R. L. (1964). Chlorination studies of unsaturated materials in nonpolar media. I. Solvent effects on radical addition of chlorine to chloroethylenes, *J. Am. Chem. Soc.*, **86**, 3807.
79. Pritchard, H. O., Pyke, J. B., and Trotnan-Dickenson, A. F. (1955). The study of chlorine atom reactions in the gas phase, *J. Am. Chem. Soc.*, **77**, 2629.

80. Ragonese, F. P., and Williams, J. A. (1971). Application of empirical rate expression and conservation equation to photoreactor design, *AIChE J.*, **17**, 1352.

81. Ré, R. M., and Cassano, A. E. (1984). Three-dimensional modelling of a tubular photoreactor when the radiation field has no angular symmetry, to be published.

82. Roger, M., and Villermaux, J. (1979). Modelling of light absorption in photoreactors, Part I. General formulation based on the laws of photometry, *Chem. Eng. J.*, **17**, 219.

83. Roger, M., and Villermaux, J. (1983). Modelling of light absorption in photoreactors. Part II. Density profile and efficiency of light absorption in a cylindrical reactor. Experimental comparison of five models, *Chem. Eng. J.*, **26**, 85.

84. Romero, R. L., Alfano, O. M., Marchetti, J. L., and Cassano, A. E. (1983). Modelling and parametric sensitivity of an annular photoreactor with complex kinetics, *Chem. Engng. Sci.*, **38**, 1593.

85. Santarelli, F., and Smith, J. M. (1974). Rate of photochlorination of liquid *n*-heptane, *Chem. Eng. Commun.*, **1**, 297.

86. Schechter, R. S., and Wissler, E. H. (1960). Photochemical reactions in an isothermal laminar-flow chemical reactor, *Appl. Sci. Res.*, **9**, 334.

87. Shirotsuka, T., and Nishiumi, H. (1971). Theoretical basis for making an apparatus for photochemical reactions, *Kagaku Kogaku*, **35**, 1329.

88. Stramigioli, C., Santarelli, F., and Foraboschi, F. P. (1977). Photosensitized reactions in an annular continuous photoreactor, *Appl. Sci. Res.*, **33**, 23.

89. Temkin, M. I. (1971). The kinetics of steady state complex reactions, *Int. J. Chem. Eng.*, **11**, 709.

90. Toor, H. L., and Marcello, J. M. (1958). Films-penetration model for mass and heat transfer, *AIChE J.*, **4**, 97.

91. Volman, D. H., and Seed, J. R. (1964). The photochemistry of uranyl oxalate, *J. Am. Chem. Soc.*, **86**, 5095.

92. Walling, C. (1957). *Free Radicals in Solution*, John Wiley, New York.

93. Williams, J. A. (1972). Radiation efficiency of elliptical reflector-photoreactors, *AIChE J.*, **18**, 643.

94. Yokota, T., Iwano, T., Deguchi, H., and Tadaki, T. (1981). Light absorption rate in a bubble column photochemical reactor, *Kagaku Kogaku Ronbunshu*, **7**, 157.

95. Zolner, W. J. III, (1971). Ph.D. Thesis, Northeastern University, Boston, Massachusetts.

96. Zolner, W. J. III, and Williams, J. A. (1972). The effect of angular light intensity distribution on the performance of tubular flow photoreactors, *AIChE J.*, **18**, 1189.

Additional Bibliography on Photochemical Reaction Engineering

97. Akehata, T., Shirai, T., Ishizoki, N., and Ito, K. (1973). Average light intensity in an annular photochemical reactor, *Kagaku Kogaku*, **37**, 1026.

98. Akehata, T., Ito, K., and Inokawa, A. (1976). Light intensity profiles in bubble-dispersed systems. An approach to the analysis of heterogeneous photoreactors, *Kagaku Kogaku Ronbunshu*, **2**, 583.

99. Bandini, E., Stramigioli, C., and Santarelli, F. (1977). A rigorous approach to photochemical reactors, *Chem. Engng. Sci.*, **32**, 89.

100. Costa, J., and Vall, F. (1972). Análisis y cálculo de reactores tubulares fotoquímicos, *Química e Industria*, **18**, 11.

101. Costa, J., and Esplugas, S. (1977). Fotorreactor anular continuo de mezcla perfecta, *Ingeniería Química*, **100**, 89.

102. Costa, J., and Esplugas, S. (1977). Descomposición fotoquímica de soluciones acuosas de acido fórmico en un fotorreactor anular continuo de mezcla perfecta, *Ingeniería Química*, **101**, 123.

103. Costa, J., Esplugas, S., and Ibarz, A. (1980). Descomposición fotoquímica de soluciones acuosas de dodecilbencenosulfonato sódico en un foto-reactor anular continuo de mezcla perfecta, *Afinidad*, **XXXVII**, 11.
104. Daniil, A., Graessley, W. W., and Dranoff, J. S. (1972). Photopolymerization of vinyl acetate in a well-stirred reactor, *Chem. React. Eng. Proc. Eur. Symp. 5th.* (*Amsterdam*), **B7**, 49.
105. Dolan, W. J., Dimon, C. A., and Dranoff, J. S. (1965). Dimensional analysis in photochemical reactor design, *AIChE J.*, **11**, 1000.
106. Dworkin, D., and Dranoff, J. S., (1978). Free radical transport in a photochemical reactor, *AIChE J.*, **24**, 1134.
107. Esplugas, S., and Costa, J. (1980). Cálculo de fotorreactores anulares continuos de mezcla perfecta, *Quaderns d'enginyeria*, **2**, 279.
108. Funayama, H., Ogiwara, K., Sugawara, and Ohashi, H. (1977). Light intensity profiles in photoreactors applied by a low pressure mercury lamp, *Kagaku Kogaku Ronbunshu*, **3**, 354.
109. Gebhard, T. J. (1978). Ph.D. Thesis, Northwestern University, Evanston, Ill.
110. Harada, J., Akehata, T., and Shirai, T. (1971). Light intensity distribution in an elliptical photoreactor, *Kagaku Kogaku*, **35**, 233.
111. Inokawa, A., and Akehata, T. (1980). Light intensity profiles in heterogeneous annular photoreactors, *Kagaku Kogaku Ronbunshu*, **6**, 178.
112. Jaeger, R. L. et al., (1968). *Engineering Compendium on Radiation Shielding*, Vol. I, 396 pp., Springer-Verlag, New York.
113. Jain, R. L., Graessley, W. W., and Dranoff, J. S. (1971). Design and analysis of a photoreactor for styrene polymerization, *Ind. Eng. Chem. Prod. Res. Develop*, **10**, 293.
114. Magelli, F., and Santarelli, F. (1978). The modelling of batch photoreactors, *Chem. Engng. Sci.*, **33**, 611.
115. Märkl, H., and Vortmeyer, D. (1971). Exact calculation of the absorption profile in an externally lighted cylinder, IInd. International Congress on Photosynthesis, Stresa.
116. Otake, T., Tone, S., Higuchi, K., and Nakao, K. (1981). Light intensity profile in gas–liquid dispersion. Applicability of effective absorption coefficient, *Kagaku Kogaku Ronbunshu*, **7**, 57.
117. Pasquali, G., and Santarelli, F. (1978). Radiant energy transfer in batch-photoreacting media, *Chem. Eng. Commun.*, **2**, 271.
118. Ragonese, F. P., and Williams, J. A., (1971). Application of empirical rate expressions and conservation equation to photoreactor design, *AIChE J.*, **17**, 1352 (1971).
119. Santarelli, F., and Stramigioli, C. (1975). Effect of the light model on conversion in a tubular photoreactor, *Ing. Chim. Ital.*, **11**, 63.
120. Santarelli, F., Stramigioli, C., Spiga, G., and Ozisik, M. N. (1982). Effects of scattering and reflection of radiation on batch photochemical reaction in a slab geometry, *Int. J. Heat Mass Transfer*, **25**, 57.
121. Shirotsuka, T., and Nishiumi, H. (1971). Design factors of photochemical reactors. Rate of absorption of light energy, *Kagaku Kogaku*, **35**, 1364.
122. Shirotsuka, T., and Nishiumi, H. (1972). Design factors of photochemical rectors. Power model of an absorption function, *Kagaku Kogaku*, **36**, 300.
123. Shirotsuka, T., and Nishiumi, H. (1972). Design factors of photochemical reactors. Photochemical reactions caused by polychromatic light, *Kagaku Kogaku*, **36**, 307.
124. Shirotsuka, T., and Nishiumi, H. (1972). Design factors of photochemical reactors. Mixing and models to absorption rate of radiation energy, *Kagaku Kogaku*, **36**, 328.
125. Shirotsuka, T., and Nishiumi, H. (1973). Kinetics of photoreduction of ferrioxalate. Deactivation and inner filter caused by ammonium molybdate, *J. Chem. Eng. Japan.*, **6**, 178.

126. Shirotsuka, T., Sudoh, M., and Nishiumi, H. (1977). Effect of optical density on n-butyraldehyde photooxidation, *Kagaku Kogaku Ronbunshu*, **3**, 231.

127. Shirotsuka, T., and Sudoh, M. (1977). Influence of mixing on light-induced chain reactions, *Kagaku Kogaku Ronbunshu*, **3**, 622.

128. Shirotsuka, T., and Sudoh, M. (1977). Influence of non-uniform profile of light intensity on photoreaction rate, *Science and Engineering Research Laboratory–Waseda University*, No. 78, 50.

129. Shirotsuka, T., and Sudoh, M. (1978). An analysis of partially illuminated photoreactor with photosensitized polymerization, *Kagaku Kogaku Ronbunshu*, **4**, 386.

130. Shirotsuka, T., and Sudoh, M. (1978). Dye-sensitized photodecomposition of aqueous phenol using visible light energy, *Kagaku Kogaku Ronbunshu*, **4**, 394.

131. Shirotsuka, T., and Sudoh, M. (1979). Scale-up of annular photochemical reactors, *Int. Chem. Eng.*, **19**, 293.

132. Skarbö, R., and Williams, J. A. (1973). The effect of internal light filtering on photoreactor performance, *Chem. Engng. Sci.*, **28**, 83.

133. Spadoni, G., Bandini, E., and Santarelli, F. (1978). Scattering effects in photosensitized reactions, *Chem. Engng. Sci.*, **33**, 517.

134. Spadoni, G., Stramigioli, C., and Santarelli, F. (1980). Influence of a reflecting boundary on a heterogeneous photosensitized reaction within a plane slab, *Chem. Eng. Commun.*, **4**, 643.

135. Spadoni, G., Stramigioli, C., and Santarelli, F. (1980). Rigorous and simplified approach to the modelling of continuous photoreactors, *Chem. Engng. Sci.*, **35**, 925.

136. Stramigioli, C., Santarelli, F., and Foraboschi, F. P. (1975). Photosensitized reactions in an annular photoreactor, *Ing. Chim. Ital.*, **11**, 143.

137. Stramigioli, C., Spadoni, G., and Santarelli, F. (1978). Photosensitized reactions in absorbing-scattering media within a plane slab, *Int. J. Heat Mass Transfer*, **21**, 660.

138. Stramigioli, C., Santarelli, F., Spiga, G., and Vestrucci, P. (1982). Sun driven photochemical reaction in a horizontal plane slab, *Chem. Eng. Commun.*, **16**, 205.

139. Sugawara, T., Tadaki, T., and Maeda, S. (1969). On the reaction characteristics in laminar flow between parallel plates. An analysis by a photochemical reaction, *Kagaku Kogaku*, **33**, 267.

140. Sugawara, T., Tadaki, T., and Maeda, S., (1971). On the analysis of plate-type photochemical reactors. Effect of light intensity distribution due to the absorption of light by the reactant on the conversion, *Kagaku Kogaku*, **35**, 461.

141. Sugawara, T., Tadaki, T., and Maeda, S., (1973). The photochlorination of chloroform in flow reactors. Kinetic studies in a tubular reactor, *Kagaku Kogaku*, **37**, 614.

142. Sugawara, T., Omori, K., and Ohashi, H. (1976). The effect of internal light filtering on the performance of plate-type photochemical reactors, *Kagaku Kogaku Ronbunshu*, **2**, 304.

143. Sugawara, T., Omori, K., and Ohashi, H., (1979). Effects of radiation profile and internal light filtering by a product on the reaction characteristics in plate type photoreactors, *J. Chem. Eng. Japan*, **12**, 143.

144. Sugawara, T., Yoneya, M., and Ohashi, H. (1981). Inactivation rate of bacillus subtilis spores irradiated by ultraviolet rays, *J. Chem. Eng. Japan*, **14**, 400.

145. Sugawara, T., Yoneya, M., and Ohashi, H. (1981). Performance of annular flow sterilizer irradiated by a germicidal lamp, *J. Chem. Eng. Japan*, **14**, 406.

146. Thiele, R., Weickert, G., Hertwig, K., and Häusler, D. (1975). Reaktionstechnische Untersuchung und mathematische Modellierung einer Fotochorierung, *Chem. Tech.*, **27**, 595.

147. Thiele, R., Voss, H., Müller, W. D., and Hertwig, K. (1980). Reaktionstechnik Photochemischer Processe. Ein und zweidimensionale Modellierung von photorohrreactoren. Modellgrundlagen, *Chem. Tech.*, **32**, 67.

148. Tournier, A., Deglise, X., Andre, J. C., and Niclause, M. (1982). Experimental determination of the light distribution in a photochemical reactor, *AIChE J.*, **28**, 156.
149. Tournier, A., Deglise, X., Andre, J. C., and Niclause, M. (1982). Industrial photochemistry I: Measurement of the absorption profile of the light in a photochemical reactor, *J. Photochem.*, **18**, 47.
150. Voss, H., Müller, W. D., Hertwig, K., and Thiele, R. (1980). Reaktionstechnik Photochemischer Processe. Ein und zweidimensionale Modellierung von photorohrreactoren. Numerische Parameteruntersuchungen, *Chem. Tech.*, **32**, 122.
151. Weickert, G., and Thiele, R. (1976). Reaktionstechnik Photochemischer Prozesse. Modellierung des Strahlungsfeldes von Photoreaktoren Tauchlampensystem, *Chem. Tech.*, **28**, 580.
152. Williams, J. A., and Ragonese, F. P. (1970). Asymptotic solutions and limits of transport equations for tubular flow photoreactors, *Chem. Engng. Sci.*, **25**, 1751.
153. Williams, J. A., and Yen, H. C. (1973). The incident wall intensity in elliptical reflector-photoreactors, *AIChE J.*, **19**, 862.
154. Williams, J. A. (1976). Experimental observations concerning the diffuse light intensity distribution model, *AIChE J.*, **22**, 811.
155. Williams, J. A. (1978). The radial light intensity profile in cylindrical photoreactors, *AIChE J.*, **24**, 335.
156. Yokota, T., Iwano, T., and Tadaki, T. (1976). Light intensity in an annular photochemical reactor, *Kagaku Kogaku Ronbunshu*, **2**, 298.
157. Yokota, T., Iwano, T., and Tadaki, T. (1977). Light absorption rate in a heterogeneous photochemical reactor, *Kagaku Kogaku Ronbunshu*, **3**, 248.
158. Yokota, T., Iwano, T., and Tadaki, T. (1978). The effect of light filtering by dispersed phase in a heterogeneous photochemical reactor, *Kagaku Kogaku Ronbunshu*, **4**, 103.
159. Yokota, T., Iwano, T., and Tadaki, T. (1981). Light absorption rate in a photochemical reactor with lamp placed at the center—The effect of specular reflection at the mirror wall of reactor, *Kagaku Kogaku Ronbunshu*, **7**, 298.
160. Yokota, T., Iwano, T., Saito, A., and Tadaki, T. (1981). Photochlorination of toluene in a bubble column photochemical reactor, *Kagaku Kogaku Ronbunshu*, **7**, 164.
161. Zolner, W. J. III, and Williams, J. A. (1971). Three-dimensional light intensity distribution model for an elliptical photoreactor, *AIChE J.*, **17**, 502.

Chapter 9

Chemical Absorption

GIANNI ASTARITA
Istituto di Principi di Ingegneria Chimica, Piazzale Tecchio, 80125 Napoli, Italy

and

DAVID W. SAVAGE
Exxon Research and Engineering Company, Corporate Research–Advancement and Transfer, Clinton Township, Route 22 East, Annandale, New Jersey 08801

Abstract

Chemical absorption is widely used in a variety of industrial gas treating processes which are briefly reviewed in the first part of this paper. Theoretical understanding of these processes has reached the stage where accurate and reliable mathematical models can be constructed. Modeling of the equilibrium behavior of these systems is discussed in the second part, and modeling of the rate behavior in the third part. Actual development of a mathematical model for any specific process requires knowledge of the values of a variety of physicochemical parameters, which can be measured with well-established laboratory techniques which are briefly reviewed. Finally, design procedures are discussed, which can be made reliable enough to allow complete scaleup from the laboratory scale of experiments to full size industrial units.

1. INDUSTRIAL GAS TREATING PRACTICE

Gas treating is the term used to describe the separation from gases of acidic impurities such as acid gases (CO_2, H_2S, SO_2), organic sulfur compounds and certain other impurities. Removal of SO_2 from gases is sometimes treated as a separate technology—flue gas desulfurization—although the separation principles and concepts are the same for both types of impurities.

In acid gas removal, the most important impurities are CO_2 and H_2S which often occur in quite large concentrations (5–50%) in industrial processes such as hydrogen manufacture, ammonia production, and natural gas purification [20, 17]. Organic sulfur contaminants tend to occur at much lower concentrations (below 0.5%). When present, the other impurities such as HCN, NH_3, SO_3, tend to occur at very low levels, but they can often be quite troublesome to remove. In flue gas desulfurization, the principal impurity, SO_2, occurs at a relatively low concentration (say 0.25 per cent), depending on the sulfur content of the coal or oil which is being burned.

The cleanup target, the allowable extent of the impurity in the treated gas to meet product specifications, varies markedly from process to process and with the nature of the acidic impurity. In general, the cleanup target for H_2S is particularly stringent, e.g., 4 ppm by volume for pipeline gas at 1000 psi, and 0.1 ppm for chemical applications such as ammonia synthesis. A stringent target is also necessary for H_2S to avoid catalyst poisoning, corrosion and environmental release. The CO_2 specification may be relatively loose for natural gas (<1%), but for ammonia and LNG manufacture CO_2 must be reduced to 10 ppm and 100 ppm, respectively. The CO_2 cleanup targets are set to avoid catalyst poisoning or to limit the amount of diluent CO_2 in the treated gas. Cleanup target plays an important role in the selection of a gas treating process.

Selective absorption of the acid impurities from the parent gas into a liquid solvent may be based on physical absorption (based on a difference in solubility) or on absorption with reaction with a chemical base. The chemical approach takes advantage of the relative acidity of the impurities. In the case of the removal of CO_2, H_2S, SO_2, and COS, acid-base chemistry is the basis for the majority of industrially important separations. Other chemical approaches, which are applicable for certain acid gases, include oxidation in the case of H_2S and hydrolysis in the case of COS and other sulfur contaminants.

The main categories of gas treating processes are shown in Table 1. Absorption into a liquid agent, either physical absorption or absorption into a solution of a chemical base, is the most commonly used approach. This type of process is used almost exclusively for bulk removal of acidic contaminants. For removal of the last traces of acid impurities from gases, adsorption onto a solid and chemical conversion to another compound, e.g., methanation, are used extensively.

The cumulative commercial use of different types of chemical and

Table 1 Categories of Gas Treating Processes

Absorption into a liquid
  Physical—polar organic solvent
  Chemical—solution of a base
Cryogenic separation
Adsorption onto a solid
  Zeolites, activated carbon, clays
	Bulk removal

Chemical conversion to another compound
  Gas phase reaction, e.g., methanation
  Reaction with solid, e.g., Fe_2O_3, ZnO
  Reaction with slurred solids (SO_2)
  Chemical oxidation (H_2S)
  Hydrolysis (COS)
	Final or trace
	purification

physical solvents for acid gas removal is shown in Table 2. Based on the number of installations, the most widely used chemical solvents are aqueous alkanolamines (aminoalcohols) and promoted hot potassium carbonate processes. These are followed by organic solvent solutions of aminoalcohols and aqueous solutions of potassium salts of amino acids. Compared to over 2000 installations using chemical solvents, less than 100 physical solvent processes are in place. Common physical solvents include organic liquids such as

Table 2 Commercial Use of Different Types of Solvent Gas Treating Processes

Solvent	Number of installations
Acid gas removal	
Aqueous alkanolamine	>1000
MEA	
DEA	
DGA	
DIPA	
Promoted hot potassium carbonate	>740
Organic promoters	
Inorganic promoters	
Organic solvent—alkanolamine	>130
Sulfolane/DIPA	
MeOH/MEA/DEA	
Aqueous solution of potassium salt	~100
of amino acids	
Physical (organic) solvents	73
Propylene carbonate	
Polyethylene glycol dialkyl ether	
N-methylpyrrolidone	
Chilled methanol	
Flue gas desulfurization	~200
Lime/limestone slurries in water	
Sodium sulfite	

alcohols, glycols, propylene carbonate and pyrrolidones. Flue gas-desulfurization is an infant technology compared to acid gas removal. However, there are over 50 installations already in the United States and a larger number in the Far East.

Acid gas removal processes, based on solvent absorption, all use variations on a basic process flowsheet which is shown in idealized form in Fig. 1. The raw gas is contacted countercurrently with regenerated (lean) solvent in an absorber tower, which usually operates at low temperature (313–373 K) and high pressure (100–1000 psig). Treated gas leaves overhead from the absorber, sometimes after water wash to prevent carryover of solvent components. The loaded (rich) solvent leaving from the bottom of the absorber is reduced in pressure, sometimes in stages, to flash off both dissolved hydrocarbons and a portion of the acid gases. The rich solvent may also be heated by the returning hot regenerated lean solvent in a rich/lean heat exchanger. The regenerator is usually a low-pressure countercurrent tower in which the remaining dissolved acid gases are stripped out with steam generated by reboiling the lean solution. In practice, the hot rich solution is usually allowed to flash into the top of the regenerator tower, i.e., no separate flash drum is provided.

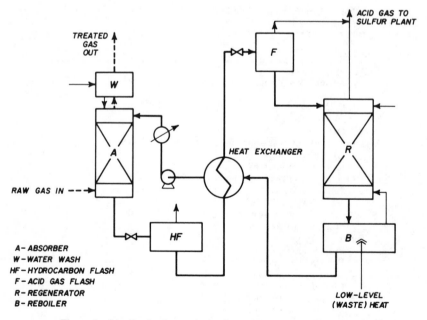

Figure 1   Idealized solvent absorption acid gas removal process.

No gas treating process is a stand-alone operation. Every gas treating unit is integrated into a larger process. Common integration features include utilization of waste heat available in the raw gas or from an associated process unit (e.g., sulfur plant or steam reformer) to reboil the stripped solvent. Mass integration is also important. When H_2S is present in the acid gas, the separated acid gas must go directly to a sulfur plant (e.g., a Claus process) to minimize emissions of H_2S to the atmosphere. In CO_2 removal, particularly in connection with ammonia manufacture, the separated CO_2 may become the feedstock for making urea, or it may be recovered as a pure product, e.g., for carbonation of beverages or for injection into the ground to enhance recovery of oil and gas.

Solvent circulation rate is the single most important factor in determining the economics of gas treating with chemical solvents [18, 22]. Solvent circulation rate, through its influence on the size of towers, pumps, lines and heat exchangers, has a large influence on the capital cost of gas treating plants. Solvent rate also has a major influence on the energy (steam) requirement for solvent regeneration, and the reboiler heat duty can often be correlated directly with liquid rate.

Rates of mass transfer, in both absorption and desorption, affect the tower heights and hence capital cost. The primary influence of rate on investment is in the absorber which is usually a high-pressure vessel. The rate of mass transfer in desorption may influence the steam required for stripping. In nonequilibrium or kinetically selective processes the rates of mass transfer are extremely important as they set the selectivity for the separation of two acid gases. Other factors that play an important role in gas treating economics include solution corrosivity, which determines materials of construction for the plant; solvent cost, especially as it relates to makeup requirements; and process reliability which determines the service factor.

The loading of chemical solvents for gas absorption is determined, in the limit, by considerations of vapor–liquid equilibrium. At any temperature, for a given solvent, there is a definite relation between the loading of acid gas in the solvent, y, and the equilibrium vapor pressure of the acid gas, p^*, over the liquid. This vapor–liquid equilibrium (VLE) relationship can be expressed mathematically by an equation of the form:

$$p^* = f(T, y) \tag{1.1}$$

Often, VLE data are available as a plot of p^* against y at constant T. Vapor–liquid equilibrium curves for CO_2 at 313 K in a physical solvent and a chemical solvent are shown in Fig. 2. Because many decades of pressure are usually covered in a gas absorption–desorption cycle, the pressure scale is logarithmic. The loading scale is linear and is expressed in mole fraction units for a physical solvent and as fractional saturation for a chemical solvent.

For the physical solvent (propylene carbonate), acid gas solubility is small at low pressures but it becomes quite large at high pressures (e.g., 200 psi). The chemical solvent (aqueous diethanolamine, DEA), on the other hand, exhibits quite high loadings at low CO_2 pressures but it quickly becomes chemically saturated at moderate pressures. Thus, when removing acid gases at low partial pressures

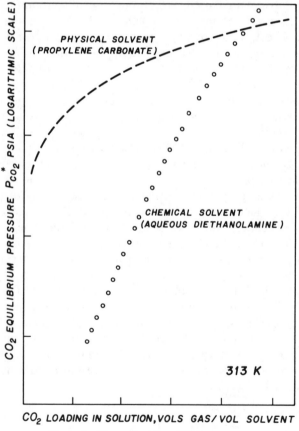

Figure 2   Vapor–liquid equilibrium curves.

in the feed gas, chemical solvents will generally be preferred, while for high partial pressures, physical solvents may have advantages over chemical solvents.

It should be borne in mind that the comparison in Fig. 2 is between a physical solvent other than water, and a chemical solvent which is an aqueous solution. In the chemical solvent, once the acid gas loading has reached the point of chemical saturation, additional absorption of acid gas can only take place by physical absorption in the solvent, which is water. As water exhibits a comparatively low physical solubility for acid gases, the chemical solvent curve in Fig. 2 eventually bends to the right and takes the shape of the physical solvent curve, but only at extremely high acid gas partial pressures. An interesting possibility is the use, as chemical solvents, of nonaqueous solutions, where loading at low acid gas pressures will be governed by chemical saturation, and at high pressure by the comparatively high physical solubility in the nonaqueous solvent.

The important parameter in practice is the cyclic capacity, C_C, of the solvent. The cyclic capacity is the maximum total number of moles of acid gas (component A) which can be extracted from the gas stream by a unit volume of the solvent. It is evident that C_C is given by (see Fig. 2):

$$C_C = (y_{max} - y_{min})m \qquad (1.2)$$

where y_{max} is the maximum or "rich" loading of the solvent leaving the absorber, y_{min} is the minimum or "lean" loading of the solvent entering the absorber (leaving the reboiler) and m is the molarity, i.e., the concentration of reactive species in the liquid solution. The importance of the concept of cyclic capacity can easily be understood from the following argument. Given a certain gas treating load, Φ_T, i.e., the number of moles per unit time of component A to be removed from the gas stream, the volumetric liquid circulation rate, L, is subject to the following obvious condition:

$$L > \frac{\Phi_T}{C_C} \qquad (1.3)$$

This equation shows that a cyclic capacity as high as possible is desirable.

The values of y_{max} and y_{min} for a given operation are essentially determined by the VLE behavior of the system. The maximum mole fraction, y_{max}, is determined by the condition that, at the gas inlet section of the absorber (absorber bottom), a finite driving force for absorption needs to be available. In the limit where this driving force is negligibly small, one may write:

$$P_B = p^*(y_{max}, T_B) \qquad (1.4)$$

where P_B is the partial pressure of A in the feed gas, and T_B is the temperature at the bottom of the absorber. The value of y_{min} is determined by the condition that at the top of the absorber there should be a finite driving force for acid gas absorption to achieve the desired degree of cleanup of acid gas. In the limit of zero driving force at the absorber top, one may write:

$$P_T = p^*(y_{min}, T_T) \qquad (1.5)$$

where P_T is the partial pressure of A in the treated gas, and T_T is the temperature at the top of the absorber. Thus, knowing P_B (from total pressure and composition of the feed gas), P_T (from total pressure and cleanup target of the treated gas) and the absorber top and bottom temperatures, the thermodynamic cyclic capacity can be estimated from VLE curves.

The actual cyclic capacity of a solvent can never be larger than the thermodynamic cyclic capacity calculated above. The actual capacity will, in fact, be smaller owing to the need to provide finite driving forces for mass transfer, and other limitations such as maximum rich solution loading allowed to avoid corrosion or precipitation.

The principal differences between physical and chemical solvents are such that physical solvents tend to be preferred when the partial pressure of acid gas in the feed is high, whereas chemical solvents are preferred when feed acid gas partial pressure is low. Both types of solvents are capable of reducing acid gas to low levels in the treated gas, but chemical solvents can do this with many fewer theoretical contacting stages owing to the different shape of the vapor–liquid equilibrium curve. Physical solvents require much taller towers to achieve a high degree of acid gas cleanup. In general, physical solvent processes require higher capital investments than chemical solvent processes owing to the taller towers, the need for refrigeration and the large solvent circulation pumps.

With chemical solvents it is generally desirable to use as high a concentration (molarity) as possible of the chemical base. However, there may be several limitations on the molarity of chemical solvents in practice. Historically, the principal limitation on molarity—has been set by corrosion considerations. However, the recent introduction of corrosion inhibitors has allowed this limitation to be lifted in certain circumstances. The practical range of molarity is usually 2–6 gmol/lt. Thus, industrial chemical solvents are solutions of high ionic strength and their behavior can be expected to be highly nonideal.

The principal advantages of chemical solvents may be listed as:

- Increased capacity, especially when removing acid gases having low partial pressure.
- High degree of removal (cleanup) of acid gas.
- Absorption mass transfer coefficient is increased (if the reaction is fast enough to occur appreciably near the interface).
- Desorption mass transfer coefficient is increased.
- There is potential for selectively absorbing one reacting gas over another, based on a difference in rates of absorption (this type of nonequilibrium selectivity is called kinetic selectivity).

Against these considerable advantages certain disadvantages must be considered, including the cost of chemical bases, high heat of absorption, corrosion of carbon steel, side reactions (base degradation) and possible environmental concerns.

2. THERMODYNAMICS

In the analysis of gas treating processes, a crucial step is the accurate calculation of vapor–liquid equilibria (VLE). This is required in order to determine the cyclic capacity of the solvent (the amount of gas it can pick up per unit volume in one absorption–regeneration cycle) and therefore the required solvent circulation rate, to establish the toughest specifications on the clean gas which can be met, and of course to design both the absorption unit and the regeneration unit. Of course, VLE data can be obtained experimentally, and in fact they are always obtained for any given system which has potential industrial use. However, a thermodynamic model is still needed for interpolation of data, as well as for extrapolation to the region of very low partial pressures where reliable data are difficult to obtain. Furthermore, a (possibly crude) thermodynamic model is a useful tool in the preliminary screening of possible solvents. In this case one generally does not have good VLE data for all the solvents under scrutiny; however, even a crude thermodynamic model based on few data can rule out several candidate solvents, so that good data need to be gathered only for a few promising solvents.

A thermodynamic model should be able to predict the equilibrium partial pressures p_n^* of the absorbable components as a function of liquid phase composition and temperature. The liquid phase composition is identfied by the molarity m (i.e., the total concentration of reactive solute in the solvent) and the fractional saturations y_n (i.e., the fraction of reactive solute which is chemically combined with the nth absorbable gas). Thus equations of the following general form need to be established:

$$p_n^* = f_n(m, T, y_1, \ldots, y_N) \qquad (2.1)$$

where N is the total number of reactive absorbable components. Absorbable components which are not reactive, such as hydrocarbons, if present in the aqueous phase will have only a very minor effect on the equilibrium partial pressures of the reactive components.

Let A_n be the reactive absorbable components, and B_m be the nonvolatile species, which may be present in the liquid phase, with $n = 1, \ldots, N$ and $m = 1, \ldots, M$. If the chemistry of the system is understood, a set of independent chemical reactions can be written, which in synthetic form is:

$$\sum_n \bar{\nu}_{jn} A_n + \sum_m \bar{\nu}_{jm} B_m = 0 \qquad (2.2)$$

In Eq. (2.2), the $\bar{\nu}$ are stoichiometric coefficients (which can be both positive and negative), and $j = 1, \ldots, R$ identifies the specific chemical reaction. It is easy to prove that $R < N + M$ [2]; the reactions (2.2) are independent if the rank of the $\bar{\nu}_{jn} \oplus \bar{\nu}_{jm}$ matrix is R.

Since the A_n's are reactive components, $R \geqslant N$. It is therefore always possible, by linear combination of reactions, to rearrange Eq. (2.2) to the following general form:

$$A_n + \sum_m \nu_{nm} B_m = 0 ; \qquad n = 1, 2, \ldots, N \qquad (2.3)$$

$$\sum_m \nu_{km} B_m = 0 ; \qquad k = 1, 2, \ldots, R - N \qquad (2.4)$$

where the set in Eq. (2.4) may be empty since it is possible that $R = N$.

The two sets of equations. (2.3) and (2.4) have a simple physical interpretation. The set given by Eqs. (2.3) corresponds to the assumption that each reactive absorbable component has undergone one (and only one) specific reaction, and that therefore it is present in the chemically combined form of one particular available chemical sink (for example, if CO_2 is the absorbable component, HCO_3^- may be chosen as the specific chemical sink for it). The set given by Eqs. (2.4) describes the possibility of redistribution of that component among all available chemical sinks (in the case of CO_2 absorption in an amine solution, it would describe the splitting of the chemically combined CO_2 between the bicarbonic, carbonic, carbamic, and possibly alkylic forms).

The procedure to be followed in order to develop a thermodynamic model is as follows. First, based on the set indicated by Eqs. (2.3), and *admissible composition* of the liquid phase is established, i.e., a chemical composition which is compatible with the assigned values of m and y_n. If b_m, a_n stand for the concentration of B_m, A_n, the admissible composition is a set of concentrations b_m^0 given by:

$$b_m^0 = m \sum_n y_n \nu_{nm} \qquad (2.5)$$

The true composition of the liquid phase must be stoichiometrically accessible from the admissible composition, through occurrence of the reactions described by the set of Eqs. (2.4). Therefore, if ξ_k is the extent of the kth reaction from the admissible to the true composition, then the true concentrations b_m are:

$$b_m = b_m^0 - \sum_k \nu_{km} \xi_k \qquad (2.6)$$

The set of equations. (2.6) reduces the unknowns of the problem from the M concentrations b_m to the $R - N$ extents of reaction ξ_k. Substitution of the set (2.6) into the $R - N$ equilibrium conditions for reactions (2.4) produces a system of $R - N$ coupled polynomial equations for the $R - N$ unknowns ξ_k. This system will produce only one solution ξ_k, $k = 1, \ldots, R - N$ such that, when substituted into (2.6), all the values of b_m are real and positive. Thus the equilibrium values b_m^* corresponding to the assigned values of m, y_n are established. These equilibrium values will, of course, be expressed by (possibly complicated) equations where the concentration-based equilibrium constants of reactions (2.4), K_k, appear as parameters:

$$b_m^* = g_m(m, y_n, K_k) \qquad (2.7)$$

Once the b_m^* values have been established, the equilibrium values a_n^* of the absorbable components in their physically dissolved form are to be calculated from the equilibrium equations for the reactions given by Eqs. (2.3):

$$a_n^* = \prod_m (b_m^{*\,-\nu_{mn}})/K_n \qquad (2.8)$$

where again the values of K_n are concentration-based equilibrium constants. Finally, the equilibrium partial pressures are calculated as follows:

$$p_n^* = H_n a_n^* \qquad (2.9)$$

where H_n is the Henry's law constant for component A_n.

The procedure as outlined above will produce equations of the form of Eq. (2.1). However, VLE data are obtained as relation between p_n^* and α_n, where α_n is the *total* content of A_n in the liquid phase, i.e., the sum of the concentration of A_n in its physically dissolved form and in all chemically combined forms. The α_n values are given by:

$$\alpha_n = a_n + m y_n \qquad (2.10)$$

and can therefore be calculated from the equations developed previously. The parameters appearing in the equations of the model are the R concentration-based equilibrium constants K_k and K_n, and the N Henry's law constants H_n. In order to have a working model, values of these $R + N$ parameters need to be estimated, and methods for doing so are discussed in the following.

Chemical solvents used in industrial gas treating operations are invariably concentrated solutions, since the cyclic capacity always increases with increasing m. Indeed, m values as large as possible are used with the upper bound being set either by the solubility of the reaction solute, or by corrosion considerations, or finally in order to avoid precipitation of reaction products. It follows that the solvents are always highly nonideal solutions, and the activity coefficients change with composition. Therefore, the equilibrium and Henry's law constants will change with composition. In general they will be functions of both m and y_n. Infinite dilution values of these constants are not to be trusted, except for order-of-magnitude analyses, since the true values may differ from the infinite dilution values by as much as 100%. However, even infinite dilution values provide some useful information. First of all, when the thermodynamic model is used simply to perform a preliminary screening of solvents, only order-of-magnitude accuracy is required, and therefore infinte dilution values can be used.

The second important result that can be obtained from infinite dilution values is the preliminary *simplification* of the thermodynamic model. Some of the reactions in the set of equations. (2.4) may have values of K_k several orders of magnitude lower than the other ones; in these cases these reactions can be excluded from consideration, and again, since only order-of-magnitude accuracy is required, infinite dilution values can be used. As an example, we may quote the case of CO_2 absorption in aqueous solutions of alkanolamines, where formation of alkylates can often be neglected if the pH range of interest is known not to exceed about 12. Once a solvent with promising VLE behavior has been selected, and the thermodynamic model has been suitably simplified, the question of estimating the values of K_k, K_n, and H_n needs to be faced. This is briefly discussed below.

The values can be extracted from VLE by different techniques of multiparameter fit. All such techniques have an inherent unreliability. Therefore, alternate techniques are preferable. At low partial pressures (and therefore low y_n values) the values of a_n^* will be negligible as compared to my_n, since the values of K_n are invariably large (if they were not, the solvent would be a poor one). Therefore, as can be seen by considering Eqs. (2.8), (2.9), and (2.10), there is no possibility of extracting from VLE data the values of H_n and K_n separately, but only the value of the ratio H_n/K_n. This is unfortunate because, in the rate analysis, the value of H_n is needed. This can sometimes be extracted from VLE data at very high partial pressures, where chemical saturation is essentially complete ($y_n \simeq 1$) and any additional uptake of absorbable components is dominated by their physical dissolution. Other techniques for measuring H_n independently are available; these techniques are all based on the assumption that the ratio H_n/H_n^0, where H_n^0 is the infinite dilution value, is independent of either the nature of the absorbing gas, or temperature, or some other parameter. In general, a Setchenov-type equation can be used:

$$\log \frac{H}{H_n^0} = K\psi \qquad (2.11)$$

where ψ is some lumped parameter used as a measure of the composition of the liquid phase. Often the ionic strength is taken to be the appropriate parameter, although for some systems use of the molarity m produces a correlation at least as good as the ionic strength.

Groupings of equilibrium constants appearing in the thermodynamic model can be estimated from their infinite dilution values, or from the VLE data themselves, by making use of some (possibily modified) form of the Debye–Huckel or the Pitzer equation. Values of K_n and K_k can also be extracted from direct chemical analysis of the liquid phase; NMR analysis is a very useful tool in this regard. Finally, since the thermodynamic model can be used to predict the alkalinity of the liquid phase, measurements of pH as function of y_n can be used to extract values of equilibrium constants.

Whenever, as often is the case (at least over some range of y_n values), one of the chemical sinks for a given absorbable component is strongly preferred thermodynamically, the model will produce equations of the following form:

$$p_n^* = Am^q f(y, \ldots, y_N) \tag{2.12}$$

It turns out that the models are in general almost surprisingly accurate in predicting the form of the function $f(\ )$. In contrast with this, the quantities A and q have to be extracted from the data; the models generally do poorly in predicting the value (through not the sign) of q, and the constant A is a grouping of equilibrium and Henry's law constants which the model is unable to predict. Fortunately, it is the form of the function $f(\ )$ which one needs to know well for the two main uses of a quantitative thermodynamic model, viz., interpolation and extrapolation of data.

3. MASS TRANSFER

The phenomena of coupled mass transfer with chemical reaction, which take place in the liquid phase in all gas treating processes, have traditionally attracted the attention of academic researchers working in the field. Indeed, the two standard reference books which appeared in the late 60's [1, 12], both written by academics, are almost exclusively concerned with such phenomena. However, it should be borne in mind that, important as these phenomena are, they are not crucial to the industrial development of gas treating processes, since lack of their understanding can easily be overcome by the comparatively inexpensive technique of overdesigning the height of absorption and desorption towers. An exception to this statement is the case of *selective* removal of one particular impurity (H$_2$S), which will be discussed in some detail below.

The theory of coupled mass transfer and chemical reaction is very briefly reviewed below, and the reader is referred to the two books quoted earlier for details. The key concept of the theory is the identification of two time scales, the diffusion time t_D and the reaction time t_R. The value of t_D is given by $D/k_L^{0^2}$, where k_L^0 is the liquid-side mass transfer coefficient in the absence of chemical reaction and D is the diffusivity in the liquid phase (both for the absorbable component under consideration). The reaction time t_R is, for a pseudo-first order chemical reaction, the inverse of the apparent kinetic constant; it can also be defined for more general kinetic equations. The relevant dimensionless group is the ratio of the two time scales:

$$\Phi = t_D/t_R \qquad (3.1)$$

As long as ϕ is significantly less than unity, mass transfer and chemical reaction are coupled only insofar as the total driving force available is split between the two phenomena, with the one which is intrinsically slower taking up a proportionally larger share. This is the situation usually referred to as "slow reaction regime". Once the splitting of the driving force has been calculated (from the requirement that the two phenomena taking place in series proceed at the same rate), the overall rate can be calculated either as a mass transfer rate or as a chemical reaction rate, as one may wish. Sometimes the regime has been subdivided into two subregimes, the kinetic and the diffusional subregimes, according to whether the chemical reaction or mass transfer takes up most of the overall driving force. In the diffusional subregime the chemical reaction is essentially at equilibrium in the bulk of the liquid phase, and in a thin concentration boundary layer near the gas–liquid interface (the thickness of which is of the order of $\sqrt{Dt_D}$) mass transfer takes place essentially unaffected by the chemical reaction (except insofar as it determines the available driving force).

However, as ϕ increases beyond the value of unity, a more direct coupling of mass transfer and chemical reaction takes place. The concentration boundary layer cannot exceed the "reaction length" $\sqrt{Dt_R}$, and therefore when $\phi > 1$ it will be thinner than it would be in the absence of chemical reactions. It follows that, *at a given driving force* (i.e., once the value of $a_i - a_0$ has been assigned, where i and 0 stand for interface and bulk-liquid conditions) the rate of mass transfer will be *larger* than it would be in the absence of chemical reactions. The ratio of the rate of mass transfer with chemical reaction to the rate one would have in their absence under the same driving force is called the rate enhancement factor, I. Since the rate of mass transfer is, other things being equal, inversely proportional to the thickness of the boundary layer, one obtains the result:

$$\phi \gg 1, \qquad I = \sqrt{\phi} \qquad\qquad (3.2)$$

This situation is the one which is called the "fast reaction regime". Of course, in the slow reaction regime ($\phi \ll 1$), the value of I is simply unity. There is of course a transition region near $\phi = 1$. Physical modeling and mathematical techniques, both of somewhat sophisticated nature, have been used to predict the $I(\phi)$ curve near $\phi = 1$. However, it turns out that the simplest possible interpolation formula:

$$I = \sqrt{1 + \phi} \qquad\qquad (3.3)$$

is more than satisfactory for all practical needs.

One may notice that, since the thickness of the concentration boundary layer in the fast reaction regime is $\sqrt{Dt_R}$, the rate of mass transfer will be independent of t_D (and therefore of k_L^0). This implies that the interfacial area per unit volume of equipment, $\bar{a}$, needs to be known, and not simply its product with the mass transfer coefficient, $(k_L^0 \bar{a})$. It is fortunate that reliable correlations for $\bar{a}$ are available [21].

As one considers progressively faster reactions, one reaches a point where Eq. (3.2) ceases to hold. One only needs to ask oneself what happens when $\phi \to \infty$, i.e., when reactions can be regarded as instantaneous ($t_R \to 0$). It is obvious that in this limit chemical equilibrium will prevail everywhere in the liquid phase. This in turn implies that the absorbable component A can diffuse both in its physically dissolved form and in its chemically combined form, with no kinetic resistance to the transformation from one form to the other. It follows that in this limit one is again limited by mass transfer (the thickness of the concentration boundary layer becomes again $\sqrt{Dt_D}$), but now the driving force is the *chemical* driving force

$\alpha_i - \alpha_0$ rather than the physical driving force $\alpha_i - \alpha_0$. Thus one has:

$$I_\infty = \lim_{\phi \to \infty} I = \frac{\alpha_i - \alpha_0}{a_i - a_0} f \qquad (3.4)$$

where α_i is in *chemical* equilibrium with the interface partial pressure of A (and therefore a thermodynamic model is required in order to calculate the value of α_i). Equation (3.4) contains a correction factor f which accounts for the fact that diffusion in the chemically combined form may occur with a diffusivity D' different from that of the physically dissolved form, D. Only when all the diffusivities are taken to be equal is the value of f unity. In practice f never lies outside the range $0.7 < f < 1.4$ (so that $f = 1$ is a reasonable approximation) and the following equation:

$$f = \sqrt{D'/D} \qquad (3.5)$$

is probably appropriate to within a very few percent.

It is important to notice that I_∞ is independent of ϕ. In other words, provided that the following condition is satisfied:

$$\Phi \gg I_\infty^2 \qquad (3.6)$$

one will have $I = I_\infty$, and further increases of the rate of reaction will have no effect on the overall observable absorption rate.

Again there is a transition region around $\phi = I_\infty^2$, where neither Eq. (3.2) nor $I = I_\infty$ can be used. In this case the prediction of the transition curve is complex. The transition may, in fact, be a rather smooth and long one and its smoothness depends on the detailed kinetics of the chemical reactions involved. This problem has a long tradition, and some recent advances have been made with new approaches; explicit algebraic equations for $I = I (\phi, I_\infty)$ have been published recently [14, 16].

What has been said above is, in extremely concise synthesis, the content of what one may call the "classical theory". Some recent conceptual advances over the classical theory are briefly reviewed in the following:

(i) *Effectiveness of rate enhancement*
If a physical solvent containing a concentration a_0 of the absorbable component A is brought in contact with a gas in which the partial pressure of A is Ha_i, the *capacity* of the liquid phase (i.e., the moles

per unit volume of A it can absorb until it is in equilibrium with the gas phase) is $a_i - a_0$. The latter quantity is also the driving force for mass transfer. Therefore, as the capacity is increased, so is the rate of mass transfer and the time required for the liquid to saturate is constant.

If one now considers the case of a chemical solvent, the capacity of the liquid has been increased to $\alpha_i - \alpha_0$, which in general will be much larger than $a_i - a_0$. Indeed, the much larger capacity is the main reason for using a chemical solvent. However, the rate of mass transfer in the chemical solvent has been enhanced by a factor I, and in general I will be *less* than the capacity enhancement $(\alpha_i - \alpha_0)/(a_i - a_0)$; consequently the chemical solvent will take a longer time to saturate than a physical solvent. The effectiveness of rate enhancement, η, is defined as:

$$\eta = \frac{\text{rate enhancement}}{\text{capacity enhancement}} = I \frac{a_i - a_0}{\alpha_i - \alpha_0} \qquad (3.7)$$

and will in general be significantly less than unity. Only when the chemical reactions are practically instantaneous $(\phi \gg I_\infty^2)$ will the effectiveness of rate enhancement approach unity as indicated by Eq. (3.4). Values of η significantly less than unity would result in excessively large absorption or desorption units. Therefore, one of the requirements for a good chemical solvent is that the reactions involved should be fast enough to result in acceptably large values of η.

(ii) *Rate promotion*
The hot carbonate process for CO_2 removal is a typical example of a chemical solvent which has very desirable features in terms of capacity and heat integration, but exhibits a rather low value of the effectiveness η. In industrial practice, additives which act as rate promoters have been used for many years [15]. The rate promoters fall into two categories: inorganic and organic.

The analysis of the effects of rate promoters has been carried out in the literature by considering two possible mechanisms. The first one is the case in which the additive simply acts as a homogeneous catalyst for the slow step of the main chemical reaction [23]. The second one is the case in which the additive reacts with the absorbable component in the interface region; the product then diffuses to the bulk of the liquid where the additive gets regenerated and can diffuse back to the interface [27]. This second mechanism is known as the shuttle mechanism.

A recent work [5] (which has generated some correspondence [13, 6]) has analyzed the rate promotion effect in general terms. The two main conclusions are as follows. The two mechanisms discussed above are in reality only quantitatively and not qualitatively different, and a single additive may act via one or the other mechanism depending on the temperature level. Also, from a chemical viewpoint, inorganic and organic additives may act in a very similar way through the reactivity of a pyramidal structure with a lone electron pair over which a weakly bonded intermediate may form.

An important recent development in the area of rate promotion is the discovery of a new class of chemicals which are very effective rate promoters, and which also increase the capacity of the carbonate solution [24]. These chemicals are amines with a sterically hindered nitrogen atom; the steric hindrance decreases the stability of the carbamate ion, which results in a higher concentration of the reactive species (the free amine). A more detailed analysis of the rate promotion effect of hindered amines is due to appear shortly [25].

(iii) *Desorption*
Although some flashing of the rich solution exiting from the absorber does take place in gas treating processes which use chemical solvents, thermal stripping of the solution is always an important part of the process. However, the literature on mass transfer with chemical reaction up to 1979 was almost entirely concerned with the analysis of chemical absorption, with little if any mention of chemical desorption. This is probably due to the fact that desorption experiments are much more difficult to carry out than absorption experiments.

A recent paper [3] presents a general analysis of chemical desorption. It turns out that several concepts of the theory of chemical absorption cannot be carried over directly to the case of desorption. The most important example is the concept of "irreversibility". Should the reactions which take place in the absorber unit really be irreversible, the desorption step would be impossible to perform. Actual values of the equilibrium constants of the reactions taking place in the absorber unit are always very large, both at the conditions prevailing during absorption and during desorption. Since the reaction takes place in the reverse direction in desorption, its equilibrium constant is in fact very small, and therefore the concept of irreversibility, if valid in any asymptotic sense, could certainly not be valid in the sense of a large equilibrium constant.

The methodology for the analysis of chemical desorption developed by ourselves [3] makes use of a nonconventional measure of the driving force, namely, the ratio of the interface and bulk values of the concentration of physically dissolved A (rather that the difference):

$$\psi = \frac{a_i}{a_0} = \frac{P_i}{p^*} \qquad (3.8)$$

where P_i is the interface partial pressure of A, and p^* is the partial pressure which equilibrates the bulk-liquid composition. One may notice that, in order to calculate the ratio p_i/p^*, one does not need to known the value of Henry's law constant H. The well-known irreversible limit in absorption occurs when $\psi \gg 1$; the analogous limit in desorption occurs when $\psi \ll 1$. Notice that both conditions may occur independently of what the value of the equilibrium constant may be, provided it is finite (which is always the case).

(iv) *Mass transfer under pinched conditions*
The identification of the parameter ψ has led to an interesting, new analysis of mass ttransfer under pinched conditions. A pinch is defined as a condition where:

$$\psi = 1 + \varepsilon , \qquad |\varepsilon| \ll 1 \qquad (3.9)$$

with $\varepsilon > 0$ corresponding to absorption and $\varepsilon < 0$ to desorption. Pinches are very likely to occur in industrial operation, and since the driving force becomes small at a pinch, most of the total packing height or number of trays is in fact used to effect pinched mass transfer. It turns out that the asymptotic behavior at pinches is significantly different from that well away from pinches, particularly as instantaneous reaction conditions are approached [4]. The enhancement factor I_∞ at pinches becomes a strongly decreasing function of temperature, and the mass transfer rate becomes proportional to $a_i - a_0$, contrary to what happens under nonpinched conditions.

(v) *Selective chemical absorption*
When a gas stream contains significant amounts of both CO_2 and H_2S, the traditional gas treating processes using an alkaline solvent

remove most of both gases from the gas stream. However, when the H_2S/CO_2 ratio in the gas stream is low, e.g. less than about 0.15, the acid gas exiting from the regeneration step will have a low H_2S content which makes operation of the Claus plant difficult if not impossible. In such cases, a step of selective removal of H_2S is needed where most of the CO_2 is left in the gas stream while essentially all the H_2S is removed.

Selectivities achieved in industrial processes are well in excess of the thermodynamic selectivity. The latter has an upper bound in aqueous alkaline solutions, which is achieved for those solvents where the only chemical sink for CO_2 is the bicarbonate ion [8] (the thermodynamic modeling has been discussed in the previous section). Since process selectivities are in excess of the thermodynamic values, they are of kinetic nature. Therefore, design of industrial units must be based on a thorough understanding of the rates of the mass transfer during simultaneous chemical absorption. An over-designed packed height will result in approaching the unfavorable conditions of thermodynamic selectivity while, of course, an under-designed height will result in the failure to meet the strick H_2S specifications in the treated gas.

Although there are comparatively many published analyses of simultaneous chemical absorption, many of these are of little value because they fail to recognize the important effect of the "shift reaction" [9]. An analysis for the case where both gases are absorbed in the instantaneous reaction regime has been published recently [7]. This analysis shows that the chemical coupling may, on occasion, reverse the *direction* of mass transfer for one of the components. Solutions of the full system of differential equations have been publisehd [11, 10], but these require a large amount of numerical computation and could not be used economically in the subroutines of a design procedure. A satisfactory approximate solution has only very recently been obtained [29].

4 LABORATORY EXPERIMENTS, PROCESS SELECTION, DESIGN ANDSCALEUP

Laboratory Experiments

Process selection and design in gas treating, as in other areas, requires the availability of reliable physico-chemical data. Often the data which are needed cannot be found in the literature and therefore they must be obtained with appropriately designed laboratory-scale experiments. Data obtained in the laboratory should be of a fundamental nature, and not purely empirical information which cannot be extrapolated confidently to conditions other than the ones prevailing when the experiment was made. The temptation of planning experiments (e.g. small countercurrent packed towers) which resemble as closely as possible the foreseeable industrial unit should be resisted: such experiments rarely if ever provide the fundamental understanding which is to be sought in laboratory-scale experiments and often they do not yield reliable predictions of the behavior of an industrial-scale unit.

Vapor–liquid equilibrium (VLE) data are gathered with the purpose of determining the partial pressure(s) of the gas components to be removed from a gas stream as a function of the relevant independent variables, namely: temperature, liquid phase molarity m and liquid phase conversions y, y'. It is useful to also obtain information on the equilibrium partial pressure of the solvent (e.g. water in the case of aqueous chemical solvents) as a function of the same variables. Vapor–liquid equilibrium data can be correlated, and often successfully extrapolated, by appropriate modeling of the kind discussed earlier.

Three different types of VLE cells have been used to obtain high quality data for pure and mixed acid gases in chemical and physical gas treating solvents. The cells all provide means for rapidly bringing vapor and liquid phases into thermal and compositional equilibrium at known temperatures and pressures. The compositions of the equilibrium phases are then determined by withdrawing samples and analyzing them. This requires the use of specialized analytical techniques such as mass spectrometry and gas chromatography. Although many types of cell are possible, only three have been used extensively: the rocking autoclave cell; the vapor recirculation cell and the gas flow-through cell. These are all sophisticated units requiring careful design, extensive instrumentation and control equipment. Great care and patience are required on the part of the operator to obtain reliable data.

The vapor recirculation closed-cell design has been used with aqueous amine and organic solvent-amine solutions with CO_2, H_2S, and mixtures of CO_2 and H_2S for partial pressures of 0.1 to 1000 psia and temperatures of 40–130°C [19]. Below 0.1 psia partial pressure the method becomes quite inaccurate owing to limitations on the size of vapor samples which can be withdrawn without disturbing equilibrium significantly. A different experimental approach is needed in the very low partial pressure region.

The vapor recirculation, closed cell is shown schematically in Fig. 3. The term "closed" refers to the fact that the cell is charged with a batch of solution and gas, which is then allowed to come to equilibrium by recirculating the vapor through the liquid. Other designs, of the open type, use a batch of liquid and gas of constant acid gas and moisture content is allowed to flow once-through the liquid and out of the cell.

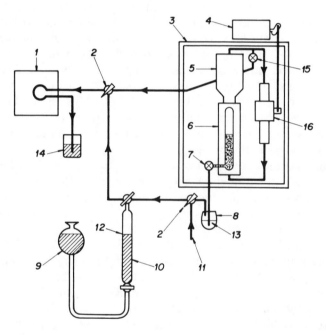

<u>LEGEND</u>

1 – GAS CHROMATOGRAPH	9 – MERCURY LEVELLER
2 – 3-WAY BALL VALVES	10 – 200 ml. GAS COLLECTING BURETTE
3 – TEMPERATURE CONTROLLED AIR BATH	11 – AIR LINE
4 – PUMP MOTOR	12 – SILICONE OIL LAYER
5 – VAPOR RESERVOIR (250 ml.)	13 – NEEDLE (0.01 INCH I.D.)
6 – HIGH PRESSURE JERGUSON GAUGE (200 mL)	14 – FLOW INDICATOR
7 – LIQUID SAMPLING VALVE	15 – VAPOR SAMPLING VALVE
8 – LIQUID SAMPLER (WITH 5N H_2SO_4)	16 – MAGNETIC PUMP

Figure 3   Vapor recirculation closed vapor–liquid equilibrium cell.

A batch of solution is contained in a pressure vessel with sight glass (Jerguson gauge). The gases CO_2 and H_2S are introduced in desired amounts and N_2 is added to bring total system pressure up to at least 1000 psig at the operating temperature. Vapor is drawn from the top of the Jerguson gauge and recirculated by a magnetic circulating pump back to the base of the Jerguson gauge. The sight glass gives a visual check that vapor is actually circulating. The vapor recirculation loop contains a 250 cc vapor reservoir to increase vapor volume. This allows many vapor samples to be withdrawn for analysis without disturbing equilibrium significantly. The Jerguson gauge, reservoir and magnetic pump are housed in an air bath for temperature control. At equilibrium, vapor and liquid samples are withdrawn for analysis. Vapor samples are taken directly off the vapor reservoir into a gas chromatograph for analysis for CO_2 and H_2S contents. Use of 10 foot column of Chromosorb W gives well-shaped, well-spaced peaks for N_2, CO_2, H_2S and H_2O. Calibration gas mixtures must be used as standards. The water vapor peak is not very reproducible and quantitative analysis for H_2O vapor is not generally possible.

Liquid samples are taken from the Jerguson gauge into a liquid sample vessel containing $5N$ sulfuric acid. All of the acid gases are evolved into a gas collecting burette which measures the volume of evolved acid gas. The evolved gases are then analyzed by gas chromatography to get the CO_2 to H_2S ratio.

Mass transfer experiments can be classified in two fundamental categories: differential and integral. In a differential experiment, the operating variables (such as temperature, pressure, composition, etc.) are fixed at some *unique* predetermined value. The value of some dependent variable, e.g. absorption rate which is presumed to depend on only the operating variables, is measured. By repeating the experiment over a range of values of the operating variables, the functional dependency of the dependent variable is determined.

In an integral experiment, the operating variables are not constant, but their values vary from point to point, or in time, within a single experiment. A pilot plant experiment is a typical example: composition, temperature and pressure have different values at different points in the pilot plant column. When some dependent variable is measured in an integral experiment, one cannot deduce any fundamental data from it, unless a theoretical model is available which relates the overall behavior of the experiment to the actual local values of physico-chemical parameters. However, if such a model is available, it would be better to determine the values of the parameters in differential experiments, and then use the model to *predict* the result of the integral experiment. If this is done, the integral experiment becomes unnecessary. Unfortunately, a reliable model is often not available and therefore integral experiments are needed; this is the main reason for performing pilot plant work.

The goal of experiments, where the rate of mass transfer is measured, is to understand the physico-chemical mechanism of absorption-reaction and desorption-reaction. For such a goal to be attained it is crucial that mass transfer experiments be of the differential type. Several classical laboratory units for mass transfer experiments are available. These may be classified as follows:

(i) Units for which the interfacial area A is known and for which the value of the physical mass transfer coefficient k_L^0 can be estimated with satisfactory accuracy from the solution of the appropriate hydrodynamic problems. The laminar jet, the short wetted-wall column and the one-sphere unit fall into this category.

(ii) Units for which the interfacial area is known, but the mass transfer coefficient k_L^0 cannot be predicted from first principles. The several variations of the stirred cell, the long wetted-wall column, the string-of-spheres unit and the string-of-disks unit fall into this category.

(iii) Units for which neither the interfacial area nor the mass transfer coefficient are known. The sparged cell, the single sieve tray and the single bubble cap plate fall into ths category.

The choice of a specific laboratory unit is based on a number of considerations; in general, units of type (i) are preferable to units of type (ii) and the latter are preferable to units of type (iii). Other important considerations for the selection of a unit include: the values of the mass transfer coefficient k_L^0, the interface are A, the ratio of liquid volume to interface area ϕ, and of the residence time of the liquid t_R. In some units it is possible to vary these quantities over a certain range by varying the operating conditions.

It is always sound policy to calibrate the unit, i.e., to measure the absorption rate of a gas in a nonreactive liquid in which the solubility of the gas is known. The calibration yields the value of the product $k_L^0 A$ for the unit. With units of types (i) and (ii), the value of k_L^0 is obtained from the calibration, since A is known.

The calibration should be carried out with a system that matches as closely as possible the physical properties of the reactive system one wishes to investigate. Usually the most important property to match is the viscosity in the case of units of types (i) and (ii).

For units of type (i), the calibration is required to check the validity of the theoretical equation for k_L^0. A check within a factor as large as two is often acceptable. The value obtained in the calibration should be used in the subsequent analysis. For these units calibration does not need to be carried out on a system with closely matching physical properties since the influence of the latter on the value of k_L^0 can be predicted from the theoretical equation for k_L^0. For units of type (ii), the calibration is indispensable in order to know the value of k_L^0, which is often needed in the subsequent analysis.

For units of type (iii) calibration is not very meaningful as it does not yield the value of k_L^0, but only of the product $k_L^0 A$. The interfacial area, A, may change very significantly for minor changes of the interfacial properties of the gas–liquid pair (not only the surface tension, but also the derivative of the latter with respect to concentration) which strongly influence the value of A. Therefore, for units of type (iii) the calibration at best offers an estimate for the actual value of $k_L^0 A$ for the reactive system one wishes to investigate. The principal characteristics of the units quoted above are reviewed briefly in the following paragraph.

The units of type (i) have certain disadvantages. First of all, they have little flexibility. The interfacial area can be varied only with the laminar jet, and not over a wide range. If the jet is too short, end effects become important; if it is too long, operation of the receiver becomes difficult. The mass transfer coefficient k_L^0 can be changed only over a limited range, due to the restricted range of values of L which are possible: in the wetted-wall and one-sphere absorbers, a lower bound on L is imposed by the requirement of completely wetting the entire solid surface, while an upper bound is imposed by the need to avoid excessive rippling and turbulence. In the laminar jet, a lower bound is imposed by the requirement of avoiding dripping and an upper bound again by the need to avoid turbulence. The residence time of the liquid and the liquid volume-to-interface area ratio, again can be varied only over a very limited range; the latter in fact not at all in the laminar jet. These shortcomings occasionally force the selection of a type (ii) unit.

The main purpose of gathering mass transfer rate data in the laboratory is to understand the chemical mechanism and the corresponding kinetics well enough to allow the development of a reliable model of the rate enhancement phenomenon. From such a model, the enhancement factor I can be calculated as a function of temperature and composition of both gas and liquid phases. This function forms the basis for the design procedures to be discussed later in this chapter.

When a set of rate data is obtained, the first step in the analysis is to ascertain which regime of mass transfer applies to the data. The quantity which is measured is the total mass transfer rate $\bar{V}$; the chemical mass transfer coefficient k_L can be extracted from the data. The key to the identification of the regime is the dependency fo $\bar{V}$ on the operating variables. The latter are: the degree of mixing as represented by k_L^0; the interface area A; the liquid volume V; temperature; and the physical driving force $a_i - a_0$.

The dependency of $\bar{V}$ on the operating variables is summarized in Table 3. The dependencies in the table are sufficiently different from one regime to another to permit identification of the regime of the available set of data (of course, data may fall in a transition region).

Once the first step of the analysis has been completed and the regime has been identified, the data can be scrutinized to gain insight into the chemical mechanisms involved. The kind of information which can be obtained is briefly reviewed in the following paragraphs.

First, consider the case where the data indicate fast reaction regime behavior:

$$\bar{V} = \sqrt{Dk_0} A \frac{p_i - p^*}{H} \qquad (4.1)$$

The value of the equivalent kinetic constant k_0 can therefore be extracted from the data, provided the VLE behavior is known (so that p^* is known), as well as the value of H. The value of k_0 depends on the liquid phase composition:

$$k_0 = g(b_{10}, \ldots, b_{N0}) \qquad (4.2)$$

If a thermodynamic model is available, the values of the bulk-liquid concentrations $b_{10}, \ldots, b_{N0}$ can be calculated as functions of the molarity m and the degree of saturation y. Therefore, the experimentally determined dependency of k_0 on m and y should yield the form of the $g(\cdot)$ function, from which the chemical mechanism and the true kinetic constant can be obtained.

Next consider the case where the data fall in the instantaneous

Table 3 Dependency of $\bar{V}$ on Operating Variables (Rate of Reaction First Order with Respect to Transferring Component)

Regimes	Slow		Fast	Instantaneous	
Operating Variables	Kinetic	Diffusional		Large Driving Force	Small Driving Force
K_L^0	$i^{(a)}$	αk_L^0	i	αk_L^0	αk_L^0
A	i	αA	αA	αA	αA
V	αV	i	i	i	i
$|a_i - a_0|$	$\alpha|a_i - a_0|$	$\alpha|a_i - a_0|$	$\alpha|a_i - a_0|$	almost i	$\alpha|a_i - a_0|$
Apparent activation energy for k_L	energy of activation of reaction	very small	1/2 energy of activation of reaction	very small	heat of reaction
	$k_L \ll k_L^0$	$k_L = k_L^0$	$k_1 \gg k_L^0$	$k_L \gg k_L^0$	$k_L \gg k_L^0$

(a)Independent.

reaction regime. The equation for $\bar{V}$ becomes in this case:

$$\bar{V} = v I_\infty k_L^0 A \frac{p_i - p^*}{H} \qquad (4.3)$$

from which values of I_∞ can be calculated and compared with predictions of models.

It is often useful to make the tentative assumption that the diffusivities of all solutes are equal to each other. If that in the case:

$$\bar{V} = v k_L^0 A (\alpha_i^* - \alpha_0) \qquad (4.4)$$

The values of α_i^* can be simply read off a VLE curve as the value of α corresponding to the interface partial pressure of the volatile component, at the bulk-liquid molarity.

Often, in actual fact, the diffusivities are not all equal and therefore the measured value of $\bar{V}$ will not be equal to the right-hand side of the above equation. However, while $\bar{V}$ itself will depend strongly on the composition of both gas and liquid phases, the ratio of $\bar{V}$ to the term on the right-hand side of the above equation will be almost constant and close to unity. Values for the ratio between 0.7 and 1.3 are typical. This ratio can be taken as an empirically determined correction factor for nonequal diffusivities.

Finally, consider the case where the data appear to fall in the transition region between fast and instantaneous reaction behavior. In this case, it is useful to perform experiments with a stirred cell, so that different values of k_L^0 (and hence of ϕ) can be obtained by changing the stirrer speed. This allows a region of ϕ values at a given liquid phase composition to be scanned. The experimentally determined I versus ϕ curve can then be compared with a theoretical model of the type discussed earlier.

Process Selection

In contrast with process analysis which is analytic and quantitative, process selection is largely an art, and the choice of any given process is often not determined uniquely by technical reasons. It is therefore difficult to develop a methodology for process selection without risking the vagueness of an essentially descriptive procedure.

No industrial gas treating process is the optimum one for all applications. Indeed, there are large numbers of processes available for the removal of acidic impurities such as carbon dioxide, hydrogen sulfide and sulfur dioxide. These are summarized in Table 4.

Since gas treating is generally integrated with at least one other process, the selection of the optimum gas treating process is always a complex problem involving both technical and economic considerations. Our attention in this section will be focused mainly on developing a methodology for selection of the solvent since the critical differences between gas treating processes are so highly dependent on the solvent.

Selection of a solvent will ultimately be based upon a thorough economic analysis of the gas treating process integrated with the associated process units and the process(es) chosen to prepare sulfur from the recovered acid gases. For example, the Claus process is the conventional, and usually the most economical, process used to prepare sulfur from H_2S-containing acid gases. However, a Claus unit generally requires an H_2S concentration in its feed greater than 15% for effective operation. When the H_2S concentration is above 50% H_2S, a Claus unit operates most efficiently and with a minimum of difficulty. Liquid phase oxidation of acid gases can be used as H_2S concentrations in the aid gas of less than 5%. However, liquid phase oxidation units are significantly more expensive from both a capital investment and operating cost point of view than the comparable Claus unit. As a generalization, it is preferable to process as much of the acid gas as possible through a Claus unit. Therefore, significant expenditures can be made for a gas treating unit that provides an acceptable feed to a Claus unit.

The essential elements of the methodology for acid gas removal processes are sketched in Fig. 4. Starting from the top, the *definition* of a process is provided by two elements: the available feed gas characteristics (composition, pressure, temperature, and so on), and the treated gas specifications, i.e., the process requirements. These two elements provide a preliminary evaluation of the solvent working capacity, which may, however, be influenced by several other elements, of which the most important ones are sketched in Fig. 4.

The controlling factors in selecting an acid gas treating system are:

– Raw feed gas pressure.
– Raw feed acid gas composition and content.
– The required treated gas purity.

To a first approximation, these factors can be used to establish the working capacity of possible solvents with the methods discussed in

Table 4 Processes Used Commercially for CO_2 and H_2S Removal from Gases

Type	Solvent	Trade Name	Typical Composition
Physical absorption system	Propylene carbonate	Fluor solvent	—
	Polyethylene glycol dimethyl ether	Selexol	—
	N-methyl-2-pyrrolidone	Purisol	—
	Methanol	Rectisol	—
Aqueous chemically-reactive systems	MEA	—	20–35% MEA
	DEA	—	30% DEA
	DGA	Econamine	60% DGA
	DIPA	ADIP	
	MDEA	ADIP	
	Promoted K_2CO_3	Benfield; Catacarb; Giammarco Vetrocoke	25–30% K_2CO_3; 5% promoters
Hybrid systems	DIPA-Sulfolane-water	Sulfinol	40%-40%-20%
	MDEA-Sulfolane-water	—	—

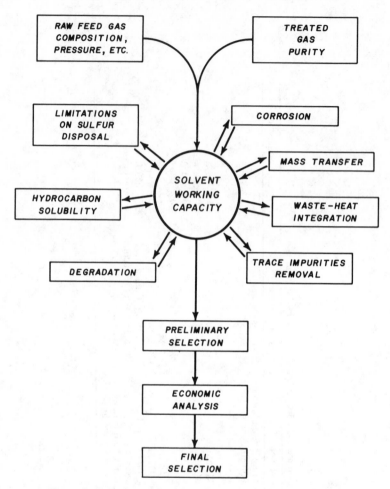

Figure 4   Selection methodology for acid gas removal solvent.

Section 2. The working capacity establishes the solvent circular rate (expressed often as a solvent/acid gas ratio) which has a major impact on both plant investment and operating cost of acid gas removal systems. It directly impacts on the size of the absorber tower, the piping, the circulation pumps, and the size of the regenerator facilities. In a solvent acid gas removal system, 50–70% of the plant investment is directly associated with the magnitude of the solvent circulation rate. Obviously, there are advantages in employing the highest possible concentration of chemical or physical solvent as limited by solubility, precipitation and corrosion considerations. While there are exceptions for a chemically-reactive solvent process to be economically attractive, the concentration of reagents must be greater than about 3 molar.

The partial pressure of acid gas in the the feed, to a first approximation, determines the solvent circulation rate. In a counter-
current column the feed gas contacts the fully loaded rich solvent at the bottom of the column. At a minimum, therefore, there must be sufficient solvent to absorb substantially all of the acid gas in the feed and provide a driving force for absorption at the bottom of the column. Therefore, the partial pressure of acid gas in the feed in controlling. The partial pressure of acid gas in the treated gas determines the degree of regeneration required. This is because the partial pressure of each acid gas in the treated gas must be higher than the equilibrium partial pressure of the same component in the regenerated solvent. Otherwise, there will not be a positive driving force for absorption at the top of the absorber.

Therefore, the two most important cost factors—circulation rate and regeneration energy duty—are established to a significant extent by the acid gas partial pressures in the feed and the product. Ten to twenty per cent of the investment of a chemically reactive acid gas removal system is dependent on the regeneration energy duty or the steam/acid gas ratio.

Over the years criteria have been developed for narrowing the choice of available processes based solely upon the partial pressure of acid gas in the feed and product [28]. For example, if the partial pressure of acid gas in the *feed* is low, below about 50–100 psi, all processes using physical solvents can be eliminated. If the partial pressure of acid gas in the *product* is very low, physical solvents can generally be eliminated unless the total pressure is quite high.

Design and Scaleup

We now turn to the principles and methodology of design for industrial processes of acid gas absorption/desorption in chemically-reactive solvents. The emphasis is on the differences between design procedures for physical solvents and for chemical solvents; the procedures for physical solvents are reviewed briefly, while those for chemical solvents are discussed in some detail.

First, we will be concerned with those aspects of design which are related mainly to the thermodynamic behavior of the system, i.e. the calculation of the minimum liquid rate and of the minimum steam rate. Secondly, we will cover sizing the height of absorption and desorption units. Tower height estimation is related to the mass transfer behavior of the system.

(i) *Minimum liquid rate*

The calculation of the minimum liquid rate is based on thermodynamics in the case of an *isothermal* absorption tower. Consider in particular the case where only one acid gas is to be removed. Let G be the molar flow rate of inert gas per unit cross-sectional area, and L be the volumetric liquid flow rate per unit cross-sectional area. A mass balance between the top of the countercurrent absorber and a generic cross section where the gas phase mole fraction of the acid gas is Y, and its total concentration in the liquid phase is α, yields:

$$G\left[\frac{Y}{1-Y} - \frac{Y_T}{1-Y_T}\right] = L(\alpha - \alpha_T) \qquad (4.5)$$

In the $Y - \alpha$ plane (see Fig. 5) the operating line, i.e. the curve representing the above equation, is concave down, since the local slope:

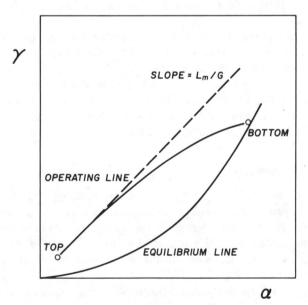

Figure 5 Determination of minimum liquid rate for isothermal countercurrent absorption,

$$\frac{dY}{d\alpha} = \frac{L(1-Y)}{G} \qquad (4.6)$$

decreases with increasing Y. Let $f(\alpha, T)$ be the equilibrium function, say

$$p^* = f(\alpha, T)$$

The equilibrium line in the $Y - \alpha$ plot is:

$$Y^* = \frac{p^*}{\pi} = \frac{f(\alpha, T)}{\pi} \qquad (4.7)$$

where π is the total pressure. For an isothermal chemical absorber (T = constant), the equilibrium is concave upwards, since $\partial^2 f / \partial \alpha^2$ is always positive.

Since the operating line is concave down, and the equilibrium line is concave upwards, the minimum liquid flow rate, L_m, corresponds to an operating line which just touches the equilibrium line at the bottom of the absorber. Therefore, the value of L_m can be calculated on the basis of only a mass balance and of knowledge of the equilibrium behavior of the system considered.

Industrial absorption towers where chemical solvents are used usually show significant deviations from isothermal behavior. The heat of absorption of the acid gas (which is large since a chemical reaction is involved) causes the liquid temperature to increase. Due to the high capacity of chemical solvents, the heat of absorption is released in a liquid which flows at a comparatively low rate, and therefore the temperature rise may be very significant. The temperature rise may be mitigated by solvent evaporation and by heat transfer to the gas phase (industrial size absorption towers can be regarded as essentially adiabatic).

Since the equilibrium function $f(\alpha, T)$ has values which strongly increase with temperature, the calculation of L_m based on Eq. (4.7) with T the liquid *inlet* temperature may result in a gross underestimate of L_m. An alternate approach is to use Eq. (4.7) with T the largest possible temperature of the liquid, i.e. the one which would result if all the heat of absorption were released in the liquid phase, and none transferred back to the gas phase by heat transfer or solvent evaporation. Again, this largest possible liquid temperature can be calculated based on only thermodynamic information. However, the calculation of L_m based on the maximum possible liquid temperature is likely to lead to a significant overestimate of L_m, since solvent evaporation and heat transfer to the gas phase take place significantly. Indeed, the maximum liquid temperature does not necessarily occur at the bottom of the absorption tower, but at some intermediate point. When such is the case, the pinch point which determines L_m may, in fact, be located at a section other than the rich end.

A calculation of L_m which takes into account the actual temperature profile in the absorber requires an analysis based on the *rates* of heat of absorption release, heat transfer to the gas phase, and solvent evaporation. It is, therefore, a calculation which cannot be performed only on the basis of thermodynamic information but requires understanding and mathematical modeling of the gas absorption rate behavior of the system.

The methodology for the calculation of L_m for nonisothermal absorbers is not completely developed even for physical solvents. A procedure which has been suggested in the literature is briefly reviewed below for the case of ammonia absorption in water [26]. The procedure is based on the assumption that the resistances to both mass and heat transfer are entirely in the gas phase; such an assumption is presumably valid only for some special cases of chemical absorption. The procedure consists essentially in writing down all the relevant differential transport equations, which can be integrated starting from one extreme section of the tower (say the top one) for different values of L/G, and a value of L_m can thus be calculated as the value which corresponds to $N \to \infty$, i.e. to an infinitely high tower. The calculation also provides temperature and concentration profiles, as well as the number of transfer units required for any assigned value of L/G; therefore, the *calculation of L_m is coupled with the sizing of the tower height*.

The results of the calculation for the ammonia–water system are given in Fig. 6. The temperature of the entering water is 61°F. At the highest liquid rate the temperature rise is about 20°F, with the maximum temperature occurring close to the base of the countercurrent absorber. As the water rate is reduced the temperature rise increases quite sharply, and the peak of the temperature maximum (bulge) moves up the column. Figure 6 gives the operating and equilibrium lines. The operating lines have slightly decreasing slopes as G/L increases while the equilibrium lines move towards the operating lines because of the temperature rise. At $G/L = 0.4515$, the operating and equilibrium lines nearly touch. It is clear that basing the minimum liquid rate on the assumption of phase equilib-

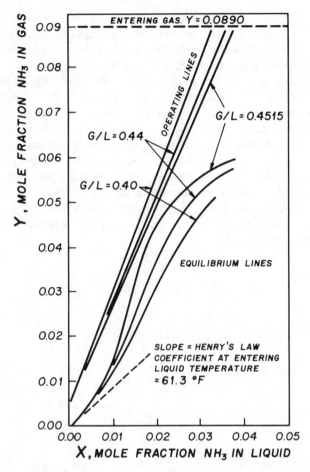

Figure 6 Operating and equilibrium lines in adiabatic ammonia absorber.

rium at the *bottom* of the absorber leads to a liquid rate which is too low.

The procedure discussed above can, in principle, be extended to chemical absorption. This is, however, not an easy task since it involves the sizing of the required packing height, which is not a straightforward procedure even in the isothermal case.

(ii) *Minimum steam rate*

The steam injected into the regeneration unit serves two purposes: it provides the sensible and latent heat required for the desorption operation, and it represents the diluent gas needed to keep the partial pressure of acid gas in the gas phase low enough to allow stripping to take place. Consequently, the required steam rate may be dictated either by the heat balance, or by the stripping operation [22]. The minimum steam rate needs to be calculated for both requirements, and the actual minimum is the larger one of the two.

First, consider the stripping requirement. In a $Y - \alpha$ plane, the equilibrium line is strongly curved upwards (see Fig. 7). The operating line also has a positive curvature, since steam condenses along the regeneration tower to supply sensible heating as well as the heat of desorption. However, the curvature of the operating line, particularly at low α values, is much less than that of the equilibrium line, and the minimum steam rate generally corresponds to an internal pinch point at which the equilibrium and operating lines are tangent to each other. The pinch point is likely to occur at a very low value of Y.

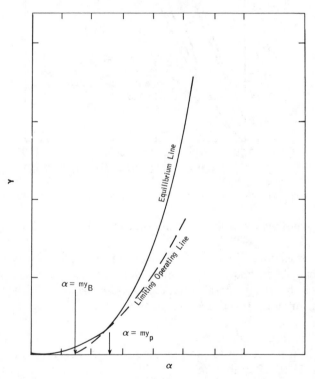

Figure 7 Equilibrium and limiting operating lines at bottom regenerator conditions.

A rather straightforward calculation yields the following equation for the minimum steam rate for stripping:

$$\frac{S_m}{L} = \frac{m^{1-q/2}}{4p_t K_0(T_2)} \exp\left[-\frac{Q_0}{R}\frac{1}{T_2} - \frac{1}{T_1}\right] \qquad (4.8)$$

Equation (4.8) shows that strong absorbents, characterized by a very low value of $K_0(T_2)$, require a large steam rate for stripping. Notice, however, that S_m increases only proportionally to the square root of $1/K_0(T_2)$. The equation also reflects the beneficial effect of a large temperature swing between regenerator bottom and absorber top which can be achieved by deeper solvent cooling or by the use of solvents with high boiling point.

We now turn attention to the heat balance limitation. The minimum steam rate required for the heat balance is:

$$\frac{S_m}{L} = \frac{\rho_L L C_L (T_1 - T_3) + m\bar{Q}\,\Delta y}{Q_s} + \left(\frac{S}{L}\right)_R \qquad (4.9)$$

where T_3 is the temperature of the liquid leaving the rich–lean heat exchanger, $\bar{Q}$ is the average heat of absorption over the absorber, Δy is the change of fractional saturation over the absorber, Q_s is the latent heat of steam, and $(S/L)_R$ is the steam rate required at the regenerator top. Equation (4.9) states that the heat supplied to the reboiler needs to: (i) supply sensible heat to raise the liquid temperature from T_3 to T_1; (ii) supply the heat of desorption of acid gases; and (iii) supply a residual steam rate at the regenerator top such that the water vapor pressure over the solution entering the regenerator is matched by the partial pressure of steam in the vapor upflow.

(iii) *Tower sizing*
The diameter of a tower is set by the flow rates of the gas and liquid phases flowing through it. The tower height is determined by the rate of mass transfer. Packings, or other tower internals, are designed to promote a large interfacial area of contact between the phases, and a high degree of mixing in both phases, with a minimum resistance to flow of the two phases (large void volume).

For estimating the diameter of packed columns with random packings, the generalized pressure drop correlation for dumped (random) packings should be used. The packing factor, F, should be taken for the appropriate size of packing. Packings smaller than 1 inch size are intended for towers 1 foot or smaller in diameter; packings 1 inch or 1.5 inches in size for towers over one foot and up to 3 feet in diameter; and 2 to 3.5 inch packings are used for towers three or more feet in diameter. The packing factor, F, is an experimentally determined quantity, obtained for the air–water system in 16-inch and 30-inch columns. In general, the largest size packing pieces that meet the packing-size-to-tower-diameter ratio criteria given above should be selected for initial designs. This will minimize tower diameter. If tower height is excessive, smaller packing pieces may provide a more optimum design, especially for the absorber and regenerator cleanup sections.

Aqueous amine and potassium carbonate solutions are generally designed to operate at 50–70% of the flooding limit (as compared to 70–90% of flood for physical solvents) owing to their tendency to foam. As far as tower diameter is concerned, the principal difference between physical and chemical solvents is that physical absorption processes operate at higher L/G ratios owing to the lower acid gas capacity of physical solvents. With chemical solvents, lower L/G ratios are used and tower diameter may be determined either by the vapor flow or by the liquid flow.

We now turn attention to the problem of sizing the required height of packing for a process of chemical absorption. The procedure for such a calculation is a rather involved one. We only discuss the *methodology* of the design procedure, and attention is therefore limited to the case of an *isothermal* absorber. The basic ideas to be developed if nonisothermal behavior needs to be taken into account have been introduced earlier.

We first briefly review the methodology of design for the case of physical solvents; furthermore, we restrict attention to the case where Henry's law applies (i.e. the equilibrium relationship between the interface partial pressure and the interface liquid-side concentration is linear). The incremental rate of absorption per unit cross-sectional area over a length dZ of packing can thus be written as:

$$N\bar{a} \, dZ = K^0_G \bar{a} \pi (Y - Y^*) \, dZ \qquad (4.10)$$

where $K^0_G \bar{a}$ is an overall, gas phase mass transfer coefficient per unit volume:

$$K^0_G \bar{a} = \cfrac{1}{\cfrac{1}{k_G \bar{a}} + \cfrac{H}{k^0_L \bar{a}}} \qquad (4.11)$$

The important assumption is that $K_G^0 \tilde{a}$ is, to a first approximation, constant along the tower axis. Its value depends on the efficiency of the packing which is being used, and on the *equilibrium* behavior of the system (through Henry's constant H).

An overall incremental number of transfer units, dN^0, can be defined as follows:

$$dZ = \frac{\bar{G}}{K_G^0 a \pi} dN^0 \qquad (4.12)$$

so that the transport equation takes the following simple form:

$$Na\, dZ = (Y - Y^*)\tilde{G}\, dN^0 \qquad (4.13)$$

When this is combined with the local mass balance equation, one obtains:

$$-d\left(\frac{Y}{1-Y}\right) = (Y - Y^*)\, dN^0 \qquad (4.14)$$

Equation (4.14) can be integrated directly to yield:

$$Z_T = \frac{\tilde{G}}{K_G^0 \tilde{a} \pi} N^0 = \left(\frac{\tilde{G}}{K_G^0 \tilde{a} \pi}\right) \int_T^B \frac{dY}{(1 - Y)^2(Y - Y^*)} \qquad (4.15)$$

where T and B identify top and bottom conditions in the absorber. The integral on the right-hand side of Eq. (4.15) represents the total number of transfer units required (and is thus a measure of the difficulty of the separation process to be carried out), while the term in round brackets represents the height of packing required for one transfer unit, or the "height of a transfer unit", HTU. The HTU is an inverse measure of the efficiency of the packing; the lower its value, the more efficient the packing.

In the special case where the gas is *dilute* ($Y \ll 1$), a considerable simplification arises. The denominator of the integrand in Eq. (4.15) simply becomes the driving force $Y - Y^*$; the latter in turn is a linear function of Y, since both the operating and the equilibrium lines are linear. Straightforward algebra leads to the following result:

$$Z_T = \frac{M}{K_G^0 \tilde{a} \pi}(Y - Y^*)_{\mathrm{lm}} \qquad (4.16)$$

where M is the total absorption rate per unit cross-sectional area:

$$M = \bar{G}(Y_B - Y_T) \tag{4.17}$$

and $(Y - Y^*)_{lm}$ is the log mean driving force:

$$(Y - Y^*)_{lm} = \frac{(Y - Y^*)_B - (Y - Y^*)_T}{\ln \dfrac{(Y - Y^*)_B}{(Y - Y^*)_T}} \tag{4.18}$$

Now consider the case of a *chemical* absorption process. The transport equation for the gas phase is unchanged, but the transport equation for the liquid phase needs to be written for chemical absorption. Thus one has:

$$N\bar{a} \, dZ = k_G \bar{a} \pi (Y - Y_i) \, dZ = I k_L^0 \bar{a} \frac{\pi}{H} (Y_i - Y^*) \tag{4.19}$$

where the enhancement factor I is a complex function of the bulk-liquid composition, interface partial pressure, and liquid-side mass transfer coefficient:

$$I = I(Y^*, Y_i, k_L^0) \tag{4.20}$$

For a nonisothermal absorber, I should be regarded as also depending on temperature.

If the right-hand side of Eq. (4.20) is known (i.e. if the *rate* behavior of the system considered has been modeled satisfactorily), Y_i can be eliminated between Eqs. (4.19) and (4.20). This elimination will never lead to any simple mathematical expression, and in fact often the resulting equation will need to be solved numerically. Formally, the result is as follows:

$$\bar{N}a \, dZ = K_g^0 \bar{a} \pi (Y - Y^*) \xi \, dZ \tag{4.21}$$

where

$$\xi = \frac{1 + P}{1 + P/I} \quad \text{and} \quad P = \frac{k_G a H}{k_L^0 a} \tag{4.22}$$

The quantity P is the ratio of the effectiveness of *physical* mass transfer in the gas phase to that in the liquid phase. Since the physical solubility of acid gases is low (large value of H), in chemical absorption processes P is likely to be a large number, say:

$$P \gg 1 \qquad (4.23)$$

(A "typical" value of $k_G a / k_L^0 a$ is 10 gmol/lt atm. Therefore, with H values of order 10^2 atm lt/gmol, values of P are of the order of 10^3.)

The crucial term in Eq. (4.21) is ξ. A brief discussion of the values that ξ may take is given in the following. First, consider the case where the enhancement factor I is not extremely large, say the following condition is fulfilled:

$$I \ll P \qquad (4.24)$$

(Condition (4.24) is likely to occur at the rich end of the absorber.) In this case, in view also of Eq. (4.23), one has:

$$\xi \simeq I \qquad (4.25)$$

Furthermore, from Eqs. (4.11) and (4.23) one obtains:

$$K_G^0 a \simeq k_L^0 a \qquad (4.26)$$

and therefore, Eq. (4.21) reduces to:

$$N\tilde{a} \, dZ = k_L^0 \tilde{a} I \pi (Y - Y^*) \, dZ \qquad (4.27)$$

The condition discussed above is the one where the enhancement factor is small enough to make the liquid-side resistance to mass transfer the rate-determining step. Therefore, $Y_i \simeq Y$, and Eq. (4.20) can be used to estimate the value of I by substituting Y for Y_i.

Conversely, consider the case where the enhancement factor is very large indeed, say:

$$I \gg P \qquad (4.28)$$

(Condition (4.28) is likely to occur at the lean end of the absorber.) In this case, one has:

$$\xi = 1 + P \qquad (4.29)$$

and Eq. (4.21) reduces to:

$$N\tilde{a}\, dZ = k_G a\pi(Y - Y^*)\, dZ \qquad (4.30)$$

This situation is the one where the enhancement factor is so large that the liquid-side resistance to mass transfer has essentially been eliminated; the local rate of mass transfer is governed by the gas phase resistance.

Combination of Eq. (4.21) with the local mass balance, and integration, yields the following design equation:

$$N^0 = \frac{K_G^0 \tilde{a}\pi}{\tilde{G}} Z_T = \int_T^B \frac{dY}{\xi(1 - Y)^2(Y - Y^*)} \qquad (4.31)$$

Although Eq. (4.31) bears a superficial analogy to the physical absorption equation, there is a crucial difference. The value of the integral on the right-hand side of Eq. (4.31) (i.e. the required number of transfer units) depends not only on the physico-chemical properties of the system considered, and on the operation to be carried out, but also on the individual mass transfer coefficients $k_G a$ and $k_L^0 a$. Furthermore, even if the gas is dilute (say $Y \ll 1$), the qantity $\xi(Y - Y^*)$ is a highly nonlinear function of Y, and therefore no analog of the simple log mean rule is ever likely to occur.

The overall mass balance between the top section and a generic one yields:

$$\frac{Y}{1 - Y} = \frac{Y_T}{1 - Y_T} + \frac{L}{G}(\alpha - \alpha_T) \qquad (4.32)$$

which can be solved for $\alpha(Y)$. If a reliable thermodynamic model is available, one can then calculate $Y^*(\alpha)$ and therefore $Y^*(Y)$. Simultaneous solution of Eqs. (4.19) and (4.20) then yields $I(Y)$ and therefore $\xi(Y)$. With the values of Y^* and ξ available at any value of Y, the integral in Eq. (4.31) can be calculated. This exhausts the design procedure.

It is important to point out that the design procedure outlined above requires both a reliable thermodynamic model and a know-ledge of the function on the right-hand side of Eq. (4.20). If, as is likely to be the case, the absorber is nearly pinched at the lean end, this information is needed particularly in the range of very low values of both Y and α, which is the range where both ther-modynamic and rate data are very difficult to obtain. Therefore, the thermodynamic model and the rate model need to have firm theoretical ground, because they both are crucially used in a range of Y and α values where they are essentially extrapolated from the range where laboratory data have been collected.

NOMENCLATURE

A, A'	the volatile component(s)
a, a'	concentrations of A, A' in liquid
$\bar{a}$	value of a that would equilibrate local composition
$\hat{a}$	interfacial area per unit volume
B	a chemical base
B_i	the ith nonvolatile component
b_i	concentration of B_i
$\hat{b}_i$	value of b_i normalized to molarity, $\hat{b}_i = b_i/m$
C_c	cyclic capacity
D	diffusivity of A
D_i	diffusivity of B_i
$\tilde{D}_i$	equivalent diffusivity of ion B_i
E	rate promotion factor
f	fugacity
g_0	value of g corresponding to the bulk-liquid composition
G	inert gas molar flow rate
$\tilde{G}$	inert gas molar superficial velocity
H, H'	Henry's Law constants for A, A' in reactive solvent
H^0	value of H is pure solvent
ΔH^v	enthalpy of vaporization
ΔH^{Diss}	enthalpy of dissolution
I	enhancement of factor
I_F	value of I in fast reaction regime
I_i	ionic strength
I_∞	value of I in instantaneous reaction regime
I_∞^s	value of I_∞ in low driving force asymptote
k	reaction rate constant
k_G	gas phase mass transfer coefficient
k_L	chemical mass transfer coefficient in liquid phase
k_L^0	physical mass transfer coefficient in liquid phase
K	thermodynamic equilibrium constant
L	solution volumetric flow rate
$\tilde{L}$	solution superficial velocity
M	number of independent chemical reactions
m	solution molarity
N, N'	rate of absorption of A, A' per unit interface area (negative in desorption)

N^0 number of transfer units

p, p' partial pressures of A, A'

$p^*, p^{*'}$ equilibrium values of p, p'

p^0 vapor pressure of pure component

q_i order of reaction with respect to B_i

r gas constant

r_0 rate of reaction at interface per unit area

$\vec{r}$ rate of forward action

$\bar{r}$ rate of reverse action

T temperature

t_D diffusion time

t_r reaction time

u partial molar volume

x distance from interface

Y gas phase mol fraction

y, y' fractional chemical saturation

$\bar{y}$ total saturation, $\bar{y} = \alpha/m$

α total concentration of A, physically dissolved plus chemically combined

α_i concentration ratio, $\alpha_0 = b_{i0}$

γ_A activity coefficient

δ film thickness

λ distance from interface to reaction plane

μ_i number of moles of A chemically combined in B_i

ν_i stoichiometric coefficient of B_i

ν_{ik} stoichiometric coefficient of B_i in kth reaction

ξ extent of reaction

π total pressure

$\bar{p}$ molar density of liquid

ϕ ratio of diffusion and reaction times

Suffixes

A of A

a activity based

B in the feed gas (bottom of countercurrent absorber)

0 in the bulk (subscript)

0 pure component (superscrpt)

I or II phase designation

i at the interface

f fugacity based (mass transfer coefficient)

i of the ith component

g gas phase

l liquid phase
k of the kth reaction
$*$ at equilibrium
ε which equilibrates the bulk
max maximum
min minimum
diss dissolution
T in the treated gas (top of countercurrent absorber)
∞ at infinite dilution

REFERENCES

1. Astarita, G. (1967). *Mass Transfer with Chemical Reaction*, Elsevier, Amsterdam.
2. Astarita, G. (1976). *Chem. Engrg. Sci.*, **31**, 1224.
3. Astarita, G., and Savage, D. W. (1980–1981). *Chem. Engrg. Sci.*, **35**, 649.
4. Astarita, G., and Savage, D. W. (1980–1982). *Chem. Engrg. Sci.*, **35**, 1513.
5. Astarita, G., Savage, D. W., and Longo, J. M. (1981). *Chem. Engrg. Sci.*, **36**, 581.
6. Astarita, G., Savage, D. W., and Longo, J. M. (1982). *Chem. Engrg. Sci.*, **37**, 1593.
7. Astarita, G., and Savage, D. W. (1982). *Chem. Engrg. Sci.*, **36**, 677.
8. Astarita, G., and Savage, D. W. (1982). In *Adv. Transport Prop.*, III (Eds. A. S. Mujumdar and R. A. Mashelkar), Wiley Eastern, New Delhi.
9. Astarita, G. (1985). *Chem. Engrg. Sci.*, forthcoming.
10. Blauwhoff, P. M. M. (1982). Ph.D. Thesis, University of Twente.
11. Cornelisse, R., Beenakers, A. A., Van Beckinn, F. P., and Van Swaoij, W. P. (1980). *Chem. Engrg. Sci.*, **35**, 1245.
12. Danckwerts, P. V. (1970). *Gas–Liquid Reactions*, McGraw-Hill, New York.
13. Danchwerts, P. V. (1981). *Chem. Engrg. Sci.*, **36**, 1741.
14. DeCoursey, W. J. (1974). *Chem. Engrg. Sci.*, **29**, 1867.
15. Giammarco, G. (1979). In *Ammonia* (Eds., A. V. Slack and G. R. James), Marcel Dekker, New York.
16. Joshi, C. V., Astarita, G., and Savage, D. W. (1981). *Chem. Eng. Prog. Symp. Series*, **77**(202), 63.
17. Kohl, A. L., and Reisenfeld, F. C. (1979). *Gas Purification*, 3rd ed., Gulf Publishing, Houston.
18. Kunkel, L. V. (1977). The fundamentals of gas sweetening, Paper I, Proceedings Gas Conditioning Conference, Norman, OK.
19. Lee, J. I., Otto, F. D., and Mather, A. E. (1973). *J. Chem. Eng. Data*, **18**(1), 71.
20. Maddox, R. N. (1977). Gas and liquid sweetening, *Campbell Petroleum Series*, *Norman, OK*.
21. Onda, K., Tagenchi, M., and Okumoto, Y. (1968). *J. Chem. Eng. Japan*, **1**, 56.
22. Ouwerkerk, C., Process selection for the treating of natural gas, Paper I-36, Natural Gas Processing and Utilization Conference, Irish Group of the Institution of Chemical Engineers, April 1976.
23. Roberts, D., and Danckwerts, P. V. (1962). *Chem. Engrg. Sci.*, **17**, 961.
24. Sartori, G., and Savage, D. W. (1983). *I.E.C. Fundam.*, **22**, 239.

25. Savage, D. W., Sartori, G., and Astarita, G. (1985). *Trans. Farad. Society.*, forthcoming.
26. Sherwood, T. K., Pigford, R. L., and Wilke, C. R. (1975). *Mass Transfer*, McGraw-Hill, New York.
27. Shrier, A. L., and Danckwerts, P. V. (1969). *Ind. Eng. Chem. Fundam.*, **8**, 415.
28. Tennyson, R. W., and Schaaf, R. P. (1977). *Oil and Gas J.*, **75**(2), 78.
29. Yu, W. C., and Astarita, G. (1985). Forthcoming.

Chapter 10

Catalysis and Catalytic Reaction Engineering

JAMES J. CARBERRY

Professor of Chemical Engineering, Laboratory of Catalysis, University of Notre Dame, Notre Dame, Indiana 46556, USA

Abstract

Some general aspects of supported metals catalysis which are of import to reaction engineering are nicely illustrated in the instance of total oxidation over Pt, Pt/Cu and Ag/Au formulations. Pretreatment, characterization as well as turnover number, sintering, alloying and surface enrichment are phenomena which deserve the close attention of those responsible for reactor design and analyses.

INTRODUCTION—THE REACTOR/REACTIONS

While signal progress has been made in our modelling prowess as regards the conventional heterogeneous catalytic *reactor* involving a single reactant bearing phase, it is safe to assert that the same level of modelling sophistication is not to be found as regards the conventional heterogeneous catalytic *reaction(s)*. To wit, we do a fair job with the gas–solid fixed bed but we can scarcely model oxidation of H_2 over Pt with naught but ambiguity.

We shall utilize supported metal and metallic alloy catalyzed total oxidation to illustrate several subtle aspects of heterogeneous catalysis which have implications with respect to reactor modelling. Reactor modelling, *per se*, is not of prime interest in this discussion—that issue having been treated in the classic text of Hougen and Watson [4] and more recent expositions [3].

In the design of the heterogeneous catalytic reactor care must be taken to respect the subtleties of the generation term, the catalytic reaction *per se*. The solid catalyzed event poses, amongst other problems, those of:

(a) kinetic modelling
(b) surface characterization
(c) pretreatment effects
(d) alloying effects and
(e) sintering

all of which can influence the generation function.

KINETICS

In modelling catalytic kinetic data we are inclined, for convenience sake, to assume one step in an assumed sequence to be rate-determining (rds). Thus a number of candidates (L–H/H–W) rate equations are generated to be compared with the rate data [4]. Yet there is no *a priori* reason why nature should be so obliging. A primitive example illustrates the point. Consider a single reversible catalytic (academic) reaction

$$A \rightleftharpoons P \tag{1}$$

e.g., simple isomerization. We assume, for sites, S

1)
$$A + S \overset{1}{\rightleftharpoons} AS$$

2)
$$AS \overset{2}{\rightleftharpoons} P + S$$

Therefore for coverge Θ

$$r_1 = k_1 A(1 - \Theta) - k_{-1}\Theta$$

$$r_2 = k_2\Theta - k_{-2}P(1 - \Theta)$$

If neither step is the rds

$$R = \frac{k_1 k_2 A - k_{-2}P}{\left[1 + K_1 A + \dfrac{k_2}{k_{-1}}(1 + P/K_2)\right]} \tag{2}$$

Had we assumed step 1 to be the rds

$$R = \frac{k_1 A - k_{-1} P / K_2}{1 + P / K_2} \tag{3}$$

or if step 2 is the rds

$$R = \frac{k_2 K_1 A - k_{-2} P}{(1 + K_1 A)} \tag{4}$$

Over a range of temperature there is no reason why either (3) or (4) should exclusively describe rate behavior save in the unlikely event that all activation energies be equal. Thus a relationship involving multi-step rate control (e.g., Eq. (2) in our example) is the more potent one.

Nor should we always assume that the catalytic steps occur in series as done above. There can be parallel steps in a reaction sequence as we shall now demonstrate.

TOTAL OXIDATION OF OLEFINS

The practical problem of elimination of automotive exhaust pollutants inspired a number of research efforts, chief amongst them being total oxidation of offending species over solid catalysts. The more poison-resistant noble metals (Pt, Pd, Rh) in alloyed form and supported on Al_2O_3 have proven to be the superior candidates at the moment. In consequence considerable research has been devoted to these systems—particularly propylene oxidation over Pt and its alloys since propylene oxidation proves to be an excellent barometer reaction for automotive exhaust catalyst activity. As is so often the case, the practical imperative gave rise to some fascinating scientific observations which in turn prove to be of import in reactor design.

For the total oxidation of propylene over 0.05% Pt/α-Al_2O_3, rate—propylene concentration data are displayed in Fig. 1. These data were secured in the Notre Dame Spinning-basket CSTR [2]. Note the pathological behavior—normal R vs. C behavior at low concentration, abnormal or negative-order kinetics at higher concentrations and ultimately zero-order behavior [5]. Conventional wisdom has it that such behavior is easily modelled by assuming surface reaction of adsorbed species to be the rds; or in excess O_2

$$R = \frac{kC\sqrt{O_2}}{(1 + KC)^2} \tag{5}$$

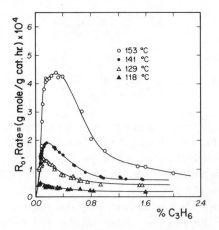

Figure 1   Global rate of total oxidation of propylene vs. propylene concentration over 0.05 wt% Pt/α-Al$_2$O$_3$ in excess O$_2$.

a model also invoked by many workers to describe CO oxidation over noble metals. In fact, in Fig. 2 the data of Fig. 1 are neatly assembled by the linear form of Eq. (5). But while Eq. (5) does exhibit positive- and then negative-order behavior, it patently does not lead to the observed zero order. And it predicts 1/2 order in O$_2$ which is not observed [6].

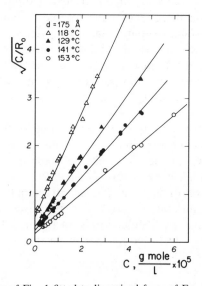

Figure 2   Data of Fig. 1 fitted to linearized form of Eq. (5) (excess O$_2$).

We postulate

1) $$O_2 + S_2 \overset{1}{\rightleftarrows} S_2O_2$$

2) $$C + S \overset{2}{\rightleftarrows} CS$$

E–R 3) $$C + S_2O_2 \rightleftarrows SO + S + P$$

L–H 4) $$CS + SO \rightleftarrows S_2 + P$$

where 3 is an Ely–Rideal step, i.e., the unadsorbed gaseous reactant 'titrates' preadsorbed O_2 and is thus oxidized to product P. Step 4 is the conventional Langmuir–Hinshelwood surface reaction step—the basis of Eq. (5). The above sequence, invoking site pairs S_2 leads to observed first order in O_2. Note that the E–R step (3) operates not in series but parallel to the L–H step. The rate then is, in excess O_2

$$R = \frac{S_0^2 k_4 KC}{(1 + KC)^2} + \frac{S_0 k_3 C}{(1 + KC)} \tag{6}$$

which exhibits positive, negative and at high KC, zero-order behavior as the data (Fig. 1) manifest. In Eq. (6), S_0 is the number of total sites. If we divide by S_0, the rate per site (or turnover number) is

$$R/S_0 = S_0(\text{L–H}) + (\text{E–R}) \tag{7}$$

so that the L–H step is total site dependent and the E–R step is not.

SURFACE CHARACTERIZATION

Knowledge of the chemistry and physics of the catalytic surface is obtained with great difficulty—particularly in the case of supported, highly dispersed crystallites. In terms of reaction kinetics, as Eq. (7) teaches, knowledge of S_0 is required to establish the turnover number or the rate per exposed site. As Boudart revealed, the rate per exposed site, or turnover number, may be invariant with changes in crystallite size or sensitive to such changes. Since it is thought that surface morphology can be crystallite size dependent, then if the reaction rate per site is sensitive to morphology we have a structure sensitive reaction. Such behavior or the lack of it (structure insensitivity) is of obvious scientific importance. As we shall see, it also has practical consequences for the reactor designer.

To secure a specific rate or turnover number requires a reliable method of exposed metal atom "counting". Termed earlier as dispersion and now more recently as "% exposed" the notion is essentially that the measuring the percentage of catalytic agents (atoms) deposited which gain surface exposure. So the Pt wire is marked by a dispersion of zero or % exposed of zero, insofar as the number of surface atoms is negligible relative to those in the bulk.

In the most elementary case one can rely upon chemisorption, e.g., H_2 or CO upon a noble metal, M

$$M + \frac{1}{2}H_2 \rightarrow MH$$

or

$$M + CO \rightarrow MCO$$

Or, more sensitively, titration

$$MO + \frac{3}{2}H_2 \rightarrow MH + H_2O$$

Whatever the method, one must assume a stoichiometry which is not always obeyed. Yet relative values of dispersion or % exposed can often be secured by chemisorption and/or titration. So the level of sintering severity can be obtained as a function of temperature, as is set forth in Fig. 3 for sintering in air of α-Al_2O_3 supported Pt [7]. A simple phenomenological second-order correlation roughly fits these sintering data and an apparent activation energy of 50–60 kal is extractable.

In Fig. 4 the specific rate, i.e., per area of exposed Pt is plotted vs. propylene concentration for diverse average crystallite sizes as induced by sintering in air. With an increase in crystallite size, S_0 is decreased. In general $S_0 \approx 1/d_c$.

It is apparent from Fig. 4 that turnover number increases with particle growth as induced by sintering. So the turnover number increases with a decrease in dispersion. Why? Examine Fig. 4 in the light of Eq. (7). That equation teaches that the L–H rate is a function of S_0 while the E–R term is independent of S_0—i.e., specific rate (per site, S_0) is predicted to be site sensitive at low and modest levels of C, but independent of S_0 at high values of KC.

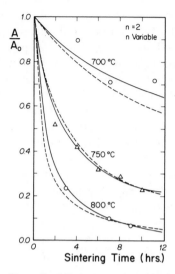

Figure 3   Normalized Pt area vs. sintering time in air.

So the specific rate (turnover number) data of Fig. 4 are in accord
with Eqs. (6) and (7). Mind that while Eq. (7) suggests that the
L–H step is proportional to S_0, the specific L–H rate coefficient is
not necessarily proportional to S_0. In fact it may be positively,
negatively or insensitive to S_0. Which is to state that the specific rate
coefficient or turnover number may increase, or decrease or be

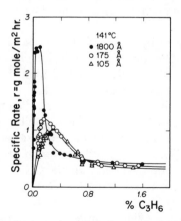

Figure 4   Specific rate (turnover number) for total oxidation of propylene vs.
propylene concentration for diverse levels of sintering. Catalyst: 0.05 wt% Pt/α-
Al$_2$O$_3$.

independent of crystallite size. What is surely evident by Eq. (7) is that structure sensitivity should or could be evident in the L–H step but absent in the E–R step. The titration step (E–R) depends solely upon O_2 equilibrium coverage, independent of S_0, while the L–H surface reaction step must depend upon S_0 and so dispersion.

What of the practical implications of structure sensitivity as induced by sintering? As has been noted the sintering data in Fig. 3 can be adequately correlated by power-law second-order kinetics (a phenomenological correlation of no fundamental significance; there being a distribution of crystallite sizes, a lumped overall order greater than intrinsic order will be manifest).

For second-order sintering, the normalized exposed area A as a function of time-on-stream is (d_c is crystallite diameter)

$$\frac{A}{A_0} = \frac{1}{1 + \alpha t} \equiv \frac{1}{d_c} \tag{8}$$

or the catalytic rate coefficient affected by sintering declines with time-on-stream

$$k_1 = \frac{k_0}{1 + \alpha t} \tag{9}$$

Note that Eq. (9), which becomes a component in the generation term of the reactor design equation network, simply declares that catalytic activity is proprotional to exposed area and thus declines with time-on-stream as that area is reduced by crystallite growth due to sintering.

The difficulty posed by Eq. (9) is the implicit assumption that k_0 is structure insensitive i.e., k_0 is independent of crystallite size. Furthermore, Eqs. (8) and (9) are linked to the assumption that the rate coefficient k is linear in exposed area.

To understand this point, consider the following argument in which we posit a growth of a catalytic crystallite while the rate per site may be invariant or increase or decrease with crystallite size.

Not to labor the point, a first-order heterogeneous rate constant is not, as taught by homogeneous kinetics, $k = \text{sec}^{-1}$; rather in a heterogeneous system it is $k_0 = \text{cm/sec}$. So

$$k_1 = k_0 \frac{\text{catalytic area}}{\text{Vol}} = k_0 \mathbf{a} = \text{sec}^{-1}$$

Now, **a** is the catalyst area per volume. If k_0 is invariant with crystallite size, k_1 may simply become linear in **a**. If, however, k_0 is sensitive to particle size (i.e., **a** which is proportional to $1/d_c$, where d_c is average crystallite size), we can anticipate rather diverse behavior of k_1. Suppose, for example, that k_0 is particle size-sensitive, i.e.,

$$k_0 \approx (d_c)^\alpha$$

d_c being average crystallite size. Now the exponent α could be zero (facile, or structure insensitivity) or other than zero (demanding or structure-sensitive behavior). If one assumes that

$$k_1 = k_0 a = k_0 (d_c)^\alpha \left(\frac{1}{d_c} \right)^\beta$$

where $(1/d_c)^\beta$ is naught but the relationship between rate and exposed area, then the rate coefficeint is

$$k_1 = k_0 (d_c)^{\alpha - \beta}$$

Note that we have abandoned the assumption of linearity between k_1 and exposed area, **a**. So for second-order sintering

$$k_1 = k_{0''}(1 + \alpha t)^{\alpha - \beta} \tag{10}$$

If the reaction is structure-insensitive ($\alpha = 0$) and $\beta = 1$ we have the conventionally assumed mode of deactivation due to sintering (Eq. (9)). Patently it is of importance in design to know α and β. Hence the practical implication of turnover number knowledge is the value of α or at least $\alpha - \beta$. Suppose one were able to alloy or promote the catalyst such that $\alpha > \beta$ or $\alpha = \beta$! In such instances activity would increase or at least remain unchanged with time-on-stream under sintering conditions.

In sum, should $\alpha > \beta$, area reduction via sintering actually increases the global rate. When $\alpha < \beta$, global rate declines. It is obvious that in the case of propylene total oxidation over Pt/α-Al$_2$O$_3$ that $\alpha < \beta$, so that sintering, i.e., an increase in d_c, sponsors a net decline in global rate in spite of the fact that specific rate (per site) is enhanced by crystallite growth.

ALLOYING EFFECTS

Alloying of one metal with another in a catalytic formulation can exert dramatic influences upon activity and selectivity as Sinfelts' classic studies teach [13].

In Fig. 5 is shown 'light-off' behavior for total oxidation of propylene over Cu and Cu/Pt alloys [9]. Note that while Cu *per se* 'lights-off' at rather high temperature, the addition of even small amounts of Pt as a Pt/Cu alloy drastically reduces 'light-off' temperature. In fact the Cu-rich Pt/Cu alloy reveals activity only slightly less than that found for Pt *per se*. These data are for a total metal loading of 1% by weight.

In the Ag–Au system Serrano [11] found a change in intrinsic kinetics of CO oxidation with an increase in Au content in a supported Ag–Au alloy.

Specifically for Au content less than 20% in Ag the well-established rate law for CO oxidation on Ag *per se* is found (in excess O_2)

$$R = \frac{kCO}{(1 + KCO_2)} \qquad (11)$$

At higher Au content the rate is described, in the limit, by (in excess O_2)

$$R = \frac{kCO}{(1 + KCO_2)^2} \qquad (12)$$

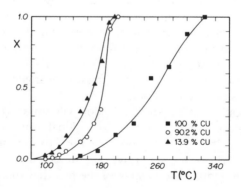

Figure 5   Oxidation of propylene "light-off" curve: conversion vs. temperature for supported Cu and Pt/Cu. Total metal loading of 1 wt%.

The role of Au in the Ag–Au alloy appears to be that of altering the rds from that of CO chemisorption as rate determining (Eq. (11)) to that of surface reaction as the rds (Eq. (12)).

Why, however, is the presence of Au without influence until present in amounts of 20% or more? We are here confronted with the issue of surface as opposed to bulk composition. Gibbs adsorption theorem predicts the possibility of surface enrichment of one component relative to its bulk value in alloys. A model which predicts surface enrichment in terms of surface tensions and free energy functions has recently been evolved (Kuczynski and Carberry—*Chemical Physics Letters*—(1984) **111** No. 4, 5; 445) which predicts that in the presence of chemisorbed O_2, Ag enriches the surface of a Ag–Au alloys—a prediction consistent with our kinetic observations [11].

PRETREATMENT

In 1925 Sir Hugh Taylor declared that the nature of the catalytic surface is determined by the nature of the reaction catalyzed. We might add—by the nature of the pretreatment. For in the Pt catalyzed oxidation of CO the rate for a CO pretreated catalyst is found to be twice as high as that for the same catalyst pretreated in O_2, (several hours at 200°C). Apparently we are dealing with two different catalytic surfaces, Pt and PtO_x [8].

OSCILLATORY BEHAVIOR

Oscillations in oxidation reactions now seem to be ubiquitous phenomena, particularly over noble metals [1, 7, 12]. A keen insight in accord with our remarks about pretreatment is provided by the University of California, San Diego group [10]. They and others postulate that oscillations are sponsored by a slow reversible surface redox reaction, i.e., slow oxidation and reduction of the surface Pt—a speculation in accord with our pretreatment observations.

CONCLUDING REMARKS

The behavior of the reactor and models thereof is critically determined by the generation or forcing function—catalysis and catalytic kinetics in the heterogeneous reaction—reactor network. Unlike homogeneous reactions, the solid catalyzed reaction system is marked by subtle complexities with respect to kinetic modelling, surface characterization, deactivation (e.g. sintering), alloying effects and pretreatment influences. Supported catalysts used in total oxidation of olefins and CO reveal these complexities and should inspire caution on the part of the catalytic reactor engineer.

REFERENCES

1. Beuch, H., Fieguth, P., and Wicke, E. (1977). *Advances in Chemistry Series*, **109**, 615.
2. Carberry, J. J. (1964). *Ind. Eng. Chem.*, **56** (Nov.), 39.
3. Carberry, J. J. (1976). *Chemical and Catalytic Reaction Engineering*, McGraw-Hill, New York.
4. Hougen, O., and Watson, K. (1943). *Chemical Process Principles*, Vol. 3, John Wiley, New York.
5. Huang, T., and Carberry, J. J. (1977). *Kinetika i Kataliz*, **18**, 562. See also Carberry, J. J. (1986), *Accounts of Chemical Research*, **18**, 358.
6. Jothi, N. (1983). Ph.D. Thesis, Univeristy of Notre Dame.
7. McCarthy, E., Zahradnik, J., Kuczynski, G. C., and Carberry, J. J. (1975). *J. Catalysis*, **39**, 29.
8. Paspek, S. (1973). Senior Project, University of Notre Dame.
9. Ping-Chau Liao (1981). Ph.D. Thesis, University of Notre Dame.
10. Sayles, B. C., Turner, J. E., and Maple, M. B. (1982). *Surf. Science*, **114**, 381.
11. Serrano, C., and Carberry, J. J. (1985). *Applied Catalysis*, **19**, 119.
12. Sheintuch, M., and Schmitz, R. A. (1977). *Cat. Reviews—Sci. and Eng.*, **15**, 107.
13. Sinfelt, J. H., Carter, J. L., and Yates, D. J. C. (1972). *J. of Catalysis*, **24**, 283.

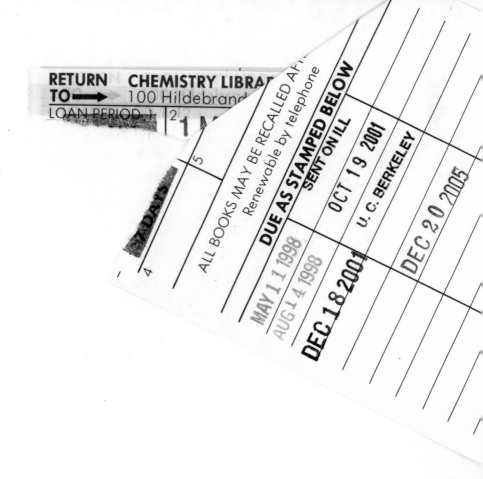